COLONY-STIMULATING FACTORS
Molecular and Cellular Biology
(Immunology Series/49)

Edited by: T. Michael Dexter, John M. Garland
and Nydia G. Testa
Published: 1990
504 pages, bound, illustrated

$135.00 (U.S. and Canada)
$162.00 (All other countries)
(Prices subject to change without notice.)

marcel dekker, inc.
270 MADISON AVENUE • NEW YORK, NY 10016 • 212-696-9000
mdi

Colony-Stimulating Factors

IMMUNOLOGY SERIES

1. Mechanisms in Allergy: Reagin-Mediated Hypersensitivity
   Edited by Lawrence Goodfriend, Alec Sehon and Robert P. Orange
2. Immunopathology: Methods and Techniques
   Edited by Theodore P. Zacharia and Sidney S. Breese, Jr.
3. Immunity and Cancer in Man: An Introduction
   Edited by Arnold E. Reif
4. *Bordetella pertussis:* Immunological and Other Biological Activities
   J.J. Munoz and R.K. Bergman
5. The Lymphocyte: Structure and Function (in two parts)
   Edited by John J. Marchalonis
6. Immunology of Receptors
   Edited by B. Cinader
7. Immediate Hypersensitivity: Modern Concepts and Development
   Edited by Michael K. Bach
8. Theoretical Immunology
   Edited by George I. Bell, Alan S. Perelson, and George H. Pimbley, Jr.
9. Immunodiagnosis of Cancer (in two parts)
   Edited by Ronald B. Herberman and K. Robert McIntire
10. Immunologically Mediated Renal Diseases: Criteria for Diagnosis and Treatment
    Edited by Robert T. McCluskey and Giuseppe A. Andres
11. Clinical Immunotherapy
    Edited by Albert F. LoBuglio
12. Mechanisms of Immunity to Virus-Induced Tumors
    Edited by John W. Blasecki
13. Manual of Macrophage Methodology: Collection, Characterization, and Function
    Edited by Herbert B. Herscowitz, Howard T. Holden, Joseph A. Bellanti, and Abdul Ghaffar
14. Suppressor Cells in Human Disease
    Edited by James S. Goodwin
15. Immunological Aspects of Aging
    Edited by Diego Segre and Lester Smith
16. Cellular and Molecular Mechanisms of Immunologic Tolerance
    Edited by Tomáš Hraba and Milan Hašek

17. Immune Regulation: Evolution and Biological Significance
Edited by Laurens N. Ruben and M. Eric Gershwin

18. Tumor Immunity in Prognosis: The Role of Mononuclear Cell Infiltration
Edited by Stephen Haskill

19. Immunopharmacology and the Regulation of Leukocyte Function
Edited by David R. Webb

20. Pathogenesis and Immunology of Treponemal Infection
Edited by Ronald F. Schell and Daniel M. Musher

21. Macrophage-Mediated Antibody-Dependent Cellular Cytotoxicity
Edited by Hillel S. Koren

22. Molecular Immunology: A Textbook
Edited by M. Zouhair Atassi, Carel J. van Oss, and Darryl R. Absolom

23. Monoclonal Antibodies and Cancer
Edited by George L. Wright, Jr.

24. Stress, Immunity, and Aging
Edited by Edwin L. Cooper

25. Immune Modulation Agents and Their Mechanisms
Edited by Richard L. Fenichel and Michael A. Chirigos

26. Mononuclear Phagocyte Biology
Edited by Alvin Volkman

27. The Lactoperoxidase System: Chemistry and Biological Significance
Edited by Kenneth M. Pruitt and Jorma O. Tenovuo

28. Introduction to Medical Immunology
Edited by Gabriel Virella, Jean-Michel Goust, H. Hugh Fudenberg, and Christian C. Patrick

29. Handbook of Food Allergies
Edited by James C. Breneman

30. Human Hybridomas: Diagnostic and Therapeutic Applications
Edited by Anthony J. Strelkauskas

31. Aging and the Immune Response: Cellular and Humoral Aspects
Edited by Edmond A. Goidl

32. Complications of Organ Transplantation
Edited by Luis H. Toledo-Pereyra

33. Monoclonal Antibody Production Techniques and Applications
Edited by Lawrence B. Schook

34. Fundamentals of Receptor Molecular Biology
Donald F. H. Wallach

35. Recombinant Lymphokines and Their Receptors
Edited by Steven Gillis

36. Immunology of the Male Reproductive Organs
Edited by Pierre Luigi Bigazzi

37. The Lymphocyte: Structure and Function
Edited by John J. Marchalonis

38. Differentiation Antigens in Lymphohemopoietic Tissues
Edited by Masayuki Miyasaka and Zdenek Trnka

39. Cancer Diagnosis In Vitro Using Monoclonal Antibodies
Edited by Herbert Z. Kupchik

40. Biological Response Modifiers and Cancer Therapy
Edited by J. W. Chiao

41. Cellular Oncogene Activation
Edited by George Klein

42. Interferon and Nonviral Pathogens
Edited by Gerald I. Byrne and Jenifer Turco

43. Human Immunogenetics: Basic Principles and Clinical Relevance
Edited by Stephen D. Litwin, with David W. Scott, Lorraine Flaherty, Ralph A. Reisfeld, and Donald M. Marcus

44. AIDS: Pathogenesis and Treatment
Edited by Jay A. Levy

45. Cell Surface Antigen Thy-1: Immunology, Neurology, and Therapeutic Applications
Edited by Arnold E. Reif and Michael Schlesinger

46. Immune Mechanisms in Cutaneous Disease
Edited by David A. Norris

47. Immunology of Fungal Diseases
Edited by Edouard Kurstak

48. Adoptive Cellular Immunotherapy of Cancer
Edited by Henry C. Stevenson

49. Colony-Stimulating Factors: Molecular and Cellular Biology
Edited by T. Michael Dexter, John M. Garland, and Nydia G. Testa

Additional Volumes in Preparation

Colony-Stimulating Factors

Molecular and Cellular Biology

edited by

T. MICHAEL DEXTER
Paterson Institute for Cancer Research
Christie Hospital and Holt Radium Institute
Manchester, England

JOHN M. GARLAND
University of Manchester Medical School
Manchester, England

NYDIA G. TESTA
Paterson Institute for Cancer Research
Christie Hospital and Holt Radium Institute
Manchester, England

MARCEL DEKKER, INC. **New York and Basel**

Library of Congress Cataloging-in-Publication Data

Colony-stimulating factors.

(Immunology series ; v. 49)
Includes bibliographical references.
1. Colony-stimulating factors (Physiology)
I. Dexter, T. Michael II. Garland, John M.
III. Testa, Nydia G.
IV. Series.
QP92.C65 1990 616.07'9 89-23380
ISBN 0-8247-8094-9 (alk. paper)

This book is printed on acid-free paper.

MARCEL DEKKER, INC.
270 Madison Avenue, New York, New York 10016

Current printing (last digit):
10 9 8 7 6 5 4 3 2 1

PRINTED IN THE UNITED STATES OF AMERICA

Series Introduction

Scientific research is often compared to an iceberg—there is much more beneath the surface than meets the eye. That description aptly fits the topic of this latest addition to the Immunology Series. Basic research on colony-stimulating factors began many years ago in the laboratories of a few devoted investigators, such as Donald Metcalf of the Walter and Eliza Hall Institute of Medical Research in Melbourne, Australia. The original objective was to learn about the factors in tissues controlling the development of blood cells from their bone marrow progenitors. The first steps were halting indeed. The tissue fractions were mixtures and the active components present in distressingly small quantities. Many years of painstaking labor were needed to separate the individual factors and identify their activities. Gradually, reliable preparation procedures were devised and reproducible assay methods developed. The introduction of gene cloning greatly accelerated the identification of colony-stimulating factors, so that they could finally be produced in acceptable amounts. It then became possible to determine their biological activities and chemical properties.

We have now entered a new stage of research on colony-stimulating factors: clinical application in the treatment of malignant infections and immunodeficiency disease. The many years of basic research, hidden from view in a few laboratories of scientists, will yield visible returns as contribution to the improvement of human health and welfare. In our admiration of recent advances, let us not forget the foundation upon which they rest.

Noel R. Rose

Preface

In very few fields can the rate of advance be greater than that concerning growth factors in general, and hemopoietic ones in particular. This advance covers the genetic cloning of the first "lymphokine" (IL-3) through to sequencing of the receptor for macrophage colony-stimulating factors (CSF) and the entry of a number of cloned, purified factors into clinical trials, all within the space of 5 years. It is unlikely that this advance will decline in any way. Therefore, any book on the subject will be subject to rapid updating. In producing this book, the authors have attempted to present the wider window of how these factors involve many different disciplines and interests: physiological observations lead to considerations of the structure of the molecules, how they are produced, how they are controlled at the genetic level, their receptors, how cells respond, and so on. In addition, such a subject demands that it should be accessible to a wide audience with very differing levels of expertise and interests.

We have chosen a format that includes information on areas of high relevance, to achieve this balance between up-to-date information and global view. Thus, the book is introduced by chapters on stem cells and marrow biology and receptors. The various CSFs are then discussed in detail followed by chapters dealing with how CSFs may relate to leukemias and the involvement of the lymphoid system. The final chapters present valuable and important information on problems faced by commercial concerns in bringing products painstakingly produced in small laboratory amounts on-stream in a cost-effective, volume-efficient manner; the use of CSFs in clinical practice; and areas critical to future understanding, namely the process of signal transduction and the control of gene activation. Thus, the reader is able to refer to different aspects of the science of the CSFs within the same book.

The authors hope that the book will be useful both to the specialist in molecular genetics who may wish to know how the end products of his constructions fit into the general plan of things cellular and to the many other disciples of cellular science who wish to relate their understandings of hematology, immunology, or medicine to things molecular. We do not presume to have covered the ground exhaustively, and for necessary omissions, differences of emphasis, or prejudice we apologize. We are grateful to all our contributors for helping us in this task, to the publisher for allowing us considerable leeway, and to all those who have helped in its production. We hope we have been successful in achieving our objectives.

T. Michael Dexter
John M. Garland
Nydia G. Testa

Contributors

Terence D. Allen Department of Ultrastructure, Paterson Institute for Cancer Research, Christie Hospital and Holt Radium Institute, Manchester, England

Nega Beru* Department of Biochemistry and Molecular Biology, The University of Chicago, Chicago, Illinois

Anna T. Brini Laboratory of Molecular Immunoregulation–BRMP, National Cancer Institute–Frederick Cancer Research Facility, Frederick, Maryland

Ruud Delwel Department of Hematology, Dr. Daniel den Hoed Cancer Center, Rotterdam, The Netherlands

T. Michael Dexter Department of Experimental Hematology, Paterson Institute for Cancer Research, Christie Hospital and Holt Radium Institute, Manchester, England

William L. Farrar Laboratory of Molecular Immunoregulation–BRMP, National Cancer Institute–Frederick Cancer Research Facility, Frederick, Maryland

Douglas K. Ferris Program Resources, Inc., National Cancer Institute–Frederick Cancer Research Facility, Frederick, Maryland

---

Present affiliation:
*Section of Hematology/Oncology, Department of Medicine, The University of Chicago, Chicago, Illinois.

John M. Garland* Division of Anatomy, Department of Cell and Structural Biology, University of Manchester Medical School, Manchester, England

Steven Gillis Immunex Research and Development Corporation, Seattle, Washington

Eugene Goldwasser Department of Biochemistry and Molecular Biology, The University of Chicago, Chicago, Illinois

Nicholas M. Gough Cancer Research Unit, The Walter and Eliza Hall Institute of Medical Research, Royal Melbourne Hospital, Melbourne, Victoria, Australia

Andrew J. Hapel Developmental Haematology Group, Division of Clinical Science, John Curtin School of Medical Research, Australian National University, Canberra, Australia

Annick Harel-Bellan† Laboratory of Molecular Immunoregulation, National Cancer Institute–Frederick Cancer Research Facility, Frederick, Maryland

Paul R. Harrison Beatson Institute for Cancer Research, Glasgow, Scotland

Ernest S. Kawasaki Department of Molecular Biology, Cetus Corporation, Emeryville, California

Marie Korner Laboratory of Molecular Immunoregulation–BRMP, Program Resources, Inc., National Cancer Institute–Frederick Cancer Research Facility, Frederick, Maryland

Martha B. Ladner Cetus Corporation, Emeryville, California

Bob Löwenberg Department of Hematology, Dr. Daniel den Hoed Cancer Center, Rotterdam, The Netherlands

Carolyn F. Morris Developmental Haematology Group, Division of Clinical Science, John Curtin School of Medical Research, Australian National University, Canberra, Australia

---

Present affiliations:
*Division of Pathology, Department of Cell and Structural Biology, University of Manchester Medical School, Manchester, England.
†Laboratoire d'Immunologie, Bâtiment de Recherches, Villejuif, France.

Nicos A. Nicola Cancer Research Unit, The Walter and Eliza Hall Institute of Medical Research, Royal Melbourne Hosptial, Melbourne, Victoria, Australia

Linda S. Park Department of Biochemistry, Immunex Research and Development Corporation, Seattle, Washington

Colin J. Sanderson National Institute for Medical Research, London, England

J. Howard Scarffe Department of Medical Oncology, Paterson Institute for Cancer Research, Christie Hospital and Holt Radium Institute, Manchester, England

P. J. Simmons Department of Transplantation Biology, Fred Hutchinson Cancer Research Center, Seattle, Washington

David Smith Joint Section of Hematology/Oncology, The University of Chicago, Chicago, Illinois

William P. Steward Department of Medical Oncology, Paterson Institute for Cancer Research, Christie Hospital and Holt Radium Institute, Manchester, England

Nydia G. Testa Department of Experimental Hematology, Paterson Institute for Cancer Research, Christie Hospital and Holt Radium Institute, Manchester, England

David R. Thatcher ICI Pharmaceuticals, Macclesfield, Cheshire, England

Ivo Touw Department of Hematology, Dr. Daniel den Hoed Cancer Center, Rotterdam, The Netherlands

David L. Urdal Immunex Research and Development Corporation, Seattle, Washington

Neil Williams Department of Physiology, University of Melbourne, Parkville, Victoria, Australia

Ian G. Young Medical Molecular Biology Group, John Curtin School of Medical Research, Australian National University, Canberra, Australia

Contents

1
Marrow Biology and Stem Cells

TERENCE D. ALLEN and T. MICHAEL DEXTER
Paterson Institute for Cancer Research, Christie Hospital and Holt Radium Institute, Manchester, England

P. J. SIMMONS
Fred Hutchinson Cancer Research Center, Seattle, Washington

I. INTRODUCTION

The hourly production of 370,000 million blood cells, required to maintain steady-state hemopoiesis in humans, may be regarded as a cellular production line involving not just the hemopoietic cells themselves but also the housing and machinery of the fixed cell populations of the bone marrow. These "microenvironmental" cells may well be intimately involved in regulating the production of the correct numbers of the respective cell types, monitoring their differentiation and development, and even recycling some of the critical components.

Specific environments for production of hemopoietic cells have been known for some time, for instance the thymus (T lymphocytes), the bursa of Fabricius (B lymphocytes), and the bone marrow (myeloid cells). Even within a single type of "environment," the distribution of cells may be heterogeneous, e.g., the distribution of developing T cells in the thymus cortex and medulla and the distribution of stem cells in mouse marrow, which show a 10-fold increase near the surface of the bone in comparison to the central regions of the marrow shaft. In 1970 Trentin suggested the term hemopoietic inductive microenvironment ("HIM") to define the local developmental restriction of hemopoiesis mediated by stromal cells, and over the years the "HER" lobby (hemopoiesis entirely random) (32), which regards blood cell production as being determined by stochastic processes, has been modified to include the role played by stromal cells and soluble regulatory molecules in determining the probability of certain developmental events. Thus, stromal-mediated hemopoiesis is now a generally accepted phenomenon among hematologists, although we are still a long way from understanding precisely how this works at the mechanistic level. Support for the view that interaction between hemopoietic cells and stroma is important in regulating hemopoiesis arose from the discovery of a culture technique whereby mouse marrow could be maintained in long term in vitro (18), displaying stem-cell renewal and pluripotential differentiation for several months. Over the surface area of the culture flasks, it was very quickly recognized that stromal and hemopoietic cells made regular, consistent, and specific associations (3). Long-term bone marrow culture (LTBMC) has remained unique in tissue culture, and its accessibility to many means of experimentation has led to its adoption worldwide as an experimental tool. In the course of this chapter we shall be concentrating mainly on our results from long-term marrow cultures, with reference to marrow in vivo wherever possible. Although this concentration on an in-vitro model system does not perhaps represent the optimal way of understanding the biology of the marrow in vivo, it does reflect the relative attention afforded to the two systems by experimenal hematologists in general. The questions that will be addressed are the following. (a) Which cells of the stroma are involved in the stimulation of hemopoietic stem cells in self renewal, commitment, proliferation, and differentiation? (b) What are the mechanisms by which these processes are brought about? (c) What stromal defects can be detected in hematological malignancies; and (d) What is the role of soluble growth factors in stromal-cell-mediated hemopoiesis? These questions will be considered initially in vivo, then usually in more detail in vitro.

II. CELLS OF THE BONE MARROW STROMA IN VIVO, THEIR POTENTIAL HEMOPOIETIC INVOLVEMENTS, AND THEIR ROLES IN VITRO

A standard histological preparation of murine femur or red marrow from larger mammals would reveal a variety of nonhematological cell types, some of which are present as routine tissue components, with no apparent active participation in hemopoiesis whatsoever. We propose, however, to briefly survey them all, with respect to potential involvement in hemopoiesis, and also with respect to their survival and proliferation in tissue culture conditions in general, and LTBMC in particular.

A. Endosteum

The endosteum forms a unicellular layer over the surface of the bone, including the trabecular regions, and possesses the potential for differentiation into osteoprogeintor cells. It is highly unlikely that these mature cells are transferred during culture initiation as trabecular bone is trimmed off mouse femurs before the contents are extruded into culture vessels. Furthermore, scanning electron microscopy of extremely well washed internal bone surfaces after marrow removal shows retention of the endosteum (T. D. Allen, unpublished data). In human cultures, internal surfaces of rib may be scrubbed during cell harvesting, possibly removing endosteal cells, but the subsequent cultures show no difference from those cultures established by liquid marrow aspirates, which perhaps would not be expected to contain endosteal cells. It should be noted, however, that "stromal" cells from rabbit marrow, also aspirated without endosteum, will subsequently generate bone when reimplanted in diffusion chambers (33), and also mouse bone marrow will form bone when reimplanted under the kidney capsule (21,22). Thus, although it appears reasonable to suggest that the mature endosteal cells are rarely removed from the bone surface (to survive and proliferate in culture conditions), osteogenic progenitor cells are indeed present in extracted marrow. One novel candidate for this role has recently been shown to be the microvessel pericytes, which will form calcifying nodules after several weeks in primary culture (A. Schor, personal communication). In terms of hemopoiesis it seems that endosteum or for that matter, osteoblasts themselves, have no direct effects on blood formation, irrespective of the finding that murine femoral bone itself may be a rich source of stem cells (29).

B. Marrow Vasculature

Hemopoietically active long bones have a nutrient artery that gives rise to radial arterioles, which in turn empty into the vascular sinuses. From here, blood reaches the venules, then the central longitudinal vein, and finally the nutrient vein (27). This vasculature has several types of cells associated with it. Lining the small arterioles are simple nonfenestrated endothelia, which are surrounded by helicoidally arranged smooth-muscle cells, themselves surrounded by a basal lamina, collagen, and reticular fibers. The capillaries have endothelia with small fenestrations, a basal lamina, and occasional pericytes, whereas the sinusoidal endothelia largely lack a basal lamina (see below) and are not fenestrated, but may be penetrated by newly formed blood cells in their constant passage into the circulation. These cells have been termed littoral cells (27) to differentiate them from other endothelia. Thus, in terms of cell types, there are five that comprise the marrow vasculature, three types of endothelia (arterial, capillary, and littoral), and two vessel-wall cells (smooth-muscle and pericytes). Both large and small-vessel endothelia from other sites (bovine aorta, retinal capillary) have been successfully grown in culture (38), indicating their ability to survive in vitro, but bone marrow sinus endothelia (littoral cells) have not been positively identified in vitro for lack of a suitable marker. Pericytes and smooth-muscle cells have also been cultured successfully from vasculature in other sites such as retina (38). It is difficult to envisage a direct role for the vasculature in hemopoietic production other than a general form of providing nutrients, although it is possible that as a final "gateway" to the entry of newly formed blood cells into the circulation, the venous sinus endothelium may possibly play a limited role.

C. Marrow Stromal Cells

Marrow stromal cells themselves are more difficult to define, because they may all be of common origin, representing phenotypic variations of a single cell type. Variations may occur dependent on their site in the marrow, or as a result of local differentiation into adipocytes. The basic stromal cell of the hemopoietic areas of the marrow is termed the reticular cell, not to be confused with reticulum cells (an early term for stem cells) or cells of the reticular endothelial series (47-49). Reticular cells are the dominant cells of the marrow stroma, with a dendritic morphology of broad sheetlike cytoplasmic processes branching into the hemopoietic spaces and forming a spongelike network similar to the reticular network of splenic cords (Fig. 1). Those reticular cells that cover areas of the abluminal surface of the sinus endothelia or littoral cells have been termed adventitial cells. Where adventitial and littoral cells come together, a basal lamina is formed. The third cell in the stroma is a lipid-laden adipocyte, typical of fatty nonhemopoietic marrow and probably derived from lipid accumulation by adventitial cells (49). Although fat cells are typically found in large numbers in the long bones of large mammals (and are associated with sites where hemopoiesis is limited or absent) they are virtually absent in the marrow of mouse femur (which, unlike that in the larger mammals, is actively hemopoietic) and thus adipocytes are not directly transferred into mouse marrow tissue culture. However, within 2-3 weeks of culture, maturing fat cells are the most obvious cell type in mouse marrow cultures, and have always represented a "bench mark" indication for hemopoietically productive cultures. This correlation between adipocyte development and hemopoiesis in the LTBMC is somewhat surprising in view of the inverse correlation seen in vivo. It should be stressed, however, that LTBMC has a finite life (normally 4-5 months) and that cessation of hemopoiesis is correlated with an increase in mature adipocytes. Thus, even in LTBMC, the presence of mature adipocytes may be contraindicatory for prolonged hemopoiesis.

The growth of reticular and adventitial cells in marrow cultures remains unconfirmed because of the absence of a convincing marker for these cell types either in vivo or in vitro. Our own working hypothesis, however, is that the bone marrow "fibroblast" as observed in stromal cell cultures [colony-forming unit F (CFU-F)] and also in haemopoietic LTBMC is the in-vitro equivalent of the reticular cell, a view shared by Freidenstein (personal communication).

All three stromal cell types appear to be involved in aspects of hemopoiesis in vivo. Fat cells are generally regarded as a "space-filling" tissue in nonhemo-

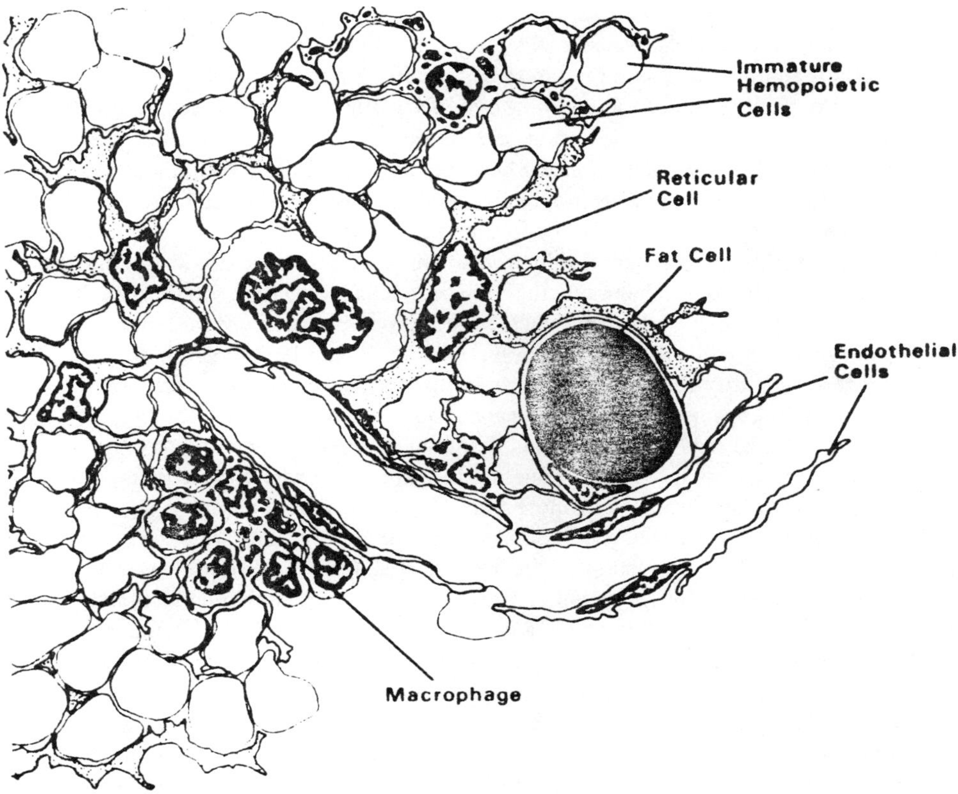

Figure 1 Diagrammatic representation of bone marrow stromal elements in vivo. The venous sinus in the center is lined with endothelial cells and is covered on its adventitial surface by an incomplete sheath of reticular cells. Processes of reticular cells branch into the extravascular space forming a scaffolding on which hemopoietic cells are arranged. An adventitial reticular cell that has accumulated fat is also shown. Macrophages occur through the marrow and are intimately associated with developing hemopoietic cells.

poietic marrow, but regress as the marrow transforms to active haemopoiesis in some circumstances. Reticular cells secrete the structural matrix of reticular fibers that support the hemopoietic tissue (along with the branched network of reticular cells themselves). Adventitial cells appear to influence the rate of delivery of blood cells into the circulation by varying the amount of cover they form on the abluminal side of the venous sinuses. Adventitial cover is increased when mice are hypertransfused, decreasing the rate of red blood cell ingress to the ciruclation (28). Conversely adventitial cell cover is reduced when erythropoietin is administered, and red blood cell production increases (14). Adventitial cover is also reduced in endotoxin-induced leukopenia (50). Associations are also reported between hemopoietic cells and reticular cells. Reticular cells appear to be directly involved in the hemopoietic mechanisms with direct association with putative stem cells in human fetal bone marrow (50) and also in the apparent sequestration of lymphocytes in fetal spleen (50).

In the highly packed dense cellularity of the hemopoietic marrow, it is clearly very difficult to comprehend whether adjacent cells are related, interacting, or merely close packed, and we may make the analogy between several families traveling on a highly crowded tube train, where it would be impossible for a stranger to identify them as interactive groups. If, however, the families were picnicking in a spacious park, where each group was discrete, then the family groups would be readily identifiable. A similar situation can be said to exist between bone marrow in vivo and bone marrow in culture where there is ample space for cell distribution. For more than 10 years, we have observed regular and consistently occurring cellular associations of cells on the "wide-open spaces" of the growing surface of the culture flask (which in mouse never becomes confluent). Thus, it is not an unreasonable inference that cellular associations in this situation are to the advantage of the cells that form them, and are necessary for the continued maintenance of stem cells and their differentiated progeny. For the remainder of this article we will survey these associations, as observed by transmission electron microscopy (TEM), scanning electron microscopy (SEM), immunocytochemistry, and time-lapse video, and compare and contrast them to the putative interactions of marrow in vivo. The majority of the work has been pursued in the mouse, but reference will be made to human and other species (cat, tupaia, rabbit) where appropriate. We shall consider all aspects of the growth of explanted bone marrow, concentrating on those in which hemopoietic production is maintained for a reasonable period.

III. GROWTH OF EXPLANTED BONE MARROW

Explanted marrow has been grown in a variety of environments, some of which result in active hemopoiesis, and some of which do not. As well as direct marrow-to-marrow transplantation, certain cells present in marrow will grow in the spleens of potentially lethally irradiated mice, and also when implanted under the kidney capsule. Explanted bone marrow consists of both hemopoietic and stromal cells, and the necessity for stromal cells can be experimentally tested in transplanted conditions. Marrow cell populations highly enriched for pluripotential stem cells using a fluorescence-activated cell sorter will produce spleen colonies and regenerate hemopoiesis in irradiated mice (31), indicating that stromal support function in the spleen is adequate for the growth and development of these cells. On an irradiated marrow stroma in vitro, however, they perform much less well, perhaps indicating a suboptimal environment (42). Whether or not stromal cells from the donor help establish transplanted marrow cells in the recipient in humans is still controversial (17). In some cases the recipient stromal tissue appears to support the transplanted marrow, and in other cases donor stroma (inoculated along with the donor marrow) has been identified by various markers for some time after transplantation (25). Recent work, however (40), supports the consensus that donor stromal cells have no role to play in facilitating hemopoiesis in vivo in a marrow transplant situation. It should be noted that this does not preclude the possibility that stromal cells are indeed transplantable. That viable stromal tissue is carried over in whole marrow transplants has been clearly demonstrated by the finding that marrow explants under the kidney capsule will produce bone first, and become hemopoietic afterwards.

Whole mouse bone marrow explanted into liquid culture with calf serum will produce a hemopoietically inactive culture (CFU-F) formed of stromal cells and macrophages only (52). In the presence of horse serum or fetal calf serum supplemented with hydrocrotisone, the bone marrow establishes an adherent layer during the first 3 weeks of culture that supports continuing hemopoiesis from several months. Initially, it was considered necessary that a second inoculum of marrow be added at 3 weeks onto the established adherent layer, but subsequently it became apparent that the hemopoietic capacity was maintained throughout the establishment period without the need for replenishment (24). An interesting and yet unanswered question in this situation is where do the stem cells reside during the 3 weeks it takes to establish the adherent layer? It is reasonable to assume that the time spent in establishing the adherent layer represents a growth and reformation of the microenvironmental architecture, and this view is supported by the observations that if a single cell suspension of the marrow inoculum is made, hemopoiesis is not established. Similarly, the use of fetal calf serum rather than horse serum supplemented with hydrocortisone will also allow cell growth (CFU-F) but obviously fails to supply the conditions necessary for maintenance and resumption of hemopoiesis. In view of the "reformation" of hemopoietic microenvironment that takes place during the establishment of the adherent layer, we have studied this process by light and electron microscopy, and also by a time-lapse video system (2), allowing minute-by-minute observations during the entire 3-week period.

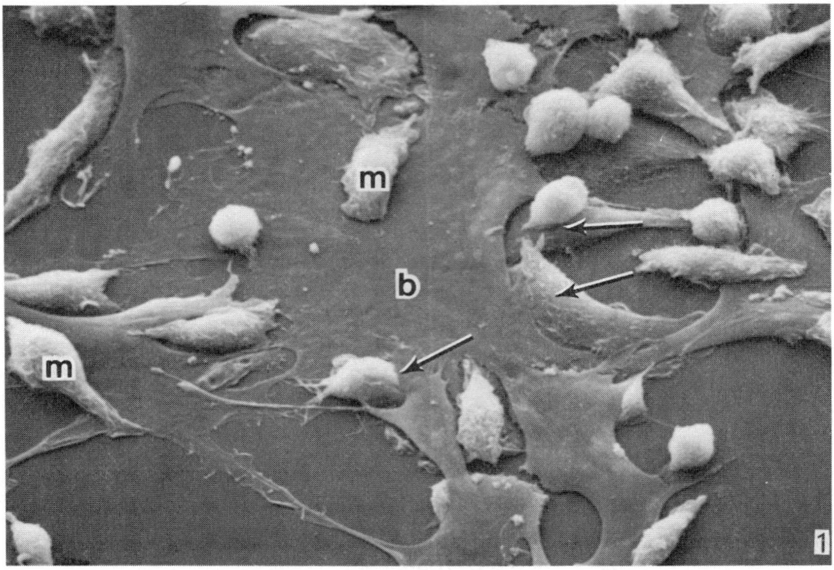

Figure 2 (1) Scanning electron micrograph (SEM) of an area of developing adherent layer cells from mouse bone marrow culture at 7 days after inoculation. Stromal precursor cells that were initially of typical fibroblastoid morphology are now adopting a less bipolar and increasingly spread morphology, typical of blanket cells (b). The majority of the nonspread cells are macrophages (m) some of which (arrowed) are migrating underneath the central blanket cell. (2) SEM detail from adherent layer of "mature" blanket cell in a 4-week culture. The cell is approximately 300 μm in diameter [compare peripheral macrophages (m) and granulocytes (g)]. Because of the highly spread morphology and veillike nature of the blanket cell cytoplasm, many macrophage profiles (three arrowed as examples) are clearly visible beneath the blanket cell, visualized by the penetration of the electron beam in the scanning electron microscope.

IV. ESTABLISHMENT OF THE ADHERENT LAYER IN LTBMC

No attempt is made to separate the roughly aspirated bone marrow into a single-cell suspension, although clumps of cells are rare, and most cells attach to the substratum as single entities. At this stage, the vast majority of cells will be relatively mature, nonadherent granulocytes, which persist until the first medium change at 7 days. Aggregates containing large numbers of granulocytes appear to be attached to the substratum by large stromal cells, and

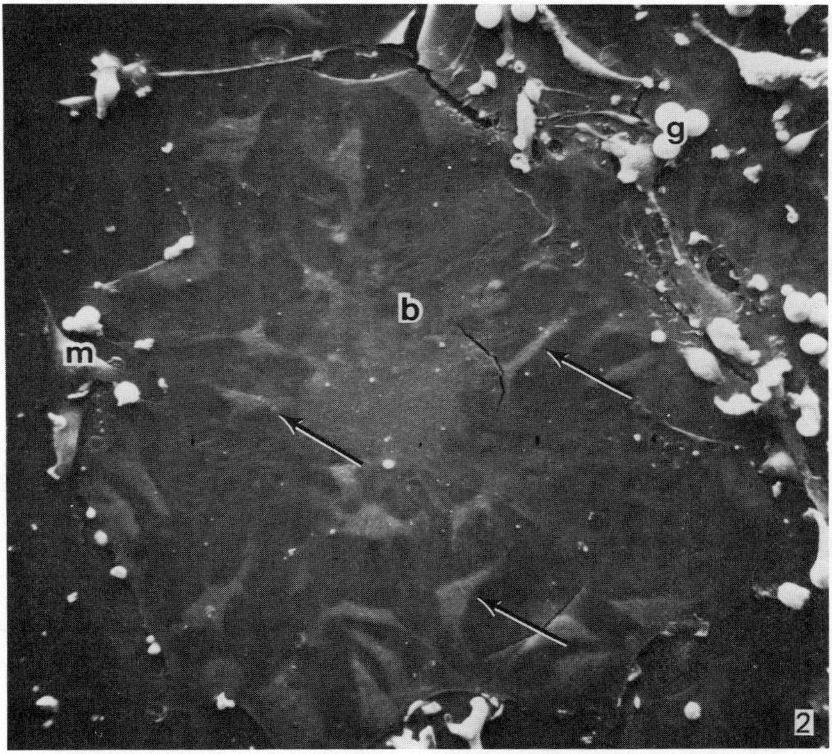

Figure 2 (continued)

after 48 hr of culture macrophages [(identified at all stages by the monoclonal antibody for mouse macrophages F4/80 (8)] appear as a well-dispersed attached population over the majority of the growing surface. At this stage (48 hr) a few bipolar or stellate cells are also apparent, characterized by a large euchromatic nucleus with prominent nucleoli. By 3 days of culture, mixed foci of macrophages and the stromal precursor cells (SPC) begin to appear, and increase in both size and number over the next 10 days. Scoring any cluster of mixed macrophage and SPC of more than 50 cells, approximately 1000 foci per flask were observed at day 4 and 1400 by day 7, after which scoring was not possible due to the overlapping and merging of the individual foci, although mouse LTBMC never become confluent in the same way as human cultures do.

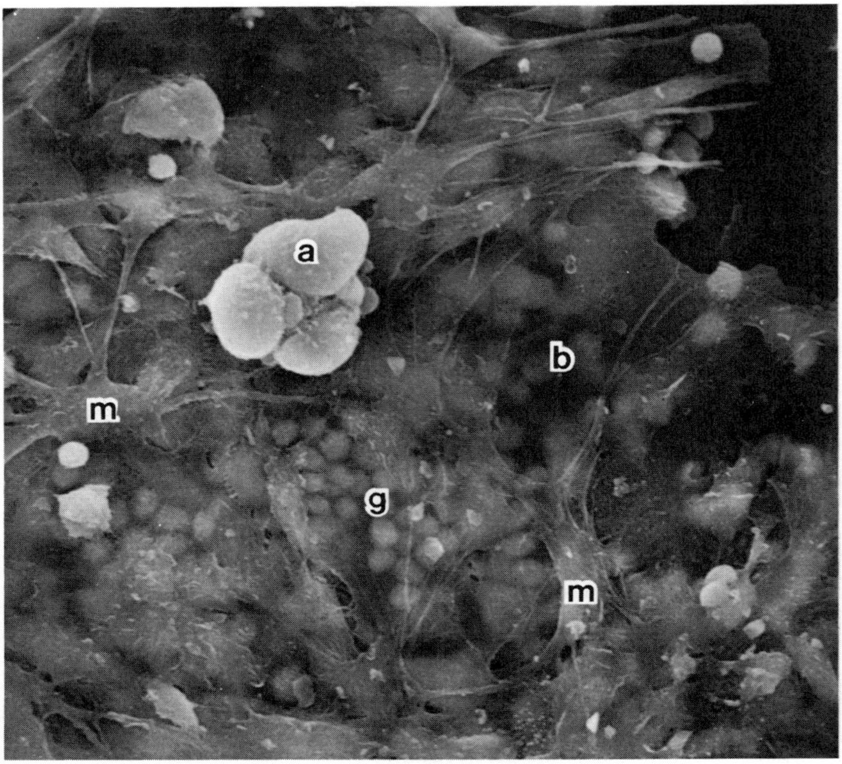

Figure 3 Scanning electron micrograph detail of a similar area to that shown in Fig. 2-2, where electron beam penetration also shows a group of developing granulocytes (g) beneath the blanket cell cytoplasm (b). Adipocytes (a) and macrophages (m) also occur on the surface of the blanket cell.

Cytochemical characterization of the SPC showed them to be uniformly positive for alkaline phosphatase as a membrane marker up to 4 days in culture, after which they became heterogeneous, reducing to 50% after 7 days, 30% after 14 days, and less than 1.5% at 6 weeks. Investigation of the nonattached cell populations with cytospin preparations showed that the SPC alkaline-phosphatase-positive cells formed a very rapidly attaching population, and on the relatively rare occasions when they were found in cytospin preparations, were often associated with maturing granulocytes. Immunocytochemical

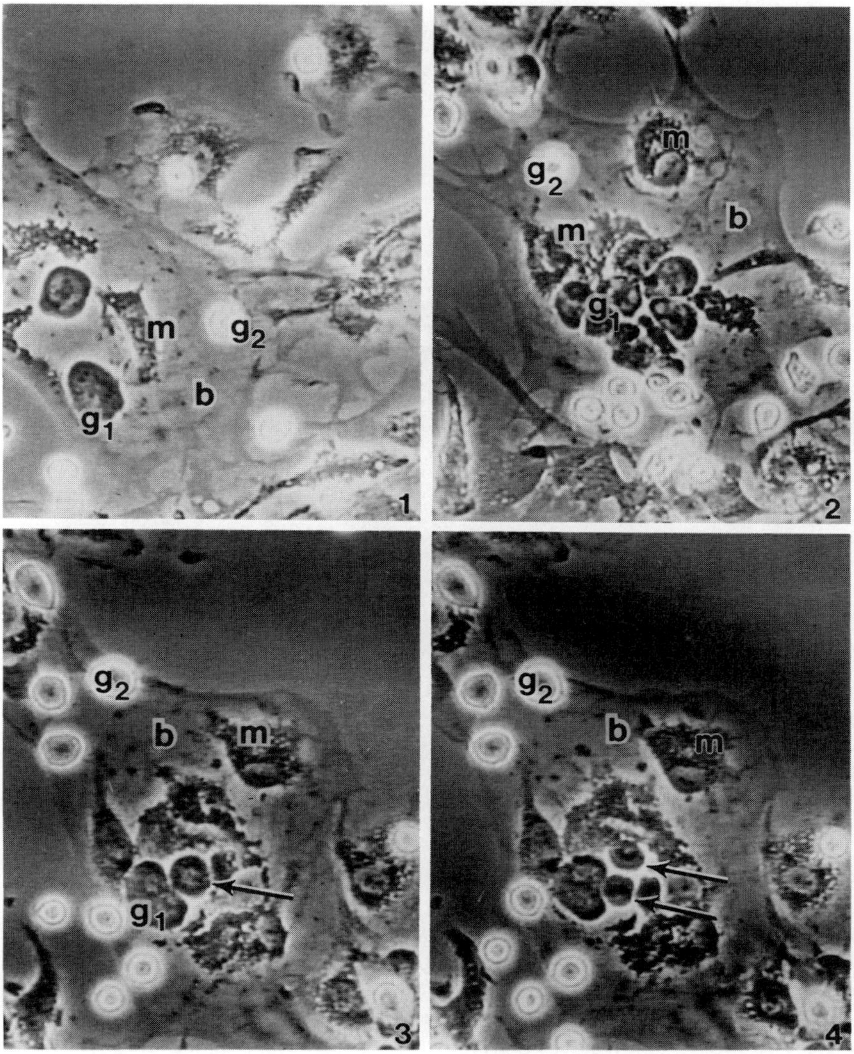

Figure 4 Sequence of time-lapse video observation recorded during approximately 24 hr at the same location in the adherent layer of a 3-week murine long-term bone marrow culture. Macrophages (m) and developing granulocytes (g_1) are beneath the cytoplasm of a blanket cell, and are both larger and more phase dense because of the flattening that occurs underneath blanket cells. In contrast, other granulocytes (g_2) on the surface of the blanket cell are spherical and phase bright, with an obvious halo. Frames 1 and 2 show the increase in numbers of both macrophages (by migration) and granulocytes during 24 hr. Frames 3 and 4 show the typical site of granulocyte proliferation, which is in association with macrophages and underneath the blanket cells, and a division of the granulocyte (arrowed in frame 4).

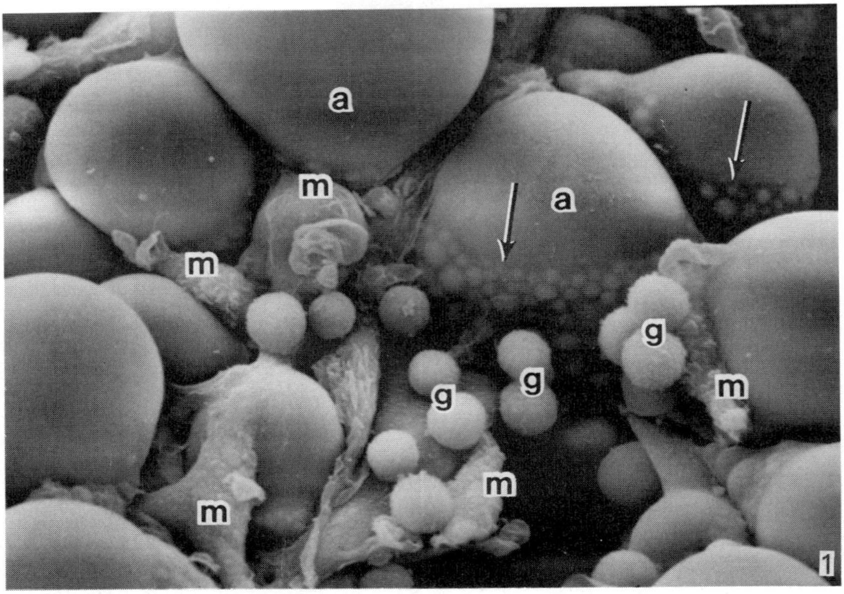

Figure 5 (1) Scanning electron micrograph of a typical cellular region of the adherent layer in a murine long-term bone marrow culture, illustrating the dense packing of adipocytes (1), macrophages (m), and granulocytes (g). The mature adipocyte (top center) consists mainly of a single lipid droplet, whereas two of the adjacent adipocytes (right) are less developed, and still display many individual lipid droplets (arrowed) (see also Fig. 6). (2) Transmission electron micrograph of a section cut parallel, and close to, the growing surface of the same flask in a similar region to that displayed in (a). Adipocytes (a), blanket cells (b), and macrophages (m) are clearly discernible. Adipocytes are characterized by one or more lipid droplets, blanket cells by their nuclear morphology and granular cytoplasm.

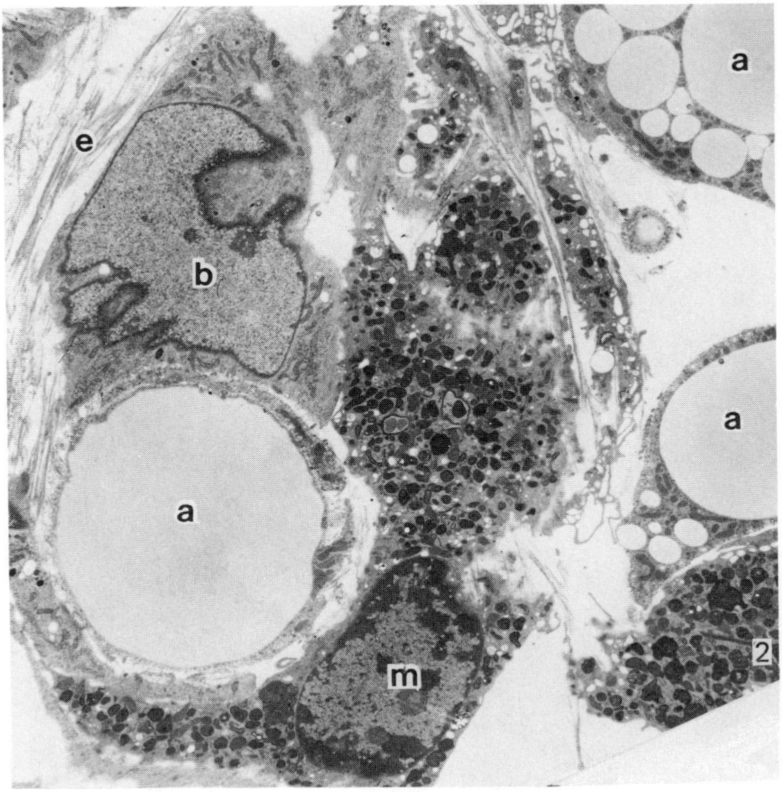

Figure 5 (continued)

characterization showed the SPC to be uniformly positive for fibronectin, collagen 1, and lamin during the first week of culture, with 60% also positive for collagen IV. By 2 weeks collagen 1 production was only apparent in 10% of the SPC, whereas lamin and collagen IV production stayed constant up to 3 weeks. Subsequently only 40-60% of the stromal cells were positive for collagen IV whereas 80% were positive for laminin.

At the same time as the numbers of cells positive for alkaline phosphatase began to decrease, the stromal precursor cells began to differentiate into two distinct subpopulations, namely, blanket cells and adipocytes (6,19,39). By 7 days blanket cells could be characterized by their increased size and highly spread cytoplasm, with characteristic euchromatic nuclear morphology with prominent nucleoli (Fig. 2). The increase in size and cytoplasmic spreading

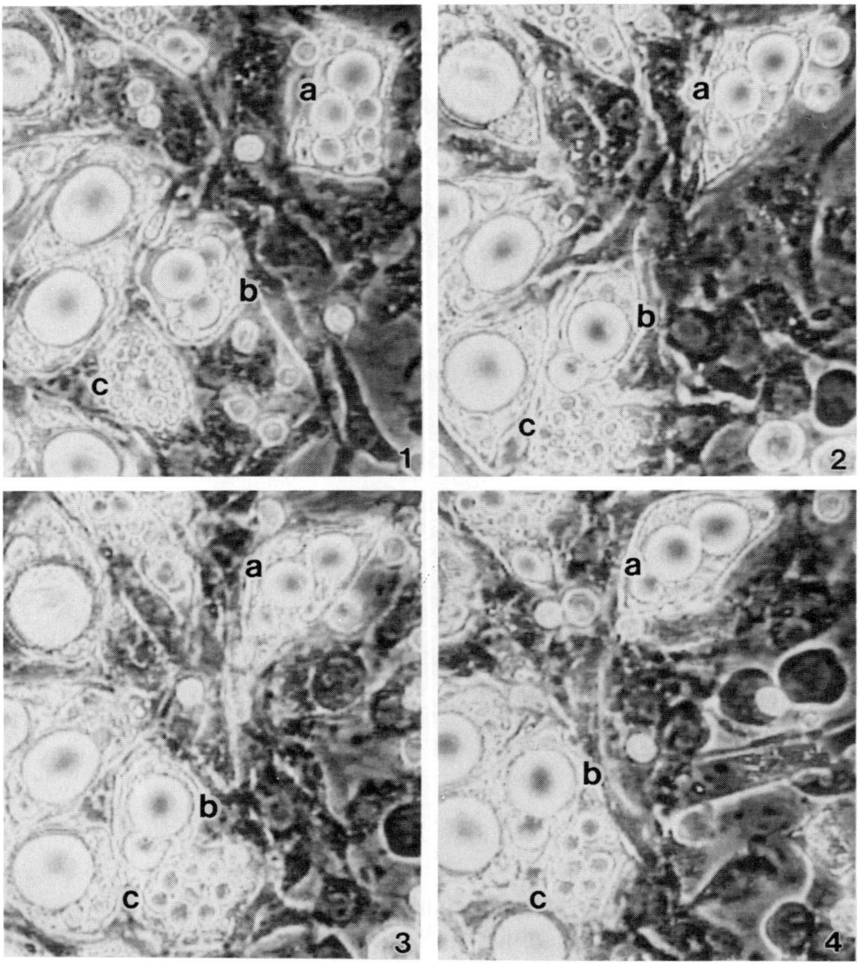

Figure 6 Sequence of time-lapse video observations of lipid accumulation in adipocytes, which occurs as initial lipid accumulation in small droplets followed by coalescence to form larger droplets, and ultimately, a single large lipid vacuole. This process is apparent over a period of 22 hr in the cells marked a, b, c. The times between frames are frame 1-frame 2: 7 hr, frame 2-frame 3: 5 hr, frame 3-frame 4: 10 hr.

was accompanied by a relative reduction in staining for alkaline phosphatase, which was lost altogether as the blanket cells matured. Adipocytes also become alkaline phosphatase negative as the earliest accumulation of lipid droplets became evident. In view of any last vestiges of misconception that might remain regarding the origin of the adipocytes, it is worth reiterating that although macrophages may occasionally possess a few small lipid droplets, they are not the adipocyte precursors, which are never positive for F4/80 antibody. At all stages of the adherent layer establishment the macrophage population is ubiquitous, but characteristically migrates underneath developing blanket cells, as shown by time-lapse video, and after dividing once or twice in this environment, remains underneath the blanket cell for several days (as directly observed by time-lapse video), and perhaps for the entire life of the culture (Fig. 3).

During the establishment of the marrow-adherent layer in mouse LTBMC, the single SPC population is in constant interaction with attached macrophages and appears to be bipotential for blanket cells and adipocytes. Our belief is that this stromal precursor cell probably represents a maintenance of the reticular/adventitial reticular cells from the explanted marrow itself, in view of the potential for lipid accumulation that mirrors the in-vivo situation for adventitial reticular cells. We have also found that these cells are consistently negative for factor VIII at the early stages of culture. This differentiation of SPC with its concomitant limitation of proliferation is probably the reason for the absence of a confluent adherent layer in mouse cultures. In human cultures, however, differentiation toward blanket cells appears to be reduced, with the whole culture becoming "overgrown" by a relatively undifferentiated "fibroblastoid" population, although adipocytes are also produced in large numbers. Although in the past we have reported the presence of endothelial cells in mouse cultures, and still believe them to occur occasionally as individual patches of pavementlike epithelia, we believe it is unlikely that they are intimately involved in the hemopoietic process in vitro. Endothelial cells have recently been positively identified in human cultures (40), but form only a very small proportion of the stromal adherent cells. No convincing marker for venous sinuous endothelium in mouse yet exists, and characterization of cell types by their basement membrane extracellular matrix constituents is fraught with difficulty when basement membranes only exist as the "jam in the sandwich" where venous sinus endothelium and adventitial reticulum cells come together in marrow in vivo (47).

V. CELLULAR INTERACTIONS IN HEMOPOIESIS

Perhaps the best known cellular association in the formation of hemopoietic elements is the erythroblastic islet in which the final stages of red blood cell maturation take place on the surface of a macrophage. Although this interaction is easy to recognize because of the differentiated nature of the cells involved, in time it may be possible to ascertain in what cellular "niche" (37) a stem cell retains or loses "stemness" and in which other "niches" it becomes committed to a particular lineage. So far, we have no marker for the stem cell itself, so we cannot pinpoint the cell or cells that provide the boundaries to the "niche" in which the stem cell resides. As a hypothesis, the "stem-cell niche" is well accepted, but the next step forward must be the identification of the cellular architecture of that niche. In the meantime we can usefully study cellular interactions between cells that are recognizable both for their own significance, and in the hope that they indicate some mechanisms that drive the self-renewal and commitment of earlier cells.

A. Stromal Precursor Cell/Macrophage Interactions

During the early establishment of the adherent layer in LTBMC (see above) there appears to be a chemotactic stimulus from the forming stroma foci to adjacent macrophages, which at that stage appear to be randomly distributed over the growing surface. Time-lapse video observations for the first few days of culture show a migration of macrophages into the foci of proliferating SPCs, where they both overlap and underlap the stromal-cell islands. Furthermore, having migrated underneath the blanket cells, the macrophages, although observed to divide, become fixed, although they may well generate peripheral cytoplasmic extensions that protude past the edge of the blanket cell. In mature regions of blanket-cell islands the subblanket-cell macrophages appear to create a network of cells. Time-lapse video shows the macrophages to be very active in terms of extending and retracting ruffling cytoplasmic extensions, but characteristically immobile with respect to cell migration, F4/80 macrophage-specific antibody staining displays a regular distribution of positive cells over large areas of the marrow cultures and also a very similar distribution in sections from bone marrow in vivo.

B. Blanket Cell/Myeloid Series/Macrophage Interactions

In mature cultures, "cobblestone" regions of granulopoiesis are formed underneath blanket cells as a mixture of macrophages and developing granulocytes (Fig. 7). This association is essentially an extention of the initial macrophage/stromal precursor association, into which the myeloid precursors migrate. Time-lapse video has shown that division of myeloid progenitors takes place almost exclusively in these regions, and that there is a constant, dynamic, membrane-to-membrane interaction between the dividing population of maturing granulocytes and the "fixed" macrophages beneath the blanket cells. Once mature, the granulocytes migrate out of the cobblestone region, passing through the blanket cells in an in-vitro "transmural passage," to rejoin the nonadherent population. In almost all observed myeloid divisions, the dividing cell had previously been in intimate membrane contact with a macrophage. Cobblestone regions are observed as transient events within the adherent layer, suggesting that a particular microenvironmental region supports a virogous but limited series of granulocytic proliferations and then becomes "exhausted," possibly requiring a period of regeneration before subsequently supporting a fresh burst of granulopoiesis.

This apparently cyclical nature of the support of granulopoiesis in vitro can be applied to marrow in vivo as an attractive hypothesis to explain the existence of the enormous reserve potential for hemopoietic production. For instance, if only a small proportion of the myeloid (or erythroid) microenvironements are in production at any one time, the observed rapid response to blood loss would merely involve switching more "production units" into active hemopoiesis.

C. Mature Granulocyte/Macrophage Interactions

"Granulocytic islets" were initially identified in the LTBMC by Simmons (39) and are associations of mature granulocytes and macrophages that occur separately from other elements in the culture (Fig. 8). The granulocytes migrate from underneath the blanket cells in the cobblestone regions, briefly join the "nonadherent" population, and then subsequently attach to the surface of free macrophages. Time-lapse video observations of these associations (Fig. 9) reveal four distinct features: (a) that the vast majority of granulocytes make a relatively permanent association with the macrophages (a few hours to a few days); (b) that this association is highly dynamic, constant movement of granulocytes and macrophages, (c) that the granulocytes involved are postmitotic; and (d) that should spontaneous granulocyte death occur, the dead cell is very rapidly phagocytosed by the macrophage. This last occurrence appears to be a regular event, because many of the macrophages contain phagocytosed cell remnants.

Speculations on these associations with respect to their in vivo significance can be made thus: Do mature granulocytes attach to macrophages deep in the hemopoietic marrow as a means of transport and delivery to the circulation via the venous sinus? Do the macrophages exert a "quality control" function over mature granulocytes in which they perceive defective granulocytes on their surface and will remove them via phagocytosis? The method of "perception" of cell death by the central macrophage may well be based on the constant

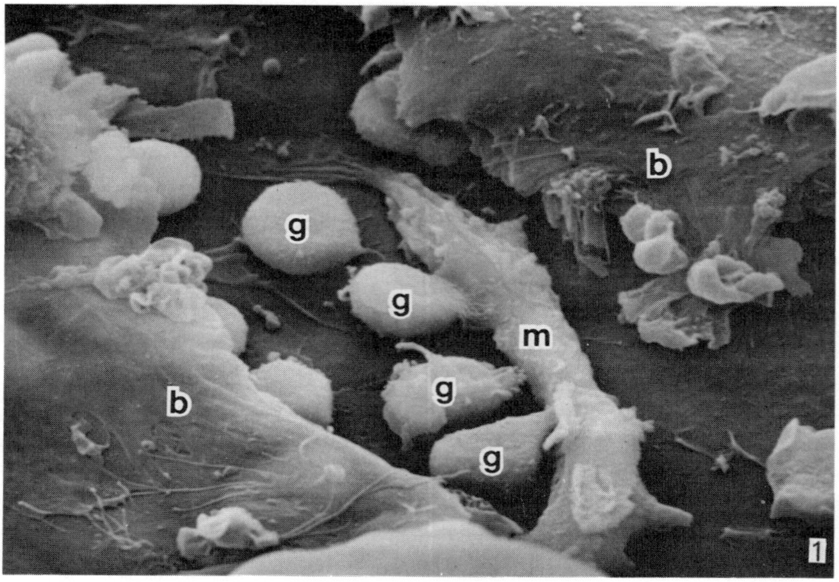

Figure 7 (1) and (2) Scanning electron microscopy of a region of granulocyte proliferation (cobblestone region) in which the normally overlying blanket cell has become split during preparation, exposing the interaction between the underlying granulocytes (g) and "fixed" macrophage (m). Note the relatively flattened appearance of the granulocytes (as indicated by their phase contrast appearance in Fig. 4), and the cellular extensions toward the macrophage by the three lower granulocytes. In the detail shown in (b), the large petallike process from the granulocyte is an extensive apposition with the macrophage membrane.

movement of the granulocytes over the surface of the macrophage. When a granulocyte dies in this situation, cell death is preceded by an intense blebbing, followed by assumption of a spherical morphology at which time membrane permeability changes take place (as the phase bright birefringence is lost and the cell becomes phase dark). The latter change often occurs between a single cycle of time-lapse exposures (i.e., less than 20 sec). Overall, from the moment of earliest alteration in the behavior and shape of the granulocyte, phagocytosis is completed within 30 min.

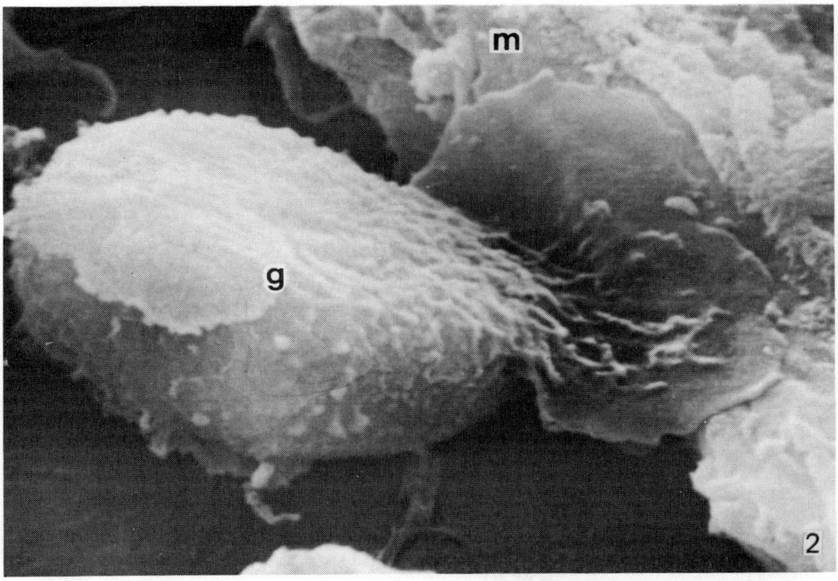

Figure 7 (continued)

D. Erythroid Series Macrophage Interactions

The erythroblastic islet is perhaps the best-documented cellular association of the hemopoietic system. Our own studies have been largely restricted to erythroid production in LTBMC generated by the addition of anemic mouse serum (AMS) (4). In these cultures, red blood cell differentiation is characterized by the production of adult hemoglobin in contrast to embryonic hemoglobin as seen in other in-vitro systems (19). Erythropoiesis in long-term cultures is particularly accessible to SEM, allowing identification of all stages by cell shape and surface morphology (4), including normoblast enucleation and the maturation of reticulocytes to the typical erythrocyte morphology. More recently, we have also studied the same events using a time-lapse video system (2).

Erythropoiesis in vitro also occurs as an association between the erythrocytic series and macrophages. In LTBMC, 2-3 days after the addition of AMS, recognizable proerythroblasts are observed on the surface of macrophages. Maturation then proceeds over four or five divisions, producing 16 or 32 erythrocytes.

Figure 8 (1) Scanning electron microscopy of a typical "granulocyte islet" as observed in murine long-term bone marrow culture, between mature (postmitotic) granulocytes (g) and a central macrophage (m). In (2) a section through a similar association shows the macrophage-specific antibody F4/80 (8) visualized using colloidal gold immunocytochemistry (39) at the surface of the macrophage.

The time-lapse video studies confirm the idea of a synchronous maturation of a cohort of red blood cells being produced as the progeny of each proerythroblast. The approximate 8-11-hr generation times oserved were in accord with those reported for mouse fetal liver erythropoiesis in short-term culture (15). Enucleation followed the last division by some 11-16 hr, and was accompanied by rapid phagocytosis of the liberated nuclei. The freshly formed reticulocytes continued to show very active movement on the surface of the macrophages for many hours, perhaps avoiding the phagocytic fate of expelled nuclei in this way (Fig. 10).

One of the most striking features of the maturation of the erythron is the high degree of synchrony maintained throughout the three or four divisions of

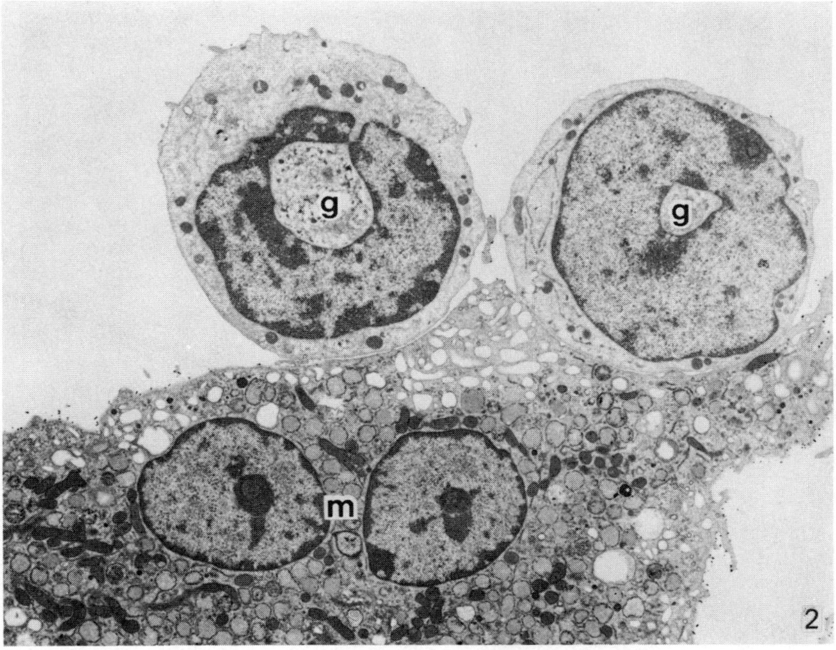

Figure 8 (continued)

the proerythroblast, as also reported in vivo (10,11). The biological significance of this synchronous maturation is intriguing. Perhaps the most obvious speculation is a simple one-namely, that if migration of the erythroblastic islet macrophage and its cohort of developing red blood cells toward the venous sinus is to be most efficient, ideally the macrophage should reach the wall of the venous sinus with *all* of its maturing reticulocytes/erythrocytes ready for delivery into the circulation, resulting in a minimal time spent at the venous sinus wall. This would be followed, perhaps, by the return of the erythroblastic islet macrophage to the hemopoietic region of the marrow for possible association with another proerythroblast and a repeat of the cycle. It is possible that the erythroblastic islet macrophage has a role to play in adjusting the cell cycle times of its attached erythron to maintain this synchrony, at the same time as it stimulates the erythron development directly. The erythroblastic islet macrophage also has the more obvious function of phagocytosis of extruded nuclei, which presumably leads to the recycling of the subsequent breakdown products. The interactions observed in all aspects of hemopoiesis in vitro are summarized diagrammatically in Figure 11.

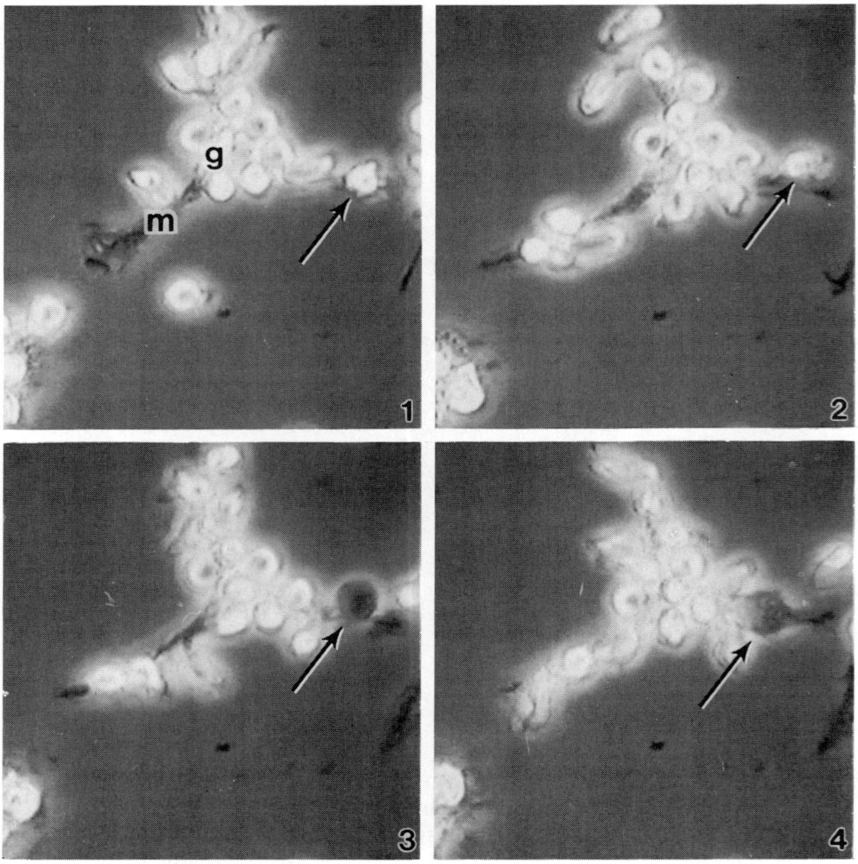

Figure 9 Sequence of time-lapse video observations recorded for 71 min in mouse long-term bone marrow culture. In frame 1 a granulocyte at the periphery of the "granulocytic islet" (g) on the surface of a macrophage (m), after undergoing characteristic blebbing for approximately 15 min, adopts a rigid spherical confirmation on the surface of the macrophage (arrowed). Thirty minutes later (frame 2) the inert granulocyte is still present on the surface of the macrophage, but after a further 18 min (frame 3) the whole cell has been engulfed in a phagocyte vacuole, which is then transported into the central region of the macrophage, during the next 24 min (frame 4 + 6 min, frame 5 + 12 min, frame 6 + 12 min).

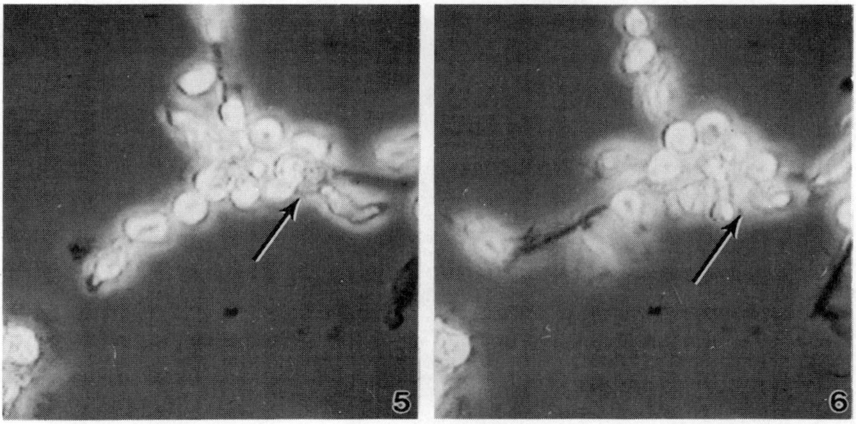

Figure 9 (continued)

VI. MECHANISMS OF CELLULAR INTERACTION AND SIGNALING IN BONE MARROW IN VIVO AND IN VITRO

Ultrastructural characterization of cell-to-cell contacts can be categorized as follows. Mechanically resistant junctions are termed desmosomes, and are typical of epithelia. They may also exist in other tissues in a slightly reduced form of differentiation as zonulae adherens junctions. Tight junctions are also typical of tissue interfaces, and provide an impermeability barrier, often adjacent to or in close collaboration with band (continuous) or spot (local) desmosomes. Communicating junctions, where membranes virtually fuse,and are spanned by small channels allowing free flow of small molecules between the connected cells, are termed gap junctions. In bone marrow in vivo, zonulae adherens, tight junctions, and gap junctions have all been reported by Watanabe (46) who also demonstrated an elegant network arrangement of stromal elements forming a three-dimensional interconnecting network of reticular cells composed of periarterial adventitial cells, intersinusoidal reticular cells, and perisinusoidal adventitial cells. The endothelial cells were connected to each other by gap junctions and tight junctions in arterioles, but by tight junctions only in sinuses and central veins. Zonulae adherens junctions occurred between adjacent adventitial and reticular cells, but there were no apparent junctions (gap or zonulae adherens) betwen any reticular cell endothelia. Gap junctions were common between adjacent reticular cells, either periarterial, intersinusoidal, or perisinusoidal. Similar findings were also reported by Campbell (13), who used a tannic-acid-supplemented fixation to show a pentalaminar structure resembling

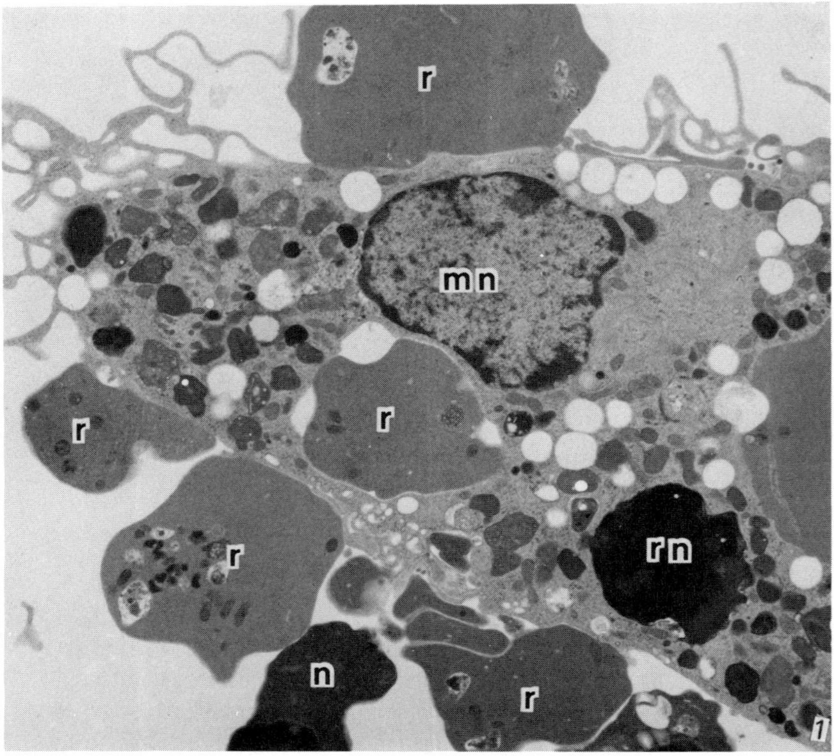

Figure 10 Macrophage/erythroblast interaction in murine long-term bone marrow culture. In (1) an erythroblastic islet has been sectioned parallel to the growing surface, showing six reticulocytes (r) attached to the surface of the central macrophage, and also an adjacent normoblast (n) probably in the process of enucleation. The macrophage nucleus (MN) is indicated, but an expelled reticulocyte nucleus (rn) is also present within a phagocytic vacuole. (2) A detail of the interfaced membranes in another erythroblastic islet reveals coated pits and vesicles (v) indicative of the process of receptor-mediated endocytosis at both the surface of the erythrocyte (e) and the macrophage (m).

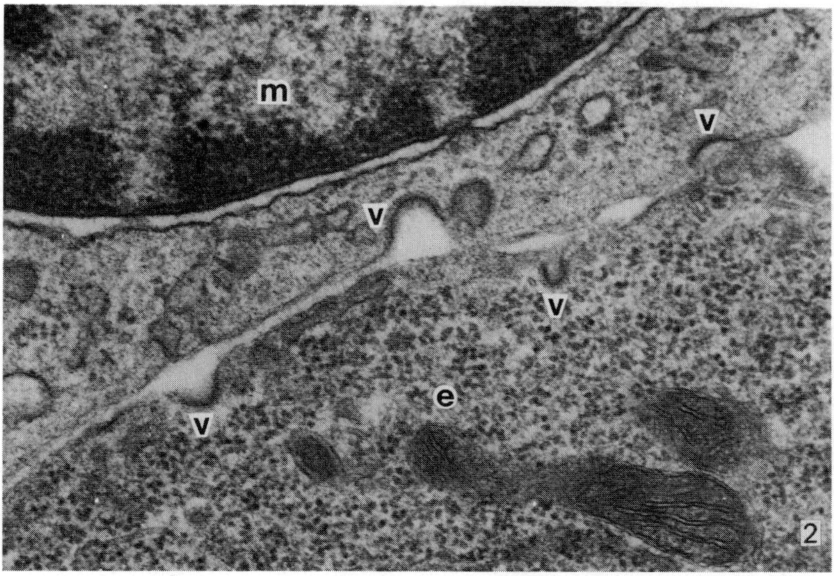

Figure 10 (continued)

gap junctions between the membranes of migrating blood cells and both endothelial and adventitial reticular cells of the sinusoidal wall. No gap junctions in these situations, however, could be confirmed by freeze fracture, and Watanabe (46) concluded that the pentalaminar junctions differed from true gap junctions. Tavassoli and Shaklai (44) also reported a failure to find gap junctions by freeze-fracture techniques in normal bone marrow, and concluded that such structures did not exist in this tissue. To add to the confusion, Porvaznik and MacVittie (34) reported typical gap junctions between adjacent macrophages in canine cultures when using freeze fracture, but were unable to visualize them by section transmission electron microscopy.

Thus, the current position for observation of ultrastructurally defined communicating structures between hemopoietic cells and stromal cells in bone marrow is relatively straightforward. They apparently do not exist. This is also the situation in marrow culture, where communicating and mechanical structures are observed between stromal cells, but not between stromal and hemopoietic cells. The finding with respect to cell communicating structure that has been consistently observed for the last few years (1,5,7) has been the presence of coated vesicles at the adjacent surfaces of cells that are in intimate membrane contact. These vesicles are indicative of the process of receptor-mediated endo-

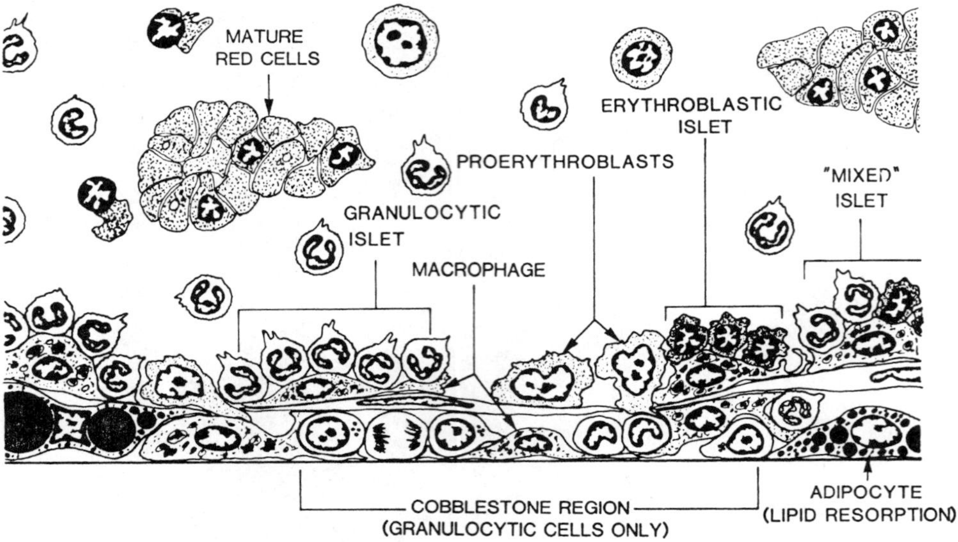

Figure 11 Diagram summarizing the cell types and their stromal/hemopoietic interactions in mouse long-term bone marrow culture. (From Ref. 39, used with permission.)

cytosis, a well-established mechanism of perception and response to stimuli in the form of membrane-bound molecules. Intimate membrane contact itself may be sufficient to allow presentation of a specific molecule from one cell to its neighbor. Membrane contact between adjacent macrophages in cobblestone regions of LTBMC has been shown to be sufficiently "tight" to exclude electron-dense tracers that are usually blocked only by tight junctions themselves (20). The requirement for both the presence and intimate contact of stromal cells with differentiation is well established (9,16,18). Thus, in the absence of any evidence (despite 3 decades of intensive investigation) it appears that the mechanism of cellular interaction between stromal and hemopoietic cells is not elaborated by the more typical cell communication mechanisms characterized by specific membrane structures in other cells. This observation should not, however, cast doubt on the role of cellular interaction in marrow, because the whole spectrum of cellular interactions in the immunological field shows no structural membrane differentiation either. Thus, the membrane mechanisms involved when a macrophage presents antigen to a T lymphocyte may well be similar to the stromal-cell presentation of a growth factor to a hemopoietic cell.

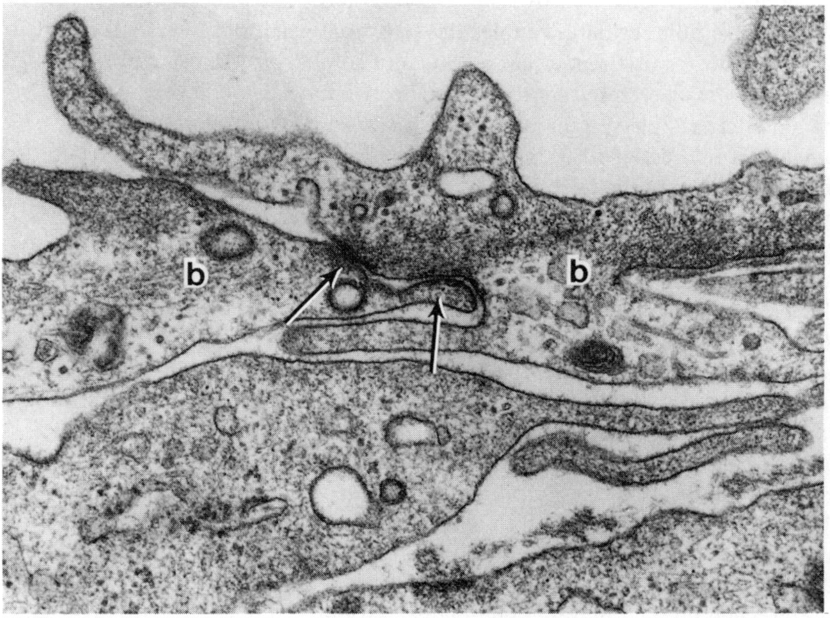

Figure 12 Section through the junction of adjacent blanket cells in mouse long-term bone marrow culture, showing membrane differentiation and microfilament involvement in an "adherens"-type cell-to-cell junction.

VII. THE ROLE OF CELLULAR ADHESION, EXTRACELLULAR MATRIX, AND GROWTH FACTORS IN STROMAL-CELL-MEDIATED HEMOPOIESIS

From the previous discussion, it is clear that stromal cells can support hemopoiesis in the absence of added growth-regulatory molecules. However, several other chapters in this volume address the role of soluble growth factors such as interleukin-3 (IL-3) and the colony-stimulating factors (CSFs) active on the granulocyte (G) and macrophage (M) lineages. Clearly, these factors also have a role to play in hemopoiesis and can stimulate the in vitro proliferation and development of multipotential stem cells, as well as cells commited to the neutrophil/macrophage, eosinophil, megakaryocyte, erythroid, and mast-cell lineages, in the absence of stromal cells. What, then, is the relationship between stromal cells and growth factor?

In the LTBMC, proliferation and development of hemopoietic cells *only* occur in intimate association with the stromal cells: if the hemopoietic stem cells are physically separated from the stromal cells using a barrier that permits diffusion of soluble regulatory molecules (e.g., agar gels), the stem cells rapidly die. This indicates that significant amounts of soluble growth regulatory molecules are not being secreted by the stromal cells. Indeed, we have also shown, using a model cell-line system (see below) that *physical proximity* with the stromal cells is insufficient to maintain the survival and proliferation of stem cells: For the stem cells to respond to the stroma, cellular *adhesion* is clearly important. These data raise three important questions. First, what is the nature of the adhesion molecules? Second, which growth factors (if any) are produced by the stromal cells? Third, how are these growth factors "presented" in a biologically active form to the stem cells?

Long-term bone marrow culture, although maintaining haemopoiesis in vitro, is in many ways unsuitable for approaching these fundamental questions, because a high level of complexity is imposed by the heterogeneity of the stromal cells. Macrophages, for example, can constitutively produce a wide range of biological response modifiers that are active on hemopoietic cells. Similarly, fibroblasts can be induced to produce and secrete various CSFs. Within the LTBMC, however, it is now known what is the relative contribution of these cell types in maintaining stem-cell self-renewal, differentiation, development, or "steady state" hemopoiesis. Furthermore, the hemopoietic cells themselves are heterogeneous, and mature granulocytes produce a range of factors (leukotrienes and eicosanoids) that may well influence the growth of other hemopoietic cells or the function of the various types of marrow stromal cells. With this in mind, an attempt to approach the role of cellular interactions between stromal cells and hemopoietic cells at the mechanistic level has been based on simplified model systems.

A. Model Systems for Cellular Interaction

Several years ago, we showed that when LTBMC are infected with a retrovirus carrying the *src* oncogene (12), the virus infects and transforms the marrow stromal cells. Presumably as a consequence of the subsequent expression of the *src* oncogene in the stromal cells, such infected cultures show a dramatic increase in the numbers of pluripoint stem cells and a block in differentiation. Significantly, however, the differentiation block is reversible when the stem cells are injected in vivo or subsequently explanted onto normal marrow stromal cells. Further examination of these hemopoietic cells demonstrated that they were not themselves infectected with the *src*-containing virus (i.e., indicating that the effects in the original cultures were indirectly mediated by the transformed stromal cells) (51) and that they could be encouraged to form continuously opening cell lines if cultured in the presence of IL-3 (41). However, of major significance was our finding that although the cells die in liquid culture if deprived of IL-3, they proliferate and differentiate in the presence of marrow stromal cells (43) to produce mature cells of multiple hemopoietic lineages (neutrophils, macrophages, eosinophils, megakaryocytes, and erythroid cells). In this respect, they behave like normal, freshly isolated stem cells. Furthermore, we have recently been able to show that the cells will also differentiate in liquid culture if supplied with an appropriate combination of growth factors (26).

When these cell lines, designated FDCP-Mix (41) are co-cultured with marrow stromal cells, they attach to the stroma, proliferate, and produce the cobblestone regions characteristic of hemopoiesis in normal LTBMC. This leads to the production of mature cells that are morphologically and functionally indistinguishable from mature cells produced in similar conditions from freshly isolated normal stem cells. The obvious benefit of the FDCP-Mix cell lines, however, is that they provide us with an unlimited number of fairly homogeneous cells.

To investigate in more detail the interaction between the FDCP-Mix cells and stromal cells, we screened a variety of stromal cell lines (derived from embryonic and normal adult tissue) for their capacity to support the survival, proliferation, and development of these cells. To our surprise, we found that embryonic mesenchymal cell lines, such as 3T3 would efficiently promote the growth and development of both the FDCP-Mix cells and normal bone marrow cells that had been highly enriched for CFU-S (35). Similar to the situation in LTBMC, the growth and development of the hemopopietic cells in association with 3T3 cells requires intimate cell contact. Indeed, variant IL-3-dependent cell lines have been characterized that show a much reduced ability to attach to stromal cells. This, in turn, is associated with lack of a survival or proliferative response.

The stem-cell adhesion molecule has so far not been characterized. Antibodies to fibronectin will not block adhesion of the FDCP-Mix on normal marrow stem cells to either marrow stroma or 3T3 cells, nor will synthetic peptides containing the classical adhesion sequence, Arg-Gly-Asp-Ser, block the ability of stem cells to interact with stroma. It is known, however, that the interaction is via a membrane-bound moiety because extracellular matrix, deposited intercellularly, is unable to support the attachment of stem cells (35).

B. Growth Factors

What about growth factors? Here the data are also confusing. Although marrow stromal cells constitutively produce an M-CSF and can be induced to secrete GM-CSF and G-CSF (as well as other interleukins such as IL-1, IL-4, IL-6, and IL-7), most of these are barely (if at all) detectable in actively hemopoietic

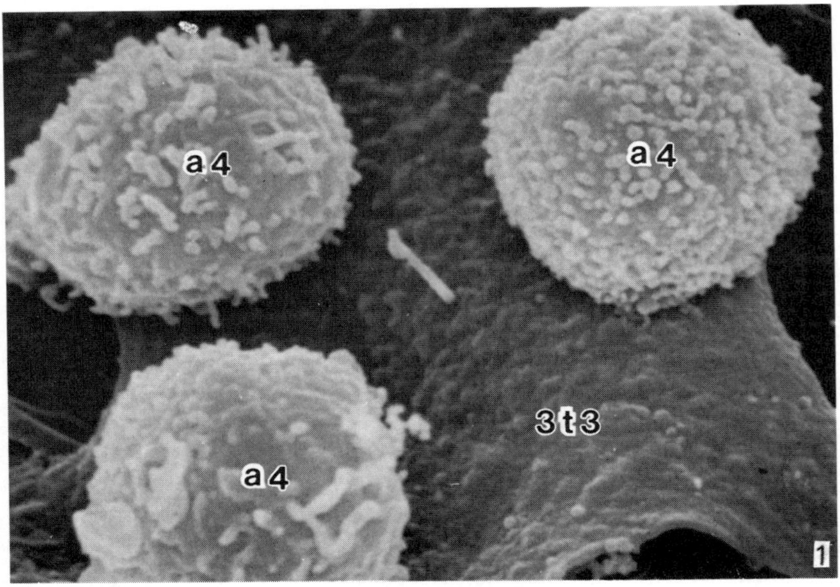

Figure 13 In (1) three factor-dependent A4 cells are seen attached to the surface of 3T3 monolayer, an association that maintains the A4 cells in the absence of their factor (IL-3) for several days. In (2), one such A4 cell has been removed from the surface of the 3T3 monolayer by a dry fracture technique and the limited areas of cellular apposition is visualized as indicated by the arrows. Despite almost constant migration by the A4 cells on the surface of the 3T3 cells (as indicated by time-lapse video) the attachment is sufficiently rigid to remove an area of 3T3 membrane during fracturing. (Courtesy of Dr. K. Yamazaki, Keio University, Tokyo.)

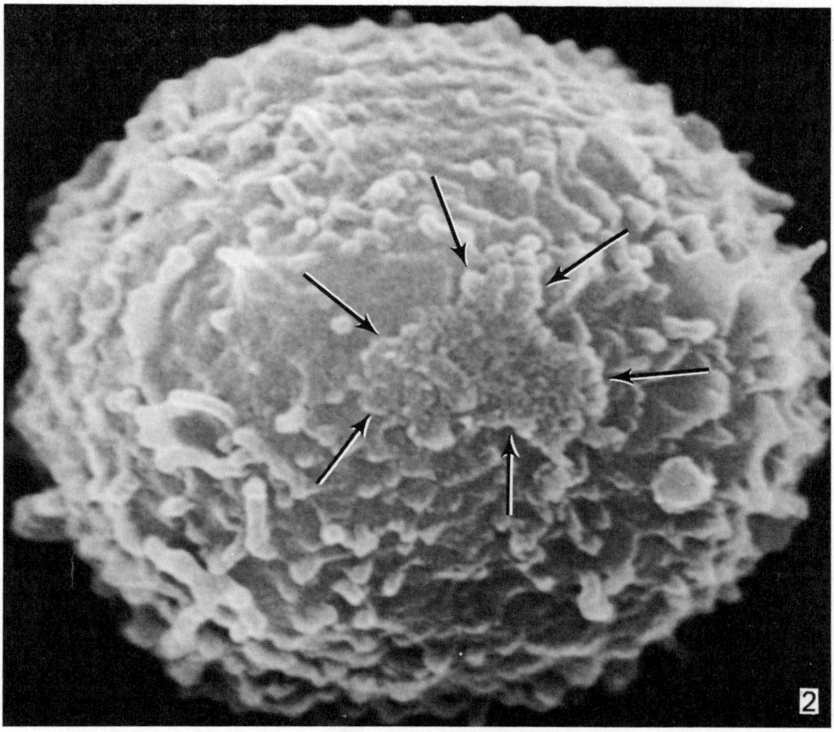

Figure 13 (continued)

LTBMC. Furthermore, IL-3 has *never* been detected in these cultures either at the biological or molecular (mRNA) level. Thus, although some of these factors may indeed be having some part to play in hemopoiesis in the LTBMC, their precise role in regulating stem-cell growth and differentiation is an enigma. Furthermore, 3T3 cells (which can support the growth and development of FDCP-Mix or normal stem cells) do so in the absence of detectable IL-3, GM-CSF, or G-CSF (again, using sensitive biological or molecular assay techniques). Clearly, myelopoiesis can occur in the apparent absence of these growth factors. This raises the intriguing question: Do stromal cells exert their effects on hemopoiesis by as yet uncharacterized growth-promoting stimuli, or is growth and development of stem cells (at least in these circumstances) regulated by another mechanism, e.g., through stimuli tranduced as a consequence of cellular adhesion?

Although the nature of the stromal cell-bound growth promoting stimuli remains unknown, recent work in both animals and humans has shown that myeloid cell growth factors (such as GM-CSF and G-CSF) can elicit dramatic responses when administered in vivo (see Chapters X and X). Clearly, high concentrations of circulating GM- and G-CSF, maintained for several days, can "override," at least in part, the growth regulatory influences exerted by the stromal cells. Because the GM- and G-CSF are administered in a soluble form, it is likely that they are exerting a direct effect on the responsive target cells via direct receptor binding. Another possible mode of action, however, is suggested by some recent experiments showing that growth factors such as IL-3 and GM-CSF can be sequestered and "compartmentalized" by molecules in the extracellular matrix (24). Furthermore, this finding has implications beyond the mere sequestration of soluble growth factors, because it provides a possible mechanism whereby growth-promoting stimuli that are themselves *produced* by the marrow stromal cell may subsequently be retained by the extracellular matrix present in the surrounding milieu. In other words, it creates the capacity for development of local "inductive" environments.

C. Extracellular Matrix

With this in mind, we have attempted to define more precisely the extracellular matrix component that is responsible for the binding of IL-3 and GM-CSF. In direct membrane-to-membrane interactions (such as occur during hemopoiesis in association with either marrow stromal cells or 3T3 cells), the most strategically placed molecules for binding and preventing growth factors to target cells are those in the extreme cell periphery. Furthermore, the finding that hemopoiesis in LTBMC or in 3T3 cells is *cell* associated suggests that the growth-promoting stimuli must be bound to membrane components. Indeed, when we analyzed the capacity of extracellular matrix to bind IL-3 and GM-CSF and to present these molecules to various target cells, we found that this function could be fulfilled by membrane-associated (but not intercellular) matrix components (36). When the membrane-associated matrix was digested with heparatinase, the ability of the matrix to bind and to present the growth factor to the target cells was much reduced. This indicated that heparan sulphate, one of the major sulphated proteoglycans present on the membrane of stromal cells, had a major role to play in growth-factor binding. This was conclusively demonstrated by isolating and purifying heparan sulphate from 3T3 cells. When this purified proteoglycan was covalently attached to a phospholipid (to facilitate adsorption onto tissue culture dishes), we found that it could bind GM-CSF and IL-3 from solution and support the growth-factor-dependent FDCP-Mix cells (36). Thus, these findings indicate a possible mecha-

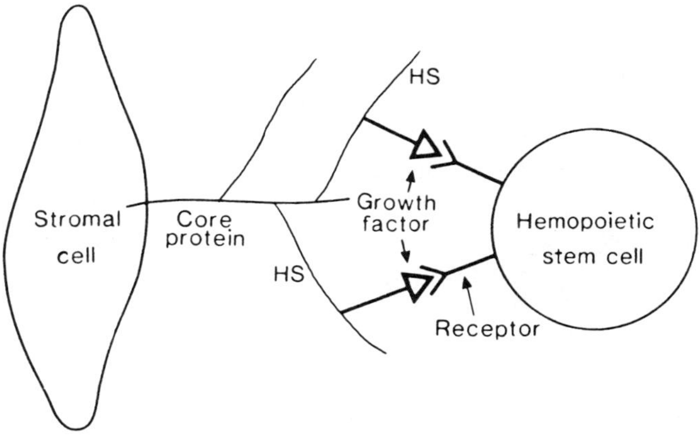

Figure 14 Diagram of putative heparan-sulphate-mediated presentation of growth factors by cellular elements of the bone marrow stroma, which may include a population of hemopoietically derived "fixed" macrophages. In this model, growth factors produced by the stromal cells of bone marrow are expressed on the stromal cell surface by binding to the heparan sulphate chains (HS) of transmembrane proteoglycans. Mitosis in hemopoietic stem cells is initiated by an interaction between specific plasma membrane receptors of the stem cells and the HS-bound growth factors in the stroma. This paracrine mechanism provides a means for restricting growth factors to specific microenvironments in the bond marrow. (From Ref. 23, used with permission.)

nisms for stromal-cell-mediated hemopoiesis involving (a) production of growth-promoting stimuli by stromal cells, (b) export to the cell surface, (c) sequestration by surface membrane-bound heparan sulphate, and (d) presentation to the "target" hemopoetic cells (Fig. 14). In this scheme, we also indicate the importance of cellular adhesion in the interactive process.

A potentially interesting scenario to this model is the finding that heparan sulphate proteoglycan inositol phosphate (which has been shown to be present on some cells) can be cleaved from the membrane and internalized by receptor-mediated endocytosis. A fragment of this molecule can then be rapidly transported to the nucleus where, in some systems, it has been shown to influence cell proliferation (23). Thus, a mechanism affecting mitotic activity can be directly associated with the ultrastructural evidence for receptor-mediated endocytosis, as suggested some time ago (1). Most of the ultrastructural evidence for receptor-mediated endocytosis in LTBMC has been observed between macrophages and their intimately associated granulocytic or erythrocytic cell populations, with evidence of receptor-mediated endocytosis (i.e., coated vesicles) apparent at the surface of both cells involved. The current evidence for heparan sulphate at macrophage surface membrane is controversial, although this may indicate that there are subsets of macrophages that bear the surface molecules specific to their particular task. Thus, a suggested role for macrophage involvement in haemopoiesis is as a "microenvironmental mediator" not of stromal origin but having integral functional signficiance with the microenvironment. Obviously, this is speculative at present but the suggestions do lend themselves to experiments at the mechanistic level.

VIII. THE STEM-CELL "NICHE"

In 1978, Schofield discussed the concept of a microenvironmental "niche" that maintained "stemness," i.e., an environment that inhibits differentiation and favors self-renewal. From the previous discussion it is perhaps opportune to ask what form such an environment may take. Stromal cells themselves are heterogenous and may therefore be able to produce and sequester (in local environments) quantitatively and qualitatively different kinds of growth-promoting, differentiation-inducing, and differentiation-inhibiting factors. It is not unreasonable to suggest, therefore, that the bone marrow consists of environmental domains (30) where certain events (self-renewal, differentiation, or lineage development) occur with a high probability. In terms of the response of *stem* cells this can be achieved in two ways. First, the stem cells may have specific receptors for relatively few growth/differentiation stimuli and the receptors for other factors are acquired during (perhaps as a consequence of) differentiation. Alternatively, the stem cells may express receptors for, and can respond to, most (perhaps all) of the growth and differentiation factors, and the cells become restricted in their ability to respond to particular stimuli as they become differentiated. Recent data from our laboratory tend to support the second model. We have shown that multipotential cells can respond to IL-3, GM-CSF, G-CSF, M-CSF, and IL-1 when used either alone (e.g., IL-3) or in combination (e.g., G-CSF plus M-CSF) (26). Of major significance, however, is that the *outcome* of the response is quite different. In the presence of IL-3, for example, the multipotential cells express all lineage potentials; in G-CSF plus M-CSF, although the same cells are recruited as in IL-3, only macrophage and neutrophil development occurs. Similarly, in M-CSF plus IL-1 (which again recruit essentially the same cells that can respond to IL-3) only macrophage development is seen.

These findings raise the possibility that stem-cell differentiation and development is deterministic in nature: that stem cells have the *potential* to respond to a variety of stimuli and that the developmental pathway taken by an individual stem cells is a reflection of the environmental milieu. This, in turn, is made up of the variety of stromal cells, extracellular matrix molecules, and soluble growth-regulatory molecules present. A fundamental problem, however, still remains to be resolved: Which environment forms self-renewal?

ACKNOWLEDGMENT

This work was supported by the Cancer Research Campaign (U.K.). T.M.D. is a Fellow of the Cancer Research Campaign.

REFERENCES

1. Allen, T. D. Haemopoietic microenvironments in vitro: ultrastructural aspects. In *Microenvironments in Haemopoietic and Lymphoid Differentiation* R. Porter, and J. Whelan (Eds.). Ciba Foundation Symposium 84, Pitman Medical, London, pp. 38-67 (1981).
2. Allen, T. D. Time lapse video microscopy using an animation control unit. *J. Microsc. 147*:129-135 (1987).
3. Allen, T. D., and T. M. Dexter. Cellular interrelationships during in vitro granulopoiesis. *Differentiation 6*:191-194 (1976).
4. Allen, T. D., and T. M. Dexter. Ultrastructural aspects of erythropoietic differentiation in long-term bone marrow culture. *Differentiation 21*:86-94 (1982).
5. Allen T. D., and T. M. Dexter. Long-term bone marrow cultures: an ultrastructural review. *Scanning Electron Microsc. 4*:1851-1866 (1983).
6. Allen, T. D., P. J. Simmons, T. M. Dexter, S. Hirsch, and S. Gordon. Stromal cell types in long-term marrow cultures. *Exp. Haematol. 11*(suppl. 14):127 (1983).
7. Allen, T. D., and T. M. Dexter. The essential cells o the haemopoietic microenvironment. *Exp. Haematol. 12*:517-521 (1984).
8. Austyn, J. M., and S. Gordon. F4/80: a monoclonal antibody directed specifically against the mouse macrophage. *Eur. J. Immunol. 10*:805 (1981).
9. Bentley, S. A. A close range cell:cell interaction required for stem cell maintenance in continuous bone marrow culture. *Exp. Hematol. 9*:303-312 (1981).
10. Bessis, M., C. Miye, and M. Prenant. Erythropoiesis: comparison of in vivo and in vitro amplification. *Blood Cells 4*:155-174 (1978).
11. Bessis, M., L. S. Lessin, and E. Beutler. Morphology of the erythron. In *Hematology*, W. J. Williams et al. (Eds.). McGraw Hill, New York, pp. 257-264 (1983).
12. Boettiger, D., S. Anderson, and T. M. Dexter. Effect of src infection on long-term marrow cultures. Increased self-renewal of haemopoietic progenitor cells without leukaemia. *Cell 36*:763-773 (1984).

13. Campbell, F. R. Gap junctions between cells of bone marrow: an ultrastructural study using tannic acid. *Anat. Res. 196*:101-117 (1980).
14. Chamberlain, J. K., P. F. Leblond, and R. I. Weed. Reduction of adventitial cell cover: an early direct effect of erythropoiesis on bone marrow ultrastructure. *Blood Cells 1*:655-674 (1975).
15. Cormack, D. Time lapse characterization of erythrocyte colony forming cells in plasma cultures. *Exp. Hematol. 4*:319-326 (1976).
16. Dexter, T. M. Cell interactions in vitro. *Clin. Haematol. 8*:453-468 (1979).
17. Dexter, T. M. Is the marrow stroma transplantable? *Nature 298*:222-223 (1982).
18. Dexter, T. M., T. D. Allen, and L. G. Lajtha. Conditions controlling the proliferation of haemopoietic stem cells in vitro. *J. Cell. Physiol. 91*:335-344 (1977).
19. Dexter, T. M., N. G. Testa, T. D. Allen, T. Rutherford, and E. Scolnick. Molecular and cell biologic aspects of erythropoiesis in long-term bone marrow cultures. *Blood 58*:699-707 (1981).
20. Dexter, T. M., E. Spooncer, P. J. Simmons, and T. D. Allen. Long-term marrow culture; an overview of techniques and experience. In *Long Term Bone Marrow Culture*, D. G. Wright and J. S. Greenberger (Eds.). Kroc Foundation Series 18, pp. 87-97 (1984).
21. Friedenstein, A. J., R. K. Chailakhjan, and K. S. Lalykina. The development of fibroblast colonies in monolayer cultures of guinea pig bone marrow and spleen cells. *Cell Tissue Kinet. 3*:393-403 (1970).
22. Friedenstein, A. J., R. K. Chailakhjan, N. V. Latsinik, A. F. Panasynk, and I. V. Keiliss-Borok. Stromal cells responsible for transferring the microenvironment of hemopoietic tissue. *Transplantation 17*:331-340 (1974).
23. Gallagher, J. G., and T. M. Dexter. Binding of growth factors to heparan sulphate: implications for the regulation of haemopoiesis by bone marrow stromal cells. In *Haematology Today* (in press).
24. Gordon, M. Y., G. P. Riley, S. M. Watt, and M. F. Greaves. Compartmentalization of a haematopoietic growth factor (GM-CSF) by glycosaminoglycans in the bone marrow microenvironment. *Nature 326*:403-405 (1987).
25. Gordon, M. Y. The origin of stromal cells in patients treated by bone marrow transplantation. *Bone Marrow Transplantation 3*:247-251 (1988).
26. Heyworth, C. M., I. L. O. Ponting, and T. M. Dexter. The response of haemopoietic cells to growth factors: developmental implications of synergistic interaction. *J. Cell. Sci. 91*:239-247 (1988).
27. Krstic, R. V. *Illustrated Encyclopedia of Human Histology*. Springer Verlag, New York (1984).
28. Leblond, P. F., J. K. Chamberlain, and R. I. Weed. Scanning electron microscopy of erythropoietin-stimulated bone marrow. *Blood Cells 1*:639-651 (1975).
29. Lord, B. I., and J. H. Hendry. Distribution of haemopoietic colony-forming units in the mouse femur and its modification by x-rays. *Br. J. Radiol. 45*:110-115 (1972).

30. Lord, B. I., N. G. Testa, and J. H. Hendry. The relative spatial distribution of CFU-S and CFU-C in the normal mouse femur. *Blood 46*:65-72 (1975).
31. Lord, B. I., and E. Spooncer. Isolation of haemopoietic spleen colony-forming cells. *Lymphokine Res. 5*:59-72 (1986).
32. McCulloch, E. A. Stem cell functions and the clonal haemopathies of man. In *Stem Cells and Tissue Homeostasis,* Brit. Soc. Cell Biology Symposium, Vol. 2, B. I. Lord, C. S. Potten, R. J. Cole (Eds.). Cambridge University Press, p. 139 (1978).
33. Owen, M. Lineage of osteogenic cells and their relationship to the stromal system. *Bone Mineral Res. 3*:1-25 (1985).
34. Porvaznik, M., T. J. MacVittie. Detection of gap junctions between the progeny of canine macrophage colony forming cell in vitro. *J. Cell Biol. 82*: 555-564 (1979).
35. Roberts, R. A., E. Spooncer, E. K. Parkinson, B. I. Lord, T. D. Allen, and T. M. Dexter. Metabolically inactive 3T3 cells can substitute for marrow stromal cells to promote the proliferation and development of multipotent haemopoietic stem cells. *J. Cell. Physiol. 132*:203-214 (1987).
36. Roberts, R., J. Gallagher, E. Spooncer, T. D. Allen, F. Bloomfield, and T. M. Dexter. Heparan sulphate bound growth factors: a mechanism for stromal cells mediated haemopoiesis. *Nature 332*:376-378 (1988).
37. Schofield, R. The relationship between the spleen colony-forming cell and the haemopoietic stem cell. A hypothesis. *Blood Cells 4*:4-7 (1978).
38. Schor, A., and S. L. Schor. The isolation and culture of endothelial cells and pericytes from bovine retinal microvasculature: a comparative study with large vessel vascular cells. *Microvasc. Res. 32*:21-28 (1986).
39. Simmons, P. J. Ph.D. Thesis. Victoria University of Manchester (1984).
40. Simmons, P. J., D. Przepiorka, E. D. Thomas, and B. Torok-Storb. Host origin of bone marrow stromal cells following allogeneic bone marrow transplantation. *Nature 328*:429-432 (1987).
41. Spooncer, E., D. Beottiger, and T. M. Dexter. Continuous in vitro generation of multipotential stem cell clones from src-infected cultures. *Nature 310*:228-230 (1985a).
42. Spooncer, E., B. I. Lord, and T. M. Dexter. Defective ability to self-renew in vitro of highly purified primitive haemopoietic cells. *Nature 316*:62-64 (1985b).
43. Spooncer, E., C. M. Heyworth, A. Dunn, and T. M. Dexter. Self-renewal and differentiation of interleukin-3 dependent multipotent stem cells are modulated by stroma cells and serum factors. *Differentiation 31*:111-118 (1986).
44. Tavassoli, M., and M. Shaklai. Absence of tight junctions in endothelium of marrow sinuses: possible significance for marrow cell egress. *Br. J. Haematol. 41*:303-307 (1979).
45. Trentin, J. J. Influence of hemopoietic organ stroma (hemopoietic micro-environment) on stem cell differentiation. In *Regulation of Hematopoiesis*, Vol. 1, A. S. Gordon (Ed.). Meredith, New York, pp. 161-186 (1970).

46. Watanabe, Y. Fine structure of bone marrow stroma. *Acta. Haematol. Jpn. 48*:1687-1700 (1985).
47. Weiss, L. The histology of the bone marrow. In *Regulation of Hematopoiesis*, Vol. 1, A. S. Gordon (Ed.). Appleton-Century-Crofts, New York, pp. 79-92 (1970).
48. Weiss, L., and L.-T. Chen. The organisation of hematopoietic cords and vascular sinuses in bone marrow. *Blood Cells 1*:617-638 (1975).
49. Weiss, L. The hematopoietic microenvironment of the bone marrow: an ultrastructural study of the stroma in rats. *Anat. Rec. 186*:161-184 (1976).
50. Weiss, L. Haemopoiesis in mammalian bone marrow. In *Microenvironments in Haemopoietic and Lymphoid Differentiation*, Ciba Foundation Symposium 84, R. Porter, and J. Whelan (Eds.). Pitman Medical, London, pp. 38-67 (1981).
51. Wyke, T. A., A. W. Stoker, S. Searle, E. Spooncer, P. J. Simmons, and T. M. Dexter. Perturbed haemopoiesis and the generation of multipotential stem cell clones in src-infected bone marrow cultures is an indirect or transient effect of the oncogene. *Mol. Cell. Biol. 6*;959-963 (1986).
52. Xu, C. X., J. H. Hendry, N. G. Testa, and T. D. Allen. Stromal colonies from mouse marrow: characterisation of cell types, optimization of plating efficiency and its effect on radiosensitivity. *J. Cell Sci. 61*:453-466 (1983).

2

Hematopoietic Growth-Factor Receptors

LINDA S. PARK, STEVEN GILLIS, and DAVID L. URDAL
Immunex Research and Development Corporation, Seattle, Washington

I. INTRODUCTION

A large body of work during the last several years has shown that a group of glycoprotein growth factors known as colony-stimulating factors (CSFs) are involved in regulating the survival, proliferation, and differentiation of hematopoietic progenitor cells, as well as affecting many functional activities of mature cells. The classical CSFs are G–CSF and M–CSF (CSF–1), which are relatively specific for end-stage progenitors of the granulocytic and monocytic lineages, respectively, and GM–CSF and IL–3, which are capable of acting on earlier progenitors to produce cells of not only granulocytic and monocytic lineages, but also eosinophils, megakaryocytes, mast cells, and erythroid cells under the right conditions (1,2). In addition to these molecules, the central role of the regulatory growth factor erythropoietin (EPO) in erythrocyte production has been known for many years (3,4). More recently, the picture of growth-factor control of hematopoiesis has become increasingly complex, with demonstrations that several other polypeptide factors are capable of acting on hematopoietic precursors. Interleukin-1 (IL-1), which has recently been shown to be identical to hemopoietin-1 (5), acts in synergy with CSFs to stimulate proliferation and differentiation of very primitive multipotent hematopoietic cells in the bone marrow (6-8). Similarly, in combination with different CSFs, IL-4 has been shown to both enhance and suppress colony formation by hematopoietic progenitor cells (9,10), whereas IL-6 (BSF-2, interferon β_2) has been shown to enhance IL-3-dependent progenitor proliferation (11). In contrast, IL-5 (eosinophil differentiation factor) functions as a lineage-specific stimulator of eosinophil production and eosinophil colony formation from bone marrow (12,13).

The availability of recombinant forms of these regulatory molecules in large quantities has allowed not only rapid progress in gaining an understanding of their biological activities, but has also allowed us to begin to study the specific cell surface receptors to which these molecules bind. Such studies are crucial to a full understanding of the mechanisms of action of the hematopoietic growth factors, and the complex interactions that must exist between them. In the last 4 years a sizable body of literature on the properties of these growth-factor receptors has begun to accumulate. Studies of the murine IL-3 (14-19), G–CSF (19-21) and CSF-1 (22-24) receptors, and the murine and human EPO (25-32), GM–CSF (33-39), and IL-4 (40-45) and IL-1 (46–52; see Ref. 53 for review of pre-1987 papers) receptors have been reported. Although some work has been reported on the human G–CSF receptor, primarily using the cross-reactive murine G–CSF (54-57), human G–CSF as well as human IL-3 have only recently become available in recombinant form for receptor studies (58). No work has yet been published on the murine or human IL-5 receptor, whereas the first article on IL-6 receptor (human) appeared recently (59). This chapter will present an overview of what is known about these hematopoietic growth-factor receptors in terms of ligand binding properties, cellular distribution, and structural characteristics. It will draw from the body of published literature as well as from recent work in our laboratory (particularly on human IL-3, G–CSF, and IL-1 receptors) and will also highlight the primary methodologies used to carry out these studies. Throughout, we will also present brief background material on principal properties of the larger family of growth-factor receptors to allow pertinent comparisons with those for the hematopoietic growth factors.

A. Historical Perspective

Due to the difficulties in purifying membrane-associated receptor proteins, information on the structural characteristics of growth-factor receptors has only begun to accumulate in the last few years. In 1984, epidermal growth factor (EGF) was the first of these receptors to be cloned and sequenced (60). Since that time the cloning and structural analysis of receptors for platelet-derived growth factor (PDGF) (61), insulin (62,63), insulin-like growth factor-1 (IGF-1) (64), nerve growth factor (NGF) (65,66), and one subunit of the IL-2 receptor (67-69) have been reported. In addition, the sequence of the protooncogene *fms*, which is believed to be closely related and possibly identical to the receptor for CSF-1 (70), has also been determined (71). All these receptors are polypeptides or multisubunit polypeptide complexes that span the plasma membrane and consist of an extracellular ligand binding domain (the amino terminus in all cases), a hydrophobic transmembrane spanning region and an intracellular domain that may participate in signal transduction after binding of hormone. The structures of the EGF, PDGF, and CSF-1 receptors exhibit many similarities. Each consists of a single polypeptide chain (M_r = 170, 180, and 165 kDa, respectively), with large intracellular and extracellular domains. The extracellular domain has a high number of hypothetical *N*-linked glycosylation sites, and the molecules are in fact heavily glycosylated. In addition, the EGF receptor has cysteine-rich regions in the extracellular domain (which may serve to stabilize regions of this domain), a feature the PDGF and CSF-1 receptors do not share. The intracellular domain of all three receptors share sequence homology with members of the tyrosine kinase family, and exhibit ligand-activated tyrosine kinase activity.

In contrast, the receptors for insulin and IGF are disulfide-linked tetrameric complexes, consisting of two α subunits of M_r 125-135 kDa, and two β subunits of M_r 95 kDa. Both these receptors have a carbohydrate- and cysteine-rich extracellular domain reminiscent of the EGF receptor and exhibit ligand-activated tyrosine kinase activity. However, additional complexity is introduced by the fact that the α subunit is responsible for ligand binding, whereas the β subunit exhibits the tyrosine kinase activity.

The receptors for NGF and IL-2 exhibit characteristics different from any of these other receptors. The NGF receptor is small (M_r = 75 kDa) with a cysteine-rich extracellular domain and low carbohydrate content. The intracellular domain is also small, consisting of 155 residues, with no protein kinase homology. Evidence from ligand binding and crosslinking studies have led to the suggestion that the receptor is present, at least partially, as a complex containing an additional M_r ~60 kDa protein (66). The subunit of the IL-2 receptor that has been cloned is also very small (M_r = 55 kDa), with a moderately carbohydrate- and cysteine-rich extracellular domain and a cytoplasmic domain too small (only 13 residues) to exhibit enzymatic activity. Although this subunit alone can bind IL-2 with low affinity, considerable evidence exists that it can also be found complexed with an additional protein of 70-75 kDa (72). This complex binds IL-2 with high affinity, whereas the 70-75 kDa subunit can itself bind IL-2 with an intermediate affinity.

In contrast, little is known about the structure of most of the hematopoietic growth factor receptors. Low expression of these receptors on most murine and human cells, as well as lack of availability of large quantities of purified ligand, has severely hampered attempts at purification and structural analysis. In addition to the receptor for CSF-1, the only other hematopoietic growth factor receptor that has been cloned is that for IL-1 (52), and that only very recently. Details on the structure of this receptor will be elaborated later in this chapter. Information about the structures of the other hematopoietic growth factor receptors has been obtained primarily through affinity crosslinking studies, and will also be discussed.

II. LIGAND BINDING CHARACTERISTICS

A. Preparation of Radiolabeled Ligands

The single most important tool for receptor characterization is a radiolabeled ligand with sufficient specific radioactivity to reproducibly detect cell-surface-associated receptors. Although it is possible to do receptor studies using [^{35}S]-methionine biosynthetically labeled recombinant proteins, almost all studies are done using radiolabeled ligands in which iodine-125 has been covalently attached to either lysine or tyrosine residues in the molecule. The principal methods employed are use of the lysine-directed Bolton-Hunter reagent, and addition of $^{125}I_2$ to tyrosine residues catalyzed by chloramine-T or lactoperoxidase (either soluble or solid phase). The success of any of these methods depends on several factors, including the absolute number of lysine or tyrosine residues present in a molecule, and sensitivity of the binding affinity of the molecule to derivatization of these particular residues.

We have employed a solid-phase lactoperoxidase method (the enzymobead radioiodination reagent from Biorad) to successfully radiolabel murine and human IL-3 (15,58), GM–CSF (33,38), and IL-4 (40,42), and human G–CSF and CSF-1 (58). This method uses a solid-phase support to which both lactoperoxidase and glucose oxidase have been immobilized to generate a low amount of hydrogen peroxide from glucose (via glucose oxidase), which is then used by lactoperoxidase to catalyze the peroxide oxidation of labeled iodide to iodine, which subsequently reacts with tyrosine (and to some extent histidine and tryptophan) residues in the protein. Using this method we have been able to generate protein-dependent specific radioactivities ranging from 5×10^{17} to 2×10^{19} cpm/mol with little loss in receptor-binding affinity. Other have also used this technique for IL-4 (41,44).

Methods using chloramine-T as an oxidizing agent to generate the tyrosine-reactive electrophilic forms of iodine (I^+) have also proven successful in many cases. Murine IL-3 (14), CSF-1 (22), human IL-1α (53), and EPO (30,32) have all been radiolabeled using modifications of the original method of Greenwood et al. (73). Use of a two-phase chloramine-T method (74) in which the protein being iodinated is kept in a separate phase from the chloramine-T iodine generating system has been even more widespread, primarily because modification of amino acid residues due to direct exposure to the oxidizing conditions generated by chloramine-T is largely avoided. Murine IL-3 (16), G–CSF (20), human EPO (31), and murine and human GM–CSF (34,37) have all been radiolabeled by this approach. Use of a solid-phase chloramine-T reagent (Iodogen from Pierce Chemical Co.) has also been reported for murine IL-4 (45) and human EPO (25,27-29). Use of the lysine-directed Bolton-Hunter reagent (75) has been less frequent, however, human IL-1β (53) and IL-6 (59) have

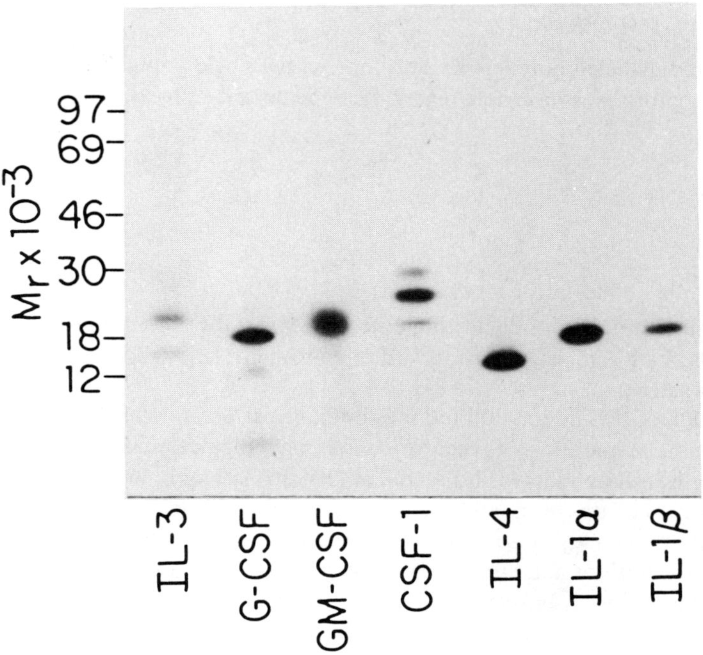

Figure 1 Characterization of ^{125}I-labeled human hematopoietic growth factors by sodium dodecyl sulfate (SDS)-polyacrylamide gel electrophoresis. ^{125}I-labeled growth factors (as indicated) were boiled for 3 min in sample buffer containing 2% SDS and 5% 2-mercaptoethanol, subjected to electrophoresis on a linear 10-20% gradient gel, and dried gels were exposed to film at –70°C.

been iodinated using this approach. Figure 1 shows a sodium dodecyl sulfate (SDS)-polyacrylamide gel comparing a number of these different iodinated hematopoietic growth factors where human IL-3, G–CSF, GM–CSF, CSF-1, and IL-4 have been iodinated using the solid-phase lactoperoxidase method, IL-1α the chloramine-T method, and IL-1β the Bolton-Hunter reagent.

Although varying methods have been employed in different laboratories to determine the specific radioactivities of iodinated hematopoietic growth factors, in general they have been high ($>1 \times 10^{18}$ cpm/mol) and the molecules have retained a substantial portion of their biological activity. Methods used to determine whether these radiolabeled molecules retain receptor-binding properties similar enough to those of their unlabeled counterparts to justify their use for receptor characterization will be addressed in Section II.C.

B. Ligand Binding Techniques

Interaction of a radiolabeled polypeptide hormone with a single population of homogeneous receptors by a reversible reaction can be described by the simple mass-action law:

$$P + R \underset{k_{-1}}{\overset{k_1}{\rightleftarrows}} PR$$

where P = radiolabeled polypeptide hormone; R = receptor; PR = hormone/receptor complex; k_1 = association or forward rate constant; k_{-1} = dissociation or reverse rate constant.

Characterization of this interaction requires both a suitable radiolabeled ligand and a suitable preparation of receptor-containing biological material (usually intact cells, cell extracts, or isolated membrane fractions). In addition, experimental analysis of this interaction requires the determination of not only appropriate incubation conditions for the two components (in terms of ligand and receptor concentrations and duration of incubation), but also a means of separating free from receptor-bound ligand at the conclusion of the incubation.

Determination of appropriate incubation conditions is influenced by requirements of the mass action law:

$$\text{initial association rate} = k_1 \; [P] \; [R]$$

$$\text{initial dissociation rate} = k_{-1} \; [PR]$$

At equilibrium, where the rate of association = rate of dissociation:

$$k_1 \; [P] \; [R] = k_{-1} \, [PR] \qquad \text{or}$$

$$\frac{[PR]}{[P]\,[R]} = k_1/k_{-1} = K_a, \text{ the equilibrium association constant}$$

The inverse of $K_a = K_d$, the equilibrium dissociation constant, which is equivalent to the concentration of polypeptide hormone at which the receptor is half-maximally occupied at equilibrium. Optimal incubation conditions must be determined empirically for each system; however, it is apparent from these equations that both the rate of attainment of equilibrium, as well as the concentration of receptor-ligand complex present at equilibrium, are dependent on the concentrations of both reactants. The time required to reach equilibrium can also be effected, often in a complex manner, by the temperature of the incubation, and must also be empirically determined.

In addition to the conditions under which a receptor and its ligand are incubated, the other main component in a ligand binding experiment is the means by which unbound ligand is separated from receptor-bound ligand at the conclusion of the incubation period. It is crucial at this stage not to disrupt the equilibrium that has been obtained between bound and unbound ligand. In almost all studies on hematopoietic growth-factor receptors done with intact cells, some variation of a rapid centrifugation technique is used. A method that we, and most others, routinely use is to layer the contents of incubation mixtures above a high density solution in a centrifuge tube, where the density is chosen to allow immediate sedimentation of cells plus bound ligand through the solution, while retaining media plus unbound ligand on its surface. The composition of the high-density solution may vary (most commonly being fetal calf serum or mixtures of phthalate oils). Tubes are then centrifuged for a short time and radioligand bound to cells is determined by cutting tubes in half and counting the top (free cpm) and tip (bound cpm) for ^{125}I in a gamma counter. Unlike earlier methods in which cells were pelleted and then washed one or more times with cold buffer to remove trapped unbound radioligand, rapid separation by spinning through a high-density solution effectively preserves the ratio of bound to unbound ligand present at the end of the incubation period.

Although the amount of unbound radioligand trapped in the cell pellet is usually small, the contribution of this component along with that of cell-associated radioligand not bound via a specific receptor (collectively known as nonspecific binding) can be determined by including an excess of unlabeled ligand in incubation mixtures to block specific binding of radioligand. This nonspecific component is then subtracted from total radioligand bound to determine the component of binding that is specific.

C. Kinetic and Equilibrium Binding Characteristics

In this section we will use the interaction of human ^{125}I-IL-3 with its receptor on human monocytes to illustrate the types of experiments used to determine kinetic and equilibrium binding characteristics, and will summarize what is known about these properties for binding of the hematopoietic growth factors (both murine and human) to their specific receptors. We would initially point out that the studies with ^{125}I-IL-3 are conducted using a recombinat form of the molecule that has been engineered to have a tyrosine-containing octapeptide fused to the amino terminus. This additional octapeptide aids in purification, and also greatly facilitates the radiolabeling of the molecule (58).

To establish the conditions needed to achieve equilibrium binding of ^{125}I-rhuIL-3, initial association experiments were performed in which one concentration of ^{125}I-rhuIL-3 was incubated with monocytes at 37°C or 4°C. Figure 2 shows that binding at 37°C was rapid and saturable, reaching equilibrium within 30 min. Binding at 4°C was much slower however, and required $>$ 6 hr to reach equilibrium. Due to extremely low site numbers observed ($<$200 at

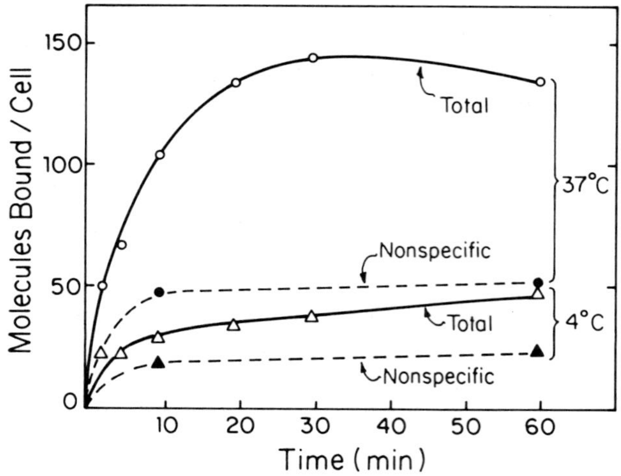

Figure 2 Time course of ^{125}I-IL-3 binding to human monocytes at 4 and 37°C. Monocytes (2.33 × 10^7 cells/ml) were incubated with 3.3 × 10^{-10} *M* ^{125}I-IL-3 at 37 or 4°C in the presence or absence of a 200-fold molar excess of unlabeled IL-3. At the times indicated aliquots were removed and assayed for binding as previously described (33).

either 4°C or 37°C), the nonspecific component of binding was a significant proportion of the total binding at either temperature. This is a problem common to other hematopoietic growth-factor receptors that are expressed at extremely low levels (particularly human GM–CSF) and contributes to the difficulty in obtaining accurate and reproducible binding data for these systems.

The observation that ^{125}I-rhuIL-3 binds to its receptor much more slowly at 4°C than 37°C is also a property shared by other members of the hematopoietic growth-factor family. Notably, all four of the classic CSFs, IL-3, GM–CSF, G–CSF, and CSF-1 exhibit this property, as does EPO. In contrast IL-1, IL-4, and IL-6 bind to their receptors almost as fast at 4°C as at 37°C (see Table 1 for summary). Extremely slow binding kinetics is not common among other growth factors, most of which reach equilibrium at 4°C within 1-4 hr. The significance of this common property among the CSF receptors is not known, but it is tempting to speculate that it may reflect some similarity in structure or mechanism of action between them.

It is worth noting here that to accurately measure kinetic and equilibrium binding constants, it is necessary to measure the initial step of ligand binding to its receptor on the cell surface, without contributions from the subsequent steps of ligand-receptor internalization and degradation. This can be accomplished by doing binding studies either at 4°C, or at 37°C in the presence of metabolic inhibitors such as sodium azide. Both conditions have pluses and minuses, as well as advocates among researchers. Binding at 37°C has the advantage over 4°C of being done at a physiological temperature, whereas prevention of internalization and degradation is more certain at 4°C than in the presence of metabolic inhibitors. Although in some systems one has the choice of picking either temperature, the extremely slow binding of the CSFs to their receptors at 4°C strongly tips the balance in favor of 37°C studies.

Further characterization studies with rhuIL-3 were therefore done at 37°C (in the presence of 0.2% sodium azide) as has been the case with most of the published studies on binding of the CSFs. To more definitively establish the conditions needed to achieve equilibrium of ^{125}I-rhuIL-3 and to determine the kinetics of binding, association experiments were performed in which different concentrations of ^{125}I-IL-3 were incubated with monocytes at 37°C. The data in Fig. 3 show that both the final equilibrium level of ^{125}I-rhuIL-3 bound to cells, and the rate at which binding achieved equilibrium, were dependent on the initial concentration of ^{125}I-rhuIL-3 in the medium. The curves passing through the data points in Fig. 3 are best-fit, single exponential, time-dependency curves (76) and are consistent with the presence of a single population of receptors on these cells. It is apparent from the data that at 37°C, equilibrium is reached within 30 min over a range of concentrations.

The experiment depicted in Fig. 3a was done under pseudo-first-order conditions, that is, where the concentration of radioligand (P) is added at concentrations in considerable excess of its receptor (R) such that the concentration of P will not change significantly over the course of the incubation. The pseudo-first-order forward rate constants (k_p) determined from fitting the curves are related to the forward or association rate constant k_1 (second-order) and the reverse or dissociation rate constant k_{-1} (first-order) by the equation:

$$k_p = k_1 [P] + k_{-1}$$

Figure 3b shows a plot of the pseudo-first order forward rate contants determined from panel a relative to the initial concentration of ^{125}I-rhuIL-3 in the medium. The data shown fit a straight line, indicative of a simple biomolecular reaction, where the slope of the line is the forward rate constant (k_1), and the intercept on the ordinate is the reverse rate constant (k_{-1}). For rhuIL-3 these values are $1.80 \pm .03 \times 10^8\ M^{-1}\ \text{min}^{-1}$ for k_1 and $4.42 \pm 0.08 \times 10^{-2}\ \text{min}^{-1}$ for k_{-1}. The ratio of these parameters also defines a value for the affinity constant (K_a) of ^{125}I-rhuIL-3 for its receptor, which in this case is $4.08 \pm .13 \times 10^9\ M^{-1}$.

In addition to mathematical determination of the reverse rate constant from association experiments, this value can also be empirically determined by the

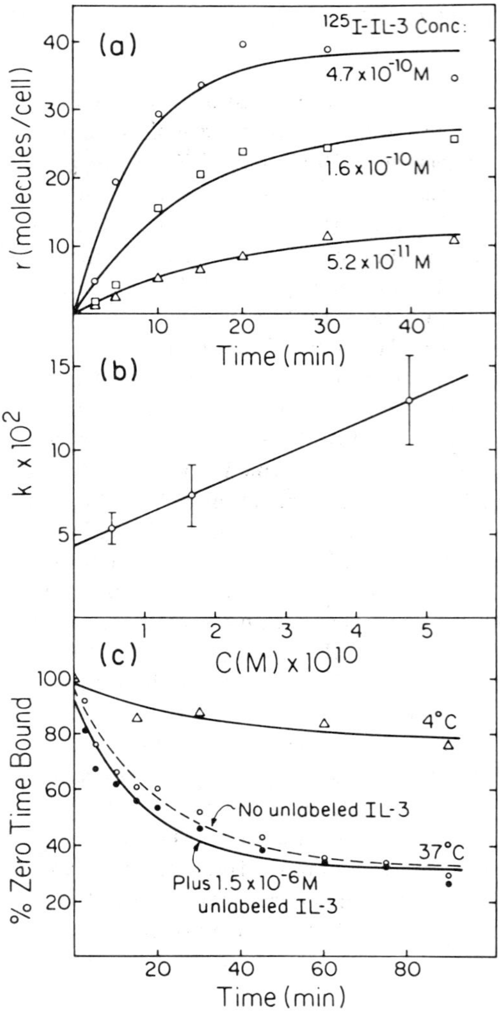

Figure 3 Association and dissociation kinetics of ^{125}I-IL-3 at 37°C. (a) Human monocytes (3.33 × 10^7 cells/ml) were incubated at 37°C with varying concentrations of ^{125}I-IL-3 and at the times indicated aliquots were removed and assayed for binding (33). Nonspecific binding has been subtracted from each curve. (b) A plot of the pseudo-first-order forward rate constants calculated from each curve versus the molar concentration of ^{125}I-IL-3 initially present in the medium. (c) Human monocytes (9.1 × 10^7 cells/ml) were incubated with ^{125}I-IL-3 (9 × 10^{-10} *M*) for 1 hr at 37°C. The cells were then divided into three pools, harvested by centrifugation, and resuspended in either medium alone at 4 or 37°C or in medium containing 1.5 × 10^{-6} *M* unlabeled IL-3 at 37°C. The cells were maintained at the indicated temperatures and aliquots were removed at various times and assayed for binding.

type of experiment shown in Fig. 3c where fixed concentrations of ^{125}I-rhuIL-3 and cell-surface receptors are allowed to come to equilibrium, after which the ^{125}I-rhuIL-3 concentration is diluted out and the rate at which bound ^{125}I-rhuIL-3 dissociates is measured either in medium alone (where only a fraction of the receptors are occupied) or in the presence of excess unlabeled rhuIL-3 (where almost all of the receptors are occupied). As originally described by Demeyts et al. (77), this experimental design is constructed to test whether receptors exhibit any cooperative properties. The data depicted in Fig. 3c reveals that at 37°C, the rate of dissociation of ^{125}I-rhuIL-3 is insensitive to occupancy of adjacent receptors suggesting that IL-3 receptors are noncooperative. The data in Fig. 2c also revealed two other features that have been generally observed for other hematopoietic growth-factor receptors. First, the dissociation rate at 4°C is extremely slow, much more so than at 37°C. As has been described in detail for CSF-1 binding (78), which is essentially irreversible at 4°C, an extremely slow dissociation rate makes it difficult to reach true equilibrium conditions at this temperature, and may result in large errors in determination of binding constants by equilibrium methods. Second, the dissociation kinetics are markedly biphasic. Complex dissociation kinetics have been observed for many cell surface receptor systems including murine IL-3 (14), GM–CSF (33,38), IL-1 (46), and IL-4 (40,42). The cause(s) for such complex kinetics remains unclear, but the complexity does not necessarily result from site heterogeneity.

In addition to obtaining values for the equilibrium association and dissociation constants from association kinetics experiments, a more direct means is through equilibrium binding experiments, such as those depicted in Fig. 4. Figure 4a shows the characteristics of both total and nonspecific binding of ^{125}I-rhuIL-3 as a function of increasing concentration of radioligand under equilibrium conditions, where specific binding is determined mathematically by subtracting nonspecific from total. Although nonspecific binding is a linear function of increasing radioligand concentration, specific binding plateaus, as expected for binding of radioligand to a saturable receptor with a single affinity (K_a). It should be noted again that in a case in which total receptor numbers are very low, such as human IL-3 or GM–CSF, nonspecific binding is a significant proportion of total binding, as illustrated in Fig. 4a.

Although many linear transformations of equilibrium binding data are available to aid in data analysis, the most popular and frequently used is the Scatchard plot (79), which allows an estimation of receptor affinity from the slope of the plot, and receptor density from the point of intersection on the x-axis. A Scatchard transformation of specific binding from Fig. 4b yields a straight line, suggesting that a single class of binding sites for ^{125}I-rhuIL-3 exists on human monocytes (Fig. 4b). The apparent K_a calculated from these data was $8.1 \pm .09 \times 10^9\ M^{-1}$ with 168 ± 4 specific binding sites per cell. Although

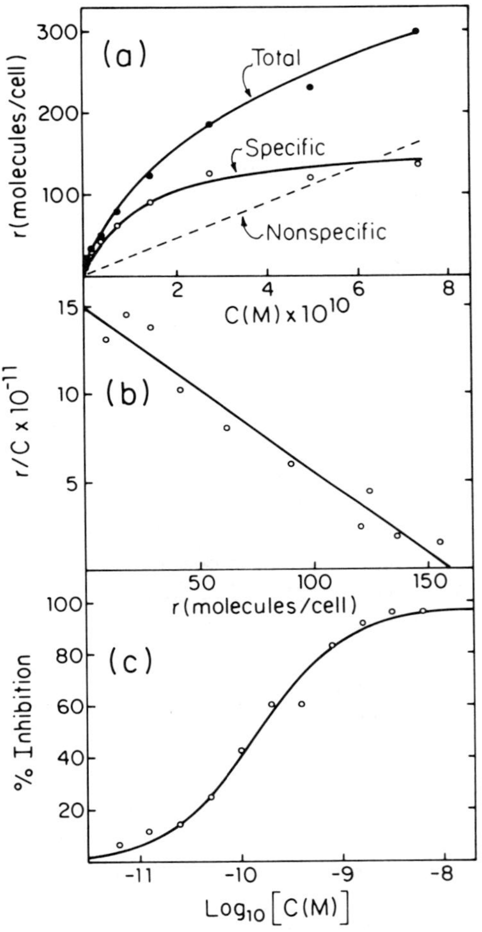

Figure 4 Equilibrium binding of ^{125}I-labeled and unlabeled human IL-3. (a) Human monocytes (3.33×10^7 cells/ml) were incubated with ^{125}I-IL-3 for 1 hr at 37°C either in the presence or absence of a 200-fold molar excess of unlabeled IL-3. Specific binding was determined by subtracting nonspecific from total binding at each point. (b) A Scatchard representation of specific binding replotted from (a). (c) Monocytes (2.33×10^7 cells/ml) were incubated with ^{125}I-IL-3 (2.1×10^{-10} *M*) and varying concentrations of unlabeled IL-3 for 1 hr at 37°C. Data were corrected for nonspecific binding and the curve passing through the data was calculated from a one-site competitive inhibition equation using a K_a value for ^{125}I-IL-3 of 6.7×10^9 M^{-1}.

Table 1 Binding Characteristics of Hematopoietic Growth-Factor Receptors

Ligand	Cells	Temp. (°C)	k_1 (M^{-1} min^{-1})	k_{-1} (min^{-1})	K_a (M^{-1})	Time to equilibrium at 4°C (hr)	Refs.
MuIL-3	Murine	37	4×10^8	.01	10^9-10^{11}	>6	14, 15, 16
HuIL-3	Human	37	1.8×10^8	.06	1×10^{10}	>6	58
MuGM-CSF	Murine	37	2.2×10^7	.048	10^9-10^{10} [a]	>6	33, 34
HuGM-CSF	Human	37	5.5×10^7	.047	10^9-10^{11}	>6	37, 38
MuG-CSF	Murine	37	2×10^9	⩽.003	10^{10}-10^{11}	>6	20, 21
HuG-CSF	Human	37	ND	ND	1-5 $\times 10^9$	ND	58
MuCSF-1	Murine	37	1.8×10^8	.07	10^9-10^{10}	>6	22, 24
HuCSF-1	Human	37	ND	ND	1-5 $\times 10^9$	ND	58
HuEpo	Murine	37	ND	ND	10^9-10^{10} [a]	>6	25, 29
HuEpo	Human	37	ND	ND	1-5 $\times 10^9$	ND	31
HuIL-1	Murine	8	2.3×10^8	.025	10^9-10^{11} [a]	3	46, 47
HuIL-1	Human	4, 20	ND	ND	10^9-10^{11}	⩾3	48, 49
MuIL-4	Murine	37	4.3×10^8	.07	10^{10}-10^{11}	<2	40, 41
HuIL-4	Human	37	3.4×10^8	.069	10^9-10^{10}	<2	42
HuIL-6	Human	4	ND	ND	10^9-10^{10}	2.5	59

[a]Receptors for MuGM-CSF (34), HuEpo (25), and HuIL-1 (47) have been reported to display two different classes of binding sites (high and low affinity).

the appearance of simple linear Scatchard plots allows reasonable estimations of affinity constants and receptor numbers, there are many examples in which equilibrium binding data do not transform to a linear plot, but instead show concave downward or concave upward plots. There are a number of possible explanations for either of these situations, many of them artifactual. For interested readers, a thorough review covering this topic, as well as other aspects of the design, execution, and interpretation of ligand-receptor binding studies has recently been published (80). Among the hematopoietic growth-factor receptors, it has been reported that murine (34) and human (39) GM–CSF and human IL-1 (47) and EPO (25) display complex concave downward Scatchard plots. We have recently observed complex Scatchard plots as well for binding of human IL-3 and human GM–CSF to certain in vitro cell lines and populations of primary cells (L. Park and D. Urdal, unpublished data), whereas others (such as ^{125}I-rhuIL-3 binding to human monocytes) are linear. The significance of this observation will be discussed in Section IV.

In addition to determining the binding affinity of a radioligand, it is essential to determine the binding affinity of the unlabeled ligand as well, to assess whether binding affinities, rate constants, and cooperative properties evaluated with the iodinated reagent may be taken as good estimates for the behavior of the unmodified ligand. This can be done most simply by a competitive binding experiment as illustrated in Fig. 4c, where inhibition of binding of ^{125}I-rhuIL-3 to human monocytes be unlabeled rhuIL-3 was determined. Analysis of the data with a single-site competitive inhibition equation derived from the mass-action law (similar to that of Cheng and Prusoff, Ref. 81), yielded an inhibition constant (K_I) of $1.68 \pm .22 \times 10^{10}$ M^{-1}, close to that determined by direct binding of ^{125}I-rhuIL-3 but somewhat higher, suggesting that a small loss in affinity may have resulted from radiolabeling the molecule.

Table 1 shows a summary of the information currently available for the forward and reverse rate constants, as well as the equilibrium association constants for binding of the murine and human hematopoietic growth factors to their specific receptors. The temperature at which these constants were determined is indicated (generally 37°C), and the estimated time required to reach equilibrium at 4°C is also listed. Although values for all three constants vary over an approximate 100-fold range, taken as a group the binding characteristics of the hematopoietic growth factors display considerable similarity. All bind rapidly at 37°C, with relatively slow dissociation rates, resulting in expression of high-affinity binding ($K_a > 10^9$ M^{-1}) in all cases. Perhaps the greatest distinguishing feature, as mentioned previously, is the extremely slow binding at 4°C exhibited by the CSFs as well as EPO, a characteristic not shared by IL-1, IL-4, and IL-6.

III. CELLULAR DISTRIBUTION

In addition to determining the characteristics of binding between individual factors and their specific receptors, it is equally important to determine the distribution of these receptors on primary cells and in vitro cell lines. This type of information makes an essential contribution to our understanding of the extent to which these factors may interact, and also allows the prediction of new activities associated with these molecules. Using equilibrium binding methods (as we have discussed in the previous section), researchers in a number of different laboratories have accumulated information on cellular distribution of hematopoietic growth-factor receptors. Although not a complete list, we have attempted to summarize the bulk of this information in Tables 2 (murine cells) and 3 (human cells) for comparative purposes. Aside from the actual patterns of distribution, the most compelling information illustrated by these tables is the extremely low levels of receptors expressed. The only excep-

Table 2 Distribution of Hematopoietic Growth-Factor Receptors on Murine Cells

Cells	Characteristics	^{125}I-Molecules bound/cell[a]				
		IL-3	G-CSF	GM-CSF	IL-4[b]	IL-1
FDC-P2	Bone marrow derived	434 ± 281	ND	<10	1740 ± 280	1490 ± 35
32D	Bone marrow derived	4500[c]	220 ± 20	ND	1590 ± 300	ND
$P388D_1$	Macrophage tumor	<10	<10	4002 ± 1998	270 ± 50	ND
J774	Macrophage tumor	90 ± 10	+[d]	1062 ± 188	ND	ND
PU5–1.8	Macrophage tumor	50 ± 20	ND	1500 ± 500	ND	ND
P815	Mastocytoma	90 ± 10	ND	1060 ± 440	1040 ± 30	ND
Wehi 3B	Myelomonocytic	<10	220 ± 20	3497 ± 303	185 ± 15	ND
LSTRA	T-lymphocytic leukemia	ND	ND	5090 ± 1910	ND	ND
LBRM-33	T-lymphoma	ND	ND	4247 ± 1507	ND	ND
EL-4	T-lymphoma	<10	ND	<10	115 ± 5	185 ± 81
CTLL	IL-2 dependent	<10	ND	<10	1130 ± 320	550 ± 570
BCL-1	B-lymphoma	ND	ND	ND	440 ± 30	ND
MST-7723	B-lymphoma	<10	ND	<10	ND	ND
SC1	Fibroblast	<10	ND	<10	575 ± 235	1850 ± 112
Murine	Bone marrow	120[c]	120 ± 40[d]	1000 ± 500	20 ± 5	280 ± 140
	Neutrophils	+[c,e]	50-500[d]	2025 ± 216	ND	ND
	Monocytes	+[c]	ND	1410 ± 490	ND	ND

[a]Molecules bound/cell were determined by Scatchard analysis of complete sets of binding data and represent total sites per cell unless noted otherwise (15, 33, 40, 46, L. Park, S. Dower, and D. Urdal, unpublished data).
[b]See Ref. 45 for binding of IL-4 to additional cell lines.
[c]Data from Refs. 14, 16. See these references for binding to additional cell lines.
[d]Data from Refs. 20, 21.
[e]+ = Binding not expressed as molecules bound/cell.

Table 3 Distribution of Hematopoietic Growth-Factor Receptors on Human Cells

		^{125}I-Molecules bound/cell[a]					
Cells	Characteristics	IL-3	G-CSF	GM-CSF	IL-4	IL-1	IL-6
Jurkat	T-lymphoma	<10	<10	<10	115 ± 15	<10	<30
HSB2	T-lymphoma	<10	<10	<10	210 ± 10	79 ± 5	<30
PEER	T-leukemia	<10	<10	<10	210 ± 50	<10	ND
RAJI	B-lymphoma	<10	<10	ND	2280 ± 210	91 ± 4	<30
ARH77	Myeloma	<10	<10	ND	200 ± 30	330 ± 3	120
CESS	B-lymphoblast	<10	<10	ND	525 ± 25	200 ± 16	2700
BMB	Early B tumor	120 ± 30	<10	<10	190 ± 30	540 ± 0	ND
Nalm-6	Pre-B lymphoma	60 ± 30	<10	ND	115 ± 15	<10	ND
JM-1	Pre-B lymphoma	90 ± 60	ND	<10	65 ± 5	<10	ND

HL-60	Promyelocytic	<10	<10	167 ± 87	125 ± 45	<10	3600
U937	Monocytic tumor	<10	700 ± 480	<10	640 ± 30	<10	2800
KG-1	Myelogenous	275 ± 60	<10	100 ± 20	475 ± 235	<10	ND
K562	Myelogenous	<10	<10	<10	ND	<10	ND
HPT	Carcinoma	<10	<10	<10	610 ± 60	<10	ND
HBT5637	Carcinoma	<10	<10	<10	495 ± 55	ND	ND
A375	Melanoma	<10	<10	<10	330	465 ± 25	ND
Human	Neutrophils	<10	50	260 ± 70	70	<10	ND
	Monocytes	125 ± 85	245 ± 5	450 ± 75	245 ± 5	<10	ND
	T cells	ND	ND	<10	140 ± 110	27 ± 1	485 ± 335
	Fibroblasts	ND	ND	<10	1950 ± 70	4900	ND

[a]Molecules bound/cell for IL-3, G-CSF, GM-CSF and IL-4 were determined by Scatchard analysis of complete sets of binding data and represent total sites/cell (38, 42, 58, L. Park and D. Urdal, unpublished data). For IL-1, molecules bound/cell were determined at 4-5 × 10^{-10} M ^{125}I-IL-1 (46, S. Dower, unpublished data). IL-6 data (59) were determined by Scatchard analysis and represent total sites/cell.

tion to this rule is the receptor for CSF-1 (data not shown in Table 2), which is expressed on a restricted subset of murine cells of monocytic phenotype, and can reach levels $>5 \times 10^4$/cell (22,23,78). The level of expression of receptors for the hematopoietic growth factors is, in general, orders of magnitude lower than expression for other growth factors such as EGF, PDGF, insulin, and NGF, which generally express in the range of 10^4-10^6 sites/cell. This would suggest that the hematopoietic growth factors must be capable of exerting a biological effect with only a few molecules bound per cell.

Several interesting observations can be made from the cellular distribution of receptors depicted in Tables 2 and 3. It would appear that receptor expression can be separated into two broad classes, those showing relatively restricted distribution (the CSF receptors) and those showing a much broader (IL-1, IL-6) or almost ubiquitous (IL-4) distribution. The CSF receptors are generally restricted to expression on primary cells or cell lines of myelomonocytic phenotype with two interesting exceptions: GM–CSF receptors have been found on two murine cell lines of T-cell lineage, and IL-3 receptors have been found on four human cell lines of early or pre-B phenotype. Although it is possible that both these cases represent aberrant expression on transformed cell lines, they nonetheless provide tantalizing evidence that GM–CSF and IL-3 may in fact be involved in development of cells of both the lymphoid and nonlymphoid compartments. In regard to the general restriction of CSF receptors to cells of myelomonocytic lineage, it should be pointed out that measurement of receptors on particular cell lines in different laboratories has produced many inconsistencies, not only in terms of level of expression, but also in terms of whether cell lines express receptors for a particular CSF at all. Although some variation between laboratories is not surprising, our own experience with myelomonocytic cell lines, particularly the popular KG-1 and HL-60 cell lines, has been that cells obtained from different sources may show completely different patterns of receptor expression. For example, we recently tested KG-1 cells from four different sources and found that two lines expressed IL-3 receptors and two did not. Much of the inconsistency in the literature may therefore be due to variants of established cell lines, and we would caution against making assumptions about receptor expression based on data reported for cell lines maintained in other laboratories.

IL-1, IL-4, and IL-6 bind to a much broader spectrum of cell types, which is in keeping with the wide range of biological activities associated with these molecules. In addition to other activities, all three have been implicated in direct effects on hematopoietic cells, as discussed in the Introduction, and evidence exists for an indirect role of IL-1 and IL-4 in hematopoiesis as well. IL-1, for example, has been shown to stimulate helper T cells, fibroblasts, and endothelial cells to synthesize CSFs, and we have recently observed that IL-4

Table 4 Distribution of CSF Receptors on Primary Human Leukemia Cells

Patient No.	FAB[a] class	^{125}I-Molecules bound/cell[b]		
		Hu-GM-CSF	Hu-IL-3	Hu-G-CSF
1	UNC	75 ± 25	20 ± 10	310
2	M2	260	30 ± 10	±[c]
3	M2	205 ± 35	20 ± 5	280 ± 80
4	M4	395 ± 65	120 ± 30	±
5	M2	180	35 ± 15	170
6	M4	200 ± 10	40	810
7	M1	270	60	10
8	M4	120 ± 30	80	80
9	M5	230 ± 40	15 ± 5	40
10	M4	130 ± 10	60 ± 20	30
11	M4	90 ± 45	40	30
12	M1	14 ± 3	±	<10[d]
13	M1	115 ± 15	70	180
14	ALL	±	±	<10
15	ALL	<10	<10	<10
16	ALL	<10	<10	ND[e]
17	ALL	<10	<10	<10
18	ALL	<10	±	ND

[a]FAB, French-American-British classification; UNC, unclassified, ALL, acute lymphocytic leukemia.
[b]Molecules bound/cell were determined by Scatchard analysis of complete sets of binding data and represent total sites/cell ± SEM.
[c]± = Binding detectable but below level of accurate quantitation.
[d]<10 = No detectable binding, where limit of accurate detection is defined as 10 molecules bound/cell.
[e]ND = not determined.

is capable of stimulating fibroblasts to produce CSFs (82, and B. Tushinski, unpublished data).

With several of the human hematopoietic growth factors in, or approaching clinical trials, knowledge of the expression of receptors for these factors on primary cells, particularly those directly involved in the disease state, may provide valuable information to the clinical trials process. We therefore recently examined the display of receptors for IL-3, GM–CSF, and G–CSF on a panel of acute nonlymphocytic leukemias (ANL) and acute lymphocytic leukemias (ALL). Table 4 shows that detectable levels of receptors for all three CSFs were expressed on essentially all ANL cells, although at consistently low levels. In contrast, expression on five T-ALL samples was either undetectable, or in a couple cases, detectable but at levels too low to be accurately estimated. It is interesting that receptor expression for all three CSFs on these primary ANL cells is much less restricted than it appears to be on continuous human cell lines of myelomonocytic lineage. As mentioned previously, this may be due to gradual selection of variant cell lines over time in culture, or may in fact be indicative of some property of these cells that allowed them to grow spontaneously in culture in the first place.

IV. RECEPTOR SPECIFICITY AND MODULATION

A major question that arises during characterization of a ligand-receptor interaction is that of specificity. Is the ligand in question capable of binding to other receptors, and conversely, can the receptor bind other ligands? The latter question is generally approached by testing the ability of unlabeled growth factors and polypeptide hormones (at a molar excess) to inhibit the binding of radiolabeled ligand to its receptor, whereas the former is generally inferred from the complexity of binding kinetics with the radioligand. The ability of a receptor to bind to more than one ligand has been reported in a few situations and is clearly the case with the IL-1 receptor, where two distantly related ligands, IL-1α and IL-1β, both bind to the same receptor molecule (53).

Modulation of the level of expression of CSF receptors by other CSFs has also been reported. Walker and Burgess (35) showed with murine bone marrow cells that the CSFs are apparently able to modulate the expression of not only their own receptors, but those for other CSFs as well. This modulation, which does not occur at 0°C, is a result of downregulation of receptors, rather than direct competition for binding. At 37°C, they observed a pattern of downregulation in which IL-3 downmodulated all CSF receptors, GM–CSF downmodulated G–CSF and CSF-1 receptors, at high concentrations CSF-1 downmodulated GM–CSF receptors, and at high concentrations G–CSF downmodulated CSF-1 receptors. Although the results are complicated by the heterogeneity of the bone marrow population, the authors suggest that downmodulation of the murine CSF receptors follows a hierarchical pattern that allows activation of lineage-specific receptors via downregulation by IL-3 or GM–CSF.

In our recent work on the receptor for human IL-3 we consistently observed that at 37°C (in the presence of sodium azide) human GM–CSF was able to partially (20-25%) compete binding of ^{125}I-rhuIL-3 to human monocytes, but not to JM1 cells. We examined this phenomenon in more detail on two variants of KG1 cells and found that at a 1000-fold molar excess, GM–CSF was able to compete 70-90% of ^{125}I-rhuIL-3 binding on one KG1 variant, and 40-60% of ^{125}I-rhuIL-3 binding to the other. This was true both at 4°C and at 37°C in the presence of 0.2% sodium azide, indicating that the effect is due to direct competition for binding to a receptor protein, rather than to a decrease in receptor number as a result of downmodulation as described (35) for the murine CSFs. An equivalent converse experiment showed that unlabeled IL-3 was also able to partially compete binding of ^{125}I-GM–CSF to KG1 cells at either 4°C or 37°C. G–CSF and CSF-1 exhibitied no inhibitor effects in either case. Further analysis on one KG1 variant showed that GM–CSF was able to compete for ^{125}I-IL-3 binding as effectively as IL-3 up to a maximum level of 40-60%; however, even at concentrations at which GM–CSF was at a 10,000-fold molar excess over ^{125}I-rhuIL-3, no further inhibition was observed.

While more data must be accumulated to fully explain these unprecedented results, they can be explained most simply by invoking heterogeneity in the binding sites for IL-3 and GM–CSF in which a subset of receptors can bind only IL-3, a subset only GM–CSF, and another subset can bind both, all with similar affinities. It is interesting to note also that KG1 cells, which strongly exhibit this cross-competition behavior, display complex biphasic binding curves, whereas human monocytes (in which the effect is minimal) do not, nor do cells that bind only GM–CSF or only IL-3. It may be that this cross-competition behavior is also responsible for the biphasic binding of ^{125}I-rhuGM–CSF observed by DiPersio et al. (39) on some cell types.

V. MOLECULAR CHARACTERIZATION

The most severe limitation to progress on the molecular characterization of hematopoietic growth-factor receptors has been low abundance of these receptors found on cells. By and large the only molecular probes directed to the receptor have been the radiolabeled growth factors to which the receptors bind. However, these limitations are being overcome by novel approaches to the molecular cloning of receptors and thus the chemical structure of many of these molecules may shortly be resolved. Studies on the characterization of the IL-1 receptor (51,52) will be discussed in detail and will serve as a paradigm on which the characterization of other hematopoietic receptors can be modeled.

A. Affinity Crosslinking

The most sensitive and usually the first means of obtaining information regarding the molecular features of the receptor has been by the technique of chemical crosslinking (15,19,33,40,42,44) (see Table 5). These studies are facilitated by first obtaining a good source of receptor and having a ligand radiolabeled to a high specific activity. Radiolabeled ligand is first bound to cells under conditions established by the equilibrium binding studies described above. The cells are washed to remove unbound radiolabeled ligand and then a bifunctional crosslinking reagent, such as dithiobis(succinimidylpropionate) is added to the cells so as to chemically crosslink the radiolabeled ligand to the receptor. Receptors and bound ligand are then extracted from the cells with detergent (commonly phosphate-buffered saline (PBS) containing 1% triton X-100 and protease inhibitors) and analyzed by SDS-polyacrylamide gel electrophoresis followed by autoradiography. The bands detected on film represent free radiolabeled ligand or radiolabeled ligand to which the receptor is chemically crosslinked. The type of results obtained by this technique are illustrated in Fig. 5a for the IL–1R. The results depicted indicate that on both T cells (El-4 6.1 C10 cells) and on fibroblasts (3T3 cells) the receptor-radiolabeled IL-1 complex exhibits a molecular mass of 97,000, suggesting that the receptor has a mass of 80,000. Analysis of the complex under reducing and nonreducing conditions allows the determination that the IL-1 receptor is not covalently associated with other subunits on the plasma membrane.

Information about the structures of the other hematopoietic growth factor receptors, summarized in Table 5, has also been obtained primarily through affinity crosslinking studies. Considerably more is known about the structure of the murine hematopoietic growth-factor receptors than their human counterparts. However, disparate results generated by different laboratories are more the rule than the exception. Several factors may contribute to these differences. Perhaps the most important is the likelihood of severely underestimating receptor size if great care is not taken to neutralize protease activity present in the detergent extracts of cells. The murine GM–CSF receptor, which was demonstrated to be exquisitely sensitive to the effects of cellular proteases, is an excellent example of this problem (33). The receptor for IL-4, which has a molecular mass of 140,000 on human cells (42,44) is another good example. We and others (40,41) initially reported that the murine IL-4 receptor was small (60-75 kDa); however, we have subsequently shown that when extraordinary care was taken to inhibit cellular proteases that a 140,000 M_r form of the murine IL-4 receptor could also be identified (82). Another factor that may contribute to variability in reported results is the cell source chosen for crosslinking studies. Differences in size may be generated by variability in the complement of cellular proteases present or in protein glycosylation patterns. Certainly it is also

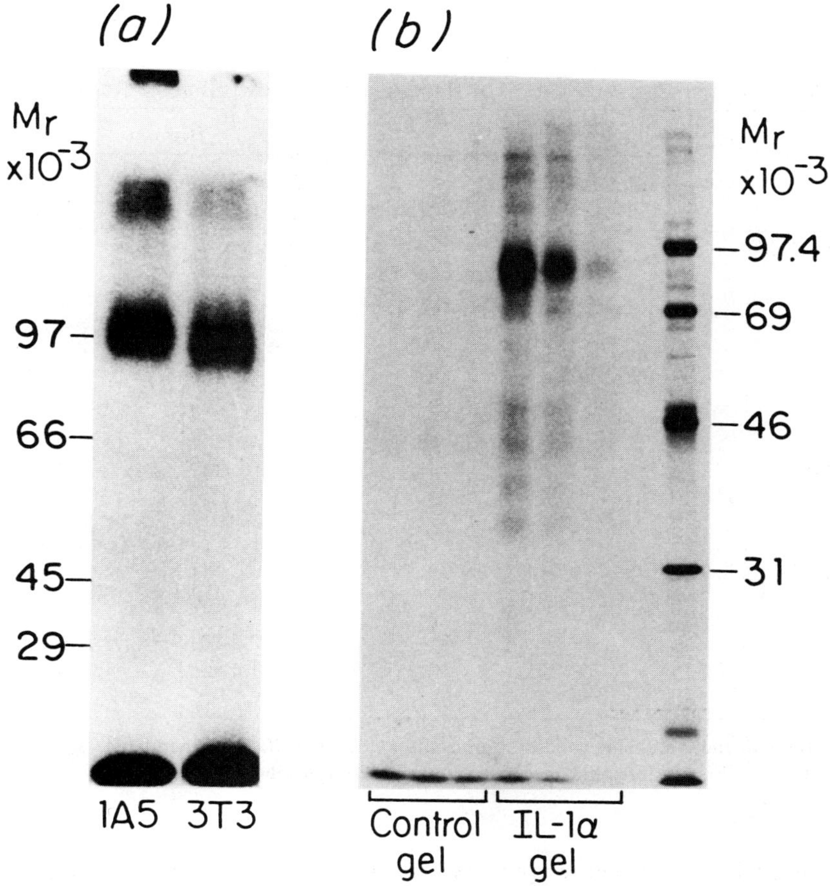

Figure 5 Characterization of the IL-1 receptor by affinity crosslinking (a) and affinity chromatography (b). (a) ^{125}I-IL-1β was bound to a murine T-cell line (LBRM–33–1A5) and to a murine fibroblast cell line (3T3) and chemically cross-linked with dithiobis(succinimidylpropionate) (83,84). The cells were extracted with detergent and analyzed by sodium dodecylsulfate (SDS) polyacrylamide gel electrophoresis under nonreducing conditions followed by autoradiography. (b) The membrane proteins present on EL4 6.1 C10 cells were iodinated by the lactoperoxidase method and then extracted from the cells with phosphate-buffered saline containing 1% Triton X-100 (51). Extracts were then applied to columns prepared by coupling GM–CSF to Affigel (control gel) or by coupling IL-1α to Affigel (IL-1α gel). The columns were washed with extraction buffer and then were eluted with 0.1 *M* glycine pH 3.0 containing 0.1% Triton X-100. Aliquots of the first three fractions collected with this buffer were analyzed by SDS polyacrylamide gel electrophoresis and autoradiography and are depicted in panel (b).

Table 5 Molecular Characterization of Hematopoietic Growth-Factor Receptors

	Murine		Human	
Factor	Molecular mass (kDa)	Ref.	Molecular mass (kDa)	Ref.
Epo	63, 94, 119	27	ND	
	85, 100	28		
	86, 106	29[a]		
IL-3	72.5 (113)	15	ND	
	65-70	17		
	67	18		
	60, 75	19		
GM-CSF	51	34	84	39
	130 (70)	33		
G-CSF	150	19	ND	
CSF-1	165	24	ND	
IL-1	80	46, 51	80	
IL-4	60	41	140	42
	75	40	140	44
	140	82		

[a]Sizes obtained under reducing conditions on sodium dodecyl sulfate-polyacrylanide gel electrophoresis. Larger sizes obtained under nonreducing conditions.

always possible that distinct receptor species exist on cells of different lineages. In most cases, although affinity crosslinking studies produce valuable preliminary information on receptor size and structure, a conclusive determination must await purification and/or molecular cloning.

B. Protein Purification and cDNA Cloning

Additional molecular characterization can be obtained if the abundance of receptors is such that direct chemical means of identification can be used. One method by which we have successfully maintained cells expressing high levels of IL-1 receptor has been by the use of fluoresceinisothiocyanate-labeled IL-1 and the fluorescence activated cell sorter. EL-4 6.1 C10 cells, for example, were found initially to express 2000-6000 IL-1 receptors per cell and by a combination of cell cloning and cell sorting gave rise to a population of cells that expressed upwards of 15,000 receptors per cell. In the absence of cell sorting the receptor display on cells would decline to presorting levels of expression.

The purification of the IL-1 receptor by IL-1 affinity chromatography was investigated as a means of confirming the results of the chemical crosslinking experiments described above (51). Purified rhu IL-1α was coupled to Affigel-10 beads. A control column was prepared by coupling rhu GM–CSF to Affigel. The plasma membrane proteins on EL-4 6.1 C10 cells were radiolabeled with ^{125}I using the lactoperoxidase procedure and then extracted from the membrane with PBS containing 1% Triton X-100 and protease inhibitors. Aliquots of this extract were then applied to an IL-1α column or a control column. The columns were washed and then eluted with a low pH (3.0) glycine buffer containing Triton X-100. Aliquots of the glycine eluate were analyzed by SDS-polyacrylamide gel electrophoresis and autoradiography as depicted in Fig. 5b. The glycine eluate from the IL-1α affinity column showed the presence of a band migrating at M_r 82,000 consistent with the mass of the protein predicted from the chemical crosslinking studies. Radiolabeled IL-1 receptor was purified as described above and subjected to digestion with the endoglycosidase, *N*-glycanase (Genzyme). After this digestion the M_r of the IL-1 receptor decreased to 62,000, suggesting that the IL-1 receptor is a glycoprotein consisting of at least 23% asparagine-linked carbohydrate.

The feasibility of receptor purification by affinity chromatography was suggested by the experiments described above and the method was applied to the goal of obtaining pure receptor for direct chemical analysis (52). To this end, EL-4 6.1 C10 cells were extracted with PBS containing 1% Triton X-100 and the extracts were applied to an IL-1α affinity column. Bound receptor was eluted with 0.1 *M* glycine, pH 3.0, and the receptor was further purified to homogeneity by chromatography on a wheat-germ-lectin column and by reversed-phase high-performance liquid chromatography. Analysis of fractions containing the receptor by SDS polyacrylamide gel electrophoresis and silver staining revealed that the receptor had a M_r of 82,000 consistent with the predictions made above using ^{125}I-labeled receptor. Fractions containing receptor were pooled and loaded onto a hexadimethrine bromide (Polybrene)-conditioned filter for amino-terminal sequencing and the following protein sequence was obtained: Leu-Glu-Ile-Asp-Val-Cys-Thr-Glu-Tyr-Pro-Asn-Glu-Ile-Val-Leu-Phe-Leu-Ser-Val-Asn-Glu-Ile-Asp-Ile-Arg-Lys.

Concurrent with attempts to purify the IL-1 receptor, direct expression techniques were examined for their potential use in the cloning of the IL-1 receptor (52). The expression vector pDC201 was designed to achieve this purpose. Within this vector, the adenovirus major late promoter and SV40 virus enhancer drive the high-level transcription of cloned cDNAs. Efficient RNA processing is assured by the presence of appropriate splicing signals upstream and a polyadenylation signal downstream from the cDNA cloning site. Additional factors that contribute to the efficient expression include the presence of the adenovirus tripartite leader and adenovirus VA genes elsewhere in the plasmid. The SV40 origin of replication is present in the plasmid and, when transfected into COS7 cells that provide the SV40 T-antigen, results in the replication of the plasmid to high copy number in this host cell. The propagation of the plasmid in *Escherichia coli* is assured by the presence in the plasmid of the pBR322 origin of replication and β lactamase gene.

The cell line EL-4 6.1 C10 was used as a source of poly A+ mRNA from which double-stranded cDNA was synthesized. The cDNA was inserted into the pDC201 vector and transformed into *E. coli* where individual cDNAs were amplified. Pools of *E. coli* transformants were grown, and plasmid DNA was prepared and transfected into COS7 cells. Seventy-two hours later, COS 7 cells that expressed the IL-1 receptor were identified by using ^{125}I-IL-1α as shown in Fig. 6a. Individual cells in which the cDNA encoding the IL-1 receptor was expressed were identified by the detection of dark foci on a light-gray background after autoradiography. The identity of the foci as IL-1 receptor-positive cells was established by the demonstration that unlabeled IL-1α or IL-1β could inhibit the binding of ^{125}I-IL-1α to the transfected COS7 cells (Fig. 6a). One hundred fifty thousand cDNA clones were screened in this fashion before one pool was found that gave a signal of approximately 25 spots per plate (52). This pool of cDNAs was broken down into individual colonies and one, clone 78, was found to encode the IL-1 receptor. The IL-1 receptor that had been cloned from EL-4 6.1 C10 cells and expressed in COS7 cells had the same specificity for IL-1α and IL-1β that had been previously determined foı the native IL-1 receptor found on EL-4 6.1 C10 cells. Moreover, the affinity with which IL-1 bound to the recombinant receptor was comparable to that observed for the native receptor. Finally, both native and recombinant receptors that had been surface iodinated and purified by IL-1α affinity chromatography migrated as M_r 82,000 proteins. Thus, all IL-1 binding activity that had been measured on EL-4 6.1 C10 cells could be explained by the activity associated with the protein encoded in cDNA clone 78.

The sequence of the cDNA isolated from clone 78 was determined (52) and allowed the molecular characterization of the receptor, which is summarized in the model depicted in Fig. 6b. The molecular mass of the protein predicted from this sequence was 64,598, which was in close agreement with the mass of 62,000 estimated from the *N*-glycanase experiment performed on the native receptor found on EL-4 6.1 C10 cells (51) and 7 potential *N*-linked glycosylation sites were identified in the first 319 amino acid residues predicted from the cDNA nucleotide sequence (52). Residues were found at the amino terminus of the cDNA sequence, which corresponded exactly to those identified by protein sequence analysis of the native EL-4 6.1 C10 receptor purified and sequenced as described above.

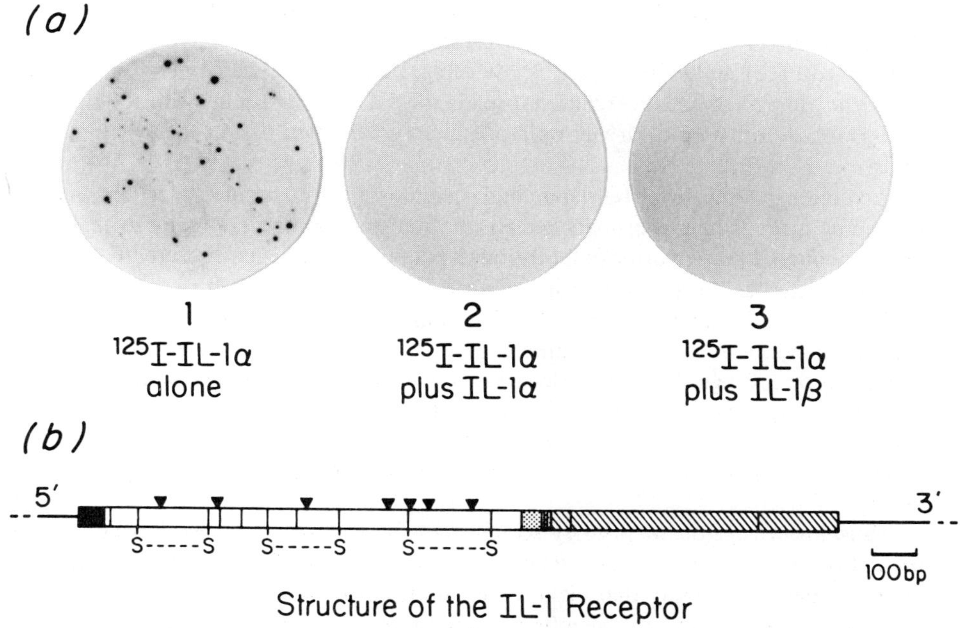

Figure 6 Cloning of the cDNA that encodes the IL-1 receptor by direct expression in COS-7 cells (a) and a model of the structure of the IL-1 receptor based on the amino acid sequence predicted from the cDNA nucleotide sequence (b). (a) cDNA cloning. Double-stranded blunt-ended cDNA was synthesized from EL-4 6.1 C10 polyadenylated RNA, ligated into the pDC201 vector and transformed into *Escherichia coli*. Plasmid DNA was prepared from pools containing 350 transformed *E. coli* colonies and transfected into COS7 cells. Pools containing cDNA that encoded the IL-1 receptor were identified using ^{125}I-labeled IL-1α and autoradiography. Panel (a), plate 1, depicts the results obtained after the transfection of COS cells with cDNA clone 78 and the detection of IL-1 binding activity by autoradiography. The results depicted in (a), plates 2 and 3, indicate that the ^{125}I-IL-1α binding shown on plate 1 could be completely inhibited by the presence in the medium of excess unlabeled IL-1α (plate 2) or unlabeled IL-1β (plate 3). (b) IL-1 receptor structure. Analysis of the primary amino acid sequence predicted from the cDNA nucleotide sequence suggests that the molecule can be divided into several regions. The signal peptide is depicted by the filled region of the box, the extracellular region is represented by the open box, the transmembrane region is stippled, and the cytoplasmic region is crosshatched. Inverted triangles above the extracellular region show potential glycosylation sites and vertical lines indicate cysteine residues. The disulfide bonds that represent the proposed Ig-like domains are indicated.

The amino-terminal 319 amino acids contain all potential *N*-glycosylation sites and thus compose the extracellular region of the IL-1 receptor. This region is followed by 21 amino acids, largely hydrophobic in nature, which most likely represent the part of the molecule that spans the plasma membrane. This domain is followed by a putative cytoplasmic region composed of 217 amino acids.

Predictions about the potential secondary and tertiary structure of the receptor can be made from the primary sequence obtained from the cDNA by searching the sequence for regions of identity with proteins of known structure. Such analysis of the IL-1 receptor revealed that the putative extracellular region could be divided into domains similar to those found in proteins that are members of the immunoglobulin gene super family. Thus, as shown in Fig. 6b, these domains can be distinguished by the location of critical Cys residues in the sequence. Comparison of the three domains with one another reveals that domain 1 and 2 are clearly related as are domains 2 and 3. Each of the three domains is similar to domains found in other members of the immunoglobulin superfamily.

Thus, knowledge of the primary sequence of the IL-1 receptor has allowed predictions to be made regarding its potential structure. It appears to be a member of a super gene family that includes, in addition to immunoglobulin, a number of well-characterized cell membrane proteins including MHC class I and class II antigens, Thy-1, CD1, CD2, CD4, CD8, LFA3, and receptors, including those specific for PDGF and CSF-1. It is intriguing to note that of the hematopoietic growth factor receptors mentioned in Table 5, the two for which the cDNA has been cloned, CSF-1 (70) and IL-1 (52), are both members of this family of genes. Similar strategies to those that have led to the cloning of the IL-1 receptor are being pursued for the cloning of other hematopoietic growth factor receptors and the identification and characterization of these molecules should occur in the near future.

VI. APPLICATION OF RECEPTOR STUDIES

Receptor studies have led to a better understanding of the biology of hematopoiesis and have provided a foundation on which a number of potential applications can be built. For example, in the course of introducing a recombinant growth factor into the clinic, rapid and quantitative assays for growth factor activity are needed to measure potency and concentration. To date most of these determinations have been made by the assessment of biological activity, which in the case of the CSFs and other cytokines depends on the capacity of the cytokines to stimulate the proliferation and differentiation of cells in vitro. This activity can be monitored by the appearance of cell colonies in agar (colony

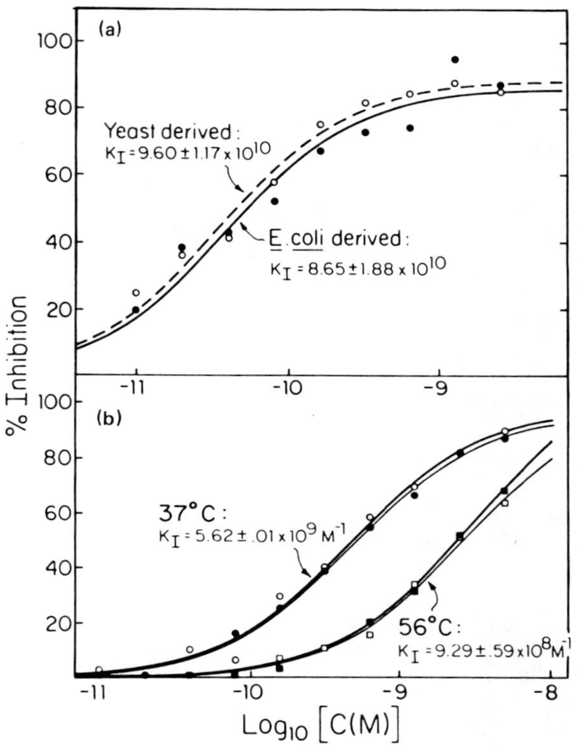

Figure 7 Practical applications of ligand binding affinity studies. (a) Binding affinity of yeast and *Escherichia coli*-derived rhuGM–CSF. HL-60 cells (3.3 X 10^7 cells/ml) were incubated for 30 min at 37°C with ^{125}I-rhuGM–CSF (3.3 X 10^{-10} *M*) and varying concentrations of unlabeled yeast or *E. coli*-derived rhuGM–CSF. Data were analyzed as described in Fig. 4a using a K_a value for ^{125}I-rhuGM–CSF of 6.9 X 10^9 M^{-1}. (b) rhuIL-4 Stability study. RAJI cells (1.3 X 10^7 cells/ml) were incubated for 1 hr at 37°C with ^{125}I-rhuIL-4 (3.3 X 10^{-10} *M*) and varying concentrations of unlabeled rhuIL-4 that had been stored for 1 month in lyophilized form at either 37 or 56°C (duplicate samples of each). Data were analyzed as described above using a K_a value for ^{125}I-rhuIL-4 of 7.0 X 10^9 M^{-1}.

assays) or by the incorporation of [^{3}H] thymidine into the DNA of proliferating cells (proliferation assays). Both in vitro assays are reasonable estimates of the activity anticipated to take place in vivo. The means by which protein growth factors stimulate cells to proliferate or differentiate is poorly understood but it is well accepted that the first event to take place is the binding of the

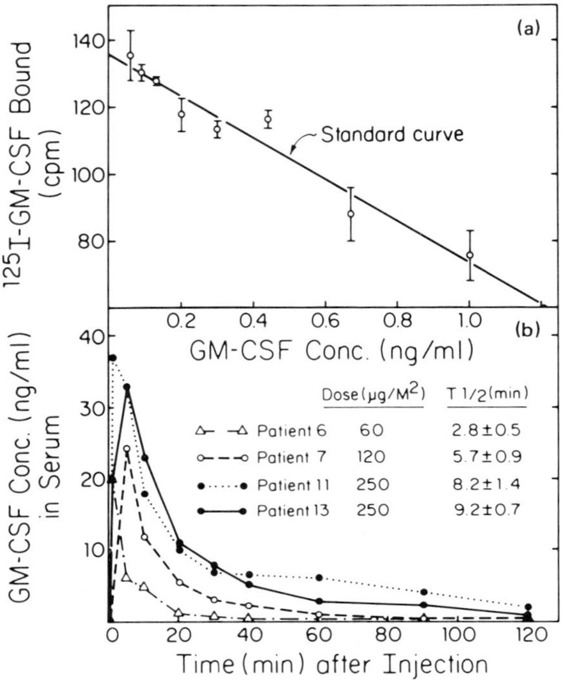

Figure 8 Pharmacokinetics of rhuGM–CSF. (a) HL-60 cells (3.3×10^7 cells/ml) were incubated for 30 min at 37°C with ^{125}I-rhuGM–CSF (3.3×10^{-10} *M*) and varying concentrations of either unlabeled rhuGM–CSF standard or serum containing unknown quantities of rhuGM–CSF collected from patients at various time points after a bolus injection (intravenous) of rhuGM–CSF. All data were corrected for nonspecific binding and a standard curve constructed by plotting ^{125}I-GM–CSF bound (cpm) versus concentration of unlabeled rhuGM–CSF (ng/ml) and the curve fit by linear least-squares analysis. (b) Concentration of rhuGM–CSF (ng/ml) in serum samples from patients receiving the indicated doses of rhuGM–CSF were determined relative to the standard curve.

growth factor to a specific plasma membrane receptor. The affinity with which a recombinant growth factor binds to a receptor is thus a physical parameter, the value of which should correlate closely with biological activity associated with the molecule. The practical advantage of receptor-based assays include the speed and reproducibility with which they can be performed.

Examples of situations in which this method of analysis has been applied are depicted in Figs. 7 and 8 (58). Figure 7a depicts the results of experiments designed to compare the potency of rhuGM–CSF expressed in *E. coli* to that expressed in yeast. Both products were found to bind to the GM–CSF receptor with comparable affinities. Similarly, the biological activity associated with GM–CSF from the two different microbial hosts was comparable. Figure 7b summarizes results that were obtained in the course of monitoring the stability of lyophilized rhuIL-4. The data revealed that storage of lyophilized material at 56°C for 1 month results in a sixfold loss in the affinity with which IL-4 bound to its receptor and corresponded to a similar loss of biological activity in samples stored at that temperature. The receptor-based assays are easy to perform and appear to accurately predict the biological activity associated with the protein. The availability of purified, cloned recombinant receptors makes it possible to contemplate assays that will predict biological potency and be free of cells.

Receptor-based assays can also be used to measure the level of growth factor present in physiological fluids by the capacity of unlabeled factor to inhibit the binding of ^{125}I-labeled factor to cells. The level of active factor, i.e., protein that has the capacity to bind receptor, can be measured at various times after administration to estimate the time of clearance from the bloodstream. As depicted in Fig. 8 (top panel), huGM–CSF levels could be accurately determined down to a level of 0.4 ng/ml huGM–CSF in the serum.

The study of hematopoietic growth factor receptors is progressing at a rapid rate and our understanding of hematopoiesis continues to grow. The number of factors involved in this process increases with each new cytokine that is cloned and the next few years will represent an exciting time during which the interactions of many of these factors and their receptors will be resolved.

REFERENCES

1. Metcalf, D. The molecular biology and functions of granulocyte-macrophage colony stimulating factors. *Blood 67*:257-267 (1986).
2. Clark, S., and R. Kamen. The human hematopoietic colony-stimulating factors. *Science 236*:1229-1237 (1987).
3. Jacobs, K., C. Shoemaker, R. Rudersdorf, S. D. Neill, R. J. Kaufmann, A. Mufson, J. Seehra, S. S. Jones, R. Hewick, E. F. Fritsch, M. Kawakita, T. Shimizu, and T. Miyake. Isolation and characterization of genomic and cDNA clones of human erythropoietin. *Nature 313*:806-810 (1985).
4. Lin, F.-K., S. Suggs, C.-H. Lin, J. K. Browne, R. Smalling, J. C. Egrie, K. K. Chen, G. M. Fox, F. Martin, Z. Stabinsky, S. M. Badrawi, P.-H. Lai, and E. Goldwasser. Cloning and expression of the human erythropoietin gene. *Proc. Natl. Acad. Sci. USA 82*:7580-7584 (1985).
5. Mochizuki, D. Y., J. R. Eisenman, P. J. Conlon, A. D. Larsen, and R. J. Tushinski. Interleukin-1 regulates hematopoietic activity, a role previously ascribed to hemopoietin 1. *Proc. Natl. Acad. Sci. USA 84*:5267-5271 (1987).

6. Stanley, E. R., A. Bartocci, D. Patinkin, M. Rosendahl, and T. R. Bradley. Regulation of very primitive, multipotent hemopoietic cells by hemopoietin-1. *Cell 45*:667-674 (1986).
7. Moore, M. A. S., and D. J. Warren. Synergy of interleukin-1 and granulocyte colony-stimulating factor: in vivo stimulation of stem-cell recovery and hematopoietic regeneration following 5-fluorouracil treatment of mice. *Proc. Natl. Acad. Sci. USA 84*:7134-7138 (1987).
8. Warren, D. J., and M. A. S. Moore. Synergism among interleukin 1, interleukin 3, and interleukin 5 in the production of eosinophils from primitive hemopoietic stem cells. *J. Immunol. 140*:94-99 (1988).
9. Peschel, C., W. E. Paul, J. Ohara, and I. Green. Effects of B cell stimulatory factor-1/interleukin 4 on hematopoietic progenitor cells. *Blood 70*:254-263 (1987).
10. Rennick, D., G. Yang, C. Muller-Sieburg, C. Smith, N. Arai, Y. Takabe, and L. Gemmell. Interleukin 4 (B-cell stimulatory factor 1) can enhance or antagonize the factor-dependent growth of hemopoietic progenitor cells. *Proc. Natl. Acad. Sci. USA 84*:6889-6893 (1987).
11. Ikebuchi, K., G. G. Wong, S. C. Clark, J. N. Ihle, Y. Hirai, and M. Ogawa. Interleukin 6 enhancement of interleukin 3-dependent proliferation of multipotential hemopoietic progenitors. *Proc. Natl. Acad. Sci. USA 84*: 9035-9039 (1987).
12. Sanderson, C. J., D. J. Warren, and M. Strath. Identification of a lymphokine that stimulates eosinophil differentiation in vitro. Its relationship to interleukin 3, and functional properties of eosinophils produced in culture. *J. Exp. Med. 162*:60-74 (1985).
13. Campbell, H. D., W. Q. J. Tucker, Y. Hort, M. E. Martinson, G. Mayo, E. J. Clutterbuck, C. J. Sanderson, and I. G. Young. Molecular cloning, nucleotide sequence, and expression of the gene encoding human eosinophil differentiation factor (interleukin 5). *Proc. Natl. Acad. Sci. USA 84*: 6629-6633 (1987).
14. Palaszynski, E. W., and J. Ihle. Evidence for specific receptors for interleukin 3 on lymphokine dependent cell line established from long-term bone marrow cultures. *J. Immunol. 132*:1872-1878 (1984).
15. Park, L. S., D. Friend, S. Gillis, and D. L. Urdal. Characterization of the cell surface receptor for a multilineage colony stimulating factor (CSF-2α). *J. Biol. Chem. 261*:205-210 (1986).
16. Nicola, N. A., and D. Metcalf. Binding of iodinated multipotential colony-stimulating factor (interleukin-3) to murine bone marrow cells. *J. Cell. Physiol. 128*:180-188 (1986).
17. May, W. S., and J. N. Ihle. Affinity isolation of the interleukin-3 surface receptor. *Biochem. Biophys. Res. Comm. 135*:870-879 (1986).
18. Sorensen, P., N. M. Farber, and G. Krystal. Identification of the interleukin-3 receptor using an iodinatable, cleavable, photoreactive cross-linking agent. *J. Biol. Chem. 261*:9094-9097 (1986).

19. Nicola, N. A., and L. Peterson. Identification of distinct receptors for two hemopoietic growth factors (granulocyte colony-stimulating factor and multipotential colony-stimulating factor) by chemical cross-linking. *J. Biol. Chem. 216*:12384-12389 (1986).
20. Nicola, N. A., and D. Metcalf. Binding of the differentiation-inducer, granulocyte-colony-stimulating factor, to responsive but not unresponsive leukemic cell lines. *Proc. Natl. Acad. Sci. USA 81*:3765-3769 (1984).
21. Nicola, N. A., and D. Metcalf. Binding of ^{125}I-labeled granulocyte colony-stimulating factor to normal murine hematopoietic cells. *J. Cell Physiol. 124*:313-321 (1985).
22. Guilbert, L. J., and E. R. Stanley. Specific interaction of murine colony-stimulating factor with mononuclear phagocytic cells. *J. Cell Biol. 85*: 153-159 (1980).
23. Guilbert, L. J., and E. R. Stanley. The interaction of ^{125}I-colony-stimulating factor-1 with bone marrow-derived macrophages. *J. Biol. Chem. 261*: 4024-4032 (1986).
24. Yeung, Y. G., P. T. Jubinsky, A. Sengupta, D. C. Y. Yeung, and E. R. Stanley. Purification of the colony-stimulating factor 1 receptor and demonstration of its tyrosine kinase activity. *Proc. Natl. Acad. Sci. USA 84*: 1268-1271 (1987).
25. Sawyer, S. T., S. B. Krantz, and E. Goldwasser. Binding and receptor-mediated endocytosis of erythropoietin in friend virus-infected erythroid cells. *J. Biol. Chem. 262*:5554-5562 (1987).
26. Sakaguchi, M., Y. Koishihara, H. Tsuda, K., Fujimoto, K. Shibuya, M. Kawakita, and K. Takatsuki. The expression of functional erythropoietin receptors on an interleukin-3 dependent cell line. *Biochem. Biophys. Res. Comm. 146*:7-12 (1987).
27. Todokoro, K., S. Kanazawa, H. Amanuma, and Y. Ikawa. Specific binding of erythropoietin to its receptor on responsive mouse erythroleukemia cells. *Proc. Natl. Acad. Sci. USA 84*:4126-4130 (1987).
28. Sawyer, S. T., S. B. Krantz, and J. Luna. Identification of the receptor for erythropoietin by cross-linking to Friend virus-infected erythroid cells. *Proc. Natl. Acad. Sci. USA 84*:3690-3694 (1987).
29. Sasaki, R., S. Yanagawa, K. Hitomi, and H. Chiba. Characterization of erythropoietin receptor of murine erythroid cells. *Eur. J. Biochem. 168*: 43-48 (1987).
30. Mufson, R. A., and T. G. Gesner. Binding and internalization of recombinant human erythropoietin in murine erythroid precursor cells. *Blood 69*: 1485-1490 (1987).
31. Fraser, J. K., F.-K. Lin, and M. V. Berridge. Expression and modulation of specific, high affinity binding sites for erythropoietin on the human erythroleukemic cell line K562. *Blood 7*:104-109 (1988).
32. Tsao, C.-J., A. Tojo, H. Fukamachi, T. Kitamura, T. Saito, A. Urabe, and F. Takaku. Expression of the functional erythropoitin receptors on interleukin 3-dependent murine cell lines. *J. Immunol. 140*:89-93 (1988).

33. Park, L. S., D. Friend, S. Gillis, and D. L. Urdal. Characterization of the cell surface receptor for granulocyte-macrophage colony-stimulating factor. *J. Biol. Chem. 261*:4177-4183 (1986).
34. Walker, F., and A. W. Burgess. Specific binding of radioiodinated granulocyte-macrophage colony stimulating factor to hemopoietic cells. *EMBO J. 4*:933-939 (1985).
35. Walker, F., and A. W. Burgess. Internalisation and recycling of the granulocyte-macrophage colony-stimulating factor (GM–CSF) receptor on a murine myelomonocytic leukemia. *J. Cell. Physiol. 130*:255-261 (1987).
36. Williams, D. E., D. C. Bicknell, L. S. Park, J. E. Straneva, S. Cooper, and H. E. Broxmeyer. Purified murine granulocyte/macrophage progenitor cells express a high-affinity receptor for recombinant murine granulocyte/macrophage colony-stimulating factor. *Proc. Natl. Acad. Sci. USA 85*:487-491 (1988).
37. Gasson, J. C., S. E. Kaufman, R. H. Weisbart, M. Tommaga, and D. W. Golde. High affinity binding of granulocyte-macrophage colony-stimulating factor to normal and leukemic human myeloid cells. Proc. Natl. Acad. Sci. USA 83:669-673 (1986).
38. Park, L. S., D. Friend, S. Gillis, and D. L. Urdal. Characterization of the cell surface receptor for human granulocyte-macrophage colony stimulating factor. *J. Exp. Med. 164*:251-262 (1986).
39. DiPersio, J., P. Billing, S. Kaufman, P. Eghtesady, R. E. Williams, and J. C. Gasson. Characterization of the human granulocyte-macrophage colony-stimulating factor receptor. *J. Biol. Chem. 263*:1834-1841 (1988).
40. Park, L. S., D. Friend, K. Grabstein, and D. L. Urdal. Characterization of the high-affinity cell surface receptor for murine B cell-stimulating factor 1. *Proc. Natl. Acad. Sci. USA 84*:1669-1673 (1987).
41. Ohara, J., and W. E. Paul. Receptors for B-cell stimulatory factor-1 expressed on cells of hematopoietic lineage. *Nature 325*:537-540 (1987).
42. Park, L. S., D. Friend, H. M. Sassenfeld, and D. L. Urdal. Characterization of the human B cell stimulatory factor 1 receptor. *J. Exp. Med. 166*:476-488 (1987).
43. Nakajima, K., Hirano, T., Koyama, K., and T. Kishimoto. Detection of receptors for murine B cell stimulatory factor 1 (BSF1): presence of functional receptors on CBA/N splenic B cells. *J. Immunol. 139*:774-779 (1987).
44. Cabrillat, H., J.-P. Galizzi, O. Djossou, N. Arai, T. Yokota, K. Arai, and J. Banchereau. High affinity binding of human interleukin 4 to cell lines. *Biochem. Biophys. Res. Comm. 149*:995-1001 (1987).
45. Lowenthal, J. W., B. E. Castle, J. Christiansen, J. Schreurs, D. Rennick, N. Arai, P. Hoy, Y. Takebe, and M. Howard. Expression of high affinity receptors for murine interleukin 4 (BSF-1) on hemopoietic and nonhemopoietic cells. *J. Immunol. 140*:456-464 (1988).
46. Dower, S. K., S. R. Kronheim, C. J. March, P. J. Conlon, T. P. Hopp, S. Gillis, and D. L. Urdal. Detection and characterization of high affinity

plasma membrane receptors for human interluekin 1. *J. Exp. Med. 162*: 501-515 (1985).
47. Lowenthal, J. W., and H. R. MacDonald. Binding and internalization of interleukin 1 by T cells. Direct evidence for high- and low-affinity classes of interleukin 1 receptor. *J. Exp. Med. 164*:1060-1074 (1986).
48. Shirakawa, F., Y. Tanaka, T. Ota, H. Suzuki, S. Eto, and U. Yamashita. Expression of interleukin 1 receptors on human peripheral T cells. *J. Immunol. 138*:4243-4248 (1987).
49. Chin, J., P. M. Cameron, E. Rupp, and J. A. Schmidt. Identification of a high-affinity receptor for native human interleukin 1β and interleukin 1α on normal human lung fibroblasts. *J. Exp. Med. 165*:70-86 (1987).
50. Paganelli, K. A., A. S. Stern, and P. L. Kilian. Detergent solubilization of the interleukin 1 receptor. *J. Immunol. 138*:2249-2253 (1987).
51. Urdal, D. L., S. M. Call, J. L. Jackson, and S. K. Dower. Affinity purification and chemical analysis of the interleukin-1 receptor. *J. Biol. Chem. 263*: 2870-2877 (1988).
52. Sims, J. E., C. J. March, D. Cosman, M. W. Widmer, H. R. MacDonald, C. J. McMahan, C. E. Grubin, J. M. Wignall, J. L. Jackson, S. M. Call, D. Friend, A. Alpert, S. Gillis, D. L. Urdal, and S. K. Dower. cDNA expression cloning of the IL-1 receptor, a member of the immunoglobulin superfamily. *Science 241*:585-589 (1988).
53. Dower, S. K., nad D. L. Urdal. The interleukin-1 receptor. *Immunol. Today 8*:46-51 (1987).
54. Nicola, N. A., C. G. Begley, and D. Metcalf. Identification of the human analogue of a regulator that induces differentiation in murine leukemic cells. *Nature 314*:625-628 (1985).
55. Souza, L. M., T. C. Boone, J. Gabrilove, P. H. Lai, K. M. Zsebo, D. C. Murdock, V. R. Chazin, J. Bruszewski, H. Lu, K. K. Chen, J. Barendt, E. Platzer, M. A. S. Moore, R. Mertelsmann, and K. Welte. Recombinanant human granulocyte colony-stimulating factor: effects on normal and leukemic myeloid cells. *Science 232*:61-65 (1986).
56. Nicola, N. A., M. A. Vadas, and A. F. Lopez. Down-modulation of receptors for granulocyte colony-stimulating factor on human neutrophils by granulocyte-activating agents. *J. Cell. Physiol. 128*:501-509 (1986).
57. Begley, C. G., D. Metcalf, and N. A. Nicola. Binding characteristics and proliferative actions of purified granulocyte colony-stimulating factor (G–CSF) on normal and leukemic human promyelocytes. *Exp. Hematol. 16*:71-79 (1988).
58. Urdal, D. L., and L. S. Park. Studies on hematopoietic growth factor receptors using human recombinant IL-3, GM–CSF, G–CSF, M–CSF, IL-1 and IL-3. *Behring Inst. Mitt. 83*:27-39 (1988).
59. Taga, T., Y. Kawanishi, R. R. Hardy, T. Hirano, and T. Kishimoto. Receptors for B cell stimulatory factor 2: quantitation, specificity, distribution, and regulation of their expression. *J. Exp. Med. 166*:967-981 (1987).

60. Ullrich, A., L. Coussens, J. S. Hayflick, T. J. Dull, A. Gray, A. W. Tam, S. Lee, Y. Yarden, T. A. Libermann, J. Schlessinger, J. Downward, E. L. V. Mayes, N. Whittle, M. D. Waterfield, and P. H. Seeburg. Human epidermal growth factor receptor cDNA sequence and aberrant expression of the amplified gene in A431 epidermal carcinoma cells. *Nature 309*:418-425 (1984).
61. Yarden, Y., J. A. Escobedo, W-J. Kuang, T. L. Yang-Feng, T. O. Daniel, P. M. Tremble, E. Y Chen, M. E. Ando, R. N. Harkins, U. Franchke, V. A. Fried, A. Ullrich, and L. T. Williams. Structure of the receptor for platelet-derived growth factor helps define a family of closely related growth factor receptors. *Nature 323*:226-232 (1986).
62. Ullrich, A., J. R. Bell, E. Y. Chen, R. Herrera, L. M. Petruzzelli, T. J. Dull, A. Gray, L. Coussens, Y.-C. Liao, M. Tsubokawa, A. Mason, P. H. Seeburg, C. Grunfeld, O. M. Rosen, and J. Ramachandran. Human insulin receptor and its relationship to the tyrosine kinase family of oncogenes. *Nature 313*: 756-761 (1985).
63. Ebina, Y., L. Ellis, K. Jarnagin, M. Edery, L. Graf, E. Clauser, J. I. Ou, F. Masiarz, Y. W. Kan, I. D. Goldfine, R. A. Roth, and W. J. Rutter. The human insulin receptor cDNA: the structural basis for hormone-activated transmembrane signalling. *Cell 40*:747-761 (1985).
64. Ullrich, A., A. Gray, A. W. Tam, T. Yang-Feng, M. Tsubokawa, C. Collins, W. Henzel, T. Le Bon, S. Kathuria, E. Chen, S. Jacobs, U. Francke, J. Ramachandran, and Y. Fujita-Yamaguchi. Insulin-like growth factor I receptor primary structure: comparison with insulin receptor suggests structural determinant that define functional specificity. *EMBO J. 5:2503-2512* (1986).
65. Johnson, D., A. Lanahan, C. R. Buck, A. Sehgal, C. Morgan, E. Mercer, M. Bothwell, and M. Chao. Expression and structure of the human NGF receptor. *Cell 47*:545-554 (1986).
66. Radeke, M. J., T. P. Misko, C. Hsu, L. A. Herzenberg, and E. M. Shooter. Gene transfer and molecular cloning of the rat nerve growth factor receptor. *Nature 325*:593-597 (1987).
67. Leonard. W. J., J. M. Depper, G. R. Grabtree, S. Rudkioff, J. Pumphrey, R. J. Robb, M. Kronke, P. B. Svetlik, N. J. Peffer, T. A. Waldmann, and W. C. Greene. Molecular cloning and expression of cDNA for the human interleukin 2 receptor. *Nature 311*:626-631 (1984).
68. Nikaido, T., A. Schimizu, N. Ishida, H. Sabe, K. Teshigawara, M. Maeda, T. Uchiyama, J. Yodoi, and T. Honjo. Molecular cloning of cDNA encoding human interleukin 2 receptor. *Nature 311*:631-635 (1984).
69. Cosman, D., D. P. Cerretti, A. Larsen, L. Park, C. March, S. Dower, S. Gillis, and D. Urdal. Cloning, sequence, and expression of human interleukin 2 receptor. *Nature 312*:768-771 (1984).
70. Sherr, C., C. Rettenmier, P. Sacca, M. Roussel, A. Look, and E. Stanley. The c-fms proto-oncogene product is related to the receptor for the mononuclear phagocyte growth factor, CSF-1. *Cell 41*:665-676 (1985).

71. Coussens, L., C. Van Beveren, D. Smith, E. Chen, R. L. Mitchell, C. M. Isacke, I. M. Verma, and A. Ullrich. Structural alteration of viral homologue of receptor proto-oncogene fms at carboxyl terminus. *Nature 320*: 277-280 (1986).
72. Smith, K. A. The two-chain structure of high-affinity IL-2 receptors. *Immunol. Today 8*:11-13 (1987).
73. Greenwood, F. C., W. M. Hunter, and J. S. Glover. The preparation of ^{131}I labeled human growth hormone at high specific radioactivity. *Biochem. J. 81*:114-123 (1963).
74. Tejedor, F., and J. P. G. Ballesta. Iodination of biological samples without loss of functional activity. *Anal. Biochem. 127*:143-149 (1982).
75. Bolton, A. E., and M. W. Hunter. The labeling of proteins to high specific radioactivities by conjugation to a ^{125}I-containing acylating agent: application to the radioimmunoassay. *Biochem. J. 133*:529-539 (1973).
76. Dower, S. K., J. A. Titus, C. Delisi, and D. M. Segal. Mechanism of binding of multivalent immune complexes to Fc receptors. 2. Kinetics of binding. *Biochemistry 20*:6335-6340 (1981).
77. Demeyts, P., J. Roth, D. M. Neville, J. R. Gavin, and M. Lesniak. Insulin interactions with its receptors: experimental evidence for ngative cooperativity. *Biochem. Biophys. Res. Commun. 55*:154-161 (1973).
78. Stanley, E. R., and L. J. Guilbert. Methods for the purification, assay, characterization and target cell binding of a colony stimulating factor (CSF-1). *J. Immunol. Meth. 42*:253-284 (1981).
79. Scatchard, G. The attraction of proteins for small molecules and ions. *Ann. N.Y. Acad. Sci. 51*:660-672 (1949).
80. Limbird, L. E. *Cell Surface Receptors: A Short Course on Theory and Methods*. Nighoff Publishing Co., Boston (1986).
81. Cheng, Y.-C., and W. H. Prusoff. Relationship between the inhibition constant (K_I) and the concentration of inhibitor which causes 50 per cent inhibition (I_{50}) of an enzymatic reaction. *Biochem. Pharmacol. 22*: 3099-3108 (1973).
82. Park, L. S., R. J. Tushinski, D. Y. Mochizuki, and D. L. Urdal. Interleukin-4 binds to murine fibroblasts: receptor characterization and induction of colony stimulating activity production. UCLA Symposia. *J. Cell Biochem.* (Suppl. 12A):111 (1988).
83. Dower, S. Characterization of the plasma membrane interleukin-1 receptor. In *Immunology Series Vol. 35, Recombinant Lymphokines and Their Receptors*, S. Gillis (Ed.). Marcel Dekker, Inc., New York, pp. 159-185 (1987).
84. Dower, S., Call, S., Gillis, S., and Urdal, D. Similarity between the IL-1 receptors on a murine T-lymphoma cell line and on a murine fibroblast cell line. *Proc. Natl. Acad. Sci. USA 83*:1060-1064 (1986).

3
Granulocyte Colony-Stimulating Factor

NICOS A. NICOLA
The Walter and Eliza Hall Institute of Medical Research, Royal Melbourne Hospital, Melbourne, Victoria, Australia

I. INTRODUCTION

Granulocyte colony-stimulating factor (G–CSF) is one member of a family of glycoprotein growth factors that control the survival, proliferation, and differentiation of hemopoietic progenitor cells and functionally activate mature hemopoietic cells. It is distinguished from the other members of this family (granulocyte-macrophage CSF, macrophage CSF, multipotential CSF or interleukin 3, interleukins 1, 2, 4, 5, 6, 7) primarily by its cellular specificity almost exclusively for neutrophilic granulocytes and their precursors and its exceptional capacity to induce terminal differentiation in certain lines of myeloid leukemic cells. The study of G–CSF has progressed from the initial description of a crude biological activity that induced differentiation of the murine myeloid leukemic cell line WEHI-3B D^+ through the purification and molecular cloning of both murine and human G–CSF, and finally to the first clinical trials of G–CSF in patients with cancer. In the present review the current knowledge of the protein and DNA structures of G–CSF, its actions on normal and leukemic cells, and its mechanism of action through specific cellular receptors will be summarized.

II. BIOLOGICAL PROPERTIES OF G–CSF

Murine G–CSF was initially described as a colony-stimulating activity that co-purified with an activity in the serum of endotoxin-injected mice that induced the terminal differentiation of the murine myelomonocytic leukemic cell line WEHI-3B D^+ (1). Subsequent purification of the WEHI-3B D^+ differentiation-inducing activity from the conditioned medium of lungs from endotoxin-injected mice showed that the colony-stimulating activity was an intrinsic property of this molecule (2). Indeed, G–CSF has now been shown to induce differentiation in at least three murine leukemic or continuous myeloid cell lines as well as the human promyelocytic leukemia cell line HL60.

A. Differentiation Induction in Myeloid Cell Lines by G–CSF

Granulocyte colony-stimulating factor accounts for nearly all the differentiation-inducing activity for WEHI-3B D^+ cells in media conditioned by various murine tissues and in endotoxin serum (1,3). Any remaining activity in these sources can be accounted for by the weaker differentiation-inducing activity of GM–CSF (4). The differentiation-inducing activity of G–CSF has most commonly been assayed by quantifying the appearance of a characteristic halo of cells around developing WEHI-3B D^+ colonies in semisolid agar cultures (1). The dose-response relationship for the action of G–CSF in causing this transformation of morphology of WEHI-3B D^+ colonies from "tight" colonies to "loose" colonies is identical to that for its colony-stimulating activity on normal neutrophil progenitor cells, with the concentration required for half-maximal effect being 3 p*M* (2). Analysis of WEHI-3B D^+ colonies developing in the presence of G–CSF has indicated that the "halo" of cells is a result of differentiation of the cells from blast cells to cells with the morphology of macrophages and neutrophils, with a greater neutrophil content at high concentrations of G–CSF (5). The motility of the latter cells in agar gives rise to the characteristic diffuse nature of colonies associated with cell differentiation.

Associated with G–CSF-induced cellular differentiation of the WEHI-3B D^+ cells is a pronounced reduction of the proliferative potential (clonogenicity) of cells within WEHI-3B D^+ colonies. Individual WEHI-3B D^+ cells growing in control agar cultures generated, during 7 days, an average of 1500 clonogenic cells per colony, whereas colonies formed in G–CSF-containing cultures contained an average of only 15 clonogenic cells (5). This reduced proliferative capacity of G–CSF-treated cells was also reflected in their behavior in vivo using a tumor transplantation model in which mice injected with G–CSF-treated WEHI-3B D^+ cells survived longer than mice injected with untreated cells (6). It should be noted that G–CSF-differentiated WEHI-3B D^+ colonies are generally of similar size to undifferentiated colonies, implying that the differentiative and clonal suppressive actions of G–CSF are not manifested until several cell divisions have occurred.

Metcalf (7) studied the mechanism of action of G–CSF on WEHI-3B D^+ cells in detail by exposing cells to G–CSF for 18-22 hr, then, by micromanipulation of washed progeny cells, determining whether the daughter and granddaughter cells could continue to proliferate. He found that such an exposure could lead to irreversible differentiation commitment in a proportion of the granddaughter cells, with cessation of proliferation most commonly after one or two cell divisions. With longer exposure of these cells to G–CSF, progressively more cells became committed to differentiation, suggesting that complete suppression of the leukemic cells could be achieved by a process of attrition. Analysis of the fate of daughter and granddaughter cells of cells exposed to G–CSF also showed that suppression of these cells was markedly asymmetric, with one progeny cell continuing to proliferate while the other differentiated and lost proliferative capacity. The irreversible action of G–CSF, sometimes being exerted within one cell cycle, and the asymmetry of its action on daughter cells suggest that G–CSF results in heritable changes in the DNA of responding cells, possibly during cell division, but that these changes accumulate slowly on a population basis so that continued exposure to G–CSF is necessary for complete suppression of a leukemic population.

More recently it has also been shown that G–CSF can induce differentiation in a different murine myelomonocytic cell line M1 (8,9). Tsuda et al. (8) showed that G–CSF inhibited the proliferation of M1 cells and caused their accumulation in the G1 phase of the cell cycle. Moreover, they showed that M1 cells arrested in early G1 (by cell density inhibition and serum removal) but not late G1 (by aphidicolin) responded rapidly (within 18 hr) and synchronously to G–CSF by differentiation as assessed by the acquisition of Fc receptors and lysozyme synthesis. Based on this and other data they suggested that cells passing through the early G1 phase of the cell cycle transiently make a decision either to proliferate or to terminally differentiate and that G–CSF influences this decision toward differentiation. In normally growing cultures, commitment to differentiation is stochastic and the cell population differentiates asynchronously, but in cells trapped in early G1 the decision to differentiate is more likely to occur in a given time-frame, resulting in synchronous differentiation. Tomida et al. (9) also showed that G–CSF could induce terminal differentiation of M1 cells to macrophages with phagocytic activity. However, they showed that considerably higher concentrations of G–CSF (500 p*M*) were required to induce differentiation in M1 cells compared with that required for WEHI-3B D^+ cells (20 p*M*). Moreover, M1 cells were also responsive to a different differentiation-inducer (D-factor) or leukemia inhibitor factor to which WEHI-3B D^+ cells do not respond (10,11).

Granulocyte colony-stimulating factor has been shown to induce differentiation in a nonleukemic immortal cell line (32D cl.3) that requires interleukin-3 (IL-3) for its continuous growth in vitro (12). Although these cells proliferate indefinitely with no differentiated cells detected in the presence of IL-3, when they are transferred to G–CSF the cells proliferate transiently but ultimately (within 10-12 days) lose all proliferative capacity and differentiate to mature neutrophilic granulocytes. The effect of G–CSF is antagonized by increasing concentrations of IL-3 while the effect of IL-3 is antagonized by G–CSF, suggesting a competitive interaction between IL-3 on the one hand, to enhance proliferation and suppress differentiation, and G–CSF on the other to induce differentiation and ultimately suppress proliferation. While growing in IL-3, 32D cl.3 cells infected with the Abelson murine leukemia virus (A–MuLV) became independent of IL-3 for their growth and could not be induced to differentiate even by very high concentrations of G–CSF. Conversely, 32D cl.3 cells that had been exposed to G–CSF, even for 24 hr, became resistant to transformation by A-MuLV as detected by the generation of IL-3-independent clones (13). This suggests that G–CSF commits the cells to terminal differentiation early, while they still have proliferative capacity, and well before the appearance of fully differentiated cells. This commitment process again appears irreversible so that it cannot be abrogated by A-MuLV. On the other hand, infection of the cells with A-MuLV in the precommitment phase interferes with the ability of the cells to respond to G–CSF by differentiation commitment.

Similar results were obtained by Crosier et al. (14) who showed that 32D Cl.23 cells were initially stimulated by but eventually lost proliferative capacity in the presence of G–CSF. Conversely, *src* transformed 32D Cl.23 cells that became factor-independent for growth were not inhibited in their growth by G–CSF.

These data suggest that certain responsive cells are transiently susceptible to the differentiating action of G–CSF during the G1 phase of the cell cycle. In a stochastic process cells may then undergo a series of irreversible events leading ultimately to the induction of terminal differentiation and cessation of proliferation. The decision to undergo these events, however, appears to be a competitive one stimulated by G–CSF but abrogated by IL-3 and various retroviral oncogens (*abl, src*, and perhaps others involved in the transformation of cells such as NFS60 and J774). It is possible that IL-3 and such oncogenes provide signals that direct the cell to traverse the cell cycle after G1 in a self-renewing process, rather than undergoing the events that lead to terminal cell differentiation (Fig. 1).

A number of murine myeloid cell lines have been shown to be unresponsive to G–CSF for differentiation induction. For some cell lines (WEHI-3B D^-, WEHI 265, PU-5, P388) this unresponsiveness is related to the absence of G–CSF receptors (15) but for others (J774, NFS60) significant numbers of G–CSF receptors have been demonstrated and, in these cases, it is presumed that the cells lack the ability to initiate differentiation programs in response to the signals delivered by the active G–CSF-receptor complex. Indeed, Bettens et al. (16) have shown that differentiation-defective WEHI-3B D^- cells that fail to respond to G–CSF can still differentiate in the presence of vitamin D_3 by activating cytosolic or nuclear events. On the other hand, the response of

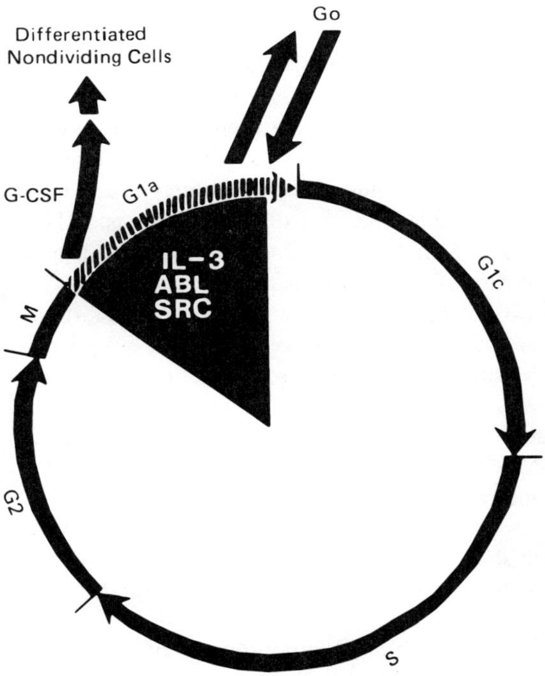

Figure 1 Model of the competing effects of G–CSF and IL-3 on differentiation commitment or self-renewal, respectively, of 32D cells. The decision between these two processes appears to be made during the early G1 phase of the cell cycle.

NFS60 cells to G–CSF appears to be, at least initially, stimulation of proliferation with no obvious differentiation (16a).

Among human myeloid cell lines it has been shown that G–CSF (as well as GM–CSF) can induce differentiation in HL60 promyelocytic leukemia cells (17-19). In clonal agar cultures G–CSF induced the appearance of colonies with a dispersed morphology (analogous to its action on WEHI-3B D^+ cells), but in most experiments there were only minor reductions in the frequency of clonogenic cells (17). In suspension cultures, however, Begley et al. (18) found that G–CSF induced the expression on HL60 cells of a variety of maturation markers including 25E11, NIMP-R10, and WEM-G11. Moreover, there was evidence for suppression of the generation of clonogenic cells by G–CSF (and GM–CSF), especially after 2-3 weeks of culture and, for some HL60 clones, complete clonal extinction was seen. By these criteria G–CSF was as effective as dimethylsulfoxide, a known differentiation inducer for HL60 cells, in suppressing the clonal proliferation of HL60 cells.

The action of G–CSF on primary human myeloid leukemias is less clear. In part, the action of G–CSF on such cells is complicated by the fact that most human leukemias are CSF dependent for growth so that there is no appropriate control to compare the action of G–CSF. Nevertheless, in preliminary studies Souza et al. (20) showed that G–CSF could induce both granulocytic and monocytic differentiation in primary human leukemic cells of the M2, M3, and M4 classification compared with unstimulated controls. On the other hand, Begley et al. (21,22) found that the majority of AML samples showed some level of proliferative stimulation with G–CSF that correlated with the degree of stimulation observed with GM–CSF but not with the number of G–CSF receptors detected. In several other studies similar results have been obtained although heterogeneous responses have been noted in all studies (23-26).

In the majority of cases of acute myeloblastic leukemia (AML) the leukemic cells proliferated with either GM–CSF or G–CSF, sometimes additively, but some cases responded only to one type of CSF and some others did not proliferate in vitro. The proliferative responses of the cases did not correspond to any of the French-American-British (FAB) classification of AML. The dose-response curves of the AML cells to G–CSF or GM–CSF were in general similar to those on normal cells.

It is clear from these data that the proliferative responsiveness of AML and CML cells to G–CSF and other factors is extremely heterogeneous. Moreover, the long term effects of G–CSF on leukemic stem cells are unknown. It follows that the use of agents such as GM–CSF and G–CSF in the treatment of patients with myeloid leukemia or preleukemic syndromes must be approached with extreme caution and on an individual basis.

Cheng et al. (27) examined the cells of 22 AML patients for rearranged G–CSF genes as a possible autocrine mechanism for leukemia. They found two patients with rearranged G–CSF genes but neither secreted detectable G–CSF and one showed no detectable G–CSF mRNA whereas the other showed only a normal-sized transcript.

B. In Vitro Actions of Hemopoietic Progenitor Cells

The first action of G–CSF on normal murine hemopoietic cells to be recognized was its ability to stimulate the formation of relatively small numbers of uniform-size granulocytic colonies from bone marrow precursor cells (1,2,28). In serum-containing cultures, G–CSF stimulated the formation of ~30 colonies/75,000 bone marrow cells with half maximal response being recorded at a concentration of 3 p*M* (2,28). At low concentrations of G–CSF the colonies were composed exclusively of neutrophilic granulocytes but at higher concentrations up to 10% of the colonies were composed of macrophages and up to 20% were composed of a mixture of granulocytes and macrophages. The colonies were uniform and generally small (100-150 cells per colony). This is in contrast to the action of GM–CSF, in which larger numbers of colonies (80–120/75,000 bone marrow cells) were stimulated to develop. They were of less uniform and generally larger size, and the inverse concentration dependence was observed (macrophage colonies were stimulated to develop at lower concentrations and granulocyte colonies at higher concentrations).

The action of G–CSF on progenitor cells was a direct one because it stimulated the proliferation and almost exclusively neutrophilic granulocyte differentiation of highly purified colony and cluster-forming cells from murine fetal liver. Somewhat surprisingly, G–CSF was able to allow the survival and/or initial proliferation of a much larger number of granulocyte, macrophage, eosinophil, and erythroid colony-forming cells although it could not sustain their proliferation and differentiation to form colonies. This was demonstrated in experiments in which cultures were initiated in G–CSF and then were transferred to GM–CSF or other CSFs (28). Granulocyte colony-stimulating factor has also been shown to stimulate the formation of blast cell colonies with multiple lineage potentials, although only the neutrophil and macrophage potentials are clearly expressed in the absence of other CSFs (29). In the presence of multi-CSF (IL-3), G–CSF also decreased the Go phase of multipotent blast cell colony-forming cells (30). Also, somewhat surprisingly, G–CSF had a synergistic effect on colony formation when added with GM–CSF or M–CSF, with the increase in colony numbers over those expected being due primarily to an increase in colonies containing both granulocytes and macrophages (28).

In serum-free cultures of murine bone marrow cells G–CSF appeared to stimulate only granulocyte formation at all concentrations of G–CSF tested, but a much smaller number of colonies were stimulated in serum-free than in serum-containing cultures (31). Because G–CSF was shown to act directly on colony-forming cells in the absence of accessory cells (28) these data suggest that either serum components are required to express the action of G–CSF or that granulocyte colony formation has a more stringent requirement for serum than other types of colony formation.

Human G–CSF, shown to be equivalent to activities previously described as CSF-β (32) or pluripoietin (33), has similar actions on human hemapoietic progenitor cells as does murine G–CSF on murine cells. Colony-stimulating-factor-β was originally shown to stimulate predominantly neutrophilic granulocyte colony formation, especially at lower concentrations, with the production of macrophage-containing colonies at higher concentrations (34). These colonies appeared earlier (day 7 of culture) and were smaller than those stimulated by human GM–CSF (maximal by day 14 of culture). Moreover, the colony-forming cell responding to CSF-β could be separated from that responding to GM–CSF by sedimentation velocity (35), suggesting that these CSFs might have a synergistic effect on colony formation. Although original reports suggested that pluripoietin was able to stimulate the formation of eosiniphil, erythroid, and mixed colonies (33), it is now thought that this effect was an indirect one mediated by the action of G–CSF on accessory cells to produce GM–CSF because it has no such actions on highly purified colony-forming cells (36,37). Granulocyte colony-stimulating factor can act as a proliferative stimulus on relatively mature granulocyte precursors, at least at the level of neutrophilic promyelocytes (37,38), and these cells can respond by at least one to two cell divisions after a single short exposure to high concentrations of G–CSF (39). Promyelocytes responded to lower doses of G–CSF than did more immature blast cells and their proliferation was transient with maximal clone numbers appearing by days 3-5 of culture (39). These results could be explained by the higher levels of G–CSF receptors seen on promyelocytes (see Section VI) and their relatively rapid differentiation to neutrophils, which then migrate out of the formed clones. As for the mouse system, G–CSF stimulated almost exclusively neutrophilic granulocyte colony formation in serum-free cultures of human bone marrow cells (40).

A surprising aspect of the actions of G–CSF is the complete two-way biological cross-reactivity of murine and human G–CSFs on murine or human cells (32), indicating the strong conservation of the active portions of both G–CSF and its receptor through evolution. These actions clearly include a direct stimulating action on the proliferation and differentiation of committed neutrophilic granulocyte progenitor cells. It is unclear whether the other actions of G–CSF on macrophage progenitor cells, progenitor cells of other cell lineages, and multipotential cells are direct ones or mediated by the action of G–CSF on accessory cells to stimulate the release of other factors (e.g, GM–CSF IL-1, etc.). However, these other actions of G–CSF, whether direct or indirect, are important to consider in evaluating the action of G–CSF injected into animals or patients.

C. In Vitro Actions on Mature Hemopoietic Cells

As for hemopoietic precursor cells, G–CSF enhanced the survival time of human peripheral blood neutrophils but did not enhance eosinophil survival in vitro (41). Although this effect was modest (increasing survival time by 5-10 hr), it occurred at considerably lower concentrations of G–CSF than that required for stimulating colony formation.

Granulocyte colony-stimulating factor was also able to stimulate several functional activities of mature neutrophils. Murine and human G–CSF stimulated both bone marrow-derived and peritoneal exudate murine neutrophils in their ability to demonstrate antibody-dependent cell-mediated cytotoxicity against antibody coated murine thymoma cells (42). Granulocyte-macrophage colony-stimulating factor also displayed this activity and, interestingly, combination of GM–CSF and G–CSF at pleateau levels caused an additive response, suggesting that different subpopulations of neutrophils respond to one or another CSF or that the combined effect of both CSFs on the same cell produced an augmented response. Granulocyte colony-stimulating factor also dramatically increased the antibody-dependent cytotoxic capacity of human blood neutrophils but not eosinophils, whereas human GM–CSF stimulated both types of cells (43,44).

G–CSF also primed human neutrophils for enhanced responsiveness to other agents. Membrane depolarization in response to the chemotactic peptide *N*-formyl Met-Leu-Phe (fMLP) was accelerated by G–CSF and the generation of superoxide anion in response to fMLP was enhanced (45,46). Similarly, arachidonic acid release in response to ionophore or chemoattractants was enhanced by G–CSF (47). Some of these effects could be mediated by the ability of G–CSF to increase the binding of fMLP to neutrophils within 30 min (48).

The ability of G–CSF to enhance differentiated functions of neutrophils was also demonstrated on the immature or defective neutrophils of patients with myelodysplastic syndromes (MDS) or chronic myelogenous leukemia (CML) (46). Granulocyte colony-stimulating factor at low concentrations (1 U/ml) was able to increase the expression of neutrophil alkaline phosphatase in MDS neutrophils and had similar but delayed effects on CML neutrophils at higher concentrations (>1000 U/ml). It also rapidly increased the responsiveness of MDS neutrophils to fMLP for superoxide anion generation. These results suggest that circulating levels of G–CSF in MDS patients may be low and that the responsiveness of neutrophils to G–CSF depends on their maturation state and their degree of prior exposure to endogenous G–CSF levels.

The functional properties of neutrophils that are enhanced by G–CSF are those related to host defense against microorganisms and, coupled with the observation that serum G–CSF levels are rapidly elevated by bacterial products (endotoxin), this suggests that a major role of G–CSF in vivo may be the activation of neutrophilic granulocytes to resolve infections.

D. Actions of G–CSF In Vivo

The dual actions of G–CSF in vitro, mentioned in the two previous sections, to stimulate neutrophil production from precursor cells and to stimulate mature neutrophil functional activity suggest that it would be useful to enhance hemopoietic recovery in situations of reduced bone marrow capacity or to enhance the ability to resolve infections in immunocompromised hosts.

In preclinical studies human G–CSF has been injected into mice (49), hamsters (50), and monkeys (51), the human molecule being active in all of these species. Using continuous i.v. infusion in golden Syrian hamsters, Cohen et al. (50) observed a selective 10-fold increase in peripheral blood neutrophils by day 3 with no effect on lymphocyte, monocyte, or eosinophil levels.

Subcutaneous injection of G–CSF (2.5 μg/injection/day) into mice caused a rapid and transient 2.5-fold elevation in blood neutrophils within 2 hr, suggesting a mobilization of preformed neutrophils from marrow or tissues. Subsequently, a sustained elevation of blood neutrophil levels (5-10-fold) was observed for at least 15 days with only minor and inconsistent effects on lymphocyte, monocyte, red blood cell, and platelet levels. However, in view of other results, the slight elevation in monocyte levels and the slight decrease in platelet levels may be significant. At the same time, spleen cellularity was increased by 50% and spleen content of all types of colony-stimulating cells (GM, Meg, CFU–E, BFU–E) and multipotential progenitor cells (CFU–S) was elevated 10-100-fold. Similar but less dramatic changes were noted in the bone marrow. After cessation of G–CSF injections all blood parameters returned to normal levels in less than 48 hr. All spleen parameters also returned to control values after cessation of G–CSF treatment but this return was much slower (10-20 days) than the return of peripheral blood cell levels (49).

Cynomolgus monkeys injected subcutaneously with human G–CSF (1 μg-100 μg/kg/day) also showed a dramatic, dose-dependent increase in blood neutrophils (up to 10-fold) that was sustained for at least 27 days. Other blood parameters were normal except that lymphocytes were also slightly increased (1.5-2.5-fold). The neutrophils in the blood of G–CSF-treated monkeys had normal or enhanced functional activities. There was evidence in the monkeys of enhanced myelopoiesis in the bone marrow, spleen, lymph nodes, and thymus but not in other organs (51).

In view of the ability of G–CSF to elevate neutrophil levels in normal animals, the capacity of G–CSF to enhance hemopoietic recovery in animals treated with myelosuppressive drugs has also been investigated. Tamura et al. (49) showed that a single intraperitoneal injection of cyclophosphamide into mice caused a dramatic depression of blood neutrophil levels and CFU–S and GM–CFC levels in the spleen. Four daily injections of G–CSF (2.5 μg/day s.c.) after cyclophosphamide treatment not only prevented the depression of blood cell neutrophil levels but actually increased them over starting levels (up to 10-fold) with increased numbers of circulating immature forms as well. Granulocyte colony-stimulating factor did not prevent the depression of spleen progenitor cell levels but significantly enhanced the rate of recovery of these progenitors to supranormal levels.

These hematological responses to G–CSF after cyclophosphamide treatment were subsequently shown to have profound consequences for the ability of the animals to resolve infections with a range of different bacteria (52). In nearly all cases tested G–CSF, at 1.0 μg/day or more, completely restored the resistance of cyclophosphamide treated mice to bacterial infections to the normal levels seen for mice untreated with cyclophosphamide. In G–CSF-injected mice there was an enhanced migration of neutrophils to the site of infection (peritoneal cavity) followed by rapid elimination of bacteria at the site as compared with untreated controls. The complete protection of these mice from otherwise lethal doses of bacteria was most striking (Fig. 2). Shimamura et al. (53) also demonstrated that G–CSF could significantly enhance the recovery of blood neutrophil, spleen CFC and, to a lesser extent, bone marrow CFC in mice treated with 5-fluorouracil. The response was dose-dependent over the range 0.5-2 $\times$ 10^5 U/day (approximately 0.25-1 μg/day). Moore and Warren (54) found similar results in 5-fluorouracil-treated mice, with G–CSF (4 μg/day) decreasing the period of neutropenia by at least 4 days and resulting in 50-fold elevations of blood neutrophil levels over normal levels by day 14. Even in the neutropenic phase, G–CSF-treated mice showed higher levels of all progenitor cells in the spleen and marrow than untreated mice. Combined treatment of these mice with G–CSF and IL-1α (0.4 μg/day) further decreased the period of neutropenia by 2 days and dramatically increased the recovery of spleen and bone marrow progenitor cells.

The advanced hemopoietic recovery seen in cyclophosphamide-treated mice treated with G–CSF has also been documented in Cynomolgus monkeys treated with cyclophosphamide (51). Granulocyte colony-stimulating factor (10 μg/kg/day) reduced the period of neutropenia in such animals from >4 weeks to 1 week with elevation of blood neutrophil levels by 10-fold after 15 days' treatment with G–CSF. This was accompanied by dramatically increased cellularity of the bone marrow by day 14, compared with animals treated only with cyclophosphamide, and recovery of all of the hemopoietic cell lineages.

E. Clinical Trials with G–CSF

The dramatic effects seen with G–CSF on recovery from chemotherapy-induced neutropenia and suceptibility to bacterial infections in animal models suggest that G–CSF would be useful in the treatment of patients who are myelosuppressed as a result of cancer chemotherapy, irradiation therpy, or undergoing bone marrow transplantation. Several clinical trials are presently under way to determine the efficacy of G–CSF in these and related conditions involving myelosuppression. In the first report published by Bronchud et al. (55) G–CSF was administered in patients receiving intensive chemotherapy for small-cell lung cancer. Granulocyte colony-stimulating factor administered to patients before chemotherapy resulted in a rapid and striking increase in blood neutrophils in all 12 patients (up to 10-fold) (1-40 μg/kg/day). No toxicities of the treatment

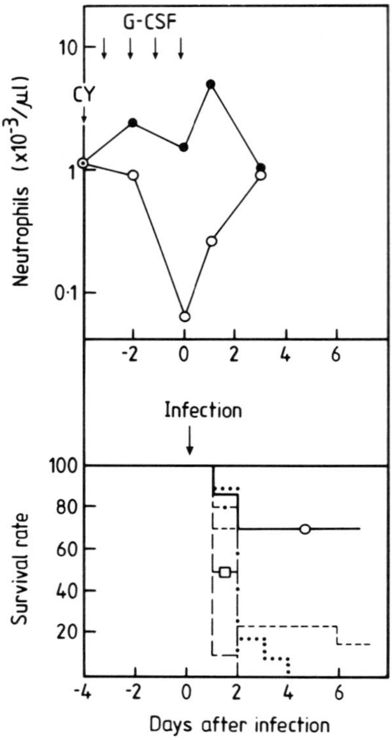

Figure 2 The effect of G–CSF on peripheral blood neutrophil counts and survival after bacterial infection of mice pretreated with cyclophosphamide (CY). Neutrophil counts are shown in the upper panel with mice treated (●——●) or untreated (○——○) with G–CSF. The lower panel shows the survival rate of mice pretreated with cyclophosphamide (day −4) and infected with bacteria on day 0. (Without G–CSF and with: - -, *Pseudomonas aeruginosa;* - • -, *Serratia marescens* . . . ., *Staphlococcus aureus*; -□-, *Candida albicans;* - - - -, *Escherichia coli*). All mice injected with G–CSF and bacteria showed 100% survival during 7 days (——) except those injected with *E. coli* (–o–). (Data redrawn from Ref. 52.)

were observed and the neutrophils appeared normal in tests of their mobility and bactericidal activity. No appreciable changes were observed in monocytes, eosinophils, platelets, lymphocytes, or hemoglobin, and neutrophil levels rapidly fell to normal levels after cessation of G–CSF infusion.

The effect of G–CSF on hemopoietic recovery after chemotherapy was assessed by infusing G–CSF for 14 days on alternate cycles of chemotherapy.

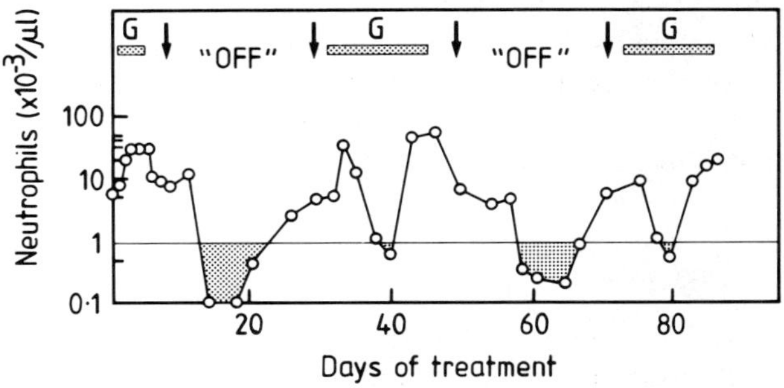

Figure 3 The effect of G–CSF on the degree and duration of peripheral blood neutropenia associated with chemotherapy in a patient with small-cell lung carcinoma. Arrows indicate the administration of chemotherapy and stippled bars indicate the period of G–CSF administration. "OFF" indicates a chemotherapy phase with no G–CSF administration. The stippled areas of the curve indicate the period in which the patient was neutropenic ($<1 \times 10^3$ neutrophils/μl). Note the decreased severity and duration of neutropenia while receiving G–CSF treatment. (Data redrawn from Ref. 55.)

Compared with the periods "off" G–CSF after chemotherapy, G–CSF was found to decrease the period of neutropenia by a median value of 80% with a return to normal or supranormal neutrophil levels within 2 weeks of initiation of chemotherapy (Fig. 3). Six severe infective episodes were recorded for patients "off" G–CSF and none for patients "on" G–CSF suggesting that, as for the animal models the reduction in neutropenic phases was protective against infections.

In a second clinical trial Morstyn et al. (56) administered G–CSF by i.v. infusion to patients being treated with melphalan as chemotherapy for a range of tumors. In the prechemotherapy phase they found a dose-related increase in peripheral blood neutrophils (2-10-fold from 1-60 μg/kg/day) and a lesser increase in blood lymphocytes and monocytes at the higher doses. This increase was preceded by a rapid (within 5 min) but transient and selective loss of neutrophils from the blood Increased immature neutrophil forms were also subsequently seen in the blood and in marrow aspirates. Granulocyte colony-stimulating factor was well tolerated by the patients with only minor side effects (notably mild bone pain).

The duration of neutropenia after melphalan depended on the dose of G–CSF and was shortened from 8 days at 1 μg/kg to 1 day at 30 μg/kg. Higher doses of G–CSF caused a more rapid depletion of blood neutrophils after melphalan treatment but shortened the neutropenic phase and increased neutrophil levels attained after cessation of melphalan.

Similar beneficial effects of G–CSF administration were observed with patients with transitional-cell carcinoma of the urothelium receiving chemotherapy. Treatment with G–CSF after chemotherapy significantly reduced the duration of neutropenia, the need for antibiotic therapy and the incidence of mucositis and increased the number of patients who qualified for further chemotherapy (56a).

F. Summary of Biological Effects

The most consistent feature of the actions of G–CSF in vitro and in vivo is its specificity for cells of the neutrophilic granulocyte cell lineage. Despite its fairly weak colony-stimulating activity in vitro it has a spectacular capacity to elevate blood neutrophil levels in all species tested in vivo. Some immediate effects of G–CSF in vivo may be due to its capacity to cause margination or demargination of preformed neutrophils but the sustained elevations seen for up to several months must also reflect increased production of neutrophils from precursors in the bone marrow and extramedullary sites. The rapid return of blood neutrophil levels to normal after cessation of G–CSF treatment suggests that the effect of G–CSF is not simply prolonged survival of existing neutrophils.

Granulocyte colony-stimulating factor is fairly rapidly cleared from the circulation in a biphasic manner ($t_{1/2}$ = 3-8 min for the first phase and 100 min for the second phase in patients). This allows for a rapid return to normal levels after G–CSF treatment but also implies that continuous infusion or subcutaneous injections may be the best way to maintain elevated G–CSF levels in patients.

It is unclear how many of the effects seen in vivo with G–CSF are due to the direct action of G–CSF or to the secondary release of other factors. Based on the results in vitro, it is likely that the effects occasionally seen on platelet and lymphocyte levels during G–CSF treatment are due to secondary events.

Regardless of the mechanisms involved it is clear that injected G–CSF is a powerful blood neutrophil elevating agent with few side effects (at least in the short-term) and that it significantly decreases the myelotoxicity associated with chemotherapeutic agents. It appears to have a promising future in the treatment of patients at risk from infections for a variety of reasons involving depressed neutrophil levels or reduced neutrophil function.

III. SOURCES OF G–CSF AND REGULATION OF ITS PRODUCTION

Almost all murine tissues produced G–CSF when cultured in vitro and the G–CSF released from all tissues had identical properties assessed by partial biochemical purification (3). The production was continuous over several days and reflected de novo synthesis rather than release of stored G–CSF because the production could be inhibited by inhibitors of protein synthesis. Injection of bacterial endotoxin into animals 3 hr before harvesting tissues resulted in the production and release of higher levels of G–CSF than from control animals. These data suggest that either all cells can produce G–CSF or that cells common to many tissues produce G–CSF and increase their production or release in response to endotoxin or its products. The latter possibility is supported by the observation that monocytes and macrophages (57,58), endothelial cells (59,60), and fibroblasts (61-64) can all be induced to express G–CSF mRNA and/or protein.

The major control of the production of G–CSF appears to be at the level of bacterial products or secondary mediators induced during infections. Mice infected with *Listeria monocytogenes* (10^4/mouse) showed a pronounced increase in circulating G–CSF levels within 48 hr that preceded the increase in colony forming cells in the bone marrow and subsequent reduction in bacterial cell numbers (65). Direct injection of a single dose (5 μg) of bacterial endotoxin into mice resulted in a rapid (peak at 6 hr) but transient ($t_{1/2} \sim 2$ hr) elevation in serum G–CSF levels (~20-fold). After three consecutive injections of endotoxin the response to further injections of endotoxin was markedly decreased (66). Whole-body irradiation did not induce G–CSF secretion but nor did it inhibit the response to endotoxin challenge. This coupled with the observation that athymic nude mice gave a normal response to endotoxin challenge suggest that radioresistant non-T cells are involved in responding to endotoxin by increased G–CSF production.

Endotoxin or endotoxin-induced products have been shown to directly stimulate macrophages, endothelial cells, or fibroblasts to increase G–CSF production. Endotoxin stimulated resident murine peritoneal cells (predominantly macrophages) to synthesize and release G–CSF (10-100-fold elevation) although a much smaller effect was seen with thioglycollate-induced macrophages (44). Endotoxin also stimulated higher levels of G–CSF mRNA expression in human monocytes (58). The other endotoxin-elicited CSF (M–CSF) has also been shown to stimulate G–CSF production by murine (59) and human macrophages (67). Two of the major mediators of endotoxin action in vivo [IL-1 and tumor necrosis factor-α (TNF-α)] activate human vascular endothelial cells (59,60) to increase transcription from the G–CSF gene, increase G–CSF mRNA levels, and increase the level of secretion of active G–CSF. The different kinetics of induction seen with IL-1 and TNF as well as the additive effects seen when both stimuli were present together suggested separate mechanisms of gene activation (59). Similar results have been obtained with human fibroblasts (61,62) and primary human bone marrow stromal cells (63,64).

Finally, products of activated T-lymphocytes (multi-CSF or IL-3, interferon-γ, and possibly GM–CSF) have also been shown to stimulate increased levels of

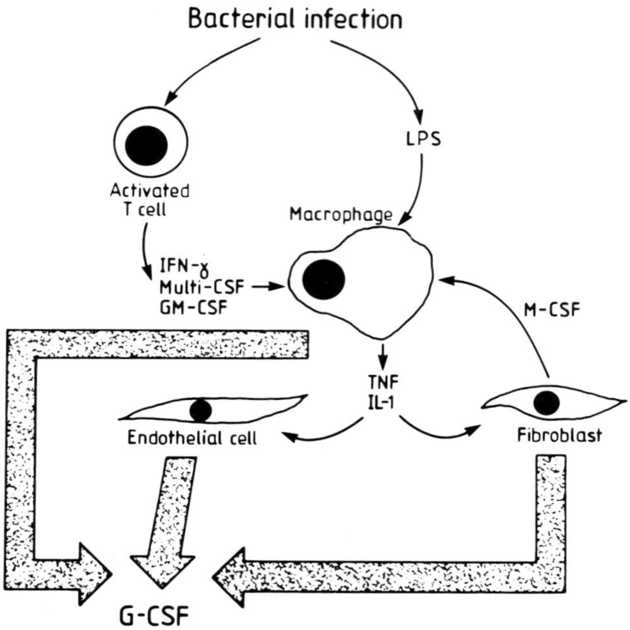

Figure 4 Sources and signals for G–CSF secretion. Macrophages, endothelial cells, and fibroblasts can be induced to secrete G–CSF by signals generated during bacterial infection or T-cell activation. Macrophages can be induced by bacterial LPS, the T-cell products IFN-γ, Multi-CSF, and GM–CSF or by M-CSF to secrete G–CSF as well as TNF and IL-1. TNF and IL-1 induce fibroblasts and endothelial cells to secrete G–CSF. The macrophage has a central role in G–CSF production.

G–CSF mRNA or increased G–CSF release from human (58) or murine (57) monocytes or macrophages.

There is thus strong evidence that bacterial infections or cell-mediated immune reactions release factors that strongly stimulate increased transcription of the G–CSF gene and the release of active G–CSF (Fig. 4). This response would result in the mobilization and activation of neutrophils to combat infections and possibly an increased production of neutrophils in the bone marrow and extramedullary sites. However, little is known of the role of G–CSF in maintaining steady-state levels of neutrophil production in the bone marrow and, in particular, whether circulating G–CSF or local production of G–CSF in bone marrow compartments is involved in this process.

IV. MOLECULAR STRUCTURE OF G–CSF

Murine G–CSF produced by various tissues with or without induction by endotoxin was shown to have essentially identical molecular properties. Those characteristics that most easily distinguished G–CSF from other CSFs were an inability to bind to *Concanavalin A* and a high degree of hydrophobicity resulting in its delayed elution from hydrophobic interaction matrices and reverse-phase high-performance liquid chromatography columns (3).

Murine G–CSF was first purified from endotoxin-induced mouse lung-conditioned medium (2) and subsequently was shown to be equivalent to a human CSF termed CSF-β (32). Human G–CSF was subsequently purified from the bladder carcinoma cell line 5637 (33) and the squamous carcinoma cell line CHU-2 (68). Murine G–CSF had a specific biological activity of 3×10^8 U/mg and human G–CSF 1×10^8 U/mg in murine or human bone marrow colony-forming assays where 50 U/ml gives 50% maximal colony formation. Thus, concentrations of 3 p*M* (mouse) 10 p*M* (human) result in 50% maximal colony formation.

The amino acid sequences of murine and human G–CSF are shown in Fig. 5. The first amino acid of the mature protein was determined in each case by direct amino acid sequencing of the purified proteins (69,70). Murine and human G–CSF mRNAs code for a 30 amino acid hydrophobic leader sequence typical of secreted proteins. Both precursor proteins have a typical signal sequence cleavage site (Glu–Ala–X–Pro–Leu) followed by 178 amino acids (M_r 19,061) in mouse and 174 amino acids (M_r 18,627) in humans. Both have two disulfide bonds formed by homologous cysteine residues (Fig. 5) and an extra cysteine residue at position 86 (murine) or 17 (human) that cannot participate in intramolecular disulfide bonds. The disulfide bonds in these molecules stabilize the structures and make them resistant to relatively harsh treatment (some proteases, high temperatures, denaturing solvents, extremes of pH), which do lead to denaturation after reduction of the disulfide bonds (71,72).

Murine and human G–CSF are glycoproteins because their isoelectric points and apparent molecular weights are altered by neuraminidase treatment (3,71). Their failure to bind to *Concanavalin A* suggested that they do not contain high mannose-attached carbohydrates and, indeed, their sequences demonstrate the absence of any potential *N*-linked glycosylation sites (Asn–X–Ser/Thr). The reduction of the apparent molecular weight of human G–CSF by *O*-glycanase from 19,200 to 18,800 (20) suggested the presence of *N*-acetyl-neuraminic acid α (2,3) [galactose β (1-3)] *N*-acetyl galactosamine-Ser/Thr and this was confirmed by direct chemical analysis of recombinant human G–CSF produced by Chinese hamster ovary cells (73). The latter study in fact showed a single *O*-glycosylation site at Thr 133, which was microheterogeneous with half the mole-

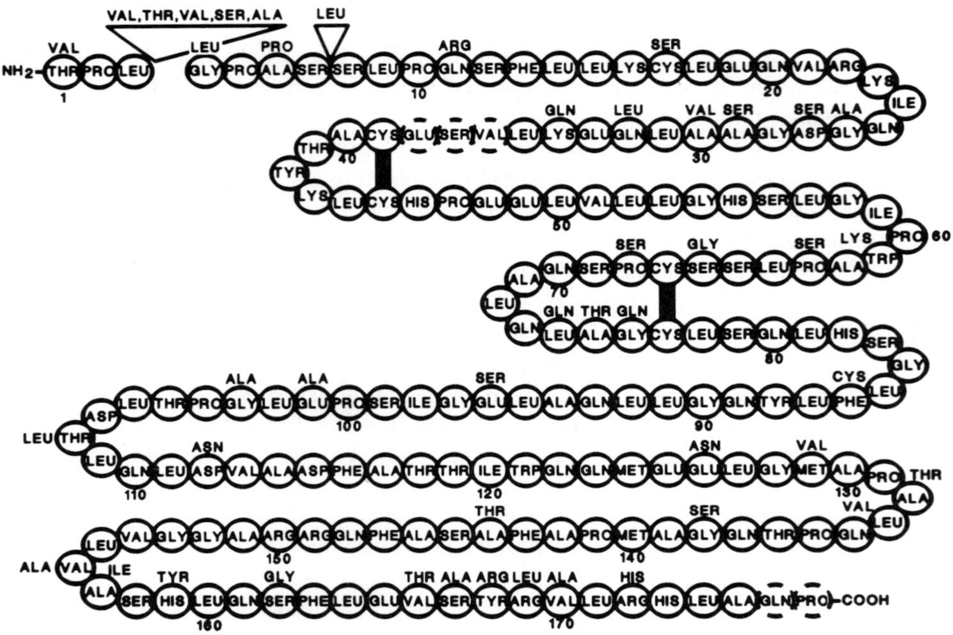

Figure 5 Amino acid sequences of human and murine G–CSF. The primary sequence shown and numbering system is for human G–CSF. Insertions in the mouse sequence are shown by triangles and deletions by dashed circles. Substitutions in the mouse sequence are shown above the human sequence. Amino acids in position 36-38 occur in only one type of human cDNA and not in the mouse. The connectivity of the disulfide bonds is shown by heavy bars.

cules glycosylated as above and the other half with a second *N*-acetyl-neuraminic acid linked α2-6 to the galactose residue.

V. MOLECULAR BIOLOGY OF G–CSF

After the purification of human G–CSF from the human bladder carcinoma cell line 5637 (33,20) or the human squamous carcinoma cell line CHU2 (68) oligonucleotides that could code for the partial amino acid sequences determined were synthesized and used to screen cDNA libraries constructed from mRNA from the appropriate cell lines. Nagata et al. (74) and Souza et al. (20) each isolated cDNA clones containing the full coding sequences of the mature protein and these were identical except that the former had a three amino acid insert

(Val–Ser–Glu) at position 35 relative to the latter cDNA. Although G–CSF produced by either cDNA was active there was some suggestion that the protein lacking the three amino acid insert had a higher specific biological activity (75).

The chromosomal gene for human G–CSF was isolated by Nagata et al. (75) and revealed the presence of 5 mRNA-encoding exons and 4 introns (Fig. 6). Nucleotide analysis of this gene revealed that alternative splicing between different 5′ splice donor sequences at the junction of exon 2 to a common acceptor sequence at the beginning of exon 3 could explain the addition or deletion of the tripeptide sequence in the two cDNAs. They showed that when the chromosomal gene for human G–CSF was expressed in monkey COS cells using the SV40 enhancer both mRNAs were produced but predominantly the mRNA containing the three amino acid insert, whereas S1 mapping analysis of CHU2 mRNA indicated that most of the mRNA (>80%) lacked the three amino acid insert. This data raises some interesting questions regarding the possibility of differential splicing in different cells and the possibility of the differential production of G–CSF with altered or reduced activities by different cells. However, there is currently little information on the differential expression of the two mRNAs or on whether both mRNAs are translated into protein.

The gene for human G–CSF exists as a single copy and has been localized by Southern hybridization analysis in human-mouse somatic cell hybrids, flow-sorted human chromosomes, and in situ hybridization on human metaphase chromosomes to the q21-q22 region of chromosome 17 (76-78). This region corresponds to a translocation between chromosomes 15 and 17 [t (15;17) (q23 q21)], which is commonly associated with acute promyelocytic leukemia (79), so that it will be interesting to determine how closely the break point is associated with control regions of the G–CSF gene. The protooncogenes c-*erb* B-2, c-*erb* A, and the nerve growth-factor receptor gene are also localized in this same segment of chromosome 17.

Subsequently, both the cDNA and chromosomal gene for murine G–CSF have been isolated and sequenced (80,81). Like the human G–CSF gene, the murine gene consists of 5 exons and 4 introns with similar spacings. However, although the coding regions of the exons are highly conserved between mouse and human (69% nucleotide homology) as is the 5′ flanking region (80% nucleotide homology for the 310 nucleotides upstream of the ATG initiation codon), there is little sequence homology in the introns. Unlike the human gene, the murine gene does not contain a second splice donor site in exon 2, so that only one spliced mRNA is produced. However, there is some evidence that two separate mRNA initiation sites (separated by 63 nucleotides) can be used in the murine gene, the major one initiating 36 nucleotides upstream of the ATG codon. Both transcripts would produce the same protein. The single copy

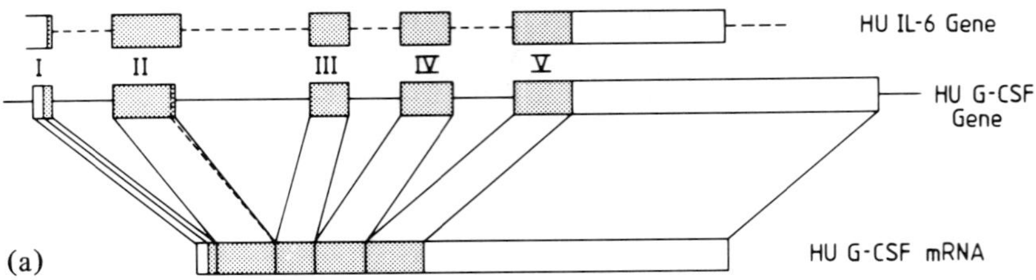

(a)

G-CSF	h	AGAGATTCCA	 147	...TATA
	m	AGAGATTCCC	 149	...
GM-CSF	h	GGAGATTCCA	 63	...
	m	GGAGATTCCA	 69	...
IL-4	h	AAGGTTTCAT	 153	...
	m	GGGGTTTCAT	 159	...
IL-2	h	GGGATTTCAC	 162	...
	m	GGGATTTCAC	 165	...
IL-5	h	TTAGTTTCAC	 65	...
IL-3	m_1	GAGATTCCA	 254	...
(b)	m_2	GAGGTTCCA	 75	...

Figure 6 (a) Genomic structure of the human G–CSF gene and mRNA. The five exons and four introns are drawn to scale and the alternate splicing pattern from exon II is shown by a dashed line. Protein coding regions are stippled and untranslated regions of the exons are blank. The exons from the human IL-6 gene are also shown to scale to emphasize their similarity to the G–CSF exons but the introns in the IL-6 gene (- - -) are not drawn to scale. (b) Conserved sequence motif in G–CSF, GM–CSF, IL-4, IL-2, IL-5, and IL-3 genes in the 5′-flanking regions (63-254 nucleotides upstream of the TATA box).

mouse gene for G–CSF has been mapped to the distal half of chromosome 11 (homologous to human chromosome 17) close to the *erb*B and *erb*B-2 loci (81a).

The highly conserved 5′ flanking regions of both human and murine G–CSF genes contain two "TATA"-like sequences 26 and 57 nucleotides upstream of the main transcription initiation site and a decanucleotide sequence GAGXTTCCXC approximately 150 nucleotides upstream of the most 3′ TATA sequence. This sequence is also present in human and murine GM–CSF and in murine and rat IL-3 genomic sequences and has been suggested to be a regulatory element for expression of the human GM–CSF gene (Fig. 6).

Human and murine G–CSF genes do not show any close homology to any other known gene sequences. However, there is some rather weak sequence homology to the human IL-6 gene (also known as B-cell stimulatory factor-2 hepatocyte stimulating factor or interferon β2) and a rather striking similarity in exon size and distribution (but not in intron size) (82) (Fig. 6). Because both IL-6 and G–CSF gene expression are inducible by IL-1β (82), this homology might extend to common regulatory elements.

VI. RECEPTOR-BINDING CHARACTERISTICS OF G–CSF

Iodinated murine G–CSF binds specifically to normal murine and hemopoietic tissues and a few cell lines (15,20,83) (Table 1). In all cases examined, however, the number of receptors detected per cell has been small, ranging from less than 100 to a few thousand. The positive murine cell lines include WEHI-3B D^+, M1, and 32 D Cl.3, which respond to G–CSF by differentiation induction, NFS60, which responds to G–CSF by proliferation (16a), and the monocytic cell lines J774 and P388D1, which do not apparently respond to G–CSF. Interestingly, sublines of WEHI-3B (D^-) and 32 D Cl.23 which, unlike the parent lines, do not respond to G–CSF, do not show detectable binding of ^{125}I-G–CSF. Among human myeloid cell lines only HL60 has been shown to exhibit G–CSF receptors (84).

Binding of ^{125}I-G–CSF to these cell lines was competed for by unlabeled G–CSF but not by any other CSF (GM–CSF, multi-CSF, M-CSF) or a variety of unrelated growth factors. In all cases examined, only a single class of saturable binding site was detected by Scatchard analysis with apparent equilibrium dissociation constant of 60-300 p*M* and apparent molecular weight (determined by chemical crosslinking experiments) of 150,000. Similar characteristics were determined for the G–CSF receptor on normal murine hemopoietic cells as for the leukemic cell lines (85).

Among murine hemopoietic tissues, ^{125}I-G–CSF bound specifically to adult bone marrow, spleen, and blood cells as well as fetal liver and cell autoradio-

Table 1 G–CSF Receptors on Normal Hemopoietic Cells and Cell Lines

	Ave. receptors/cell		Affinity [K_D (apparent)]	
	Murine	Human	Murine	Human
1. Normal cells				
Blast cells	160	300	65-90	
Promyelocytes	180	500		
Metamyelocytes	120	400		
Polymorphs	300	300		
Promonocytes	60	100		
Monocytes	30	50		
Eosinophils	0	0		
Lymphocytes	0	0		
Erythroid cells	0	0		
Blood neutrophils		500-3000		~300 pM
2. Cell lines				
WEHI-3B (D^+)	300		90pM	
M1	50		–	
J774	320		100 pM	
NFS-60	400		120 pM	
32D.Cl.3	500		100 pM	
HL60		50		
3. Human leukemias				
Acute myeloid leukemias (M1, M3 > M2, M4, M5)		100-3000		~100 pM
Chronic myeloid leukemias		100-2000		~100 pM
Acute lymphoid leukemias		0		
Chronic lymphoid leukemias		0		

graphic analysis indicated that binding was restricted to cells within the neutrophilic granulocytic cell lineage and, to a lesser extent, the monocyte-macrophage cell lineage (Table 1). Essentially all cells within the neutrophil lineage displayed specific G–CSF receptors with numbers increasing from an average of ~150 on promyelocytes and myelocytes to ~300 on mature polymorphs. On the other hand, only a proportion (60-80%) of promonocytes and monocytes displayed a relatively small number (30-60) of G–CSF receptors, whereas all other cell types were negative (83). ^{125}I-Granulocyte colony-stimulating factor also bound specifically to human bone marrow and blood cells with the same cellular specificity as observed for murine hemopoietic cells. The only difference observed was that neutrophilic promyelocytes displayed the highest level of G–CSF receptors (500-1000 receptors/cell) with decreased levels of expression on mature polymorphs (up to 500/cell) (32).

In agreement with the cross-species biological reactivity of G–CSFs, iodinated murine G–CSF bound to human hemopoietic cells with the same affinity as human G–CSF and iodinated human G–CSF bound to murine hemopoietic cells, indicating strong conservation of receptor-binding sequences in murine and human G–CSF as well as their receptors (20,32). Granulocyte colony-stimulating factor bound specifically to all primary myeloid leukemia blast cells tested but not to lymphoid leukemias. Receptor numbers appeared to correlate with the FAB classification of leukemias with low levels on monocytic leukemias (M4) and high levels on promyelocytic leukemias (M3) (up to 2000/cell). In all cases tested the apparent binding affinity for G–CSF as determined by Scatchard analysis was comparable to that observed for normal human or murine hemopoietic cells (kDa ≅ 100 p*M*) (21,22,32).

For all cell systems G–CSF binding was characterized by a rapid kinetic association constant (ka ~ 10^8-10^9 M^{-1} min^{-1}) and a slow kinetic dissociation constant. On normal hemopoietic cells dissociation was barely measurable over several hours at 0°C whereas on WEHI-3B D^+ cells dissociation was measurable but slow ($t_{1/2}$ ~ 5 hr). Kinetic dissociation rates at 37°C could not be measured directly because of ligand-receptor internalization. However, analysis of competing kinetic interactions at 37°C by a curve-fitting program suggested that the kinetic dissociation rate at 37°C was much faster than at 0°C (~0.02-0.04 min^{-1} for several cell types) (86).

Analysis of the kinetics of receptor and ligand processing at 37°C also revealed some consistent patterns in different cell types. In all cell types G–CSF-occupied receptors were internalized 5-10 times more rapidly than unoccupied receptors so that G–CSF resulted in a net loss of G–CSF receptors from the cell surface (downregulation). The rate of dissociation of bound G–CSF was also slower than the rate of G–CSF-induced receptor internalization so that the majority of G–CSF bound to receptors was internalized rather than dissociated

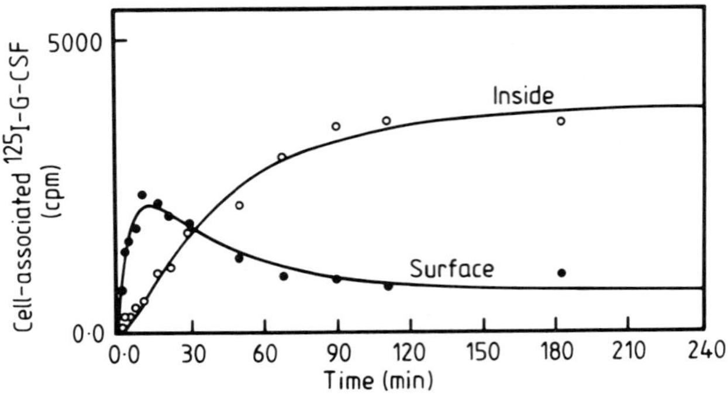

Figure 7 Internalization of ^{125}I-G–CSF by NFS60 cells at 37°C. NFS60 cells were incubated with ^{125}I-G–CSF (3×10^{-10} *M*) at 37°C and at the indicated times cell-surface radioactivity (●——●) and internalized radioactivity (○——○) were determined by acid dissociation.

back into the medium. Finally, the rate of G–CSF degradation was slower than the rate of internalization, resulting in an accumulation of G–CSF inside the cell so that, in the steady state, considerably more G–CSF (5-50 times more) was inside the cell than at the cell surface (N. A. Nicola and L. Peterson, unpublished data) (Fig. 7).

These data indicate that the primary G–CSF interaction with its receptor results in the rapid internalization of receptor and ligand followed by the slow degradation of internalized G–CSF. It is difficult to determine whether this process is required for G–CSF action or is merely required to deactivate the activation signal. However, there are other ways that G–CSF receptors can be lost from the cell surface.

When murine bone marrow cells were incubated at 37°C with multi-CSF or GM–CSF, there was a rapid ($t_{1/2} \sim 15$ min) dose-dependent loss of G–CSF receptors from the surface of bone marrow cells (trans-downmodulation) (87). Similarly, when human blood neutrophils were incubated at 37°C with the chemotactic peptide fMLP, bacterial lipopolysaccharide, or GM–CSF, there was a rapid, dose-dependent loss of G–CSF receptors (47,88). Those factors that trans-downmodulated G–CSF receptors do not antagonize the action of G–CSF on these cells but rather synergize with it, and this has led us to speculate that trans-downmodulation which, like G–CSF action itself, could be mediated by internalization of G–CSF receptors, may serve to functionally activate the G–CSF receptor in the absence of G–CSF. In this type of model, internalization of the G–CSF receptor by whatever mechanism it occurs is crucial for signal transduction by the receptor (87,88).

The mechanism of trans-downmodulation is unknown in this system and it is by no means clear that the above model is correct. Nevertheless, it provides one framework within which it is possible to analyze the synergistic actions of different growth factors.

VII. SUMMARY

Granulocyte colony-stimulating factor was discovered during attempts to define the normal regulators present in cell supernatants that could induce terminal differentiation of the murine myeloid leukemic cell line WEHI-3B D^+. The purification and subsequent cloning of both murine and human G–CSF allowed the normal functions of this molecule to be elucidated and indicated that it was a relatively lineage-specific stimulator of the survival, proliferation, and differentiation of precursor cells of the neutrophilic granulocyte cell lineage as well as an activator of mature neutrophil function. Murine and human G–CSFs as well as the ligand-binding domains of G–CSF receptors have been strongly conserved so that the biological activities and receptor-binding characteristics of G–CSFs are completely species cross-reactive. The evolutionary conservation of G–CSFs is suggestive of an important role for the molecule in maintaining neutrophil levels and activity in vivo, a role amply supported by a series of animal experiments and clinical trials that have been performed recently. These experiments and trials have suggested that G–CSF will be clinically useful in augmenting patients' resistance to certain infections, in enhancing the neutrophil responses of some patients with reduced granulocyte counts or defective granulocytes, and in reducing the decline in granulocytes that occurs during chemotherapy or irradiation therapy and bone marrow transplantation. The special capacity of G–CSF among the CSFs to induce terminal differentiation in some myeloid leukemias is not understood in molecular terms. However, the capacity of G–CSF to stimulate proliferation in some myeloid leukemias as well argues that caution needs to be exercised in the timing of G–CSF administration and patient selection before considering this form of therapeutic approach.

ACKNOWLEDGMENT

The author's work reviewed here was supported by the National Health and Medical Research Council, Canberra, the Anti-Cancer Council of Victoria, Australia, the J. D. and L. Harris Trust Fund, and the National Institutes of Health, Bethesda, Maryland, U.S.A., grant CA-22556.

REFERENCES

1. Burgess, A. W., and D. Metcalf. Characterization of a serum factor stimulating the differentiation of myelomonocytic leukemic cells. *Int. J. Cancer 39*:647-654 (1980).
2. Nicola, N. A., D. Metcalf, M. Matsumoto, and G. R. Johnson. Purification of a factor inducing differentiation in murine myelomonocytic leukemia cells: identification as granulocyte colony-stimulating factor (G–CSF). *J. Biol. Chem. 258*:9017-9023 (1983).
3. Nicola, N. A., and D. Metcalf. Biochemical properties of differentiation factors for murine myelomonocytic leukemic cells in organ conditioned media. Separation from colony-stimulating factors. *J. Cell. Physiol. 112*: 257-264 (1981).
4. Metcalf, D. Clonal analysis of the action of GM–CSF on the proliferation and differentiation of myelomonocytic leukemic cells. *Cancer 24*:616-623 (1979).
5. Metcalf, D., and N. A. Nicola. Autoinduction of differentiation in WEHI-3B leukemia cells. *Int. J. Cancer 30*:773-780 (1982).
6. Metcalf, D. Regulatory control of the proliferation and differentiation of normal and leukemia cells. *Natl. Cancer. Inst. Monogr. 60*:123-131 (1982).
7. Metcalf, D. Regulator-induced suppression of myelomonocytic leukemic cells: clonal analysis of early cellular events. *Int. J. Cancer 30*:203-210 (1982).
8. Tsuda, H., L. M. Neckers, and D. H. Pluznik. Colony stimulating factor (CSF)-induced differentiation of murine M1 myeloid leukemic cells is permissive in early G_1 phase. *Proc. Natl. Acad. Sci. USA 83*:4317-4321 (1986).
9. Tomida, M., Y. Yamamoto-Yamaguchi, M. Hozumi, T. Okabe, and F. Takaku. Induction by recombinant human granulocyte colony-stimulating factor of differentiation of mouse myeloid leukemic M1 cells. *FEBS Lett. 207*:271-275 (1986).
10. Tomida, M., Y. Yamamoto-Yamaguchi, and M. Hozumi. Purification of a factor inducing differentiation of mouse myeloid leukemic M1 cells from conditioned medium of mouse fibroblast L929 cells. *J. Biol. Chem. 259*: 10978-10982 (1984).
11. Hilton, D. J., N. A. Nicola, N. M. Gough, and D. Metcalf. Resolution and purification of three distinct factors produced by Krebs ascites cells which have differentiation-inducing activity on murine myeloid leukemic cell lines. *J. Biol. Chem. 263*:9238-9243 (1988).
12. Valtieri, M., D. J. Tweardy, D. Caracciola, K. Johnson, F. Mavilio, S. Attmann, D. Santoli, and G. Rovera. Cytokine-dependent granulocytic differentiation. Regulation of proliferative and differentiative responses in a murine progenitor cell lines. *J. Immunol. 138*:3829-3835 (1987).
13. Rovera, G., M. Valtieri, F. Mavilio, and E. P. Reddy. Effect of Abelson murine leukemia virus on granulocytic differentiation and interleukin-3 dependence of a murine progenitor cell line. *Oncogene 1*:29-35 (1987).

14. Crosier, P. S., J. M. Bungey, K. E. Barber, and J. D. Watson. Evidence for distinct hemopoietic cell development mechanisms for IL2, IL3 and granulocyte colony-stimulating factor. *J. Immunol.* (in press).
15. Nicola, N. A., and D. Metcalf. Binding of the differentiation-inducer, granulocyte colony-stimulating factor, to responsive but not unresponsive leukemic cell lines. *Proc. Natl. Acad. Sci. USA 81*:3765-3769 (1984).
16. Bettens, F., E. Schlick, W. Farrar, and F. Ruscetti. 1, 25-Dihydroxy cholecalciferol-induced differentiation of myelomonocytic leukemic cells unresponsive to colony stimulating factors and phorbol esters. *J. Cell. Physiol. 129*:295-302 (1986).
16a. Hara, K., T. Suda, J. Suda, M. Eguchi, J. N. Ihle, S. Nagataka, Y. Miura, and M. Saito. Bipotential murine hemopoietic cell line (NFS60) that is responsive to IL03, GM–CSF, G–CSF and erythropoietin. *Exp. Hematol. 16*:256-261.
17. Metcalf, D. Clonal analysis of the response of HL60 human myeloid leukemia cells to biological regulators. *Leuk. Res. 7*:117-132 (1983).
18. Begley, C. G., D. Metcalf, and N. A. Nicola. Purified colony-stimulating factors (G–COF and GM–COF) induce differentiation in human HL60 leukemic cells with suppresion of clonogenicity. *Int. J. Cancer 39*: 99-105 (1987).
19. Hara, T., T. Umeda, T. Niijima, and T. Okabe. Cytostasis of tumor cell lines by promyelocytic leukemic cell line HL60 differentiated to granulocyte lineage. *J. Cancer Res. Clin. Oncol. 109*:103-106 (1985).
20. Souza, L. M., T. C. Boone, J. Gabrilove, P. H. Lai, K. M. Zsebo, D. C. Murdock, V. R. Chazin, J. Bruszewski, H. Lu, K. K. Chen, J. Barendt, E. Platzer, M. A. S. Moore, R. Mertelsmann, and K. Welte. Recombinant human granulocyte colony-stimulating factor: effects on normal and leukemic myeloid cells. *Science 232*:61-65 (1986).
21. Begley, C. G., D. Metcalf, and N. A. Nicola. Primary human myeloid leukemia cells: comparative responsiveness to proliferative stimulation by GM–CSF or G–CSF and membrane expression of CSF receptors. *Leukemia 1*:1-8 (1987).
22. Begley, C. G., D. Metcalf, and N. A. Nicola. Binding characteristics and proliferative action of purified granulocyte colony-stimulating factor (G–CSF) on normal and leukemic human promyelocytes. *Exp. Hematol. 16*: 71-79 (1988).
23. Vellenga, E., D. Young, K. Wagner, D. Wiper, D. Ostapovids, and J. D. Griffin. The effects of GM–CSF and G–CSF in promoting growth of clonogenic cells in acute myeloblastic leukemia. *Blood 69*:1771-1776 (1987).
24. Kelleher, C., J. Miyauchi, G. Wong, S. Clark, M. D. Minden, and E. A. McCulloch. Synergism between recombinant growth factors, GM–CSF and G–CSF, acting on the blast cells of acute myeloblastic leukemia. *Blood 69*:1498-1453 (1987).
25. Miyauchi, J., C. A. Kelleher, Y-C. Yang, G. G. Wong, S. C. Clark, M. D. Minden, S. Minkin, and E. A. McCulloch. The effects of three recombinant

growth factors, IL-3, GM–CSF and G–CSF, on the blast cells of acute myeloblastic leukemia maintained in short-term suspension culture. *Blood 70*:657-663 (1987).
26. Asano, Y., T. Shibuya, S. Okamura, S. Yamaga, T. Otsuka, and Y. Niho. Effect of human recombinant granulocyte/macrophage colony-stimulating factor and native granulocyte colony-stimulating factor on clonogenic leukemic blast cells. *Cancer. Res. 47*:5647-5648 (1987).
27. Cheng, G. Y. M., C. A. Kelleher, J. Miyauchi, C. Wang, G. Wong, S. C. Clark, E. A. McCulloch, and M. D. Minden. Structure and expression of genes of GM–CSF and G–CSF in blast cells from patients with acute myeloblastic leukemia. *Blood 71*:204-208 (1988).
28. Metcalf, D., and N. A. Nicola. Proliferative effects of purified granulocyte colony-stimulating factor (G–CSF) on normal mouse hemopoietic cells. *J. Cell. Physiol. 116*:198-206 (1983).
29. Suda, T., J. Suda, S. Kajigawa, S. Nagata, S. Asano, M. Saito, and Y. Miura. Effects of recombinant murine granulocyte colony-stimulating factor on granulocyte-macrophage and blast colony formation. *Exp. Hematol. 15*: 958-965 (1987).
30. Ikebuchi, K., S. C. Clark, J. N. Ihle, L. M. Souza, and M. Ogawa. Granulocyte colony-stimulating factor enhances interleukin 3-dependent proliferation of multipotential hemopoietic progenitors. *Proc. Natl. Acad. Sci. USA 85*:3445-3449 (1988).
31. Eliason, J. F. Granulocyte-macrophage colony formation in serum-free culture: effects of purified colony-stimulating factors and modulation by hydrocortisone. *J. Cell. Physiol. 128*:231-238 (1986).
32. Nicola, N. A., C. G. Begley, and D. Metcalf. Identification of the human analogue of a regulator that induces differentiation in murine leukaemic cells. *Nature 314*:625-628 (1985).
33. Welte, K., E. Platzer, L. Lu, J. L. Gabrilove, E. Levi, R. Mertelsmann, and M. A. S. Moore. Purification and biochemical characterization of human pluripotent hematopoietic colony-stimulating factor. *Proc. Natl. Acad. Sci. USA 82*:1526-1530 (1985).
34. Nicola, N. A., D. Metcalf, G. R. Johnson, and A. W. Burgess. Separation of functionally distinct human granulocyte-macrophage colony-stimulating factors. *Blood 54*:614-627 (1979).
35. Nicola, N. A., and D. Metcalf. The colony-stimulating factors and myeloid leukaemia. *Cancer Surv. 4*:789-815 (1985).
36. Strife, N. A., C. Lambek, D. Wisniewski, S. Gulati, J. C. Gasson, D. W. Golde, K. Welte, J. L. Gabrilove, and B. Clarkson. Activities of four purified growth factors on highly enriched human hematopoietic progenitor cells. *Blood 69*:1508-1523 (1987).
37. Begley, C. G., D. Metcalf, A. F. Lopez, and N. A. Nicola. Fractionated populations of normal human marrow cells respond to both human colony-stimulating factors with granulocyte-macrophage activity. *Exp. Hematol. 13*:956-962 (1985).

38. Metcalf, D., C. G. Begley, and N. A. Nicola. The proliferative effects of human GM–CSFα and β and murine G–CSF in microwell cultures of fractionated human marrow cells. *Leuk. Res. 9*:521-527 (1985).
39. Begley, C. G., N. A. Nicola, and D. Metcalf. Proliferation of normal human promyelocytes and myelocytes after a single pulse stimulation by purified GM–CSF or G–CSF. *Blood 71*:640-645 (1988).
40. Ohara, A., T. Suda, M. Saito, Y. Miura, T. Okabe, and F. Takakyu. Effect of recombinant human granulocyte colony-stimulating factor on hemopoietic cells in serum-free culture. *Exp. Hematol. 15*:695-699 (1987).
41. Begley, C. G., A. F. Lopez, N. A. Nicola, D. J. Warren, M. A. Vadas, C. J. Sanderson, and D. Metcalf. Purified colony-stimulating factors enhance the survival of human neutrophils and eosinophils in vitro: a rapid and sensitive microassay for colony-stimulating factors. *Blood 68*:162-166 (1986).
42. Lopez, A. F.,N. A. Nicola, A. W. Burgess, D. Metcalf, F. L. Battye, W. A. Sewell, and M. A. Vadas. Activation of granulocyte cytotoxic function by purified mouse colony-stimulating factors. *J. Immunol. 131*:2983-2988 (1983).
43. Vadas, M. A., N. A. Nicola, and D. Metcalf. Activation of antibody-dependent cell-mediated cytotoxicity of human neutrophils and eosinophils by separate colony-stimulating factors. *J. Immunol. 130*:795-799 (1983).
44. Vadas, M. A., and A. F. Lopez. Regulation of granulocyte function by colony-stimulating factors and monoclonal antibodies. *Lymphokines 12*:179-200 (1985).
45. Sullivan, R., J. P. Fredette, J. D. Griffin, and E. R. Simons. Effect of GM–CSF_{rh} and G–CSF_{rh} on the granulocyte's membrane potential. [Abstract 601]. *Blood 70*(Suppl 1).186a (1987).
46. Yuo, A., S. Kitagawa, T. Okabe, A. Urabe, Y. Komatsu, S. Itoh, and F. Takaku. Recombinant human granulocyte colony-stimulating factor repairs the abnormalities of neutrophils in patients with myelodysplastic syndromes and chronic myelogenous leukemia. *Blood 70*:404-411 (1987).
47. Avalos, B. R., C. Hedzat, G. C. Baldwin, D. W. Golde, J. C. Gasson, and J. F. Di Persio. Biological activities of human G–CSF and characterization of the human G–CSF receptor. [Abstract 517]. *Blood 70*(suppl 1):165a (1987).
48. Platzer, E., K. Welte, J. Gabrilove, L. Lu, P. Harris, R. Mertelsmann, and M. A. S. Moore. Biological activities of a human pluripotent hemopoietic colony stimulating factor on normal and leukemic cells. *J. Exp. Med. 162*: 1788-1801 (1985).
49. Tamura, M., K. Hattori, H. Nomura, M. Oheda, N. Kubota, I. Imazeki, M. Ono, Y. Ueyama, S. Nagata, N. Shirafuji, and S. Asano. Induction of neutrophilic granulocytosis in mice by administration of purified human native granulocyte colony-stimulating factor (G–CSF). *Biochem. Biophys. Res. Commun. 142*:454-460 (1987).
50. Cohen, A. M., K. M. Zsebo, H. Inoue, D. Mines, T. C. Boone, V. R. Chazin,

L. Tsai, T. Ritch, and L. M. Souza. In vivo stimulation of granulopoiesis by recombinant human granulocyte colony-stimulating factor. *Proc. Natl. Acad. Sci. USA 84*:2484-2488 (1987).
51. Welte, K., M. A. Bonilla, A. P. Gillio, T. C. Boone, G. K. Potter, J. L. Gabrilove, M. A. S. Moore, R. J. O'Reilly, and L. M. Souza. Recombinant human granulocyte colony-stimulating factor: effects on hematopoiesis in normal and cyclophosphamide-treated primates. *J. Exp. Med. 165*:941-948 (1987).
52. Matsumoto, M., S. Matsubara, T., Matsuno, M. Tamura, K. Hattori, H. Nomura, M. Ono, and T. Yokota. Protective effect of human granulocyte colony-stimulating factor on microbial inefection on neutropenic mice. *Infec. Immun. 55*:2715-2720 (1987).
53. Shimamura, M., Y. Kobayashi, A. Yuo, A. Urabe, T. Okabe, Y. Komatsu, S. Itoh, and F. Takaku. Effect of human recombinant granulocyte colony-stimulating factor on hematopoietic injury in mice induced by 5-fluorauracil. *Blood 69*:353-355 (1987).
54. Moore, M. A. S., and D. J. Warren. Synergy of interleukin 1 and granulocyte colony-stimulating factor: in vivo stimulation of stem-cell recovery and hematopoietic regeneration following 5-fluorouracil treatment of mice. *Proc. Natl. Acad. Sci. USA 84*:7134-7138 (1987).
55. Bronchud, M. H., J. H. Scarffe, N. Thatcher, D. Crowther, L. M. Souza, N. K. Alton, N. G. Testa, and T. M. Dexter. Phase I/II study of recombinant human granulocyte colony-stimulating factor in patients receiving intensive chemotherapy for small cell lung cancer. *Br. J. Cancer 56*:809-813 (1987).
56. Morstyn, G., L. Campbell, L. M. Souza, N. K. Alton, J. Keech, M. Green, W. Sheridan, D. Metcalf, and R. Fox. Effect of granulocyte colony stimulating factor on neutropenia induced by cytotoxic chemotherapy. *Lancet 1*:667-672 (1988).
56a. Gabrilove, J. L., A. Jakubowski, H. Scher, C. Sternberg, G. Wong, J. Grous, A. Yagoda, K. Fain, M. A. S. Moore, B. Clarkson, H. F. Oettge, K. Alton, K. Welte, and L. Souza. Effect of granulocyte colony-stimulating factor on neutropenia and associated morbidity due to chemotherapy for transitional-cell carcinoma of the urothelium. *N. Engl. J. Med. 319*:1414-1422 (1988).
57. Metcalf, D., and N. A. Nicola. Synthesis by mouse peritoneal cells of G–CSF, the differentiation inducer for myeloid leukemia cells: stimulation by endotoxin, M–CSF and Multi-CSF. *Leukemia Res. 9*:35-50 (1985).
58. Rambaldi, A., D. C. Young, and J. D. Griffin. Expression of the M–CSF (CSF–1) gene by human monocytes. *Blood 69*:1409-1413 (1987).
59. Seelentag, W. K., J-J. Mermod, R. Montesano, and P. Vassalli. Additive effects of interleukin-1 and tumor necrosis factor-α on the accumulation of the three granulocyte and macrophage colony-stimulating factor mRNAs in human endothelial cells. *EMBO J. 6*:2261-2265 (1987).
60. Zsebo, K. M., V. N. Yuschenkoff, S. Schiffer, D. Chang, E. McCall, C. A.

Dimarcello, M. A. Brown, B. Altrock, and G. C. Bagby Jr. Vascular endothelial cells and granulopoiesis: Interleukin-1 stimulates release of G–CSF and GM–CSF. *Blood 73*:99-103 (1988).
61. Koeffler, H. P., J. Gasson, J. Ranyard, L. Souza, M. Shephard, and R. Munker. Recombinant human TNFα stimulates production of granulocyte colony-stimulating factor. *Blood 70*:55-59 (1987).
62. Yang, Y-C., S. Tsai, G. Wong, and S. Clark. Interleukin-1 regulation of hematopoietic growth factor production by human stromal fibroblasts. *J. Cell. Physiol. 134*:292-296 (1988).
63. Febbe, W. E., J. van Damme, A. Billiau, H. M. Goselink, P. J. Voogt, G. van Eeden, P. Ralph, B. W. Altrock, and J. H. F. Falkenburg. Interleukin 1 induces human marrow stromal cells in long-term culture to produce granulocyte colony-stimulating factor and macrophage colony-stimulating factor. *Blood 71*:430-435 (1988).
64. Rennick, D., G. Yang, L. Gemmell, and F. Lee. Control of hemopoiesis by a bone marrow stromal cell clone: lipopolysaccharide and interleukin-1-inducible production of colony-stimulating factors GM–CSF and G–CSF. *Blood 69*:682-691 (1987).
65. Cheers, C., A. M. Haigh, A. Kelso, D. Metcalf, E. R. Stanley, and A. M. Young. Production of colony-stimulating factors (CSFs) during infection: separate determinations of macrophage-, granulocyte-, granulocyte-macrophage-, and Multi-CSFs. *Infect. Immun. 56*:247-251 (1988).
66. Metcalf, D. Sources and biology of regulatory factors active on mouse myeloid leukemic cells. *J. Cell. Physiol.* (suppl. 1):175-183 (1982).
67. Ishizaka, Y., K. Motoyoshi, K. Hatake, M. Saito, F. Takaku, and Y. Miura. Mode of action of human urinary colony-stimulating factor. *Exp. Hematol. 14*:1–8 (1986).
68. Nomura, H., I. Imazeki, M. Oheda, N. Kubota, M. Tamura, M. Ono, Y. Ueyama, and S. Asano. Purification and characterization of human granulocyte colony-stimulating factors (G–CSF). *EMBO J. 5*:871-876 (1986).
69. Simpson, R. J., E. C. Nice, and N. A. Nicola. Structural studies on the murine granulocyte colony-stimulating factor. *Biol. Chem. Heppe-Seyler 368*:1327-1331 (1987).
70. Nagata, S., M. Tsuchiya, S. Asano, Y. Kaziro, T. Yamazaki, O. Yamamoto, Y. Hirata, N. Kubota, M. Oheda, H. Nomura, and M. Ono. Molecular cloning and expression of cDNA for human granulocyte colony-stimulating factor. *Nature 319*:415-418 (1986).
71. Nicola, N. A., and D. Metcalf. Differentiation induction in leukemic cells by normal growth regulators: molecular and binding properties of purified granulocyte colony-stimulating factor. In *Genes and Cancer*, J. M. Bishop, J. D. Rowley, and M. Greaves (Eds.). Alan R. Liss, New York, pp. 591-610 (1984).
72. Nicola, N. A. Granulocyte-colony-stimularing factor. *Methods Enzymol. 116*:600-619 (1987).

73. Oheda, M., S. Hase, M. Ono, and T. Ikenaka. Structures of the sugar chains of recombinant human granulocyte-colony-stimulating factor produced by Chinese hamster ovary cells. *J. Biochem.* (*Tokyo*) *103*:544-546 (1988).
74. Nagata, S., M. Tsuchiya, S. Asano, Y. Kaziro, T. Yamazaki, O. Yamamoto, Y. Hirata, N. Kubota, M. Oheda, H. Nomura, and M. Ono. Molecular cloning and expression of cDNA for human granulocyte colony-stimulating factor. *Nature 319*:415-418 (1986).
75. Nagata, S., M. Tsuchiya, S. Asano, O. Yamamoto, Y. Hirata, N. Kubota, M. Oheda, H. Nomura, and T. Yamazaki. The chronosomal gene structure and two mRNAs for human granulocyte colony-stimulating factor. *EMBO J. 5*:575-581 (1986).
76. Tweardy, D. J., L. A. Cannizzaro, A. P. Palumbo, S. Shane, K. Heubner, P. Vantuinen, D. H. Ledbetter, J. B. Finan, P. C. Nowell, and G. Rovera. Molecular cloning and characterization of a cDNA for human graulocyte colony-stimulating factor (G–CSF) from a glioblastoma multiforme cell line and localization of the G–CSF gene to chromosome band 17q21. *Oncogene Res. 1*:209-220 (1987).
77. Simmers, R. N., L. M. Webber, M. F. Shannon, O. M. Garson, G. Wong, M. A. Vadas, and G. R. Sutherland. Localization of the G–CSF gene on chromosome 17 proximal to the breakpoint in the t (15;17) in acute promyelocytic leukemia. *Blood 70*:330-332 (1987).
78. Kanda, N., S-I. Fukushige, T. Murotsu, M. C. Yoshida, M. Tsuchiya, S. Asano, Y. Kazior, and S. Nagata. Human gene coding for granulocyte-colony stimulating factor is assigned to the q21-q22 region of chromosome 17. *Somat. Cell. Mol. Gent.13*:679-684 (1987).
79. Larson, R. A., K. Kondo, J. W. Vardiman, A. E. Butler, H. M. Golombo, and J. D. Rowley. Evidence for a 15,17 translocation in every patient with acute promyelocytic leukemia. *Am. J. Med. 76*:827-841 (1984).
80. Tsuchiya, M., S. Asano, Y. Kaziro, and S. Nagata. Isolation and characterization of the cDNA for murine granulocyte colony-stimulating factor. *Proc. Natl. Acad. Sci. USA 83*:7633-7637 (1986).
81. Tsuchiya, M., Y. Kaziro, and S. Nagata. The chromosomal gene structure for murine granulocyte colony-stimulating factor. *Eur. J. Biochem. 165*: 7-12 (1987).
81a. Buckberg, A. M., H. G. Bedigan, B. A. Taylor, E. Brownell, J. N. Ihle, S. Nagata, N. A. Jenkins, and N. G. Copeland. Localization of Evi-2 to chromosome 11: linkage to other photo-oncogene and growth factor locii using interspecific backcross mice. *Oncogene Res. 2*:149-165 (1988).
82. Yasukawa, K., T. Hirno, Y. Watanabe, K. Muratani, T. Matsuda, S. Nakai, and T. Kishimoto. Structure and expression of human B cell stimulating factor-2 (BSF-2/IL-6) gene. *EMBO J. 6*:2939-2945 (1987).
83. Nicola, N. A., and D. Metcalf. Binding of ^{125}I-labeled granulocyte colony-stimulating factor to normal murine hemopoietic cells. *J. Cell. Physiol. 124*:313-321 (1985).

84. Nicola, N. A. Murine granulocyte colony-stimulating factor: Actions on normal and leukemic cells. *Behring Inst. Mitt. 83*:207-215 (1988).
85. Nicola, N. A., and L. Peterson. Identification of distinct receptors for two hemopoietic growth factors (granulocyte colony-stimulating factor and multipotential colony-stimulating factor) by chemical cross-linking. *J. Biol. Chem. 261*:12384-12389 (1986).
86. Nicola, N. A. Kinetic aspects of the interaction of colony-stimulating factors with cellular receptors. In *Recent Advances in Leukemia and Lymphoma*, R. P. Gale and D. W. Golde (Eds.). Alan R. Liss, New York, pp. 215-228 (1987).
87. Walker, F., N. A. Nicola, D. Metcalf, and A. W. Burgess. Hierarchical down-modulation of hemopoietic growth factor receptors. *Cell 43*:269-276 (1985).
88. Nicola, N. A., M. A. Vadas, and A. F. Lopez. Down-modulation of receptors for granulocyte colony-stimulating factor on human neutrophils by granulocyte-activating agents. *J. Cell. Physiol. 128*:501-509 (1986).

4
Granulocyte-Macrophage Colony-Stimulating Factor

NICHOLAS M. GOUGH and NICOS A. NICOLA
The Walter and Eliza Hall Institute of Medical Research, Royal Melbourne Hospital, Melbourne, Victoria, Australia

I. INTRODUCTION

Granulocyte-macrophage colony-stimulating factor (GM–CSF) was one of the first hemopoietic growth factors to be described, purified, and molecularly cloned. Its name derives from its ability to stimulate individual progenitor cells in semisolid cultures of bone marrow cells to proliferate and differentiate to form colonies of neutrophils, macrophages, or a mixture of the two cells. It later became apparent that GM–CSF had a broader specificity than suspected, with the ability to stimulate eosinophil and megakaryocyte progenitors as well as the initial proliferation of early erythroid and some multipotential precursor cells. It also was discovered that, in addition to its actions on the production of mature hemopoietic cells, GM–CSF had a powerful action in enhancing the responsiveness of mature cells to bacterial and parasitic infections.

Despite the potential usefulness of such a molecule in clinical medicine, the molecular heterogeneity of GM–CSF, its production by almost all tissues, and the lack of data in vivo meant that GM–CSF was considered by some as an in vitro curiosity. The present review will detail the advances that have been made over the last few years that have elevated GM–CSF from a curiosity to a drug being currently evaluated in a large series of clinical trials.

II. BIOLOGICAL ACTIVITIES OF GM–CSF

A. Proliferative Actions of GM–CSF on Normal Hemopoietic Progenitor Cells

The primary action of GM–CSF on normal hemopoietic progenitor cells, as its name suggests, is to enhance the survival as well as to stimulate the proliferation and differentiation of cells committed to the neutrophilic granulocyte and macrophage cell lineages (1).

When hemopoietic progenitor cells or GM–CSF-dependent cell lines are cultured in vitro without a source of GM–CSF, they lose viability, with a half-life varying from 6 to 24 hr (2,3). The concentration of GM–CSF needed to maintain cell viability is significantly less than that required to stimulate cell proliferation (4).

Exposure of GM progenitors to GM–CSF causes rapid entry of cells into cell cycle (5), and both the rate and extent of cell division depend on the concentration of GM–CSF over the concentration range of 1-100 p*M* (6). However, the onset of proliferation of GM progenitor cells remains highly asynchronous, and different kinds of murine bone marrow progenitor cells vary in their responsiveness to GM–CSF. For example, larger progenitor cells separated by sedimentation velocity from mouse bone marrow are highly responsive to GM–CSF and generate relatively small macrophage colonies, whereas smaller progenitor cells are less responsive to GM–CSF but generate larger GM colonies (7). With human bone marrow cells, similar studies have separated day 14 colony-forming cells (CFC) from day 7 CFC and shown a differential responsiveness to GM–CSF and G–CSF (8).

With mouse bone marrow cells, the types of progenitor cells recruited to proliferation and differentiation with GM–CSF increase with increasing concentrations. Macrophage colonies are formed at a half-maximal dose of GM–CSF of 2.7 p*M*, granulocyte colonies at 8 p*M*, GM colonies at 23 p*M*, eosinophil colonies at 23 p*M*, and megakaryocyte colonies at 270 p*M* (9,10). In human bone marrow cultures, human GM–CSF shows a similar but less marked concentration-dependence in recruiting different kinds of progenitor cells (macrophage, neutrophil, and eosinophil) (11) including megakaryocytes (12) and early erythroid cells (13).

In addition to these apparently direct effects on the proliferation and differentiation of precursor cells, GM–CSF also shows additive or synergistic effects with other hemopoietic growth factors to allow complete differentiation to mature cells. Both murine and human GM–CSF have been shown to have a direct burst-promoting activity in that they can initiate the proliferation or at least survival of early erythroid precursor cells (BFU–E), which then require erythropoietin for full terminal differentiation (13,14). This effect also extends to multipotential precursor cells (15,16). Granulocyte-macrophage colony-stimulating factor has additive effects with multi-CSF or interleukin (IL)-3 in stimulating maximal numbers of eosinophil, erythroid, and megakaryocyte colonies to develop (12,17). In the human system, extremely low levels of GM–CSF (1-10 p*M*) synergized with M–CSF to dramatically elevate macrophage colony numbers (18), and in the murine system, GM–CSF synergized with M–CSF in stimulating maximal growth of blood monocytes and peritoneal macrophages (19,36).

A number of immortal factor-dependent myeloid cell lines derived from long-term murine bone marrow cultures (FDCP-1), SV40-transformed macrophage cell lines (BAC1) or from myeloid leukemic mice (DA1, NFS60), (20-22) proliferate in response to GM–CSF (or to multi-CSF) with no associated cell differentiation, although this proliferation is sometimes transient (23). Somewhat unexpectedly, a number of murine and human T-cell lines have also been shown to respond to GM–CSF by proliferation: These include the TALL101 cell line derived from a patient with undifferentiated acute T-cell leukemia (24) and the murine IL-2-dependent HT2 cell line (25,26). In addition, GM–CSF receptors have been detected on two other murine T-cell lines (LSTRA and LBRM33) (27), although GM–CSF receptors or biological responses to GM–CSF have not been observed with normal T cells.

B. Proliferative Activity of GM–CSF on Primary Human Leukemia Cells

A number of recent studies have also investigated whether GM–CSF stimulates the proliferation of primary human myeloid leukemic cells (28-32). Most of these studies have monitored these effects in cultures of acute myeloblastic leukemic (AML) cells in short-term colony-forming cultures or suspension cultures thought to detect blast-cell "self-renewal." These cultures do not necessarily detect the leukemic stem cells that maintain the leukemic population in vivo. Nevertheless, most studies have documented a heterogeneous response of AML cells to GM–CSF: ~20% of cases fail to grow in vitro, 25% respond to GM–CSF only, 15% respond to G–CSF only, and 40% respond to either GM–CSF or G–CSF (31). A majority of the latter class shows synergistic or additive effects with the GM–CSF + G–CSF combination (29-31). Granulocyte-macrophage colony-stimulating factor has also been shown to stimulate the proliferation of erythroid leukemic cells in a few cases of early erythroblastic leukemia (30). In most cases no evidence was found of enhanced differentiation of leukemic cells treated with GM–CSF.

In a very small number of cases of AML, autonomous growth of leukemic cells was observed in vitro. In two of these cases normal-size GM–CSF mRNA was detected in the leukemic cells and the growth of the leukemic cells was inhibited by antibodies to GM–CSF, suggesting an autocrine growth mechanism (33). In a second study, 2 of 22 AML patients showed rearranged GM–CSF genes and expressed abnormally large transcripts (34), whereas in a third study 11 of 12 of AML patients, but no pre-B-cell acute lymphocytic leukemia or chronic myeloid leukemia patients expressed GM–CSF transcripts and six of these secreted GM–CSF into the medium (35).

These studies suggest caution in the use of GM–CSF in patients with myeloproliferative disorders because of the danger of increasing the leukemic burden or of precipitating leukemia in preleukemic syndromes. The data do not, however, indicate that altered GM–CSF expression is a necessary prerequisite for the development of leukemia.

C. Differentiation-Inducing Actions of GM–CSF

Granulocyte-macrophage colony-stimulating factor alone appears to be sufficient to allow complete maturation of at least subpopulations of neutrophil, macrophage, eosinophil and, at higher concentrations, megakaryocyte progenitors. However, although GM–CSF can initiate proliferation of erythroid progenitor cells, it does not allow complete differentiation of erythroid cells in the absence of erythropoietin, and other factors may also be involved in further maturation events associated with eosinophils and megakaryocytes.

The increased colony size observed when bone marrow cultures are stimulated with high concentrations of GM–CSF may suggest that the onset of terminal differentiation can be delayed by increasing concentrations of GM–CSF (1). The concentration-dependent development of different types of colonies by GM–CSF already mentioned could be explained by differential responsiveness of different types of committed progenitor cells to varying concentrations of GM–CSF. However, there is also evidence, using paired daughter cells derived from a single cell, that high or low concentrations of GM–CSF may selectively commit cells to the neutrophilic or monocytic cell lineages, respectively (6). Similar paired daughter-cell analyses with cultures initiated in either GM–CSF or M–CSF and then transferred to the alternate stimulus also suggested that GM–CSF could irreversibly commit progenitor cells to a particular differentiation pathway (37).

The ability of GM–CSF to irreversibly commit normal myeloid progenitor cells to terminal differentiation suggests that it may also have the capacity to induce the differentiation and thereby suppress the proliferation of myeloid leukemic cells. In fact, two murine myeloid leukemic cells lines (WEHI-3B D^+ and a subclone of M1) (38,39) can be induced to differentiate mainly to macrophages by GM–CSF and this results in suppression of their proliferative capacity. The human promyelocytic leukemic cell line HL60 responds to GM–CSF by increased proliferation initially followed by differentiation and clonal suppression (40,41). There is also some evidence that differentiated HL60 cells display an enhanced dependence or responsiveness to GM–CSF for viability and growth (42). Although GM–CSF alone had little effect on the human monocytic cell line U937, it synergized with 1,25-dihydroxyvitamin D_3 in inducing the expression of a number of functional surface molecules on both U937 and normal human monocytes (43).

As well as enhancing the differentiated functions of normal and leukemic hemopoietic cells, GM–CSF can modulate the viability or functional capacity of apparently nonhemopoietic cells. These activities include a capacity to induce myeloid markers on small-cell lung carcinoma cell lines and inhibit their proliferation (44), a capacity to inhibit the replication of osteoclast precursors (45), and a capacity to increase the antigen-presenting ability of epidermal Langerhans' cells (46).

D. Actions of GM–CSF in Stimulating Functional Activities of Mature Hemopoietic Cells

In addition to its effects in stimulating the production of neutrophils, macrophages, and eosinophils, GM–CSF also has a pronounced capacity to increase the functioning of these cells in immune reactions. Part of this capacity is due to the ability of GM–CSF to enhance the lifespan of human blood neutrophils and eosinophils in vitro (47), and increase their metabolic activity (48).

Granulocyte-macrophage colony-stimulating factor enhances the ability of both murine and human neutrophils to kill antibody-coated tumor cells (antibody-dependent cellular cytotoxicity) (49,50), phagocytose bacteria, parasites, and yeast (1,51,52,53), and kill them intracellularly (53). Granulocyte-macrophage colony-stimulating factor also inhibits the migration of neutrophils (54), increases their cellular adhesiveness (55), and is directly chemotactic for both neutrophils and monocytes (56). These effects are expected to result in the recruitment of neutrophils from the circulation to sites of GM–CSF secretion where they would be immobilized and activated to fight infections. In addition to these direct activations, GM–CSF is also able to prime neutrophils so that they can be more readily activated by bacterial products such as formyl peptides or complement and leukotrienes (57,58) to produce reactive oxygen intermediates (superoxide anion) involved in bacterial killing. Induction of IL-1 secretion by neutrophils stimulated with GM–CSF may further recruit and activate accessory immune cells (59).

Monocytes and macrophages can be stimulated in many ways by GM–CSF. These include the stimulated release or prostaglandin E, arachidonic acid, plasminogen activator, interferons, IL–1 (1), tumor necrosis factor (60), and M–CSF (61). Similarly to its effects on neutrophils, GM–CSF also stimulates antibody-dependent cytotoxicity of macrophages (62), the intracellular killing of parasites (63-65), and suppresses the intracellular proliferation of the acquired immune deficiency syndrome (AIDS) virus (human T-lymphotropic virus type III) (66). The direct effects of GM–CSF on macrophages as well as the indirect effects mediated by the stimulated release of inflammatory monokines would be expected to provide a powerful response to bacterial or parasitic infections.

Granulocyte-macrophage colony-stimulating factor stimulates the calcium ionophore-induced release of leukotriene C4 by human eosinophils (67), increases the autofluorescence, superoxide anion production, and protein synthesis of eosinophils (68), antibody-dependent cytotoxicity for tumor cells (69), and adherence to *Schistosomula* as well as their antibody-dependent killing (67,70).

It is clear from these data that GM–CSF plays a pronounced immunomodulatory role in bacterial and parasitic infections and in generating a local inflammatory response. Bacterial products or T-cell activation would provide a localized release of GM–CSF (possibly via IL-1 and tumor necrosis factor action on local cells) that would recruit neutrophils, macrophages, and eosinophils (in the case of parasitic infections) to the site of infection. These cells would be immobilized at the site and be activated either directly by GM–CSF or be primed to respond more aggressively to other stimulatory signals (bacterial peptides, antibody coating of parasites) to phagocytose the infectious agent and kill it intracellularly or extracellularly. At the same time, the GM–CSF-induced release of cytotoxic agents and cellular recruiting stimuli [IL-1, tumor necrosis factor (TNF)] would ensure that a generalized reaction took place that would resolve the local infection. This may well constitute the major role of GM–CSF in vivo because, in general, the circulating levels of GM–CSF are low, GM–CSF-producing cells are widely scattered throughout various tissues, and the major GM–CSF-inducing signals are bacterial products or T-lymphocyte immune reactions.

E. Actions of GM–CSF In Vivo

The effects of GM–CSF on hemopoietic cells in vivo have been assessed by the injection of GM–CSF into normal and immunocompromised mice or monkeys, by the development of a transgenic strain of mice containing a GM–CSF gene under the control of a viral promoter, and by reconstituting irradiated mice with bone marrow cells that contain a GM–CSF-expressing retrovirus.

Injection of GM–CSF (200 ng) into the peritoneal cavity of normal mice resulted in a marked elevation of macrophages (15-fold), neutrophils, and eosinophils (up to 100-fold) in the peritoneal cavity within 6 days (71). These peritoneal macrophages demonstrated enhanced phagocytic capacity toward antibody-coated erythrocytes and other signs of activation (71,72). Peripheral blood neutrophils were only slightly elevated (twofold), but increased numbers of infiltrating neutrophils and macrophages in the liver and lung were observed (71). The spleen weight and content of monocytes, megakaryocytes, and myeloid progenitor cells were increased whereas bone marrow cellularity and content of myeloid cells were decreased (71).

Granulocyte-macrophage colony-stimulating factor continuously infused into monkeys or administered subcutaneously caused a rapid leukocytosis with elevated levels of circulating granulocytes (sixfold), eosinophils (fivefold), monocytes (fourfold) and lymphocytes (threefold) but not platelets or erythrocytes (73,74). Circulating neutrophils obtained from GM–CSF-treated monkeys were primed for enhance *N*-formyl Met-Leu-Phe (FMLP)-induced superoxide anion generation and capacity to kill *Escherichia coli* compared with control cells (74). The effects of GM–CSF on leukocytosis were readily reversed by cessation of GM–CSF treatment (73,74).

Low doses of GM–CSF were shown to have a synergistic effect when injected with other CSFs (M–CSF and multi-CSF or IL-3) into mice pretreated with lactoferrin. Lactoferrin suppresses hemopoiesis but the percentage of cycling colony-forming cells (of all types) and total number of colony-forming cells can be increased in the bone marrow by GM–CSF or combinations with the other CSFs (75,76).

Although GM–CSF itself has little radioprotective effect on lethally irradiated mice, it does synergize with suboptimal doses of IL-1α and tumor necrosis factor-α in conferring radioprotection (77). This may be due to the stimulation of hemopoietic recovery after irradiation damage. Indeed, in models of allogeneic bone marrow transplantation in mice using T-cell-depleted bone marrow, pretreatment in vitro of the bone marrow before transplantation or treatment of transplanted mice in vivo with GM–CSF resulted in an increased degree of bone marrow engraftment in recipients (78). In autologous models of bone marrow transplantation using irradiated monkeys, infusion of GM–CSF resulted in a significant shortening of the period of neutropenia (9 days compared with 24 days) and thrombocytopenia as well as elevated levels of circulating neutrophils compared with controls (79,80).

All of these effects, with the exception of elevated lymphocyte levels in monkeys, are consistent with the activities of GM–CSF in vitro in stimulating the production, functional activity, and chemotaxis and migration inhibition of neutrophils, eosinophils, and macrophages. They suggest that GM–CSF would be useful in immunodeficiency, anemias, infection, and bone marrow transplantation. However, there are also long-term experiments of chronic GM–CSF administration that suggest that some caution needs to be exercised in using GM–CSF clinically.

In transgenic mice carrying the GM–CSF gene expressed from a retroviral promoter (81), elevated levels of GM–CSF were observed in the serum, peritoneal cavity and eye. Accumulations of apparently activated macrophages were observed in the eye, leading to retinal damage and blindness; in striated muscle, leading to muscle wastage, and in the peritoneal cavity, where the macrophages were large multinucleated cells, as well as in the pleural cavity. By contrast, peripheral blood cell counts and bone marrow cellularity and progenitor cell content were near normal, whereas the spleen showed variable responses. The hemopoietic progenitor cells retained a normal responsiveness to GM–CSF. The result of these pathological effects was a dramatically reduced lifespan of transgenic mice compared with littermate controls, with death resulting from a wasting disease.

The pathological effects of chronic GM–CSF exposure in transgenic mice are likely to be due to effects of GM–CSF produced by and acting on hemopoietic cells rather than insertional mutagenesis because two independent lineages of transgenic mice display similar symptoms, and mice repopulated by bone marrow cells infected with a GM–CSF expressing retrovirus show similar or more dramatic effects (82). Such mice express much higher levels of GM–CSF in the serum and show the same pathological effects, albeit more dramatically, as transgenic mice. In addition, infiltration of neutrophils and eosinophils as well as macrophages was observed in the liver, lungs, and other organs. The pathological effects of excess GM–CSF expression seem to be related to chronic activation of inflammatory cells with lesser effects on hemopoietic cell production. This again suggests that the major role of GM–CSF in vivo may be the transient modulation of inflammatory responses rather than the steady-state production of hemopoietic cells.

F. Clinical Trials with GM–CSF

Phase I/II clinical trials of GM–CSF have been initiated in patients with leukopenia and impaired host defense mechanisms as a result of AIDS (83), myelodysplastic syndromes (84), malignancy (85,86), cytotoxic chemotherapy for cancer (87,88), or chemotherapy followed by bone marrow transplantation (89).

Patients with AIDS often show decreased levels of circulating functional neutrophils and monocytes (as well as lymphocytes) and this, in part, contributes to the high incidence of opportunistic infections in such patients and limits the use of toxic drugs used to treat infections. Infusion of GM–CSF into AIDS patients caused a dose-dependent increase in circulating mature neutrophil and band forms (3-15-fold), monocytes (two to fourfold), and eosinophils (10-20-fold), reaching a maximum by days 10-17 after treatment initiation. Marrow cellularity increased in some patients, especially associated with an eosinophilia and shift to more immature myeloid forms in the marrow. There were no consistent effects of GM–CSF on AIDS virus recovery from mononuclear cells.

Continuous infusion of GM–CSF (30-500 $\mu g/m^2$) into patients with myelodysplastic syndromes (refractory anemia with excess blasts) showed a rapid and sustained elevation in circulating neutrophils (10-100-fold), monocytes (10-100-fold), eosinophils (100-fold), reticulocytes (2-10-fold) and, less consistently, hematocrit and platelets. Marrow cellularity was also increased with a decreased ratio of blast cells to mature myeloid cells, suggesting that GM–CSF induced differentiation of the blast cells. No transformation to acute leukemia was observed with the treated patients despite the ability of GM–CSF to stimulate the proliferation of leukemic and preleukemic myeloid cells.

In patients receiving chemotherapy for solid tumors, GM–CSF had little effect in the prechemotherapy phase on bone marrow CFC but increased peripheral blood GM–CFC (18-fold) and erythroid CFC (BFU-E) (eightfold). Chemotherapy abolished these CFC in the peripheral blood but GM–CSF increased the rate and extent of CFC recovery and leukocyte recovery in the blood after chemotherapy. Higher levels of CFC were attained on recovery after chemotherapy in the presence of GM–CSF than the prechemotherapy levels in the presence of GM–CSF. In cancer patients receiving autologous bone marrow transplantation after high-dose chemotherapy, GM–CSF infused for 14 days after transplantation caused a dose-dependent increase in total leukocyte and neutrophil recovery by day 14 and decreased the number of neutropenic days from 17 to 12 compared with historical controls.

All studies have thus demonstrated some beneficial effect of GM–CSF in elevating leukocyte levels and CFC in the peripheral blood of leukopenic patients, as well as increasing the rate of recovery from leukopenia-inducing treatments. This occurs after a rapid and transient fall in circulating neutrophil and monocyte levels (nadir by 30 min) due to sequestration in the lungs (86,90). Similar clinical effects were observed with either glycosylated yeast-derived GM–CSF or nonglycosylated *E. coli*-derived GM–CSF. However, all the studies have also noted some mild to severe side effects of GM–CSF treatment including local phlebitis, edema, skin rashes, bone pain, fever, and chills. The modes and sites of GM–CSF administration as well as the dose and duration still need to be optimized to achieve maximal hemopoietic responses with minimal side effects.

III. BINDING OF GM–CSF TO CELLULAR RECEPTORS

Specific binding of iodinated murine GM–CSF has been observed on a variety of murine myeloid and some lymphoid cell lines (91,92) as well as to normal murine hemopoietic tissue including bone marrow, spleen, and peritoneal exudate cells (91). However, the two reports describing these binding characteristics differ considerably. Walker and Burgess (91) found two affinity classes of GM–CSF receptor on most cell types—a very small number (10-100) of high-affinity (K_D 20-60 p*M*) receptors and a larger number (300-500) of low-affinity (K_D 0.7-1.2 n*M*) receptors. Chemical crosslinking suggested that the receptor had a molecular weight of 51,000. On the other hand Park et al. (92) found a single class of GM–CSF receptor on all cells (1000-5000 per cell, K_D 0.9-3.7 n*M*) and a receptor molecular weight of 130,000 in most cells.

Murine GM–CSF receptors directly recognize only the cognate ligand but the GM–CSF receptor can be "downmodulated" (loss of binding sites from the cell surface) on bone marrow cells by multi-CSF (IL-3) or high concentrations of M–CSF (CSF-1) (91,93). This indirect or "transmodulation" of GM–CSF receptors is rapid, dose dependent, and occurs only at biological temperatures (~37°C). The mechanisms underlying it are not understood and its biological consequences are also unclear. However, those CSFs that cause rapid GM–CSF receptor downmodulation (M–CSF and multi-CSF) show synergistic growth and differentiating activities on bone marrow cells with GM–CSF (94) and the transmodulation phenomenon is not apparent on some leukemic cell lines such as WEHI-3B D^+ (91). Possible interpretations of these phenomena have been described in detail elsewhere (94).

Granulocyte-macrophage colony-stimulating factor, itself, causes internalization and downregulation of GM–CSF receptors in a variety of murine cells at 37°C (95,96). Internalization was followed by degradation of GM–CSF, probably in lysozymes because degradation was inhibited by ammonium chloride or monensin (95) and ligand degradation was balanced by receptor reexpression at the cell surface by both new receptor synthesis and receptor recycling (95). Considerable variation in the intracellular processing events associated with GM–CSF/receptor interactions was observed with different cell types. For example, WEHI-3B D^+ cells showed a slower rate of internalization of GM–CSF receptors than did most other cell types and peritoneal macrophages showed a much faster ligand degradation rate than did most other cells (96). These differences resuted in variations of the steady state internalized to surface GM–CSF ratio from 0.5 in peritoneal cells to 3.3 in factor-dependent cells (96).

Autoradiographic analysis of murine bone marrow cells incubated with radioiodinated GM–CSF showed that almost all cells within the neutrophil, eosinophil, and monocyte cell lineages displayed GM–CSF receptors whereas lymphocytes and erythroid cells did not. In general, receptor numbers decreased with cellular differentiation but even the most mature cells displayed significant numbers (up to 100 per cell) of GM–CSF receptors (94). The pattern of receptor distribution was consistent with the known cellular specificity of GM–CSF biological actions on both immature and mature cells. Furthermore, the lack of species cross-reactivity of the biological actions of GM–CSF was matched by the inability of murine GM–CSF to bind to human hemopoietic cells and vice versa.

The binding of human GM–CSF to specific cellular receptors on human myeloid cells, myeloid leukemias, and myeloid cell lines has also been studied (97-100). In all cases GM–CSF receptor expression was restricted to normal neutrophils, eosinophils, and monocytes and to the myeloid cell lines HL60 and KG1. All authors agree that only small numbers (50-500/cell) of GM–CSF receptors are expressed in the various cell systems but, again, there is disagreement on the binding affinity (K_D = 20 p*M* versus 600 p*M*). Although these authors have all observed a single affinity class of GM–CSF receptors, we (N. A. Nicola and L. Peterson, unpublished data, 1986) consistently observe both high-affinity (K_D = 20-50 p*M*) and low-affinity (K_D = 0.8-1.2 n*M*) GM–CSF receptors on various cell types with the low-affinity receptors being rapidly dissociated under normal washing conditions. Consistent with the observation that nonglycosylated GM–CSF has a higher specific biological activity than glycosylated GM–CSF, Kelleher et al. (100) found that nonglycosylated GM–CSF also had a much higher affinity for the GM–CSF receptor. Chemical crosslinking of iodinated GM–CSF to its receptor on neutrophils or HL60 cells indicates a receptor molecular weight of 84,000 (99).

As for murine GM–CSF receptors, the human receptor levels can be transmodulated. The chemotactic peptide, FMLP, causes a dramatic loss of GM–CSF receptors from the surface of neutrophils without altering their apparent affinity (99). On the other hand, long-term treatment of HL60 cells with dimethylsulfoxide caused an increase in the number of high-affinity GM–CSF receptors (99). Dimethylsulfoxide causes differentiation of HL60 cells to neutrophils and the increased number of receptors may reflect the differentiated state of the cells. Kelleher et al. (100) found that AML blast cells incubated in media conditioned by the 5637 bladder carcinoma cell line (a source of a number of different hemopoietic growth and differentiation factors) showed an increased number and lower apparent affinity of GM–CSF receptors. In both cases (HL60 and AML blasts) the pretreatment that resulted in increased receptor numbers also resulted in an increased responsiveness to GM–CSF.

IV. CLONING OF MURINE AND HUMAN GM–CSF

The isolation of murine and human GM–CSF cDNA clones exemplified two of the more elegant strategies for identifying clones corresponding to low abundance mRNAs: hybridization with degenerate oligonucleotides corresponding to short stretches of protein sequence and functional expression.

Murine GM–CSF was purified to homogeneity from the lungs of mice injected with bacterial endotoxin (101,102), and amino acid sequence analysis yielded an amino terminal sequence of some 34 residues (101). Using this sequence, short oligonucleotides complementary to the predicted GM–CSF mRNA were synthesized and used as hybridization probes to screen a library of cDNA clones complementary to mRNA from the same source (103). Despite the apparent richness of GM–CSF in lung-conditioned medium, the abundance of the corresponding mRNA was quite low, and only 2 GM–CSF clones from a library of ~100,000 were isolated. Neither of these clones was full-length, but could be used as hybridization probes to identify a richer source of GM–

CSF mRNA: the murine CD4$^+$ T lymphocyte line LB3, in which high levels of GM–CSF mRNA (as well as multi-CSF and IFN-γ mRNA) are inducible with the lectin *Concanavalin A* (103). Full-length murine GM–CSF cDNA clones were thus isolated from a cDNA library synthesized using stimulated LB3 mRNA (104).

At the time that Wong et al. (105) and Lee et al. (106) isolated cDNA clones encoding human GM–CSF, no amino acid sequence data were available. Both groups therefore resorted to an elegant "functional expression" strategy. cDNA libraries complementary to mRNA from the human T-lymphoblastoid cell line, Mo, or from a *Concanavalin A*-stimulated human helper T-cell clone, T7, were constructed in vectors designed to allow efficient expression of the cloned sequences after transfection into simian COS cells. The libraries of cDNA clones were divided into pools, the DNA from each pool were transfected into COS cells, and the conditioned medium was assayed for CSF activity using either the KG1 human myeloid leukemic cell line or agar cultures of myeloid progenitor cells. By subdivision of positive pools, both groups isolated several independent clones, each containing sequences homologous to the previously cloned murine GM–CSF (103).

Subsequently, several other groups have isolated murine and human GM–CSF clones, using the published sequences to construct hybridization probes (107, 108). Particularly noteworthy were the reports of isolation of human GM–CSF clones by cross-hybridization with murine probe (108,109). Recombinant GM–CSF has been produced in a number of host/vector systems, including *E. coli* (107,110), yeast (108,111,112), and mammalian cells (104-106,109, 113,114) with no gross differences in the spectrum of biological activities displayed by the factor produced, but evidence that nonglycosylated GM–CSF has a 10-20-fold higher specific biological activity than the glycosylated forms (100,115,116).

V. GM–CSF STRUCTURE

The structures of murine, human, and gibbon GM–CSF have been determined both by partial amino acid sequence analysis (101,105,117) and by deduction from the nucleotide sequence of cDNA clones (103-105,118) (Fig. 1). In each species, the native protein is preceded by a hydrophobic leader sequence, 25 amino acids in length. Mature murine GM–CSF comprises 124 amino acids (molecular weight = 14 kD) whereas human and gibbon GM–CSF are three residues longer (127 residues). The additional three residues in the primate sequences are located between residues 21 and 22 of the murine sequence. Interestingly, this is within one of two strongly prediced α-helices that appear to be essential for GM–CSF activity (119) (see below).

```
                                   +1
MOUSE      MWLQNLLFLGIVVYSLSAPTRSPITVTRPWKHVEAIKEA---LNLLDDMPVTLNEEVEVV   40
           **** ** ** *  * *** ***   * ** ** ** **   ***  *     ** ***
HUMAN      MWLQSLLLLGTVACSISAPARSPSPSTQPWQHVNAIQEARRLLNLSRDTADEMNETVEVI   43
           ******************* ****** *** ******************* * ******
GIBBON     MWLQSLLLLGTVACSISAPSRSPSPSRQPWEHVNAIQEARRLLNLSRDTAAEINETVEVV   43
                                                      +++       +++

                                     +++      +++
MOUSE      SNEFSFKKLTCVQTRLKIFEQGLRGNFTKLKGALNMTASYYQTYCPPTPETDCETQVTTY   100
           *  *     ** ****    *****  ***** * * ** *   ******* * ** **
HUMAN      SEMFDLQEPTCLQTRLELYKQGLRGSLTKLKGPLTMMASHYKQHCPPTPETSCATQTITF   103
           *********************** ******************************** ***
GIBBON     SEMFDLQEPTCLQTRLELYKQGLQGSLTKLKGPLTMMASHYKQHCPPTPETSCATQIITF   103


MOUSE      ADFIDSLKTFLTDIPFECKKPVQK   124
             *   ** **  *** *  ***
HUMAN      ESFKENLKDFLLVIPFDCWEPVQE   127
           ************* **********
GIBBON     ESFKENLKDFLLVTPFDCWEPVQG   127
```

Figure 1 Comparison of the amino acid sequence of murine (104), human (105), and gibbon (118) pre-GM–CSF. Identities between the mouse and human and human and gibbon sequences are indicated with asterisks. The *N*-terminal residue of native GM–CSF is +1. *N*-linked glycosylation sites are indicated with +.

Murine and human GM–CSF bear only ~56% amino acid sequence identity, and thus GM–CSF is one of the less well conserved of the myeloid growth factors (Fig. 1). Given the relatively low sequence identity between murine and human GM–CSF, the lack of cross-species activity (1) or receptor binding (91) displayed by this factor is perhaps not surprising. Despite the rather low primary amino acid squence identity between GM–CSF from mouse and humans, far-ultraviolet circular-dichroic spectral data suggest that the two proteins in fact have closely similar backbone conformations (120).

The murine and human GM–CSF polypeptide sequences both contain two potential *N*-linked glycosylation sites but interestingly they are at different locations within the respective molecules (Fig. 1). Granulocyte-macrophage colony-stimulating factor isolated from most murine and human sources is glycosylated to ~23,000 Da (121). Higher-molecular-weight forms, presumed to represent glycosylation or aggregation variants, have also been observed (121). Although natural GM–CSF is glycosylated, nonglycosylated recombinant GM–CSF derived from *E. coli* has high biological activity (107,110,115, 116), supporting earlier work that demonstrated that deglycosylated authentic GM–CSF is biologically active (101). Moreover, GM–CSF produced in yeast cells, bearing different carbohydrate structures to the mammalian protein, is also active (108,111,112). The data indicate that attached carbohydrate is not necessary for biological activity of GM–CSF (and probably decreases the affinity of GM–CSF for its receptor), but rather leads to enhanced stability and solubility of GM–CSF.

The primary amino acid sequences of GM–CSF strongly predict two α-helices close to the *N*-terminus (positions 13 to 27 and 31 to 46 in the murine molecule), which appear to be required for biological activity: Truncation of either helix by introduction of a helix-breaking glycine residue markedly reduced the activity of the molecule (119). Removal of up to seven residues from the *N* terminus, which does not include any of the first helix, did not, however, appear to impair the molecule (101).

The pattern of disulphide bonding has been determined, with the first and third and second and fourth cysteine residues, respectively, being paired (122). These intramolecular disulfide bonds enhance the stability of GM–CSF and are required for its correct folding because GM–CSF is denatured in chaotropic salts containing reducing agents (102).

VI. CELLULAR SOURCES OF GM–CSF

Granulocyte-macrophage colony-stimulating factor can be produced by a number of different cell types, under a number of different circumstances. Injection of bacterial endotoxin into mice results in a rapid release of GM–CSF into the serum (1), probably from macrophages and endothelial cells. Almost all tissues and organs derived from endotoxin-primed mice and cultured in vitro release GM–CSF into the culture medium (121). Indeed, conditioned medium from lungs of endotoxin-primed mice was the source from which murine GM–CSF was first purified (102). Production from organ cultures is continuous over several days and reflects de novo synthesis rather than release of preformed material because the production can be inhibited by protein synthesis inhibitors (121). Studies with purified primary cell cultures, with cloned cell lines and with tumors have identified a number of the cell types that synthesize GM–CSF. In all cases, GM–CSF production requires stimulation of the producer cell, for example, by other cytokines, antigen (in the case of T lymphocytes), inflammatory agents, etc. (see below). Only very low levels of GM–CSF are produced in the absence of stimulation. Figure 2 schematically illustrates the network of signals and stimuli that can give rise to GM–CSF production (discussed in detail in next section), and highlights and central role of T lymphocytes and macrophages in regulating GM–CSF production.

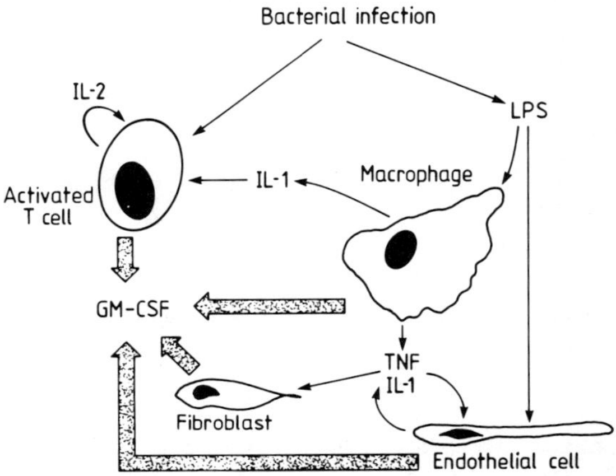

Figure 2 Potential signals stimulating GM–CSF production after bacterial infection (see Section VI).

A. Expression of GM–CSF in T Cells

Of the several cell types known to produce GM–CSF, T lymphocytes are the most well studied, due largely to the availability of cloned cell lines. T lymphocytes can be induced to express at least eight different hemopoietic regulators, including GM–CSF, IL 2-6, interferon (IFN)-γ, and TNFβ (123). These molecules have diverse and, in some cases, opposing effects on cells of the hemopoietic and lymphoid systems.

The major pathway for the activation of GM–CSF synthesis in T cells is induced by ligands of the T-cell receptor for antigen (TCR), such as monoclonal antibodies, lectins, and antigen itself, and results in the synthesis of a number of lymphokines, including GM–CSF. Work from a number of groups has identified some of the early events that follow TCR binding (see 124,125 for reviews; and 126-129). T-cell-receptor-dependent activation of phospholipase C results in the hydrolysis of phosphatidyl inositol bisphosphate to yield inositol trisphosphate, which induces an increase in intracellular free Ca^{2+}, and diacylglycerol which, in turn, activate protein kinase C. The two major pathways so activated can also be stimulated individually by Ca^{2+} ionophores, which cause an influx of extracellular Ca^{2+}, and by analogues of diacylglycerol, such as phorbol myristate acetate (PMA), which in the presence of Ca^{2+} directly activates protein kinase C. Ca^{2+} ionophores and PMA have thus been found to act synergistically to induce lymphokine synthesis analogous to that induced by ligands of the TCR. TCR-induced lymphokine expression, as well as that induced by Ca^{2+}

ionophore, is substantially inhibitable by the immunosuppressive drug cyclosporin A (CSA) (130-139). Cyclosporin A appears to inhibit Ca^{2+}-dependent activation pathways at a point distal to the early rise in intracellular-free Ca^{2+} (134,139,140), perhaps as a result of binding to the cytosolic protein, cyclophilin (141). The TCR-induced pathway does not involve de novo protein synthesis, because cytoplasmic accumulation of lymphokine mRNA can ensue in the presence of the protein synthesis inhibitor cycloheximide (123).

Other stimuli that can induce GM–CSF synthesis in T cells have also been observed, including IL-1 (142) and the T-cell growth factor IL-2 (143). Induction by IL-2 results in synthesis of lower levels of GM–CSF and IFN-γ than with TCR stimulation, with little or no multi-CSF production. As with the TCR pathway, de novo protein synthesis is not required for mRNA induction (N. M. Gough and A. Kelso, unpublished data, 1988). By contrast with the TCR-induced pathways, binding of IL-2 to its receptor has been shown to activate protein kinases, including those of the C-type (144-146) without detectable breakdown of inositol phosphate or rise in intracellular-free Ca^{2+} (147,148). The IL-2-induced production of GM–CSF appears to be resistant to CSA (137).

The consequences of activation of any of the kinases implicated in TCR or IL-2-dependent stimulation and the later nuclear events that lead to lymphokine gene expression remain largely undefined. An increased rate of transcription of the GM–CSF gene after lectin stimulation of certain murine T-cell clones has been observed (R. James, personal communication) and progress has been made in identifying the DNA sequences and nuclear DNA binding factors implicated in this process (see Section VIII.B).

In addition to mitogen-stimulated production of GM–CSF by T lymphocytes, there is evidence that mitogen-stimulated B cells are also a source of GM–CSF (149).

B. Expression of GM–CSF in Macrophages

The mononuclear phagocyte system is well documented to participate in immunoregulatory functions an in host defense against infectious disease (150). In addition, macrophages may play a role in controlling hemopoietic development: It is the first hemopoietic cell to be identified during ontogeny (151,152) and appears to be the sole source of erythropoietin during the fetal liver gestational period (153), a role that appears to be carried over into the adult (154, 154a). Macrophage populations can be induced to produce GM–CSF in conjunction with a range of cytokines different to that coexpressed with GM–CSF in T cells, including G–CSF, TNFα, and IL-1. Among the most common signals to induce cytokine production by macrophages is the bacterial cell-wall product, lipopolysaccharide (155), which, like ligands of TCR in activation of T cells (see Section VI.A), appears to generate inositol trisphosphate and diacylglycerol as second messengers.

Using short-term cultures of peritoneal exudate cells, substantially enriched for macrophages, Thorens et al. (156) demonstrated that a number of stimuli could induce secretion of GM–CSF, including LPS, latex beads, thioglycolate, and adherence. In this system, increased GM–CSF mRNA stability appeared to be a major contributor to the elevation of cytoplasmic GM–CSF mRNA evident after stimulation. By contrast with induction of GM–CSF expression in T cells (123), Thorens et al. (156) reported that in their peritoneal macrophage populations, accumulation of cytoplasmic GM–CSF mRNA was cycloheximide sensitive. Because the macrophage populations used in these studies comprised a few percent Thy-1^+ cells, the possibility that at least a fraction of the GM–CSF produced in these cultures was derived by indirect stimulation of T cells (perhaps by IL-1; 142) cannot be discounted. However, using longer-term cultures of bone-marrow-derived macrophages devoid of T lymphocytes, Rich (157) has also documented GM–CSF production by macrophages.

C. GM–CSF Production by Endothelial Cells and Fibroblasts

Endothelial cells, which separate blood from the various tissues, are involved in a number of inflammatory reactions. Indeed, inflammation is characterized by an enhanced passage of white blood cells across the endothelium. It has been noted for some time that various factors that act on the growth and differentiation of hemopoietic progenitor cells are produced by stimulated endothelial cells (158-159a) and more recently it has been observed that TNF-α and IL-1, which are released by activated monocytes and macrophages, induce various phenotypic changes in cultured endothelial cells, including increased adhesiveness for leukocytes (160,161) and enhanced expression of plasminogen activator inhibitor (162-164).

By Northern blot analysis it has now become clear that IL-1, TNF-α, and LPS can induce the expression of three of the CSFs in cultured endothelial cells: GM–CSF, G–CSF, and M–CSF (165-168). Although LPS is a potent inducer of IL-1 and TNF-α production by macrophages, it seems clear from the data of Seelentag et al. (167) that LPS, like IL-1 and TNF-α, can act directly on endothelial cells. Enhanced expression of GM–CSF, G–CSF, and M–CSF in these cells is due, at least in part, to an increased rate of transcription (167). Interestingly, TNF-α also stimulates production of IL-1 by endothelial cells (167) and thus autocrine stimulation of endothelial cells by IL-1 at the site of inflammation is a possibility.

The stimulation by inflammatory mediators of endothelial cells to release IL-1 and the CSFs may constitute an important step in the inflammatory process. The CSFs could not only functionally activate mature white blood cells that migrate to an inflammatory site and, by inhibiting further migration, ensure their retention in the region of inflammation, but also could enhance proliferation and differentiation of progenitor cells. Thus, an initial release of IL-1 and TNF-α, probably from macrophages stimulated by an inflammatory or immunological stimulus, would mediate an amplification of the local inflammatory response.

Fibroblasts, which can also be induced by a variety of agents to produce GM–CSF, including TNF-α (166), IL-1 (169), inflammatory and tumor-promoting diterpene esters (170), and retroviruses (171), may also contribute to the development of a local inflammatory focus or, in the case of bone marrow fibroblastoid stromal cells (172), may play a role in controlling development of granulocyte/macrophage progenitors.

The particular location of endothelial cells not only predisposes them to a role in local inflammatory reactions, but also to a role in acute systemic reactions. Thus, the very rapid increase in serum GM–CSF, G–CSF, and M–CSF after intravenous injection of LPS (1) may occur through stimulation of a large number of endothelial cells, which could lead to a generalized increase in the number of mature granulocytes and macrophages.

VII. GM–CSF GENE LOCALIZATION

Granulocyte-macrophage colony-stimulating factor is encoded by a unique gene in both mouse (103) and humans (105,108,109,173) with no kindred genes revealed by hybridization analysis at low stringency. In the mouse, the GM–CSF gene was initially localized by analysis of a panel of mouse/Chinese hamster somatic cell hybrids, which indicated that this gene is located on chromosome 11 (103). The specific location on chromosome 11, subband B1, was derived by a genetic approach in which tight linkage of the GM–CSF gene to the previously localized SPARC gene was established by analysis of a set of *Mus musculus* × *M. spretus* F1 blackcross mice (174). This localization was subsequently confirmed using recombinant inbred strains of mice (175).

In humans, the GM–CSF gene was localized to chromosome 5 at band q21-q31 by in situ chromosome hybridization (176,177). This band is within a region of chromosome 5 frequently deleted in secondary neoplasias arising after chemotherapy and radiotherapy—the so-called $5q^-$ syndrome (176-178). However, it is difficult to evaluate the relevance to this clinical syndrome of loss of one GM–CSF allele, because a large segment of the long arm of chromosome 5 is deleted in such situations. This region of chromosome 5 undoubtedly contains many genes, and interestingly it contains several other growth factor (M–CSF, IL-3, IL-4, IL-5, ECGF) and receptor (c-fms, platelet-derived growth factor receptor, β_2-adrenergic receptor) loci (see Ref. 178 for review).

The most surprising revelation arising from the gene localization of GM–CSF is its close proximity to the multi-CSF gene—as little as 10 kbp in humans (179) and between 19 and 230 kbp in the mouse (174). The most obvious explanation for this tight linkage is that the two genes are ancestrally related, and indeed, multi-CSF is a functionally related lymphokine. However, there is no apparent structural homology between the two proteins at either the primary nucleotide or amino acid sequence level, suggesting that either the two proteins are not descendants of a common ancestral gene or alternatively that they are products of quite an ancient duplication event. Either way there is no reason a priori to expect tight linkage of the two genes to be conserved, particularly because ancient duplicates tend to be on separate chromosomes (e.g. , α and β globins and immunoglobulin heavy and light chains). Thus, it is likely that there are selective pressures to maintain the tight linkage of these two genes. One level at which selective pressure might be exerted is at the control of expression of the two genes: the tight linkage of these genes may in some way facilitate or modulate their coexpression in activated T cells (123).

VIII. GM–CSF GENE AND mRNA STRUCTURE

A. The Transcriptional Unit

The unique murine and human GM–CSF genes have been cloned and analyzed in a number of laboratories (109,173,180). The GM–CSF gene comprises four exons, spread over approximately 2.5 kbp of DNA (Figs. 3 and 4).

The transcriptional initiation site of murine GM–CSF has been determined to be at either position 1136 (173) or 1133 (180) in Fig. 3. The slight discrepancy may reflect either microheterogeneity in the start site of transcription, or differences in the technique used to derive the location (S1 nuclease mapping versus primer extension). A similar level of microheterogeneity is evident at the 3′ end of the transcriptional unit, with most reported mRNA sequences terminating with a poly A tail after positin 3506 (103,104,173) (Fig. 3) but with that reported by DeLamarter (107) terminating after position 3513. Processing of the primary transcript yields a native mRNA of some 780 nucleotides, with a 5′ untranslated region of 32-35 residues, a 3′ untranslated region of 319-326 residues, and a coding region of 423 nucleotides.

The human GM–CSF gene is arranged similarly to that of the mouse. The transcriptional start site once again displays a level of microheterogeneity, having been variously assigned to position 620-622 (173) and 625 and 627 (181) in Fig. 4. Note that the latter location is analogous to the murine start site as determined by Stanley et al. (180). The 3′ end of the mature transcript is at position 2998 (105,106,173). The mature human GM–CSF mRNA is also ~780 nucleotides in length, with 5′ and 3′ untranslated regions of 39-42 and 316 nucleotides, respectively, and a coding region of 432 nucleotides.

A number of sequence motifs within and around the GM–CSF gene have been noted as potential control elements, either by virtue of homology be-

Figure 3 Nucleotide sequence of the murine GM–CSF gene (173,180). Flanking and intervening sequences are in lower-case letters. Various sequence motifs (see text) are highlighted.

```
ttctcagagtggctgcagtctcgctgctggatgtgcacatggtggtcattccctctgctcacaggggcaggggtccccccttactggactgaggttgcccccctgctccaggtcctgggtgggagc  125
ccatgtgaactgtcagtggggcaggtctgtgagagctcccctcacactcaagtctctctcacagtggccagagaagaggaaggctggagtcagaatgaggcaccagggcgggcatagcctgccca  250
aaggcccctgggattacaggcaggatggggagccctatctaagtgtctcccacgccccaccccagccattccaggccaggaagtccaaactgtgcccctcagagggagggggcagcctcaggccc  375
attcagactgcccagggagggctggagagccctcaggaaggcgggtgggtgggctgtcggttcttggaaaggttcattaatgaaaaccccaagcctgaccacctagggaaaaggctcaccgttc  500
ccatgtgtggctgataagggccaggagattccacagttcaggtagttccccgcctccctggcatttttgtggtcaccattaatcatttcctctgtgtatttaagagctcttttgccagtGAGCCC  625
                                          MetTrpLeuGlnSerLeuLeuLeuLeuGlyThrValAlaCysSerIleSerAlaProAlaArgSerProSerProSerThrGlnProTrp
AGCTACACAGAGAGAAAGGCTAAAGTTCTCTGGAGGATGTGGCTGCAGAGCCTGCTGCTCTTGGGCACTGTGGCCTGCAGCATCTCTGCACCCGCCCGCTCGCCCAGCCCCAGCACGCAGCCCTG  750
 GluHisValAsnAlaIleGlnGluAlaArgArgLeuLeuAsnLeuSerArgAspThrAlaAlaGluMet
GGAGCATGTGAATGCCATCCAGGAGGCCCGGCGTCTCCTGAACCTGAGTAGAGACACTGCTGCTGAGATGgtaagtgagagaatgtgggcctgtgctaggcaccagtggccctgactggccacgc  875
                                AsnGluThrValGluValIleSerGluMetPheAspLeuGln
ctgtcagcttgataacatgacattttccttttctacagAATGAAACAGTAGAAGTCATCTCAGAAATGTTTGACCTCCAGgtaagatgcttctctctgacatagctttccagaagcccctgccct  1000
ggggtggaggtggggactccatttagatggcaccacacagggttgtccactttctctccagtcagctggctgcaggaggagggggtagcaactgggtgctcaagaggctgctggccgtgcccct  1125
atggcagtcacatgagctcctttatcagctgagcggccatgggcagacctagcattcaatggccaggagtcaccaggggacaggtggtaaagtgggggtcacttcatgagacaggagctgtgggt  1250
ttggggcgctcactgtgccccgagaccaagtcctgttgagacagtgctgactacagagaggcacagaggggtttcaggaacaacccttgcccacccagcaggtccaggtgaggccccacccccct  1375
ctccctgaatgatggggtgagagtcacctccttccctaaggctgggctcctctccaggtgccgctgagggtggcctgggcggggcagtgagaagggcaggttcgtgcctgccatggacagggcag  1500
ggtctatgactggacccagcctgtgcccctcccaagccctactcctgggggctgggggcagcagcaaaaaggagtggtggagagttcttgtaccactgtgggcacttggccactgctcaccgacg  1625
                  GluProThrCysLeuGlnThrArgLeuGluLeuTyrLysGlnGlyLeuArgGlySerLeuThrLysLeuLysGlyProLeuThrMetMetAlaSerHisTyrLysGln
aacgacattttccacagGAGCCGACCTGCCTACAGACCCGCCTGGAGCTGTACAAGCAGGGCCTGCGGGGCAGCCTCACCAAGCTCAAGGGCCCCTTGACCATGATGGCCAGCCACTACAAGCAG  1750
HisCysProProThrPro
CACTGCCCTCCAACCCCGgtgagtgcctacggcagggcctccagcaggaatgtcttaatctaggggtggggtcgacatggggagagatctatggctgtggctgttcaggaccccagggggtttc  1875
tgtgccaacagttatgtaatgattagccctccagagaggaggcagacagcccatttcatcccaaggagtcagagccacagagcgctgaagcccacagtgctccccagcaggagctgctcctatcc  2000
tggtcattattgtcattacggttaatgaggtcagaggtgagggcaaacccaaggaaacttggggcctgcccaaggcccagaggaagtgcccaggcccaagtgccaccttctggcaggactttcct  2125
ctggccccacatggggtgcttgaattgcagaggatcaaggaagggaggctacttggaatggacaaggacctcaggcactccttcctgcgggaagggagcaaagtttgtggccttgactccactcc  2250
ttctgggtgcccagagacgacctcagcccagctgccctgctctgccctgggaccaaaaaggcaggcgtttgactgcccagaaggccaacctcaggctggcacttaagtcaggcccttgactctgg  2375
ctgccactggcagagctatgcactccttggggaacacgtgggtggcagcagcgtcacctgacccaggtcagtgggtgtgtcctggagtggcctcctggcctctgagttctaagaggcagtagag  2500
                                                                                                    GluThrSerCysAlaThrGlnIleIleThrPheGluSerPheLysGlu
aaacatgctggtgcttccttcccccacgttacccacttgcctggactcaagtgtttttatttttctttttttaaagGAAACTTCCTGTGCAACCCAGATTATCACCTTTGAAAGTTTCAAAGAG  2625
AsnLeuLysAspPheLeuLeuValIleProPheAspCysTrpGluProValGlnGlu***
AACCTGAAGGACTTTCTGCTTGTCATCCCCTTTGACTGCTGGGAGCCAGTCCAGGAGTGAGACCGGCCAGATGAGGCTGGCCAAGCCGGGGAGCTGCTCTCTCATGAAACAAGAGCTAGAAACTC  2750
AGGATGGTCATCTTGGAGGGACCAAGGGGTGGGCCACAGCCATGGTGGGAGTGGCCTGGACCTGCCCTGGGCACACTGACCCTGATACAGGCATGGCAGAAGAATGGGAATATTTTATACTGACA  2875
GAAATCAGTAATATTTATATATTTATATTTTTAAAATATTTATTTATTTATTTATTTAAGTTCATATTCCATATTTATTCAAGATGTTTTACCGTAATAATTATTATTAAAAATATGCTTCTACt  3000
tgtccagtgttctagtttgtttttaaccatgagcaaatgccat  3043
```

Figure 4 Nucleotide sequence of the human GM–CSF gene (173). Flanking and intervening sequences are in lower-case letters. Various sequence motifs (see text) are highlighted.

Table 1 A Family of Nuclear Protein Binding Sites in the Promoter Region of Various Hemopoietic Regulator Genes

Gene	Ref.	Sequence	Position
Murine GM–CSF	173,180	GAGATTCCAC–TCAGGTA	-78
Human GM–CSF	180	GAGATTCCAC–TCAGGTA	-73
Murine multi-CSF	182,183	GAGGTTCCAT–TCAGATA	-84
Rat multi-CSF	184	GAGGTTCCAT–TCAGATA	-79
Human multi-CSF	185	GAGGTTCCAT–TCAGATA	-97
Murine IL-2	186	GGGATTTCAC	-175
Human IL-2	187	GGGATTTCAC	-172
Human IFN-γ (R)	188	GAGTTTCCTT	-137
Murine G–CSF	189	GAGATTCCCC	-130
Human G–CSF	190	GAGATTCCAC	-125
Human IL-5	191	AAGATTCCTT	-183
Human IL-5	192	AAGATTCTTC	-183
Human IL-6	193	AGGTTTCCAA	-87
Mouse IL-4 (R)	194	GAAATTACAC	-129

Position corresponds to the distance between the 5′ nucleotide of the region of homology and the 5′ nucleotide of the TATA box element of the promoter. R indicates that the decanucleotide occurs in reverse orientation.

tween different lymphokine genes, or homology between the human and murine genes, or by comparison with other functionally characterized sequence elements. More recently a number of these elements have also been more directly implicated in controlling the expression of the gene.

B. Nuclear Factor Binding Sites

A family of closely related decanucleotides located 100-300 nucleotides upstream of the transcriptional initiation sites of various lymphokine genes, including GM–CSF, was noted several years ago (123,173,180) (Table 1). It was estimated (123) that the probability of chance occurrence of this set of sequences at this location was extremely low (3×10^{-6}) and thus this represented a candidate sequence to be involved in the coordinate regulation of these genes in activated T cells. Homologous decanucleotides were subsequently found in the 5′ flanking region of other lymphokine genes (IL-4, IL-5, and IL-6) and also the G–CSF gene, thus extending its putative role to other cell types (e.g., macrophages) (Table 1). In addition to this highly conserved decanucleotide common to many lymphokine genes, the GM–CSF and multi-CSF genes share an additional heptameric sequence immediately 3′ of the decamer (173,195) (Table 1). These two sequence motifs have been dubbed CK1 and CK2, respectively (195).

Recently this region has been more directly implicated in the control of GM–CSF gene expression with the demonstration by gel retardation experiments that the region spanning the decamer and the heptamer interacts with two specific nuclear proteins (195). One, NFa, appears to bind to the decamer sequence, and the second NFb, possibly binds to the heptamer (196). Interestingly, although NFa has a reasonably broad cellular distribution, NFb is much more restricted, and very importantly is inducible in the bladder carcinoma 5637 with PMA, an agent that stimulates transcription of the GM–CSF gene in these cells. It will thus be extremely interesting to see if this factor is also induced in other cell types in which GM–CSF expression is inducible, e.g., T cells, endothelial cells and macrophages (see Section VI).

DNAase I footprinting studies (197) have also implicated a third, adjacent region as binding a specific nuclear protein. Extracts from unstimulated MLA144 and S-LB-1 cells, protect a region of 34 bp (573-591 in Fig. 4) that is located immediately upstream of the transcriptional initiation site of the human GM–CSF gene and contains a few nucleotides of the TATA box. Interestingly, when extracts from PHA plus PMA-stimulated MLA144 cells were used, the protected (protein-bound) region contracted to 19 bp, implying a change in the nature, conformation, or context of the protein or protein-complex binding to this region on stimulation (197). The presence or otherwise of this nuclear factor has not been assessed in cells that do not express GM–CSF, and it will be of interest to see its cellular distribution and whether the change in the footprint evident on stimulation of MLA144 cells is also characteristic of other inducible cell types.

Experiments in which the 5′ flanking region plus promoter of the human GM–CSF gene are linked to a chloramphenicol acetyltransferase "reporter" gene and the constructs introduced into MLA144 cells indicate that sequences located between –53 and +37 of the transcriptional initiation site are sufficient to confer PHA/PMA-inducible expression on the CAT gene (197). This region contains most of the "footprint" region identified using MLA144 nuclear extracts, but does not include the CK1 and CK2 sequences. Thus, of the three nuclear factor binding sites so far identified–CK1, CK2, and the "footprint" region encompassing the TATA box–the latter would appear to be the more likely to be directly involved in controlling the inducible expression of GM–CSF in T cells.

C. A/U-Rich Sequence in the 3′ Untranslated Region

The most striking structural motif in the GM–CSF mRNA is the extensive U_nA region in the 3′ untranslated region (103-106). A sequence of 51 nucleotides close to the 3′ end of the GM–CSF mRNA comprising entire A and U residues (predominantly UUUA) is in fact the portion of the mRNA sequence most conserved between mouse and humans (198). A similar situation was noted for human and mouse TNF-α mRNA, in which the most highly conserved region is an A/U rich segment toward the 3′ end of the mRNA (199). Indeed, a large number of transiently expressed mRNAs appear to contain variants of this sequence motif (198-200).

It is likely that this sequence plays a role in modulating mRNA stability: Shaw and Kamen (198) demonstrated that inclusion of the 51 nucleotide A/U rich segment from the human GM–CSF mRNA on the otherwise stable β-globin mRNA markedly reduced its stability in fibroblasts. Moreover, Wreschner and Rechavi (200) have demonstrated that in in vitro reticulocyte lysate systems, relative mRNA stability inversely correlates with the extent of U_nA sequences within the 3′ untranslated region.

D. Other Sequence Motifs

In addition to these regions and elements, for which there is direct molecular evidence of a role in gene control, a number of motifs have been identified purely on the basis of sequence homologies. Approximately 1 kbp 5′ of the promoter in the murine gene is a sequence composed of 14 contiguous GT dinucleotides (173; position 112-139 in Fig. 3). Such a sequence is capable of adopting a Z-DNA conformation characteristic of many enhancers and that Rich et al. (201) have suggested could have either a positive or negative influence on expression from nearby promoters. A similar sequence is present immediately 5′ of the human M–CSF gene promoter (202). A candidate SP1 transcription factor binding site (203-205) (5′GGGCGG) is located, in this instance in reverse orientation, immediately 3′ of the CK1 and CK2 motifs in the mouse gene sequence (Fig. 3). Note, however, that this motif in the human (Fig. 4) does not conform well to the consensus sequence (203,204) (5′AGGCGG compared with the canonical 5′GGGCGG) raising some doubt as to whether it does actually represent an SP1 binding site. There are a number of direct repeats within the second intron of the mouse gene (173), although the biological significance of these is unclear because they are not represented in the human gene.

IX. AUTOCRINE GM–CSF PRODUCTION AND LEUKEMOGENESIS

Myeloid leukemia is a monoclonal proliferation of hemopoietic cells, probably arising through a succession of genetic alterations. One such change appears to be autocrine growth factor production. The autocrine model envisages that an affected cell produces an inappropriate level of a growth factor to which it can respond, thus leading to sustained self-stimulated proliferation (206). Because proliferation and differentiation are tightly coupled in normal hemopoietic cells (e.g., 8), autonomous production of CSFs by otherwise normal cells should not in itself lead to the emergence of a leukemic clone. Indeed, autocrine production of GM–CSF by macrophages in either transgenic mice expressing GM–CSF (81) or in irradiated mice repopulated with hemopoietic stem cells infected with a retroviral construct expressing GM–CSF (72) does not appear to represent a preleukemic state. However, leukemogenicity could ensue if autocrine stimulation were a secondary event occurring in a preleukemic clone that had already incurred a lesion causing a differentiation block. Continuous hemopoietic cell lines can be derived from various long-term murine bone marrow cultures, which are absolutely dependent for growth on the continuous presence of one or other of the CSFs (generally GM–CSF and/or multi-CSF). Such cell lines have a block to differentiation–CSF exposure stimulates proliferation but not differentiation–and are generally nonleukemogenic. However, in accordance with the autocrine model, they can be made leukemogenic by rendering them independent of exogenous CSF for growth (207,208). At least two distinct mechanisms can be envisaged by which this could occur: by direct autonomous factor production and by lesions that circumvent the requirement for CSF by precipitating those intracellular events normally induced by the CSF-receptor complex; examples of each type of mechanism have been documented.

Direct autonomous growth factor production has been demonstrated to contribute to leukemogenicity in a number of different experiments. In a direct experimental test of the autocrine hypothesis, Lang et al. (208) constructed a retroviral vector expressing both GM–CSF and a gene conferring resistance to the drug G418. This construct was used to infect FDCP1 cells, which are dependent for proliferation on GM–CSF or multi-CSF and which are nonleukemogenic. One hundred percent of drug-resistant infectants produced significant quantities of GM–CSF and were independent of exogenous growth factor for proliferation. When injected into syngeneic recipients they were tumourigenic.

More recently, in an attempt to directly select for factor independence and to determine the frequency with which the acquisition of autonomous growth was due to CSF production, Stocking et al. (209) isolated a set of spontaneously arising factor-independent variants of another factor-dependent murine myeloid cell line, D35. Of the 11 mutants examined, 10 had achieved factor independence by autonomous growth factor production and only one had achieved factor-independence by a nonautocrine mechanism. Of the 10 producing factors, 6 were producing GM–CSF, 3 multi-CSF (the two factors known to stimulate D35 proliferation), and 1 appeared to be producing a hitherto unidentified factor. The lesions in the GM–CSF genes in four of the six GM–CSF producers were examined in some detail. In all cases one allele of the GM–CSF gene was altered, either by insertion of a partial retroviral genome derived from the viral complex used in the original derivation of D35 (Friend SFFV and Rauscher MuLV) or by insertion of an endogenous intracisternal A particle genome. In three cases the retroviral inserts were located immediately 5′ of the GM–CSF gene, within a region implicated in controlling the expression of the gene (see Section VIII). The retroviral genomes are all inserted in the opposite transcriptional orientation to that of the GM–CSF gene, and the deregulated expression of the GM–CSF gene evident in these cells is presumably a result of insertion of enhancer elements within the retroviral genomes close to the promoter of the GM–CSF gene. In one case, the IAP genome was inserted in the 3′ untranslated region of the GM–CSF gene, giving rise to a hybrid mRNA terminating within the IAP genome. This insertion removes the U_nA sequence normally found in the 3′ untranslated region of the GM–CSF mRNA (see Section VIII.C) and thus in this instance, increased mRNA stability is implicated in conjunction with enhancer insertion as an additional means to aberrant GM–CSF expression.

It would seem, therefore, that the GM–CSF gene should be viewed as a second oncogene required to transform myeloid cells from an immortalized to a leukemic state. It could be argued, however, that the immortalized parental cells are already fully transformed, but fail to grow as leukemias on transplantation simply because there is insufficient CSF at the site of injection. Either way, however, it must be said that acquisition of growth-factor independence allows the cell to escape the normal growth control mechanisms in the body and establish a frank leukemia.

The experiments of Dührsen (210,211) are particularly interesting in this regard: tumorigenic variants of the nontumorigenic (growth-factor-dependent) line FDCP1 were isolated by selection in vivo for tumor formation in irradiated recipient mice. Although tumor-derived cell lines frequently were growth-factor-independent (and were in many instances producing GM–CSF), not all leukemic derivatives were factor-independent in vitro. This situation is remarkably similar to human AML, in which only a proportion of leukemias display evidence of autonomous growth factor production (34,35) and in which apparently unlimited proliferation in vivo often contrasts with extremely poor growth in vitro (212).

ACKNOWLEDGMENTS

Work in the authors' laboratories reviewed here was supported by the National Health and Medical Research Council, Canberra, The Anti-Cancer Council of Victoria, The J. D. and L. Harris Trust Fund (Victoria), and the National Institutes of Health, Bethesda, Maryland, grant CA-22556.

REFERENCES

1. Metcalf, D. *The Hemopoietic Colony Stimulating Factors*. Elsevier, Amsterdam (1984).
2. Metcalf, D., and S. Merchav. Effects of GM–CSF deprivation on precursors of granulocytes and macrophages. *J. Cell. Physiol. 112*:411-418 (1982).
3. Metcalf, D. Multi-CSF-dependent colony formation by cells of a murine hemopoietic cell line: specificity of action of Multi-CSF. *Blood 65*:357-362 (1985).
4. Burgess, A. W., N. A. Nicola, G. R. Johnson, and E. C. Nice. Colony-forming cell proliferation: a rapid and sensitive assay system for murine granulocyte and macrophage colony-stimulating factors. *Blood 60*:1219-1223 (1982).
5. Moore, M. A. S., and N. Williams. Functional, morphological and kinetic analysis of the granulocyte-macrophage progenitor cell. In *Hemopoiesis in Culture*, W. A. Robinson (Ed.). DHEW Publication No. 74-205, Washington, D. C., pp. 17-27 (1973).
6. Metcalf, D. Clonal analysis of proliferation and differentiation of paired daughter cells: action of granulocyte-macrophage colony-stimulating factor on granulocyte-macrophage precursors. *Proc. Natl. Acad. Sci. USA 74*: 5327-5330 (1980).
7. Metcalf, D., and H. R. MacDonald. Heterogeneity of in vitro colony- and cluster-forming cells in the mouse marrow. Segregation by velocity sedimentation. *J. Cell. Physiol., 85*:643-654 (1975).
8. Nicola, N. A., and D. Metcalf. The colony-stimulating factors and myeloid leukemia. *Cancer Surveys 4*:789-815 (1985).
9. Metcalf, D., A. W. Burgess, G. R. Johnson, N. A. Nicola, E. C. Nice, J. DeLamarter, D. R. Thatcher, and J-J. Mermod. In vitro actions on hemopoietic cells of recombinant murine GM–CSF purified after production in *E. coli*: comparison with purified native GM–CSF. *J. Cell. Physiol. 128*:421-431 (1986).
10. Robinson, B. E., H. E. McGrath, and P. J. Quesenberry. Recombinant murine granulocyte macrophage colony-stimulating factor has megakaryocyte colony-stimulation activity and augments megakaryocyte colony stimulation by interleukin 3. *J. Clin. Invest. 79*:1648-1652 (1987).
11. Metcalf, D., C. G. Begley, G. R. Johnson, N. A. Nicola, M. A. Vadas, A. F. Lopez, D. J. Williamson, G. G. Wong, S. C. Clark, and E. A. Wang. Biological properties in vitro of recombinant human granulocyte-macrophage colony-stimulating factor. *Blood 67*:37-45 (1986).
12. Mazur, E. M., J. L. Cohne, G. G. Wong, and S. C. Clark. Modest stimulating effect of recombinant human GM–CSF on colony growth from peripheral blood human megakaryocyte progenitor cells. *Exp. Hematol. 15*:1128-1144 (1987).
13. Migliaccio, A. R., M. Bruno, and G. Migliaccio. Evidence for direct action of human biosynthetic (recombinant) GM–CSF on erythroid progenitors in serum-free culture. *Blood 70*:1867-1871 (1987).

14. Metcalf, D., G. R. Johnson, and A. W. Burgess. Direct stimulation by purified GM–CSF of the proliferation of multipotential and erythroid precursor cells. *Blood 55*:138-147 (1980).
15. Koike, K., M. Ogawa, J. N. Ihle, T. Miyake, T. Shimizu, A. Miyajima, T. Yokota, and A. Arai. Recombinant murine granulocyte-macrophage (GM) colony-stimulating factor supports formation of GM and multipotential blast cell colonies in culture: comparison with the effects of interleukin-3. *J. Cell. Physiol. 131*:458-464 (1987).
16. Sieff, C., S. Emerson, R. Donahue, and D. Nathan. Human recombinant granulocyte-macrophage colony-stimulating factor: a multilineage hemopoietin. *Science 230*:1171-1174 (1985).
17. Messner, H. A., K. Yamasaki, N. Jamal, M M. Minden, Y-C. Yang, G. G. Wong, and S. C. Clark. Growth of human hemopoietic colonies in response to recombinant gibbon interleukin 3: comparison with human recombinant granulocyte and granulocyte-macrophage colony-stimulating factor. *Proc. Natl. Acad. Sci. USA 84*:6765-6769 (1987).
18. Caracciolo, D., N. Shirshat, G. G. Wong, B. Lange, S. Clark, and G. Rovera. Recombinant human macrophage colony-stimulating factor (M–CSF) requires subliminal concentrations of granulocyte/macrophage (GM)–CSF for optimal stimulation of human macrophage colony formation in vitro. *J. Exp. Med. 166*:1851-1860 (1987).
19. Chen, B.D-M., C. R. Clark, and T-h. Chou. Granulocyte/macrophage colony-stimulating factor stimulates monocyte and tissue macrophage proliferation and enhances their responsiveness to macrophage colony-stimulating factor. *Blood 71*:997-1002 (1988).
20. Dexter, T. M., J. Garland, D. Scott, E. Scolnick, D. Metcalf. Growth of factor-dependent hemopoietic precursor cell lines. *J. Exp. Med. 152*:1036-1047 (1980).
21. Ihle, J. N. Immunological regulation of hematopoietic stem cell function by interleukin 3 and its role in leukemogenesis. *Lymphokines 15*:127-161 (1988).
22. Morgan, C., J. W. Pollard, and E. R. Stanley. Isolation and characterisation of a cloned growth factor dependent macrophage cell line, BACl.2FS. *J. Cell. Physiol. 130*:420-427 (1987).
23. Branch, D. R., J. M. Turc, and L. J. Guilbert. Identification of an erythropoietin-sensitive cell line. *Blood 69*:1782-1785 (1987).
24. Valtieri, M., D. Santoli, D. Caracciolo, B. L. Kreider, S. W. Altmann, D. J. Tweardy, I. Gemperlein, F. Mavilio, B. Lange, G. Rovera. Establishment and characterization of an undifferentiated human T leukemia cell line which requires granulocyte-macrophage colony stimulating factor for growth. *J. Immunol. 138*:4042-4050 (1987).
25. Woods, A., J. West, R. Rasmussen, and K. Bottomly. Granulocyte-macrophage colony stimulating factor produced by cloned L3T4a$^+$, class II-restricted T cells induces HT-2 cells to proliferate. *J. Immunol. 138*:4293-4297 (1987).

26. Kupper, T., P. Flood, D. Coleman, and M. Horowitz. Growth of an interleukin 2/interleukin 4-dependent T cell line induced by granulocyte-macrophage colony-stimulating factor (GM–CSF). *J. Immunol. 138*:4288-4292 (1987).
27. Park, L. S., D. Friend, S. Gillis, and D. Urdal. Characterization of the cell surface receptor for granulocyte-macrophage colony-stimulating factor. *J. Biol. Chem. 261*:4177-4183 (1986).
28. Begley, C. G., D. Metcalf, and N. A. Nicola. Primary human myeloid leukemia cells: comparative responsiveness to proliferative stimulation by GM–CSF to G–CSF and membrane expression of CSF receptors. *Leukemia 1*:1-8 (1987).
29. Asano, Y., T. Shibuya, S. Okamura, S. Yamaga, T. Otsuka, and Y. Niho. Effect of human recombinant granulocyte/macrophage colony-stimulating factor and native granulocyte colony-stimulating factor on clonogenic leukemic blast cells. *Cancer Res. 47*:5647-5648 (1987).
30. Mitjavila, M. T., J-L. Villeval, P. Cramer, A. Henri, J. Gasson, G. Krystal, M. Tulliez, R. Berger, J. Breton-Gorius, and W. Vainchenker. Effects of granulocyte-macrophage colony-stimulating factor and erythropoietin on leukemic erythroid colony formation in human early erythroblastic leukemias. *Blood 70*:965-973 (1987).
31. Vellenga, E., D. C. Young, K. Wagner, D. Wiper, D. Ostapovicz, and J. D. Griffin. The effects of GM–CSF and G–CSF in promoting growth of clonogenic cells in acute myeloblastic leukemia. *Blood 69*:1771-1776 (1987).
32. Kelleher, C., J. Miyauchi, G. Wong, S. Clark, M. D. Minden, and E. A. McCulloch. Synergism between recombinant growth factors, GM–CSF and G–CSF, acting on the blast cells of acute myeloblastic leukemia. *Blood 69*:1498-1503 (1987).
33. Young, D. C., and J. D. Griffin. Autocrine secretion of GM–CSF in acute myeloblastic leukemia. *Blood 68*:1178-1181 (1986).
34. Cheng, G. Y., C. A. Kelleher, J. Miyauchi, C. Wang, G. Wong, S. C. Clark, E. A. McCulloch, and M. D. Minden. Structure and expression of genes of GM–CSF and G–CSF in blast cells from patients with acute myeloblastic leukemia. *Blood 71*:204-208 (1988).
35. Young, D. C., K. Wagner, and J. D. Griffin. Constitutive expression of the granulocyte-macrophage colony-stimulating factor gene in acute myeloblastic leukemia. *J. Clin. Invest. 79*:100-106 (1987).
36. Hamilton, J. A., G. Vairo, N. A. Nicola, A. W. Burgess, D. Metcalf, and S. Lingelbach. Activation and proliferation signals in murine macrophages. Synergistic interaction between the hemopoietic growth factors and with phorbol ester for DNA synthesis. *Blood 71*:1574-1580 (1988).
37. Metcalf, D., and A. W. Burgess, Clonal analysis of progenitor cell commitment to granulocyte or macrophage production. *J. Cell. Physiol. 111*:275-283 (1982).
38. Metcalf, D. Clonal analysis of the action of GM–CSF on the proliferation and differentiation of myelomonocytic leukemic cells. *Int. J. Cancer 24*: 616-623 (1979).

39. Lotem, J., Y. Shabo, and L. Sachs. Role of different normal hematopoietic regulatory proteins in the differentiation of myeloid leukemic cells. *Int. J. Cancer 41*:101-107 (1988).
40. Tomonaga, M., D. W. Golde, and J. C. Gasson. Biosynthetic (recombinant) human granulocyte-macrophage colony-stimulating factor: effect on normal bone marrow and leukemia cell lines. *Blood 67*:31-36 (1986).
41. Begley, C. G., D. Metcalf, and N. A. Nicola. Purified colony stimulating factors (G–CSF and GM–CSF) induce differentiation in human HL60 leukemic cells with suppression of clonogenicity. *Int. J. Cancer 39*:99-105 (1987).
42. Brennan, J. K., K. S. Lee, C. N. Abboud, C. L. Erickson-Miller, and P. C. Keng. Interactions of dimethyl sulfoxide and granulocyte-macrophage colony-stimulating factors on the growth and maturation of HL-60 cells. *J. Cell. Physiol. 132*:246-254 (1987).
43. Zuckerman, S. H., Y. M. Surprenant, and J. Tang. Synergistic effect of granulocyte-macrophage colony-stimulating factor and 1,25-dihydroxy vitamin D3 on the differentiation of the human monocytic cell line U937. *Blood 71*:619-624 (1988).
44. Ruff, M. R., W. L. Farrar, and C. B. Pert. Interferon gamma and granulocyte/macrophage colony-stimulating factor inhibit growth and induce antigens characteristic of myeloid differentiation in small-cell lung cancer cell lines. *Proc. Natl. Acad. Sci. USA 83*:6613-6617 (1986).
45. Lorenzo, J. A., S. L. Sousa, J. M. Fonseca, J. M. Hock, and E. S. Medlock. Colony-stimulating factors regulate the development of multinucleated osteoclasts from recently replicated cells in vitro. *J. Clin. Invest. 80*:160-164 (1987).
46. Heufler, C., F. Koch, and G. Schuler. Granulocyte/macrophage colony-stimulating factor and interleukin 1 mediate the maturation of murine epidermal Langerhans cells into potent immunostimulatory dendritic cells. *J. Exp. Med. 167*:700-705 (1988).
47. Begley, C. G., A. F. Lopez, N. A. Nicola, D. J. Warren, M. A. Vadas, C. J. Sanderson, and D. Metcalf. Purified colony-stimulating factors enhance the survival of human neutrophils and eosinophils in vitro: a rapid and sensitive microassay for colony-stimulating factors. *Blood 68*:162-166 (1986).
48. Stanley, I. J., and A. W. Burgess. Granulocyte-macrophage colony stimulating factor stimulates the synthesis of membrane and nuclear proteins in murine neutrophils. *J. Cell. Biochem. 23*:241-258 (1983).
49. Lopez, A. F., N. A. Nicola, A. W. Burgess, D. Metcalf, F. L. Battye, W. A. Sewell, and M. A. Vadas. Activation of granulocyte cytotoxic function by purified mouse colony-stimulating factors. *J. Immunol. 131*:2983-2988 (1983).
50. Vadas, M. A., N. A. Nicola, and D. Metcalf. Activation of antibody-dependent cell-mediated cytotoxicity of human neutrophils and eosinophils by separate colony stimulating factors. *J. Immunol. 130*:795-799 (1983).
51. Fleischmann, J., D. W. Golde, R. H. Weisbart, and J. C. Gasson. Granulo-

Granulocyte-macrophage colony-stimulating factor enhances phagocytosis of bacteria by human neutrophils. *Blood 68*:708-711 (1986).
52. Williamson, D. J., and T. C. Brown. Enhancement of neutrophil-mediated phagocytosis by human granulocyte-macrophage colony-stimulating factor demonstrated using a novel mathematical model. *Immunol. Cell. Biol. 65*: 329-335 (1987).
53. Villalta, F., and F. Kierszenbaum. Effects of human colony-stimulating factor on the uptake and destruction of a pathogenic parasite (*Trypanosoma cruzi*) by human neutrophils. *J. Immunol. 137*:1703-1707 (1986).
54. Weisbart, R. M., D. W. Golde, S. C. Clark, C. G. Wong, and J. C. Gasson. Human granulocyte-macrophage colony-stimulating factor is a neutrophil activator. *Nature 314*:361-363 (1985).
55. Wang, J. M., S. Colella, P. Allavena, and A. Mantovani. Chemotactic activity of human recombinant granulocyte-macrophage colony-stimulating factor. *Immunology 60*:439-444 (1987).
56. Burgess, A. W., C. G. Begley, G. R. Johnson, A. F. Lopez, D. J. Williamson, J-J. Mermod, R. J. Simpson, A. Schmitz, and J. F. DeLamarter. Purification and properties of bacterially synthesized human granulocyte-macrophage colony stimulating factor. *Blood 69*:43-51 (1987).
57. Weisbart, R. H., L. Kwan, D. W. Golde, and J. C. Gasson. Human GM–CSF primes neutrophils for enhanced oxidative metabolism in response to the major physiological chemoattractants. *Blood 69*:18-21 (1987).
58. Lindemann, A., D. Riedel, W. Oster, S. C. Meuer, D. Blohn, R. M. Mertelsmann, and F. Hermann. Granulocyte/macrophage colony-stimulating factor induces interleukin 1 production by human polymorphonuclear neutrophils. *J. Immunol. 140*:837-839 (1988).
59. Arnaout, M. A., E. A. Wang, S. C. Clark, and C. A. Sieff. Human recombinant granulocyte-macrophage colony-stimulating factor increases cell-to-cell adhesion and surface expression of adhesion-promoting surface glycoproteins on mature granuloctyes. *J. Clin. Invest. 78*:597-601 (1986).
60. Cannistra, S. A., A. Ramboldi, D. R. Spriggs, F. Herrmann, D. Kufe, and J. D. Griffin. Human granulocyte-macrophage colony-stimulating factor induces expression of the tumor necrosis factor gene by the U937 cell line and by normal human monocytes. *J. Clin. Invest. 79*:1720-1728 (1987).
61. Horiguchi, J., M. K. Warren, and D. Kufe. Expression of the macrophage-specific colony-stimulating factor in human monocytes treated with granulocyte-macrophage colony-stimulating factor. *Blood 69*:1259-1261 (1987).
62. Grabstein, K. H., D. L. Urdal, R. J. Tushinski, D. Y. Mochizuki, V. L. Price, M. A. Cantrell, S. Gillis, and P. J. Conlon. Induction of macrophage tumoricidal activity by granulocyte-macrophage colony-stimulating factor. *Science 232*:506-508 (1986).
63. Handman, E., and A. W. Burgess. Stimulation by granulocyte-macrophage colony stimulating factor of *Leishmania tropica* killing by macrophages. *J. Immunol. 122*:1134-1137 (1979).
64. Reed, S. G., C. F. Nathan, O. L. Pihl, P. Rodricks, K. Shanebeck, P. J. Con-

lon, and K. H. Grabstein. Recombinant granulocyte/macrophage colony-stimulating factor activates macrophages to inhibit *Trypanosoma cruzi* and release hydrogen peroxide—comparison with interferon gamma. *J. Exp. Med. 166*:1734-1746 (1987).

65. Weiser, W. Y., A. Van Niel, S. C. Clark, J. R. David, and H. G. Remold. Recombinant human granulocyte/macrophage colony-stimulating factor activates intracellular killing of *Leishmania donovani* by human monocyte-derived macrophages. *J. Exp. Med. 166*:1436-1446 (1987).
66. Hammer, S. M., J. M. Gillis, J. E. Groopman, and R. M. Rose. In vitro modification of human immunodeficiency virus infection by granulocyte-macrophage colony-stimulating factor and gamma interferon. *Proc. Natl. Acad. Sci. USA 83*:8734-8738 (1986).
67. Silberstein, D. S., W. F. Owen, J. C. Gasson, J. F. Di Persio, D. W. Golde, J. C. Bina, R. Soberman, K. F. Austen, and J. R. David. Enhancement of human eosinophil cytotoxicity and leukotriene synthesis by biosynthetic (recombinant) granulocyte-macrophage colony-stimulating factor. *J. Immunol. 137*:3290-3294 (1986).
68. Vadas, M. A., G. Varigos, N. Nicola, S. Pincus, A. Dessein, D. Metcalf, and F. L. Battye. Eosinophil activation by colony-stimulating factor in man: metabolic effects and analysis by flow cytometry. *Blood 61*:1232-1241 (1983).
69. Vadas, M. A., N. A. Nicola, and D. Metcalf. Activation of antibody-dependent cell-mediated cytotoxicity of human neutrophils and eosinophils by separate colony-stimulating factors. *J. Immunol. 130*:795-799 (1983).
70. Dessein, A. J., M. A.Vadas, N. A. Nicola, D. Metcalf, and J. R. David. Enhancement of human blood eosinophil cytotoxicity by semi-purified eosinophil colony-stimulating factors. *J. Exp. Med. 156*:90-103 (1982).
71. Metcalf, D., C. G. Begley, D. J. Williamson, E. C. Nice, J. DeLamarter, J-J. Mermod, D. Thatcher, and A. Schmidt. Hemopoietic responses in mice injected with purified recombinant murine GM—CSF. *Exp. Hematol. 15*: 1-9 (1987).
72. Morrisey, P. J., L. Bressler, K. Charrier, and A. Alpert. Response of resident murine peritoneal macrophages to in vivo administration of granulocyte-macrophage colony-stimulating factor. *J. Immunol. 140*:1910-1915 (1988).
73. Donahue, R. E., E. A. Wang, D. K. Stone, R. Kamen, G. G. Wong, P. K. Sehgal, D. G. Nathan, and S. C. Clark. Stimulation of haematopoiesis in primates by continuous infusion of recombinant human GM—CSF. *Nature 321*:827-875 (1986).
74. Mayers, P., C. Lam, H. Obenbhaus, E. Lichl, and J. Bessemer. Recombinant human GM—CSF induces leukocytosis and activates peripheral blood polymorphonuclear neutrophils in non-human primates. *Blood 70*:206-213 (1987).
75. Broxmeyer, H. E., D. E. Williams, S. Cooper, R. K. Shadduck, S. Gillis, A. Waheed, D. L. Urdal, and D. C. Bicknell. Comparative effects in vivo of recombinant murine interleukin 3, natural murine colony-stimulating factor-

1, and recombinant murine granulocyte-macrophage colony-stimulating factor on myelopoiesis in mice. *J. Clin. Invest.* *79*:721-730 (1987).
76. Broxmeyer, H. E., D. E. Williams, G. Hangoc, S. Cooper, S. Gillis, R. K. Shadduck, and D. C. Bicknell. Synergistic meylopoietic actions in vivo after administration to mice of combinations of purified natural murine colony-stimulating factor-1, recombinant murine interleukin 3, and recombinant murine granulocyte/macrophage colony-stimulating factor. *Proc. Natl. Acad. Sci. USA* *84*:3871-3875 (1987).
77. Neta, R., J. J. Oppenheim, and S. D. Douches. Interdependence of the radioprotective effects of human recombinant interleukin alpha, tumor necrosis factor alpha, granulocyte colony-stimulating factor, and murine recombinant granulocyte-macrophage colony-stimulating factor. *J. Immunol.* *140*:108-111 (1988).
78. Blazar, B. R., M. B. Widmer, C. C. Soderling, D. L. Urdal, S. Gillis, L. L. Robison, D. A. Vallera. Augmentation of donor bone marrow engraftment in histoincompatible murine recipients by granulocyte/macrophage colony-stimulating factor. *Blood* *71*:320-328 (1988).
79. Monroy, R. L., R. R. Skelly, T. J. MacVittie, T. A. Davis, J. J. Sauber, S. C. Clark, and R. E. Donahue. The effect of recombinant GM–CSF on the recovery of monkeys transplanted with autologous bone marrow. *Blood* *70*:1696-1699 (1987).
80. Neinhuis, A W., R. E. Donahue, S. Karlsson, S. C. Clark, B. Agricola, N. Antinoff, J. E. Pierce, P. Turner, W. F. Anderson, and D. G. Nathan. Recombinant human granulocyte-macrophage colony-stimulating factor (GM–CSF) shortens the period of neutropenia after autologous bone marrow transplantation in a primate model. *J. Clin. Invest.* *80*:573-577 (1987).
81. Lang, R. A., D. Metclaf, R. A. Cuthbertson, I. Lyons, E. Stanley, A. Kelso, G. Kannourakis, D. J. Williamson, G. K. Klintworth, T. J. Gonda, and A. R. Dunn. Transgenic mice expressing a hemopoietic growth factor gene (GM–CSF) develop accumulations of macrophages, blindness, and a fatal syndrome of tissue damage. *Cell* *51*:675-686 (1987).
82. Johnson, G R., T. J. Gonda, D. Metcalf, I. K. Hariharan, and S. Cory. A lethal myeloproliferative syndrome in mice transplanted with bone marrow cells infected with a retrovirus expressing granulocyte-macrophage colony-stimulating factor. *EMBO J.* *8*:441-448 (1989).
83. Groopman, J. E., R. T. Mitsuyasu, M. J. DeLeo, D. H. Oette, and D. W. Golde. Effect of recombinant human granulocyte-macrophage colony-stimulating factor on myelopoiesis in the acquired immunodeficiency syndrome. *N. Engl. J. Med.* *317*:593-598 (1987).
84. Vadhan-Raj, S., M. Keating, A. Le Maistre, W. N. Hittelman, K. McCredie, J. M. Trujillo, H. E. Broxmeyer, C. Henney, and J. U. Gutterman. Effects of recombinant human granulocyte-macrophage colony-stimulating factor in patients with myelodysplastic syndromes. *N. Engl. J. Med.* *317*:1545-1552 (1987).

85. Vadhan-Raj, S., S. Buescher, A. Le Maistre, M. Keating, R. Walters, C. Ventura, W. Hittelman, H. E. Broxmeyer, and J. U. Gutterman. Stimulation of hematopoiesis in patients with bone marrow failure and in patients with malignancy by recombinant human granulocyte-macrophage colony-stimulating factor. *Blood 72*:134-141 (1988).
86. Lieschke, G. J., D. Maher, J. Cebon, M. O'Connor, M. Green, W. Sheridan, A. Boyd, M. Rallings, E. Bonnem, D. Metcalf, A. W. Burgess, K. McGrath, R. M. Fox, and G. Morstyn. Effects of subcutaneously administered bacterially-synthesized recombinant human granulocyte-macrophage colony-stimulating factor in patients with advanced malignancy. *Ann. Intern. Med. 110*:357-364 (1989).
87. Socinski, M. A., S. A. Cannistra, A. Elias, K. H. Antman, L. Schnipper, and J. D. Griffin. Granulocyte-macrophage colony stimulating factor expands the circulating haemopoietic progenitor cell compartment in man. *Lancet 1*:1194-1198 (1988).
88. Dührsen, U., J-L. Villeval, J. Boyd, G. Kannourakis, G. Morstyn, and D. Metcalf. Effects of recombinant human granulocyte-colony stimulating factor on hemopoietic progenitor cells in cancer patients. *Blood 72*:2074-2081 (1988).
89 Brandt, S. J., W. P. Peters, S. K. Atwater, J. Kurtzberg, M. J. Borowitz, R. B. Jones, E. J. Shpall, R. C. Bast, Jr., C. J. Gilbert, and D. H. Oette. Effect of recombinant human granulocyte-macrophage colony-stimulating factor in hematopoietic reconstitution after high-dose chemotherapy and autologous bone marrow transplantation. *N. Engl. J. Med. 318*:869-876 (1988).
90. Devereux, S., D. C. Linch, D. Campos Costa, M. F. Spittle, and A. M. Jelliffe. Transient leucopenia induced by granulocyte-macrophage colony-stimulating factor. *Lancet 2*:1523-1524 (1987).
91. Walker, F., and A. W. Burgess. Specific binding of radioiodinated granulocyte-macrophage colony stimulating factor to hemopoietic cells. *EMBO J. 4*:933-939 (1985).
92. Park, L. S., D. Friend, S. Gillis, and D. L. Urdal. Characterization of the cell surface receptor for granulocyte-macrophage colony-stimulating factor. *J. Biol. Chem. 261*:4177-4183 (1986).
93. Walker, F., N. A. Nicola, D. Metcalf, and A. W. Burgess. Hierarchical down-modulation of hemopoietic growth factor receptors. *Cell 43*:269-276 (1985).
94. Nicola, N. A. Why do hemopoietic growth factor receptors interact with each other? *Immunol. Today 8*:134-140 (1987).
95. Walker, F., and A. W. Burgess. Internalization and recycling of the granulocyte-macrophage colony-stimulating factor (GM–CSF) receptor on a murine myelomonocytic leukemia. *J. Cell. Physiol. 130*:255-261 (1987).
96. Nicola, N. A., L. Peterson, D. J. Hilton, and D. Metcalf. Cellular processing of murine colony-stimulating factor (Multi-CSF, GM–CSF, G–CSF) receptors by normal hemopoietic cells and cell lines. *Growth Factors 1*: 41-49 (1988).

97. Gasson, J. C., S. E. Kauffman, R. H. Weisbart, M. Tomonaga, and D. W. Golde. High affinity binding of granulocyte-macrophage colony-stimulating factor to normal and leukemic human myeloid cells. *Proc. Natl. Acad. Sci. USA 83*:669-673 (1986).
98. Park, L. S., D. Friend, S. Gillis, and D. L. Urdal. Characterization of the cell surface receptor for human granulocyte/macrophage colony-stimulating factor. *J. Exp. Med. 164*:251-262 (1986).
99. Di Persio, J., P. Billing, S. Kaufman, P. Eghtesady, R. E. Williams, and J. C. Gasson. Characterization of the human granulocyte-macrophage colony-stimulating factor receptor. *J. Biol. Chem. 263*:1834-1841 (1988).
100. Kelleher, C. A., G. G. Wong, S. C. Clark, P. F. Schendel, M. D. Minden, and E. A. McCulloch. Binding of iodinated recombinant human GM–CSF to the blast cells of acute myeloblastic leukemia. *Leukemia 2*:211-215 (1988).
101. Sparrow, L. G., D. Metcalf, M. W. Hunkapiller, L. E. Hood, and A. W. Burgess. Purification and partial amino acid sequence of asialo murine granulocyte-macrophage colony-stimulating factor. *Proc. Natl. Acad. Sci. USA 82*:292-296 (1985).
102. Burgess, A. W., J. Camakaris, and D. Metcalf. Purification and properties of colony stimulating factor from mouse lung-conditined medium. *J. Biol. Chem. 252*:1978-2003 (1977).
103. Gough, N. M., J. Gough, D. Metcalf, A. Kelso, D. Grail, N. A. Nicola, A. W. Burgess, and A. R. Dunn. Molecular cloning of cDNA encoding a murine haemotopoietic growth regulator, granulocyte-macrophage colony stimulating factor. *Nature 309*:763-767 (1984).
104. Gough, N. M., D. Metcalf, J. Gough, D. Grail, and A. R. Dunn. Structure and expression of the mRNA for murine granulocyte-macrophage colony stimulating factor. *EMBO J. 4*:645-653 (1985).
105. Wong, G. G., J. S. Witek, P. A. Temple, K. M. Wilkens, A. G. Leary, D. P. Luxenberg, S. S. Jones, E. L. Brown, R. M. Kay, E. C. Orr, C. Shoemaker, D. W. Golde, R. J. Kaufmann, R. M. Hewick, E. A. Wang, and S. C. Clark. Human GM–CSF: molecular cloning of the complementary DNA and purification of the natural and recombinant proteins. *Science 228*:810-815 (1985).
106. Lee, F., T. Yokota, T. Otsuka, L. Gemmell, N. Larson, J. Luh, K. Arai, and D. Rennick. Isolation of a cDNA for a humans granulocyte-macrophage colony stimulating factor by functional expression in mammalian cells. *Proc. Natl. Acad. Sci. USA 82*:4360-4364 (1985).
107. DeLamarter, J. F., J-J. Mermod, C-M. Liang, J. F. Eliason, and D. R. Thatcher. Recombinant murine GM–CSF from *E. coli* has biological activity and is neutralized by a specific antiserum. *EMBO J. 4*:2575-2581 (1985).
108. Cantrell, M. A., D. Anderson, D. P. Cerretti, V. Price, K. McKereghan, R. J. Tushinski, D. Y. Mochizuki, A. Larsen, K. Grabstein, S. Gillis, and D. Cosman. Cloning, sequence and expression of a human granulocyte/macrophage colony stimulating factor. *Proc. Natl. Acad. Sci. USA 82*:6250-6254 (1985).

109. Kaushansky, K., P. J. O'Hara, K. Berkner, G. M. Segal, F. S. Hagen, and J. W. Adamson. Genomic cloning, characterization, and multilineage growth-promoting activity of human granulocyte-macrophage colony-stimulating factor. *Proc. Natl. Acad. Sci. USA 83*:3101-3105 (1986).
110. Libby, R. T., G. Braedt, S. R. Kronheim, C. J. March, D. L. Urdal, T. A. Chiaverotti, R. J. Tushinski, D. Y. Mochizuki, T. P. Hopp, and D. Cosman. Expression and purification of native human granulocyte-macrophage colony-stimulating factor from an *Escherichia coli* secretion vector. *DNA 6*:221-229 (1987).
111. Price, V., D. Mochizuki, C. J. March, D. Cosman, M. C. Deeley, R. Klinke, W. Clevenger, S. Gillis, P. Baker, and D. Urdal. Expression, purification and characterization of recombinant murine granulocyte-macrophage colony-stimulating factor and bovine interluekin-2 from yeasts. *Gene 55*: 287-293 (1987).
112. Ernst, J. F. Efficient secretion and processing of heterologous proteins in *Saccharomyces cerevisiae* is mediated solely by α-factor precursor. *DNA 7*:355-360 (1988).
113. Dunn, A. R., D. Metcalf, E. Stanley, D. Grail, J. King, E. C. Nice, A. W. Burgess, and N. M. Gough. Biological characterization of regulators encoded by cloned hematopoietic growth-factor gene sequences. In *Cancer Cells*, Vol. 3, J. Feramisco, B. Ozanne, and C. Stiles (Eds.). Cold Spring Harbor, New York, pp. 227-234 (1985).
114. Seiff, C A., S G. Emerson, R. E. Donahue, D. G. Nathan, E. A. Wang, G. G. Wong, and S. C. Clark. Human recombinant granulocyte-macrophage colony stimulating factor: a multilineage hematopoietin. Science 230:1171-1173 (1985).
115. Moonen, P., J-J. Mermod, J. F. Ernst, M. Hirshi, and J. F. DeLamarter. Increased biological activity of deglycosylated recombinant human granulocyte/macrophage colony stimulating factor produced by yeast or animal cells. *Proc. Natl. Acad. Sci. USA 84*:4428-4431 (1987).
116. Kaushansky, K., P. J. O'Hara, C. E. Hart, J. W. Forstrom, and F. S. Hagen. Role of carbohydrate in the function of human granulocyte-macrophage colony-stimulating factor. *Biochemistry 26*:4861-4867 (1987).
117. Burgess, A. W., D. Metcalf, L. G. Sparrow, R. J. Simpson, and E. C. Nice. Granulocyte/macrophage colony-stimulating factor from mouse lung conditioned medium. *Biochem. J. 235*:805-814 (1986).
118. Wong, G. G., J. S. Witek, P. A. Temple, K. M. Wilkens, A. G. Leary, D. P. Luxenberg, S. S. Jones, E. L. Brown, R. M. Kay, E. C. Orr, C. Shoemaker, D. W. Golde, R. J. Kaufmann, R. M. Hewick, S. C. Clark, and E. A. Wang. Molecular cloning of human and gibbon T-cell-derived GM–CSF cDNAs and purification of the natural and recombinant human proteins. In *Cancer Cells*, Vol. 3, J. Feramisco, B. Ozanne, and C. Stiles (Eds.). Cold Spring Harbor, New York, pp. 235-242 (1985).
119. Gough, N. M., D. Grail, D. P. Gearing, and D. Metcalf. Mutagensis of murine granulocyte/macrophage-colony stimulating factor reveals critical residues near the N terminus. *Eur. J. Biochem. 169*:353-358 (1987).

120. Wingfield, P., P. Graber, S. Craig, and R. H. Pain. The conformation and stability of recombinant-derived granulocyte-macrophage colony stimulating factors. *Eur. J. Biochem. 173*:65-72 (1988).
121. Nicola, N. A., A. W. Burgess, and D. Metcalf. Similar molecular properties of granulocyte-macrophage colony stimulating factors produced by different mouse organs in vitro and in vivo. *J. Biol. Chem. 254*:5290-5299 (1979).
122. Shrimser, J. L., K. Rose, M. G. Simona, and P. Wingfield. Characterization of human and mouse granulocyte-macrophage colony-stimulating factors derived from *Escherichia coli*. *Biochem. J. 247*:195-199 (1987).
123. Kelso, A., and N. Gough. Expression of hemopoietic growth factor genes in murine T lymphocytes. In *The Lymphokines*, Vol. 3, D. R. Webb and D. V. Goeddel (Eds.). Academic Press, New York, pp. 209-238 (1987).
124. Imboden, J. B., A. Weiss, and J. D. Stobo. Transmembrane signalling by the T3-antigen receptor complex. *Immunol. Today 6*:328-331 (1985).
125. Hadden, J. W. Transmembrane signals in the activation of T lymphocytes by mitogenic antigens. *Immunol. Today 9*:235-239 (1988).
126. Taylor, M. V., J. C. Metcalfe, T. R. Hesketh, G. A. Smith, and J. P. Moore. Mitogens increase phosphorylation of phosphoinositides in thymocytes. *Nature 312*:462-465 (1984).
127. Weiss, A., J. Imboden, D. Shoback, and J. Stobo. Role of T3 surface molecules in human T-cell activation: T3-dependent activation results in an increase in cytoplasmic free calcium. *Proc. Natl. Acad. Sci. USA 81*: 4169-4173 (1984).
128. Imboden, J. B., and J. D Stobo. Transmembrane signalling by the T cell antigen receptor. *J. Exp. Med. 161*:446-456 (1985).
129. Kozumbo, W. J., D. T. Harris, S. Gromkowski, J-C. Cerottini, and P. A. Cerutti. Molecular mechanisms involved in T cell activation. II. The phosphatidylinositol signal-transducing mechanism mediates antigen-induced lymphokine production but not interleukin 2-induced proliferation in cloned cytotoxic T lymphocytes. *J. Immunol. 138*:606-612 (1987).
130. Britton, S., and R. Palacios. Cyclosporin A—usefulness, risks and mechanisms of action. *Immunol. Rev. 65*:5-22 (1982).
131. Dunn, D. E., K. C. Herold, G. R. Otten, D. W. Lancki, T. Gajewski, S. N. vogel, and F. W. Fitch. Interleukin 2 and concanavalin A stimulate interferon-γ production in a murine cytolytic T cell clone by different pathways. *J. Immunol. 139*:3942-3948 (1987).
132. Elliott, J. F., Y. Lin, S. B. Mizel, R. C. Bleackley, D. G. Harnish, and V. Paetkau. Induction of interleukin 2 messenger RNA is inhibited by cyclosporin A. *Science 226*:1439-1441 (1984).
133. Granelli-Piperno, A., K. Inaba, and R. M. Steinman. Stimulation of lymphokine release from T lymphoblasts. Requirement for mRNA synthesis and inhibition by cyclosporin A. *J. Exp. Med. 160*:1792-1802 (1984).

134. Harris, D. T., W. J. Kozumbo, P. A. Cerutti, and J. C. Cerottini. Mechanism of cyclosporin A-induced immunosuppression. Cyclosporin inhibits receptor-mediated and non-receptor-mediated lymphokine production as well as interleukin-2-induced proliferatin in cloned T lymphocytes. *Cell. Immunol. 109*:104-114 (1987).
135. Herold, K. C., D. W. Lancki, R. L. Moldwin, and F. W. Fitch. Immunosuppressive effects of cyclosporin A on cloned T cells. *J. Immunol. 136*: 1315-1321 (1986).
136. Krönke, M., W. J. Leonard, J. M. Depper, S. K. Arya, F. Wong-Staal, R. C. Gallo, T. A. Waldmann, and W. C. Greene. Cyclosporin A inhibits T-cell growth factor gene expression at the level of mRNA transcription. *Proc. Natl. Acad. Sci. USA 81*:5214-5218 (1984).
137. Kelso, A., and N. M. Gough. Differential inhibition by cyclosporin A reveals two pathways for activation of lymphokine synthesis in T cells. *Growth Factors 1*: (in press) (1989).
138. Orosz, C. G., D. C. Roopenian, M. B. Widmer, and F. H. Bach. Analysis of cloned T cell function. II. Differential blockade of various cloned T cell functions by cyclosporine. *Transplantation 36*:706-711 (1983).
139. Wiskocil, R., A. Weiss, J. Imboden, R. Kamin-Lewis, and J. Stobo. Activation of a human T cell line: a two-stimulus requirement in the pretranslational events involved in the co-ordinate expression of interleukin 2 and γ-interferon genes. *J. Immunol. 134*;1599-1603 (1985).
140. Metcalfe, S. Cyclosporine does not prevent cytoplasmic calcium changes associted with lymphocyte activation. *Transplantation 38*:161-164 (1984).
141. Quesniaux, V. F. J., M. H. Schreier, R. M. Wenger, P. C. Hiestand, M. W. Harding, and M. H. V. Van Regenmortel. Cyclophilin binds to the region of cyclosporine involved in its immunosuppressive activity. *Eur. J. Immunol. 17*:1359-1365 (1987).
142. Hermann, F., W. Oster, S. C. Meuer, A. Lindermann, and R. H. Mertelsmann. Interleukin 1 stimulates T lymphocytes to produce granulocyte-monocyte colony stimulating factor. *J. Clin. Invest. 81*:1415-1418 (1988).
143. Kelso, A., D. Metcalf, and N. M. Gough. Independent regulation of granulocyte-macrophage colony-stimulating factor and multi-lineage colony stimulating factor production in T lymphocyte clones. *J. Immunol. 136*:1718-1725 (1986).
144. Benedict, S. H., G. B. Mills, and E. W. Gelfand. Interleukin 2 activates a receptor-associated protein kinase. *J. Immunol. 139*:1694-1697 (1987).
145. Farrar, W. L., and M. Taguchi. Interleukin 2 stimulation of protein kinase C membrane association: evidence for IL-2 receptor phosphorylation. *Lymphokine Res. 4*:87-93 (1985).
146. Farrar, W. L., and W. B. Anderson. Interleukin-2 stimulates association of protein kinase C with plasma membrane. *Nature 315*:233-237 (1985).

147. Mills, G. B., R. K. Cheung, S. Grinstein, and E. W. Gelfand. Interleukin 2-induced lymphocyte proliferation is independent of increases in cytosolic-free calcium concentrations. *J. Immunol. 134*:2431-2435 (1985).
148. Mills, G. B., D. J. Stewart, A. Mellars, and E. W. Gelfand. Interleukin 2 does not induce phosphatidylinositol hydrolysis in activated T cells. *J. Immunol. 136*:3019-3024 (1986).
149. Merchav, S., A. Nayler, and I. Tatarsky. Human stem cell colony-stimulating activity (CFU–GEMM) in medium conditioned by leukemic B-lymphocytes. *Exp. Hematol. 15*:115-118 (1987).
150. Adams, D. O., and T. A. Hamilton. The cell biology of macrophage activation. *Annu. Rev. Immunol. 2*:283-318 (1984).
151. Cline, M. J., and M. A. S. Moore. Embryonic origin of the mouse macrophage. *Blood 39*:843-849 (1972).
152. Kelemen, E., and M. Janossa. Macrophages are the first differentiated blood cells formed in human embryonic liver. *Exp. Hematol. 8*:996-1000 (1980).
153. Grober, D. F., J. R. Zucali, and E. A. Mirand. Identification of erythropoietic producing cells in fetal mouse liver cultures. *Exp. Hematol. 5*: 392-398 (1977).
154. Paul, P., S. A. Rothmann, J. T. McMahon, and A. S. Gordon. Erythropoietin secretion by isolated rat Kupffer cells. *Exp. Hematol. 12*:825-830 (1986).
154a. Rich, I. N., W. Heit, and B. Kubanek. Extrarenal erythropoietin production by macrophages. *Blood 60*:1007-1017 (1982).
155. Hamilton, T. A., and D. O. Adams. Molecular mechanisms of signal transduction in macrophages. *Immunol. Today 8*:151-158 (1987).
156. Thorens, B., J.-J. Mermod, and P. Vassalli. Phagocytosis and inflammatory stimuli induce GM–CSF mRNA in macrophages through post-transcriptional regulation. *Cell 48*:671-679 (1987).
157. Rich, I. N. A role for the macrophage in normal hemopoiesis. I. Functional capacity of bone-marrow-derived macrophages to release hemopoietic growth factors. *Exp. Hematol. 14*:738-745 (1986).
158. Ascensao, J. L., G. M. Vercellotti, H. S. Jacobs, and E. D. Zanjani. Role of endothelial cells in human hematopoiesis: modulation of mixed colony growth in vitro. *Blood 63*:533-558 (1984).
159. Quesenberry, P. J., and M. A. Gimbrone. Vascular endothelium as a regulator of granulopoiesis. Production of colony-stimulating activity by cultured human endothelial cells. *Blood 56*:1060-1067 (1980).
159a. Bagby, G. C., E. McCall, K. A. Bergstrom, and D. Burger. A monokine regulates colony-stimulating activity production by vascular endothelial cells. *Blood 62*:663-668 (1983).
160. Bevilacqua, M. P., J. S. Prober, M. E. Wheeler, R. S. Cotran, and M. A. Gimbrone Jr. Interleukin 1 acts on cultured endothelial cells to increase the adhesion of polymorphonuclear leukocytes, monocytes and related leukocyte cell lines. *J. Clin. Invest. 76*:2003-2011 (1985).

161. Gamble, J. R., J. M. Harlan, S. J. Klebanoff, and M. A. Vadas. Stimulation of the adherence of neutrophils to umbilical vein endothelium by human recombinant tumor necrosis factor. *Proc. Natl. Acad. Sci. USA 82*:8667-8671 (1985).
162. Bevilacqua, M. P., R. R. Schleef, M. A. Gimbrone Jr., and D. J. Loskutoff. Regulation of the fibrolytic system of cultured human vascular endothelium by interleukin 1. *J. Clin. Invest. 78*:587-591 (1986).
163. Emeis, J. J., and T. Kooistra. Interleukin 1 and lipopolysaccharide induce an inhibitor of tissue-type plasminogen activator in vivo and in cultured endothelial cells. *J. Exp. Med. 163*:1260-1266 (1986).
164. Nachman, R. L., K. A. Hajjar, R. L. Silverstein, and C. A. Dinarello. Interleukin-1 induces endothelial cell synthesis of plasminogen activator inhibitor. *J. Exp. Med. 163*:1595-1600 (1986).
165. Broudy, V. C., K. Kaushansky, G. M. Segal, J. M. Harlan, and J. W. Adamson. Tumor necorsis factor type α stimulates human endothelial cells to produce granulocyte/macrophage colony stimulating factor. *Proc. Natl. Acad. Sci. USA 83*:7467-7471 (1986).
166. Munker, R., J. Gasson, M. Ogawa, and H. P. Koeffer. Recombinant human TNF induces production of granulocyte-monocyte colony-stimulating factor. *Nature 323*:79-82 (1986).
167. Seelentag, W. K., J-J. Mermod, R. Montesano, and P. Vassalli. Additive effects of interleukin 1 and tumour necrosis factor-α on the accumulation of the three granulocyte and macrophage colony-stimulating factor mRNA's in human endothelial cells. *EMBO J. 6*:2261-2265 (1987).
168. Seiff, C. A., S. Tsai, and D. V. Faller. Interleukin-1 induces cultured human endothelial cell production of granulocyte-macrophage colony stimulating factor. *J. Clin. Invest. 79*:48-51 (1987).
169. Zsebo, K. M., V. N. Yuschenkoff, S. Schiffer, D. Chang, E. McCall, C. A. Dinarello, M. A. Brown, B. Altrock, and G. C. Bagby Jr. Vascular endothelial cells and granulopoiesis: interleukin-1 stimulates release of G–CSF and GM–CSF. *Blood 71*:99-103 (1988).
170. Koury, M. J., A. Balmain, and I. B. Pragnell. Induction of granulocyte-macrophage colony-stimulating activity in mouse skin by inflammatory agents and tumor promoters. *EMBO J. 2*:1877-1882 (1983).
171. Koury, M. J., and I. B. Pragnell. Retroviruses induce granulocyte-macrophage colony stimulating activity in fibroblasts. *Nature 299*:638-640 (1982).
172. Broudy, V. C., K. S. Zuckerman, S. Jetmalani, J. H. Fitchan, and G. C. Bagby. Monocytes stimulate fibroblastoid bone marrow stromal cells to produce multilineage hematopoietic growth factors. *Blood 68*:530-534 (1986).
173. Miyatake, S., T. Otsuka, T. Yokota, F. Lee, and K. Arai. Structure of the chromosomal gene fro granulocyte-macrophage colony stimulating factors: comparison of the mouse and human genes. *EMBO J. 4*:2561-2568 (1985).

174. Barlow, D. P., M. Bucan, H. Lehrach, B. L. M. Hogan, and N. M. Gough. Close genetic and physical linkage between the murine haemopoietic growth factor genes GM–CSF and Multi-CSF (IL3). *EMBO J. 6*:617-623 (1987).
175. Buchberg, A. M., H. G. Bedigian, B. A. Taylor, E. Brownell, J. N. Ihle, S. Nagata, N. A. Jenkins, and N. G. Copeland. Localization of *Evi-2* to chromosome 11: linkage to other proto-oncogene and growth factor loci using interspecific backcross mice. *Oncogene Res. 2*:149-165 (1988).
176. Huebner, K., M. Isobe, C. M. Croce, D. W. Golde, S. E. Kaufman, and J. C. Gasson. The human gene encoding GM–CSF is at 5q21-q32, the chromosome region deleted in the $5q^-$ anomaly. *Science 230*:1282-1285 (1985).
177. Le Beau, M. M., C. A. Westbrook, M. O. Diaz, R. A. Larson, J. D. Rowley, J. C. Gasson, D. W. Golde, and C. J. Sherr. Evidence for the involvement of GM–CSF and FMS in the deletion (5q) in myeloid disorders. *Science 231*:984-987 (1986).
178. Nimer, S. D., and D. W. Golde. The $5q^-$ abnormality. *Blood 70*:1705-1712 (1987).
179. Yang, Y-C., S. Kovac, R. Kriz, S. Wolf, S. C. Clark, T. E. Wellems, A. Nienhuis, and N. Epstein. The human genes for GM–CSF and IL-3 are closely linked in tandem on chromosome 5. *Blood 71*:958-961 (1988).
180. Stanley, E., D. Metcalf, P. Sobieszczuk, N. M. Gough, and A. R. Dunn. The structure and expression of the murine gene encoding granulocyte-macrophage colony-stimulating factor: evidence for utilization of alternative promoters. *EMBO J. 4*:2569-2573 (1985).
181. Chan, J. Y., D. J. Slamon, S. D. Nimer, D. W. Golde, and J. C. Gasson. Regulation of expression of human granulocyte/macrophage colony-stimulating factor. *Proc. Natl. Acad. Sci. USA 83*:8669-8673 (1986).
182. Miyatake, S., T. Yokota, F. Lee, and K-i. Arai. Structure of the chromosomal gene for murine interleukin 3. *Proc. Natl. Acad. Sci. USA 82*:316-320 (1985).
183. Campbell, H. D., S. Ymer, M-C. Fung, and I. G. Young. Cloning and nucleotide sequence of the murine interleukin-3 gene. *Eur. J. Biochem. 150*:297-304 (1985).
184. Cohen, D. R., A. J. Hapel, and I. G. Young. Cloning and expression of the rat interleukin 3 gene. *Nucleic Acids Res. 14*:3641-3658 (1986).
185. Yang, Y-C., R. Kriz, S. Kovacic, A. C. Leary, A. Ciarletta, J. Witek-Giannatti, R. G. Donahue, and S. C. Clark. Characterization and functional expression of the human gene for interleukin-3 (multilineage-CSF) (in press).
186. Fuse, A., T. Fujita, H. Yasumitsu, N. Kashima, K. Hasegawa, and T. Taniguchi. Organization and structure of the mouse interleukin 2 gene. *Nucleic Acids. Res. 12*:9323-9331 (1984).
187. Fujita, T., C. Takaoka, H. Matsui, and T. Taniguchi. Structure of the human interleukin 2 gene. *Proc. Natl. Acad. Sci. USA 80*:7437-7441 (1983).

188. Gray, P. W., and D. V. Goeddel. Structure of the human immune interferon gene. *Nature 298*:859-863 (1982).
189. Nagata, S., M. Tsuchiya, S. Asano, O. Yamamoto, Y. Hirata, N. Kubota, M. Oheda, H. Nomura, and T. Yamazaki. The chromosomal gene structure and two mRNA's for human granulocyte colony-stimulating factor. *EMBO J. 5*:575-581 (1986).
190. Tsuchiya, M., Y. Kaziro, and S. Nagata. The chromosomal gene structure for murine granulocyte colony-stimulating factor. *Eur. J. Biochem. 165*: 7-12 (1987).
191. Campbell, H. D., W. Q. J. Tucker, Y. Hort, M. E. Martinson, G. Mayo, E. J. Clutterbuck, C. J. Sanderson, and I. G. Young. Molecular cloning, nucleotide sequence and expression of the gene encoding human eosinophil differentiation factor (interleukin 5). *Proc. Natl. Acad. Sci. USA 84*: 6629-6633 (1987).
192. Tanabe, T., M. Konishi, T. Mizuta, T. Noma, and T. Honjo. Molecular cloning and structure of the human interleukin-5 gene. *J. Biol. Chem. 262*:16580-16584 (1987).
193. Yasukawa, K., T. Hirano, Y. Watanabe, K. Muratani, T. Matsuda, S. Nakai, and T. Kishimoto. Structure and expression of human B cell stimulatory factor-2 (BSF-2/IL-6) gene. *EMBO J. 6*:2939-2945 (1987).
194. Otsuka, T., D. Villaret, T. Yokota, Y. Takebe, F. Lee, N. Arai, and K.-I. Arai. Structural analysis of the mouse chromosomal gene encoding interleukin 4 which expresses B cell, T cell and mast cell stimulating activities. *Nucleic Acis Res. 15*:333-334 (1987).
195. Shannon, M. F., J. R. Gamble, and M. A. Vadas. Nuclear proteins interacting with the promoter region of the human granulocyte/macrophage colony-stimulating factor gene. *Proc. Natl. Acad. Sci. USA 85*:674-678 (1988).
196. Shannon, M. F., M. A. Vadas, and J. R. Gamble. Transcriptional regulation of GM–CSF production: two nuclear proteins bind to the human GM–CSF promoter region. *J. Cell. Biochem*. (in press).
197. Nimer, S. D., E. A. Morita, M. J. Martis, W. Washsman, and J. C. Gasson. Characterization of the human granulocyte-macrophage colony-stimulating factor promoter region by genetic analysis: correlation with DNase I footprinting. *Mol. Cell. Biol. 8*:1979-1984 (1988).
198. Shaw, G., and R. Kamen. A conserved AU sequence from the 3′ untranslated region of GM–CSF mRNA mediates selective mRNA degradation. *Cell 46*:659-667 (1986).
199. Caput, D., B. Beutler, K. Hartog, R. Thayer, S. Brown-Shimer, and A. Cerami. Identification of a common nucleotide sequence in the 3′-untranslated region of mRNA molecules specifying inflammatory mediators. *Proc. Natl. Acad. Sci. USA 83*:1670-1674 (1986).
200. Wreschner, D. H., and G. Rechavi. Differential mRNA stability to reticulocyte ribonucleases correlates with 3′ non-coding $(U)_nA$ sequences. *Eur. J. Biochem. 172*:333-340 (1988).

201. Rich, A., A. Nordheim, and A. Wang. The chemistry and biology of left-handed Z–DNA. *Annu. Rev. Biochem.* *53*:792-846 (1984).
202. Ladner, M. B., G. A. Martin, J. A. Noble, D. M. Nikoloff, R. Tal, E. S. Kawasaki, and T. J. White. Human CSF-1: gene structure and alternative splicing of mRNA precursors. *EMBO J.* *6*:2693-2698 (1987).
203. Dynan, W. S., and R. Tijan. Control of eukaryotic messenger RNA synthesis by sequence-specific DNA binding proteins. *Nature 316*:774-778 (1985).
204. Dynan, W. S., and R. Tijan. The promoter-specific transcription factor SP1 binds to upstream sequences in the SV40 early promoter. *Cell 35*: 79-87 (1983).
205. Kadonaga, J. T., K. R. Carner, F. R. Masiarz, and R. Tijan. Isolation of cDNA encoding transcription factor SP1 and functional analysis of the DNA binding domain. *Cell 51*:1079-1090 (1987).
206. Sporn, M. B., and A. B. Roberts. Autocrine growth factors and cancer. *Nature 313*:745-747 (1985).
207. Cook, W. D., D. Metcalf, N. A. Nicola, A. W. Burgess, and F. Walker. Malignant transformation of a growth factor-dependent myeloid cell line by Abelson virus without evidence of an autocrine mechanism. *Cell 41*:677-683 (1985).
208. Lang, R. A., D. Metcalf, N. M. Gough, A. R. Dunn, and T. J. Gonda. Expression of a haemopoietic growth factor cDNA in a factor-dependent cell line results in autonomous growth and tumorigenicity. *Cell 43*:531-542 (1985).
209. Stocking, C., C. Löliger, M. Kawai, S. Suciu, N. Gough, and W. Ostertag. Identification of genes involved in growth autonomy of hematopoietic cells by analysis of factor-independent mutants. *Cell 53*:869-879 (1988).
210. Dührsen, U., and D. Metcalf. A model system for leukemic transformation of immortalized hemopoietic cells in irradiated recipient mice. *Leukaemia 2*:329-333 (1988).
211. Dührsen, U. In vitro growth patterns and autocrine production of hemopoietic colony stimulating factors: analysis of leukaemic populations arising in irradiated mice from cells of an injected factor-dependent continuous cell line. *Leukaemia 2*:334-342 (1988).
212. Moore, M. A. S., G. Spitzer, N. Williams, D. Metcalf, and J. Buckley. Agar culture studies in 127 cases of untreated acute leukemia: the prognostic value of reclassification of leukemia according to in vitro growth characteristics. *Blood 44*:1-18 (1974).

5
Molecular Biology of Macrophage Colony-Stimulating Factor

ERNEST S. KAWASAKI and MARTHA B. LADNER
Cetus Corporation, Emeryville, California

I. INTRODUCTION

As of this review, four major myeloid growth factors have been molecularly cloned. Three of the factors, discussed elsewhere in this volume, include granulocyte-macrophage colony-stimulating factor (GM–CSF), granulocyte colony-stimulating factor (G–CSF), and interleukin-3 (IL-3). In this section we review the molecular biology of M–CSF.

Macrophage colony-stimulating factor, when purified from human urinary or tissue culture sources is a heterogeneous, heavily glycosylated dimer with molecular weights ranging from 45 to >70 kd (1-3). This factor promotes the growth and differentiation of bone marrow progenitor cells in the mononuclear phagocytic lineage and can stimulate the proliferation of mature macrophages via specific cellular receptors (1,4,5). Macrophage colony-stimulating factor also has a variety of stimulatory effects on the function of monocytes and mature macrophages (6). Recent studies have shown that M–CSF is much more complex than other hemopoietic CSFs at the DNA, RNA, and protein levels. These studies include the isolation and characterization of human M–CSF genomic (7,8) and human complementary DNA (cDNA) clones (7,9), and murine M–CSF cDNA clones (10-12). We summarize these findings and include some of our unpublished results.

II. HUMAN M–CSF GENE

In 1984, M–CSF was purified from human urine (7) and 12 residues were obtained by sequencing of the amino-terminal end. Using labeled oligonucleotides derived from this sequence, we screened a Maniatis human genomic library (13) and identified several positively hybridizing clones. Most clones contained a 3.8-kilobase (kb) Hind III fragment that hybridized strongly to a consensus oligonucleotide probe designated EK 14. DNA sequence analysis of this Hind III fragment revealed an open reading frame around the EK 14 binding site that encoded the amino-terminus of M–CSF. Finding an authentic M–CSF genomic sequence enabled us to synthesize new, M–CSF-specific oligonucleotides that were used to screen for and isolate M–CSF cDNA clones (7).

Complete DNA sequence analysis revealed that only a portion of M–CSF was present in the 3.8-kb Hind III genomic fragment. For further characterization of the total genomic clone a large number of subclones were generated and analyzed. Extensive DNA sequence analysis, coupled with data obtained from M–CSF cDNA and oligonucleotide hybridizations to Southern blots of enzyme-restricted DNA, allowed us to put together a map of the entire human M–CSF gene (Fig. 1).

The gene, which spans 21 kb, is comprised of 10 exons and 9 introns. The introns range in size from 0.3 to 2.7 kb and the exons vary from 52 base pairs (bp) to >2 kb. The majority of the coding sequence exons (exons 1-8) are relatively small and conform to the earlier observation that most exon sizes are clustered around means of 52, 140, 223, and 299 bp (14). However, exon 6 is either 131 or 1025 bp, depending on which splice acceptor site is used (see below). The larger size appears to be unusual in genes comprised of several exons. The exception to small exons are usually those that contain 3′ untranslated sequences, and these can be quite large. For example, the hamster *HMG* CoA reductase gene contains 19 coding region exons of 85-248 bp; the 20th exon contains the last 52 bp of coding sequence and 2 kb of 3′ untranslated sequence (15). In the M–CSF gene, the largest exon (exon 10) codes for a 3′ untranslated sequence.

Southern analysis of genomic DNA as well as an exhaustive study of recombinant genomic clones indicates that M–CSF exists as a single-copy gene (7,8).

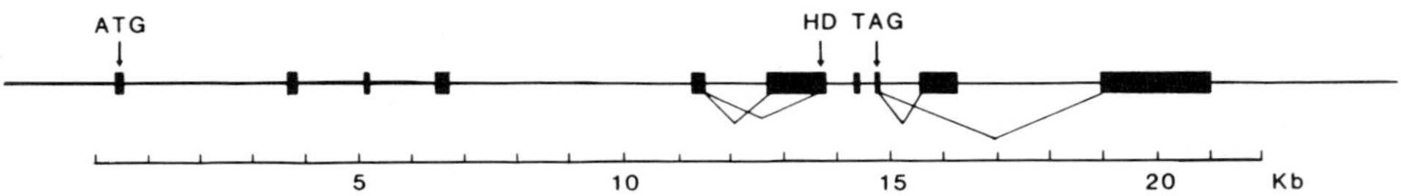

Figure 1 Human M–CSF gene structure. The solid boxes represent the 10 exons, with introns denoted by the straight lines. The positions found for alternative splicing are indicated. ATG represents the beginning of the translated sequence, HD refers to the area encoding the 23 amino acid hydrophobic domain, and TAG indicates the position of the stop codon.

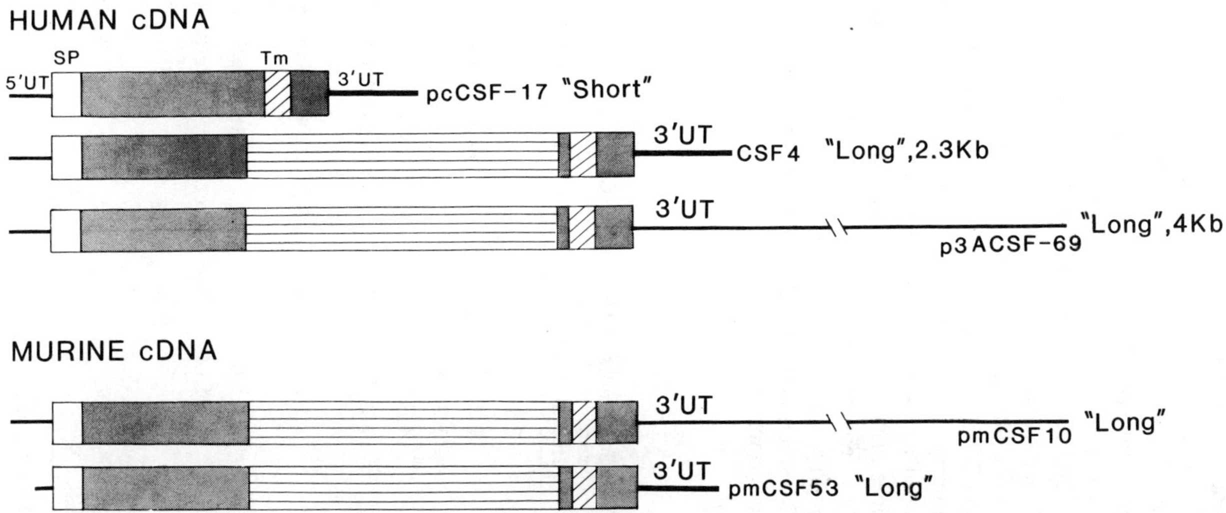

Figure 2 Structure of the human and murine cDNA clones. The postulated 32-amino-acid signal peptide (SP) is indicated by the clear block. The coding sequences in both the short [pcCSF-17 (7)] and long [CSF4 (8), p3ACSF-69 (9), pMCSF10 and pMCSF53 (12)] protein forms of M–CSF are indicated by the dark areas. The coding sequence unique to the long protein forms is indicated by the blocks with horizontal lines, and the transmembrane domain (Tm) is depicted by boxes with slanted lines. The short 3′ untranslated (UT) sequences found in pcCSF-17, CSF4, and pmCSF53 are shown by thick lines and the 2-kb 3′ UT regions in p3ACSF-69 and pmCSF10 are depicted by the long, thin lines.

However, this gene produces several distinct M–CSF mRNA species (Fig. 2) through alternative splicing mechanisms, and several of these alternatively spliced mRNAs have been cloned (7-9). Of particular interest are the mRNAs that are derived from the alternatively spliced exon 6 described above. The most common M–CSF mRNAs use the more upstream acceptor site and contain 1662 bp (554 amino acids) of coding sequence whereas other transcripts, which contain 768 bp (256 amino acids), use the downstream acceptor site for exon 6 (8).

All the donor and acceptor boundary sites conform to the most important aspects of the consensus sequences compiled by Mount (16) (Table 1). That is, all the introns begin with the GT dinucleotide and end with the AG dinucleotide. All of the introns also contain a T/C nucleotide-rich sequence immediately upstream of the acceptor boundary, with the exception of the more 3′ intron 5 acceptor, which does not have this T/C-rich stretch. The alternative splice site in intron 5, further 5′ward, does conform to this rule, and is strikingly rich in T/C nucleotides. The more 5′ acceptor site is responsible for formation of the large, 1025 bp exon 6, the long T/C nucleotide stretch preceding this acceptor site may be the reason why this site is preferentially used over the downstream site in cells that synthesize both transcripts.

Table 1 Exon-Intron Boundary Sequences of Human M–CSF

Intron	Donor boundary	Intron length	Acceptor boundary
1	ACGgtaagc	3.1 Kb	ccctctctgtcacagACA
2	CTGgtgagt	1.3 Kb	cctgctctctcttacagATT
3	TTGgtgagt	1.3 Kb	ttcctgtttttcagAAA
4	AAGgtagga	4.6 Kb	ccctgatctccttccagGCC
5	AAGgtaagc	1.2 Kb	tatctgtggttctttcagATG
			c c
		2.1 Kb	cgtaatgttcctttgactgacacagGCC
			c
6	CGGgtgagt	0.6 Kb	ctctttcctacctcagAGC
7	CAGgtgaga	0.3 Kb	ccctgtgctctgcctgcagCCC
8	AAGgtaagg	0.8 Kb	ttttcagACC
Consensus:			
	CAGgtaagt		t ncagG
	A g		c n

The consensus sequence is from Mount (16).

Exons 9 and 10 contain the sequence for the two different 3′ untranslated regions that have been found in M–CSF cDNA clones. The shorter 3′ untranslated sequence (679 bp), comprising exon 9, is present in pcCSF-17 (7) and CSF4 (8). The alternate 3′ sequence, present in the 4-kb cDNA, p3ACSF-69 (9), is contained in exon 10. Although complete sequencing of the genomic sequence comprising exon 10 has not been reported, two groups working independently have sequenced enough of the area to confirm that the sequence must reside in a single 2-kb exon (8,17). Curiously, the last 16 bp of untranslated sequence of p3ACSF-69 diverges from the genomic sequence, as shown below.

CCGTTGACGTAGCTCGAG p3ACSF-69 (9)

CCATGTGAGTCGTGGGAA genomic sequence (8)

At present, the basis for this difference is unknown.

Within 400 bp upstream of the putative M–CSF mRNA start site are several elements that may be involved in control of M–CSF expression (Fig. 3). There are two possible Goldberg and Hogness (or TATA) boxes. CATAAA is 54 bases upstream of the beginning of pcCSF-17 (the cDNA clone with most 5′ untranslated sequences), and TTAAA is 26 bases upstream. The TATA sequence is usually found within 30 residues of the transcriptional start site and is the most ubiquitous element in the promoter region of eukaryotic genes (18).

```
ctgcagaggaagaagggggctgccggcaaacctgctgactcaggctccacgagggagcaa -301
gtaacactggactcctttcggcactccgagaatggggtgggggcgtcttcaaaggatttc -271
cctccccttcccagtgcttgtccctgctctcggtccgttttctgctaagatttggggattt -241
tcagggcctggagggaaagtcccttgggacgatcatagagcgctagcactgaatcagcct -211
ggagagcgcggaaggaaagggtcggtccgcagagggcgcggggaaggcagggtggggacg -181
cggtggagcccgcgctcgtttgctgaaggcttggaagtgcagcgcagaagacagagggtg -151
actaggaagacgcgcgaacggggctggccggccggcgggtgggggaggggaggcgggggga -121
aggcggctgagtgggcctctggagtgtgtgtgtctgtgtcagtgtgtgtgtgtgtgtgtg -91
tatgtgtgtgtctggcgcctggccagggtgatttcccataaaccacatgccccccagtcc -31
tctcttaaaaggctgtgccgagggctggcc -1
```

Figure 3 Human M–CSF gene 5′ flanking sequence. The two possible TATA boxes are underlined an the long TG repeat is boxed. The GC-type boxes are overlined and the two sequences similar to enhancer core consensus sequences are boxed.

The residues -80 to -126, relative to the pcCSF-17 sequence, contain a long series of alternating thymine and guanine bases. A conformational form of DNA, known as Z-DNA, has been shown to occur along such stretches of regularly repeating nucleotides. The structural changes occurring along the DNA by these repeating elements have been postulated to have effects on transcription (19).

Upstream of the TG repeats are two GGCGGG sequences (-159 and -177). Recently, it has been shown that a similar sequence, GGGCGG, binds the transcription factor, SP1 (20). Identical GC boxes have been found in the promoter region of several eukaryotic genes, although not all have yet been shown to have an effect on transcription (15,21-23). The same GGCGGG sequence is represented five times between -167 and -214 in the human hypothanthine phosphoribosyl-transferase gene. Their homology with the SP1 binding site and location within 200 residues upstream of transcriptional start sites have led to suggestions that these sequences are involved in transcriptional regulation (24).

Finally, there are two sequences in the 5′ flanking region that are similar to the consensus enhancer core sequence, TGGAAAG (25). These two sequences are AGGAAAG at position -317, and GGGAAAG at -377. True enhancer sequences can regulate the transcriptional level of the gene with which they are associated, independently of position or orientation.

The importance of any of the abovementioned sequences in the regulation of M–CSF expression is unknown at this time. However, ongoing studies include using the 5′ flanking sequences attached to a reporter gene to determine which elements might have specificity for the induction of M–CSF.

III. HUMAN M–CSF COMPLEMENTARY DNAs

To isolate human M–CSF cDNA clones (7), we chose as starting material mRNA from the pancreatic tumor line, MIA-PaCa-2 (26). This cell line constitutively secretes moderate amounts of M–CSF, and methods were developed for inducing M–CSF in serum-fee media (27). Oligonucleotides derived from the M–CSF gene sequence described above were used to screen and isolate clones from a cDNA bank constructed from sucrose gradient fractionated mRNA. Only one clone encoding biologically active M–CSF was found. Subsequently, different forms of M–CSF cDNA clones were isolated from an SV40 transformed trophoblast cell line (9), and MIA-PaCa-2 (8). The various types of cDNA clones encoding biologically active M–CSF are illustrated (Fig. 2) and their DNA and protein sequences given (Fig. 4). These different forms of M–CSF are generated by alternative splicing of the primary RNA transcripts, and they contain large differences in both coding and noncoding regions. The exact splicing mechanisms were described in the previous section covering the M–CSF gene and will not be discussed in detail here.

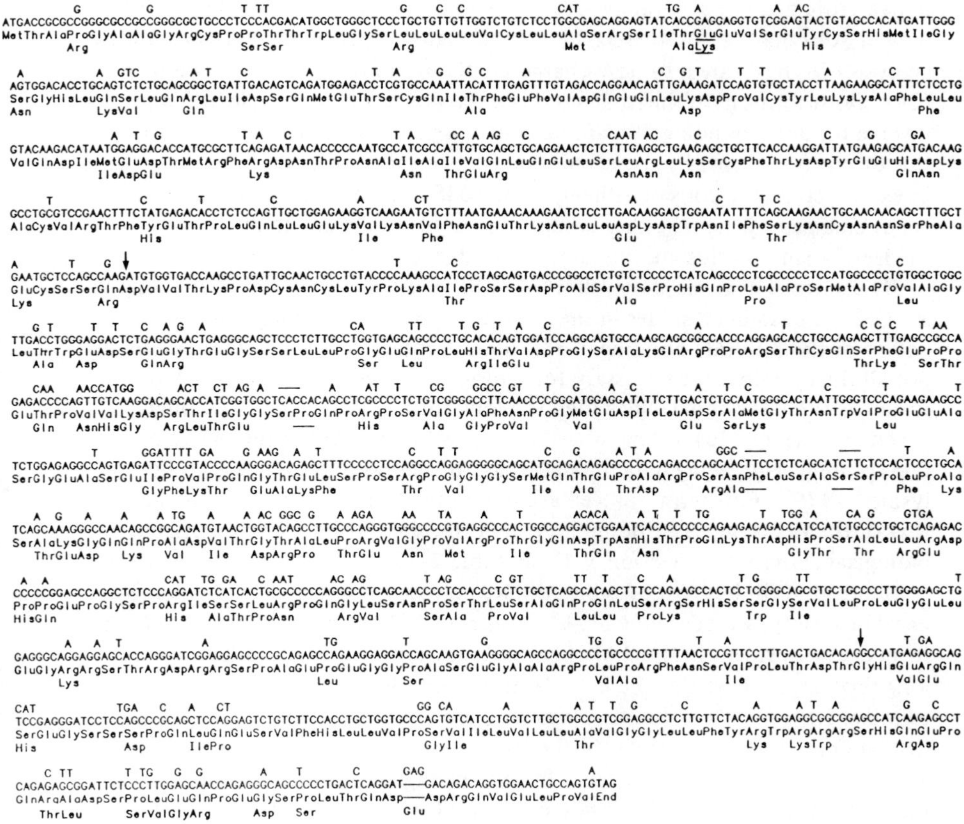

Figure 4 Human and murine DNA and protein sequences. The complete human M–CSF DNA and protein sequences are shown. Only the differences found for the murine nucleotide and amino acid sequences are given; the murine nucleotide changes are shown above the human DNA sequence and the amino acid changes are given below the human protein sequence. The first amino acid residue of mature M–CSF is underlined. The two arrows delineate the human M–CSF sequence not contained in pcCSF-17 (7).

By cDNA cloning, two biologically active forms of M–CSF have been found; a 256 (7) and a 554 amino acid form (8,9). For simplicity we shall refer to these two types as the "short" and "long" forms of M–CSF. The amino acid numbering used here will refer to amino acids present in the "mature" protein, minus the signal peptide. The term mature is in quotes because the exact size of the native protein is not accurately known, as will be discussed later. Both short and long M–CSF have a putative 32 amino acid signal peptide, which hydrophilicity analysis (28) has shown to have characteristics of a functional signal sequence (29). Both forms have the same coding sequence until amino acid 149 and then diverge for an astounding 298 amino acids. As described in the previous section, this divergence occurs by use of two alternative splice acceptor sites that results in exon 6 coding regions of either 131 or 1025 bp. Because the reading frame is not changed by use of either acceptor site, the sequences converge again and the last 74 amino acids are the same for both the short and long forms.

By sequence analysis alone, short M–CSF could have 7 cysteine and long M–CSF 10 cysteine residues in the "mature" protein. Although the exact nature of the disulfide bonding of this dimer protein is not known, it seems that not all the cysteines are necessary for biological activity, because the short form has three fewer cysteines than the long form. As evidence, the final cysteine (residue 226) of long M–CSF may not be present in the native protein; protein sequencing data indicates that the carboxy-terminus ends before lysine 218 (9). Furthermore, we have determined by site-directed mutagenesis that of the first nine cysteines in long M–CSF, the last two are dispensable for biological activity (see Section VII and Table 2) (30).

Short M–CSF has two possible *N*-linked glycosylation sites of the Asn-X-Ser/Thr variety, whereas long M–CSF has four. The last two sites in the long form are not present in short M–CSF and may not be present in the mature long protein, because they lie outside of the putative boundary of the carboxy-terminal end (9). Thus, the first two sites present in both short and long M–CSF are probably responsible for the large amount of glycosylation and heterogeneity seen in the native protein. In spite of the heavy carbohydrate attachment, the sugars are not absolutely necessary for biological function (Table 2) (30).

Both forms of M–CSF cDNAs have an identical coding sequence for the last 75 amino acids. This region is of interest because it contains a very hydrophobic stretch of 23 amino acids (residues 166-188 of short M–CSF), followed by the sequence Arg-Trp-Arg-Arg-Arg and 31 more amino acids to the carboxy-terminal end. This type of structure resembles membrane proteins, which have a hydrophobic transmembrane domain followed by positively charged residues and a cytoplasmic tail (31). Thus, M–CSF may first be synthesized and function as a membrane protein (32,33). Further processing at carboxy-terminus would then release the mature form into the circulation (7,28,34).

Characteristics of the 5′ and 3′ untranslated regions of the long and short forms of M–CSF were described in the previous section on the human gene. Other interesting highlights are discussed below in relation to the murine M–CSF cDNA clones.

Table 2 Biological Activity of M–CSF Mutants

Mutant sample	Activity
1. Wild type, short or long M–CSF	+
2. Tyr-59 to Asp-59	+
3. Lys-52 to Pro-52	–
4. Lys-52 to Gln-52	+
5. Cys-7 to Ser-7 or Ala-7	–
6. Cys-31 to Ser-31 or Ala-31	–
7. Cys-48 to Ser-48 or Ala-48	–
8. Cys-90 to Ser-90 or Ala-90	–
9. Cys-102 to Ser-102 or Ala-102	–
10. Cys-139 to Ser-139 or Ala-139	–
11. Cys-146 to Ser-146 or Ala-146	–
12. Cys-157 to Ser-147	+
13. Cys-159 to Ser-149	+
14. Asn-122 to Gln-122	+
15. Asn-140 to Gln-140	+
16. Asn-122 and 140 to Gln-122 and 140	+
17. 66 aa C-terminal deletion	+
18. 74 aa C-terminal deletion	+
19. 301 aa C-terminal deletion	+

M–CSF from wild-type and mutant short or long forms were assayed in the mouse bone marrow colony assay (7,54). Mutants 2-11 and 14-18 were tested with the short M–CSF sequence. Mutants 12-13 and 19 were made from the long M–CSF cone. +: activity within a factor of 3 of wild-type; –: <10% of wild-type activity.

IV. MURINE M–CSF cDNAs

Murine L-cells have traditionally been the source of purified M–CSF (35). All of the described murine cDNA clones have been isolated from banks synthesized from L-cell mRNA. The clones were identified by either hybridization to a human cDNA probe (10) or by binding an antibody against murine M–CSF to clones expressing M–CSF, followed by a second screen using oligonucleotides derived from the amino-terminal sequence of murine M–CSF (11).

All of the murine M–CSF sequences encode a protein that is similar to the 554 amino acid form of human M–CSF (Figs. 2 and 4). There are 10 Cys residues in the murine protein, which are all conserved in the human sequence, and four potential *N*-linked glycosylation sites. The protein contains a 32 amino acid leader peptide that is 81% similar to the human sequence. The signal sequence is followed by a 520 amino acid reading frame, and has, like its human counterpart, a 23 amino acid hydrophobic domain.

Overall, the murine amino acid sequence is very similar to human M–CSF, but certain areas are more highly conserved than others, as is clearly shown by the comparison of the two sequences (Fig. 4). From the beginning of the mature polypeptide to amino acid residue 227, the two proteins exhibit an 82% similarity. However, the middle 174 amino acids share only a 53% similarity. From amino acid 402 in the human to the end of the protein, the similarity increases again to 76%.

The high degree of similarity through the amino-terminal-half of the two proteins reflects the fact that this portion of the protein is responsible for biological activity. The evidence for the importance of the amino-terminus in M–CSF activity comes from two different lines of research: (a) Sequencing of the human protein has indicated that the carboxy-terminus of the secreted protein lies around amino acid residue 191-223 (9,17). (b) Both the smaller, 256-amino-acid form of human M–CSF and the larger 554-amino-acid form are secreted from cells after cleavage of the hydrophobic domain, and both forms exhibit biological activity. This result also indicates that the active domain of the molecule is in the amino-terminus, because the amino acid sequence of the two proteins diverge after the first 149 amino acids (Fig. 4) (8,9).

The last 120-amino-acid residues of the human and murine sequences both contain a 23-amino-acid hydrophobic region, and the sequence similarity through this last portion of the two proteins probably indicates the necessity of maintaining this sequence. Whether or not the short cytoplasmic tail has any biological role is unknown at this time.

Northern blots of murine L-cell RNA hybridized with an M–CSF probe show two bands of hybridization (Fig. 5). One band corresponds to a 4-kb mRNA, and the other to a 2.3-kb transcript. cDNA clones have been isolated for both transcripts (Fig. 6) (12), and contain essentially the same coding sequence. The only differences in the coding sequence is that the 4-kb clone has an Asp at amino acid 260 and a Ser at position 313 while the 2-kb clone encodes a Gly and a Pro at the corresponding positions. One other complete murine M–CSF sequence has been published (10), and this 2-kb cDNA clone codes for Asp and a Ser at the positions in question. The 2-kb cDNA clones obtained by both groups contain a short, 235 bp 3′ untranslated sequence, while the 4 kb type contains a 2.1 kb 3′ untranslated sequence, which is not related to the short sequence found on the 2-kb clone (Figs. 4 and 6).

Thus, the 2.3-kb and 4-kb murine L-cell transcripts correspond to the 2.3-kb and 4-kb human M–CSF cDNAs. In both the human and the murine trans-

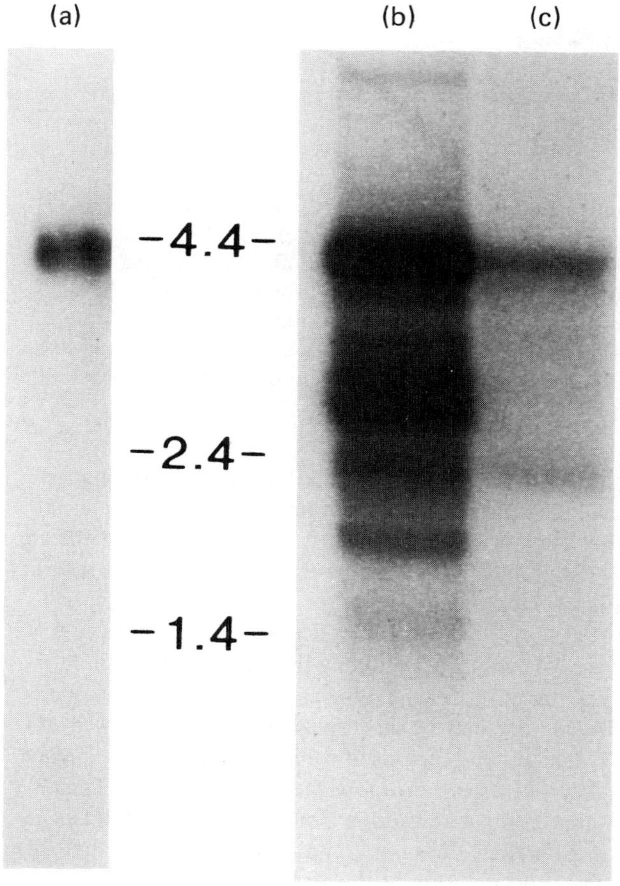

Figure 5 Northern gel analysis of M–CSF mRNA transcripts from human and murine cells. Human monocyes (a) were adhered to plastic for 24 hr in the presence of 10 μg/ml LPS. MIA-PaCa-2 human cells (b) were induced (27) before RNA isolation. Untreated murine L929 cells (c) were used for obtaining RNA. RNA was electrophoresed in 1% agarose gels containing 2.2 *M* formaldehyde and transferred to either nylon (monocyte RNA) or nitrocellulose membranes. The filters were probed with labeled, M–CSF-specific oligonucleotides.

(a)

```
ATTCTAAG↓ACCCCTCACCATCCTGGACACACTCGTTTGTCAATGTCCCTCTGAAAATGTGACGCCCAGCCCCGGACACAGTACTCCAGATGTTGTCTGAC
       T                                                            G            T

CAGCTCAGAGAGAGTACAGTGGGACTGTTACCTTCCTTGATATGGACAGTATTCTTCTATTTGTGCAGATTAAGATTGCATTAGTTTTTTCTTAACAAC
          ----                G   GT           C            C        C C                CA          -------  G

TGCATCATACTGTTGTCATATGTTGAGCCTGTGGTCT[ATTTAAA]ACCCCTAGTTCCA
         - A                 T   A           A   C      T
```

Figure 6 Human and murine cDNA 3′ untranslated regions. (a) The short 3′ untranslated sequence (exon 9) found on the human cDNA clones pcCSF-17 (7), CSF4 (8), and the murine sequence pmCSF53 (12). (b) The long 3′ untranslated sequence (exon 10) found in the human cDNA clone p3ACSF-69 (9), and the murine clone pmCSF10 (12). The human sequence is depicted below. The arrow indicates where the splice point is between exon 8 and exon 9 (sequence a) or exon 10 (sequence b). The polyadenylation signals are boxed and the AU sequences (55) are underlined.

cripts, the 3′ untranslated sequences diverge 16 bp after the TAG stop codon. Because it is known that there is an intron in the human genomic sequence at this point, we assume that the murine counterpart is the same, and that the two 3′ untranslated sequences contained in the murine transcripts reflect an alternative choice of exons, as has been shown to occur for the human transcripts. These findings indicate that these 3′ alternative splice sites have been conserved over many millions of years of evolution, implying that there may be some critical, perhaps regulatory, function for the different 3′ untranslated exons.

The control of mRNA stability may be partly regulated by a conserved AU sequence found in the 3′ untranslated regions of some mRNAs. This sequence has been found in several cytokines and protooncogenes (55,56). The 2-kb 3′ untranslated sequence found in the 4-kb murine clone contains this sequence three times (Fig. 6). The 2-kb clone does not contain this motif. Whether this AU sequence leads to a faster degradation of the 4-kb mRNA or whether the presence of guanine and cytosine bases between the sequences in the long 3′ end abrogates this effect is not known.

V. EXPRESSION OF M–CSF IN CELLS, TISSUES, AND ORGANS

Because the first human and murine cDNA clones for M–CSF were isolated from MIA-PaCa-2 and L-cells, respectively, these two cell lines were used to estimate the molecular size of M–CSF mRNA. MIA-PaCa-2 cells produce several different-size transcripts, depending on the incubation conditions (27). Cells grown with or without serum contain a 4 kb mRNA. If phorbol myristic acetate (PMA) is added to the cells growing in serum-containing medium, they

(b)

```
          ↓
GAATTCTAAGCTGGACGCACAGAACAGTCTCTTCGTGGGAGGAGACATTATGGGGCGTCCACCACCACCCCTCCCTGGCCATCCTCCT–GGAATGTGGTC
G       T       G  C  G  TA      TA   A           ATG  AA A         T –      T TTA     C    G       C

TGCCCTCCACCAGAGCTCCTGCCTGCCAGGACTGGACCA——GAGCAGCCAGGCTGGGGCCCCTCTGTCTCAACCCGCAGACCCTTGACTGAATGAGAGA
–A  A T                  A   A      TG AAGA      TTT A      T TT   CA  CTC     TT     T  CA  CA  GAGA AG

GGCCAGAGGATGCTCCC—CATGCTGCCACTATTTATTGTGAGCCCTGGAGGCTCCCATGTGCTTGAGGAA–GGCTGGTGAGCCCGGCTCAGGACCCTCT
   TG        C   CA    A                    G                  –  CAT G      G   A C C  — A              –

TCCCTCAGGGGCTGCACCCTCCTCTCACTCCCTTCCATGCCGGAACCCAGGCCAGGGACCCACCGGCCTG—TGGTTTGTGGGAAAGCAGGGTGGACGCT
      T          TT       C       ————————— AA  A   –  T   A        T   TA      GA    CCCA       G C    ACA  T AG

GAGGAGTGAAAGAACCCTGCACCCAGAGGGCCTGCCTGGTGCCAAGGTATCCCAGCCTGGACAGGCATGGACCTGTCTCCAGAGAGAGGAGCCTGAAGTT
A      CAG     G       TG        A A  CAA                 A          A          G           T   CA    AGA        TA

CGTGGGGCGGGACAGCGTCGGCCTGATTTCCCGTAAAGGTGT—————————GCAGCCTGAGAGACGGCAAGAGGAGGCCTCTGGACCTGCTGGTCTGCACT
AAA     T          A   T    C   C                  CA AAAGGCAC        CA A        G                  T    —        TG G ——

GACAGCCTGAA–GGGTCTACACCCTCGGCTCACCTAAGT–CCCTGTGCTGGTTGCCAGGCGCAGAGGGGAGGCCAGCCCTGC–CCTCAGGACCTGCCTGA
          T A   G                       AA    TG T        G      C        A A        — A   GA        A                 C                    A ——

CCTGCCAGTGATGCCAAGAGGGGGATCAAGCACTGGCCTCTGCCCCTCCTCCTTCCAGCACCTGCCAGAGCTTCTCCAGGAGGCCAAGCAGAGGCTCCCC
C  G    TCA                    AA   CAG      T   A          GT  TT          G  CCA         C                   T  CT ———— TT       C                  A

TCATGAAGGAAGCCATTGCACTGTGAACACTGTACCTGCCTGCTGAACAGCCTGCCCCCGTCCATCC—ATGAGCCAGCATCCGTCCG————TCCTCCA
                                          T       A                                      –T      A             CT            TG  C                AATGT   TC

CT—CTCCAGCCTCTCCCCAGCCTCCTGCACTGAGCTGGCCTCACCAGTCGACTGAGGGAGCCCCTCAGCCCTGACCTTCTCCTGAC–CTGCCTTTGACT
  TC  T               T  G   T   T                             GT   T                             G          CA           C         C  T A                  T

CCCCGGAGTGGAGTGGGGTGGGAGAACCTCCTGGGCCGCCAGCCAGAGCCGGTCTTTAGGCTGTGTTGTTCGCCCAGGTTTCTGCATCTTGCACTTTGAC
GT    A G      A — CT                         G                 TA          T            T               G                 CC  T

ATTCCCAAGAGAGGGAAGGGACTAGTGGGAGAGAGCAAGGGAGGGGAGGGCACAGACAGAGAGGCTACAGGGCGAGCTCTGACTGAAGATGGGCCTTTGAAA
            G                T                A   G      AG     A            ——      G C A         C          CA           A    A      GA          T     GA

TATAGGTATGCACCTGAGGTTGGGGGAG–GGTCTGCACTCCCAAACCCCAGCGCAGTGTCCTTTCCCTGCTGCCGACAGGAACCTGGGGCTGAGCAGGTT
     GT     GCA  C     A   C       AAATT  ATC        AT          TAGCC     T        –         CC          ———    C      TT  T  T  A                A    A

ATCCCTGTCAGGAGCCCT—GGACTGGGCTGCATCTCAGCCCCACCTGCATGGTATCCAGCTCCCATCCACTTCTCACCCTTCTTTCCT——CCTGACCT
G –                CA       T    GG             A     C        —–  T    A ——————————————        T   T           G      T     G    CT  CAGA               C

TGGTCAGCAGTGATGACCTCCAACTCTCACCCACC–CCCTCTACCATCACCTCTAACCAGGCAAGCCAGGGTGGGAGAGCAATCAGGAGAGCCA————
A     AG       AG  G  AC           C  G              T      ATA    G  G          ——            G      A                     T  TA          AGGG          AA  A  AAAAAAAAAG

–GGCCTCAGCTTCCAA—TGCCTGGAGGGCCTCCACTT–TGTGGCAGCCTGTGGTGGTGGCTCTGAGGCCTAGGCAACGAGCGACAGGGCTGCCAGTTG
G  TTG  TTA              CT      TTCT      T  C          TGT       CCCA               G  ————        T     G    GT  A  CT     GC  T  G  T                   CT     T

CCCCTGGGTTCCTTTGTGCTGCTGTGTGCCTCCTCTCCTGCCGCCCTTTGTCCTCCGCTAAGAGACCCTGCCCTACCTGGCCGCTGGGCCCCGTGACTTT
       T      CT        A             A   ——————————    C              ——————————                C            TA           T  AATAAT

CCCTTCCTG–CCCAGGAAAGTGAG–GGTCGGCTGGCCCCACCTTCCCTGTCCTGATGCCGACAGC–TTAGGGAAGGGCAGTG—AACTTGCATATGGGG–
   AGGC  A     AAT  T     G  G  A        T         T  TAA      T       T  AG                      ——               AGGG                A    C    GAG         T  T  G           C

—CTTAGCCTTCTAGTCACAGCCTCTATATTTGATGCTAGAAAACACATATTTTTAAATGGAAG————————————AAAAATAAAAAGGCATTCC
TTA                  G    T        A    A —–      CC                          T                    A        A           GAAAAACACAGAAACA      C  A    C                  —

CCCTTCATCCCCCTACCTTAAACATATAATATTTTAAAGGTCAAAAAAGCAATCCAACCCACTG—CAGAAGCTCTTTTTGAGCACTTGGTGGCATCAGA
—    CT  C      T  C                       T  —            ACAG    G  GAA                        T   —A                     G    G  TG  C          TA         G

GCAGGAGGAGCCCCAGAGCCACCTCTGGTGTCCCCCAGGCTACCTGCTCAGGAACCCCTTCTGTTCTCTGAG–AAGTCAAGAGAGGACATTGGCTCACGC
        –            T   A  T              T      A  C  G        —–        GTG          ATTA  G  A                  C  C              A          CTC           G  AG     C    C             AA

ACTGTGAGATTTTG–TTTTTATACTTGGAAGTGGTGAATTATTTTTTATAAAGTCATTTAAATATCTATTTAAAAGATAGGAAGCTGCTTATATATTTAA
                 C  GC  T                                                            A          G                          A                              T

TAATAAAAGAAGTGCACAAGCTGC
```

Figure 6 (continued)

continue to synthesize the same large-molecular-weight M–CSF mRNA. However, if the cells are grown without serum in the presence of PMA, after 48 hr they produce approximately seven different-size M–CSF transcripts, ranging in size from 4 to 1.6 kb (Fig. 5). These different mRNAs differ in size depending on the splice site used for exon 6 and which 3′ untranslated region they contain (see Section II). A second human pancreatic cell line, PANC1, also produces multiple M–CSF transcripts similar to MIA-PaCa-2 (27). A particularly abundant source for the large M–CSF mRNA is the SV40-transformed trophoblast cell line TPA30-1 (9). In the same study (9) lectin-stimulated peripheral blood lymphocytes and the human T-cell line, C10-MJ2 (36), synthesized several M–CSF mRNAs, with the stimulated lymphocytes having predominant bands at approximately 4 and 2 kb.

Murine L-cells constitutively express at least two transcripts for M–CSF, which are approximately 2.3 and 4 kb (Fig. 5). Others have reported that L-cells also produce low levels of 3.8 and 1.4 kb M–CSF mRNAs (11), but these have not been detected in other L-cell lines. Another murine cell line, W20 (37), synthesizes both the 4 and 2.3 kb transcripts at similar levels (9).

To date, human monocytes have been the most extensively studied cell type for the expression of M–CSF transcripts. Circulating blood monocytes do not contain M–CSF mRNA, but adherence for 24 hr in human AB serum or adherence followed by a 4-hr incubation under serum-free conditions results in the production of the high-molecular-weight M–CSF mRNA (Fig. 5) (38). Phorbol myristic acetate addition induces higher levels of the large mRNA within 4-6 hr (38-40).

In both MIA-PaCa-2 cells and human monocytes, it appears that synthesis of the mRNA for M–CSF is not always correlated with high levels of protein expression. Although one signal may be enough to induce the expression of M–CSF transcripts, a second signal is necessary to allow the appearance of high levels of M–CSF protein in the medium (27,38). This observation may be related to the postulated membrane bound and circulating forms of M–CSF; the membrane form of M–CSF would not normally be detected unless first processed by a proteolytic cleavage to allow release into the medium.

Tissues and organs which have been shown to produce M–CSF transcripts include human liver and placenta (9), and murine liver, lung, brain, heart, spleen, and uterus (11). From the published data, most produce M–CSF mRNA of several sizes. However, human liver produces almost exclusively, perhaps only, the 4-kb variety (9). Intriguingly, mouse liver produces large amounts of a small, 1.4-kb M–CSF transcript (11). One tissue that has been studied extensively at both the protein and mRNA levels is the murine uterus. During pregnancy, there is a 10,000-fold increase mouse uterine M–CSF (41). Northern blots of uterine RNA from pregnant mice showed a preponderance of the

2.3-kb M–CSF mRNA and only a small amount of the large M–CSF RNA transcript. RNA from uteri of nonpregnant mice contained undetectable amounts of M–CSF message. In addition, in-situ hybridization of M–CSF RNA probes to uterine sections showed a localization of the signal to the luminal and glandular secretory epithelial cells of the endometrium (42).

Human endothelial cells, which normally synthesize very low levels of M–CSF mRNA, increase their transcription of M–CSF (as well as G- and GM–CSF) when stimulated with LPS, TNF-α or IL-1 (43). Finally, it has been shown that a clonal cell line derived from murine bone marrow stroma, phenotypically similar to a population of fibroblastoid cells termed adventitial cells, can synthesize M–CSF mRNA and produce M–CSF protein (44).

The foregoing has shown that M–CSF can be produced by many different cell types, tissues, and organs. There should be a note of caution in interpreting some of the results, because many organs or tissues are heavily infiltrated with other cells, such as macrophges, which can produce large amounts of M–CSF mRNA. However, the evidence presented above suggests that there is more to the function of M–CSF than just the stimulation of monocytes, macrophages, and their precursors.

VI. CHROMOSOMAL LOCALIZATION OF THE M–CSF GENE

In-situ hybridization studies (45) have located the human M–CSF gene on the long arm of chromosome 5 at 5q33.1. This region is of particular interest because deletions in this area are known to be associated with several hematological disorders (46). In addition, the C–FMS protooncogene, which recently has been identified as the M–CSF receptor (47), has also been localized to 5q33 (48). In fact the 5q arm is becoming crowded with growth factor genes and receptors. So far, the 5q23-33 region is known to contain the genes for M–CSF (45), C–FMS (48), GM–CSF (48), IL-3 (49), endothelial growth factor (ECGF) (50), and the receptor for platelet-derived growth factor (PDGF) (51). Thus, it is not surprising that deletions or translocations in this region can cause hematopoietic dysfunction. In the evolutionary sense, there may be some positive selection for this array of genes, because this linkage group may have been conserved in the mouse; the murine GM–CSF and IL-3 genes lie within 230 kb of each other on the murine chromosome 11 (52).

VII. STRUCTURE-FUNCTION STUDIES OF HUMAN M–CSF

A variety of mutations were engineered into the human short and long M–CSF sequence to study their effects on biological activity (28,30). Standard methods were applied for site-directed mutagenesis (53), and the expressed mutated proteins were tested for biological activity in the mouse bone marrow colony assay (Table 2) (7,54).

The original short M–CSF clone (7) encoded a Tyr at residue 59, while the genomic clone encoded an Asp. To determine whether this difference affected the biological activity of M–CSF expressed from this clone, we changed the Tyr to Asp. As shown (mutant 2), changing the Tyr to Asp had no effect on activity. We now believe that the presence of the Tyr codon was due to a single base reverse transcriptase error in the synthesis of the cDNA, because all other human and mouse M–CSF sequences contain an Asp at this position.

To aid in bacterial expression of M–CSF, a potentially protease-sensitive, dibasic site was eliminated (Lys-51-Lys-52). Lys-52 was changed to Pro or Gln to help inhibit trypsinlike protease activity at this position (Table 2). The Lys and Pro mutation (mutant 3) completely abolished activity, whereas the change to Gln (mutant 4) had little effect. This finding suggests that the amino-terminal end of M–CSF is important for biological activity, because large changes in the carboxy-terminal end have little effect (see below).

All seven Cys in the short form of M–CSF were mutated to Ser or Ala to define which are involved in intra- or intermolecular disulfide bonds. None of these mutants was active (mutants 5-11), implying that they may all be necessary for monomer folding and/or dimer formation. However, when cysteines eight and nine of the mature sequence of the long protein were changed to Ser, the biological activity remained unchanged (mutants 12 and 13). This result is evidence that these residues are not involved in critical disulfide bonds.

Because native M–CSF is such a heavily glycosylated protein, we wished to determine whether the attachment of carbohydrate is an absolute requirement for biological activity. The sugar attachments sites at Asn-122 and Asn-140 of short M–CSF were changed to Gln. Surprisingly, neither the single mutants (mutants 14 and 15) nor the double mutants (mutant 16) had much effect on biological activity. Of course, this result only means that removal of the carbohydrates does not disrupt binding of the protein to its receptor; the sugars may have some other function or functions in vivo.

The coding sequence of the short and long forms of M–CSF predict a monomer size of 26 and 61 kDa, respectively. These sizes are much larger than originally predicted (14.5 kDa) for the subunits after removal of the carbohydrate using glycosidases (1). A well-defined amino-terminal sequence, starting 33 amino acids from the predicted start codon, has been found for M–CSF protein from a variety of human and murine sources. This fact implies that any large-scale posttranslational processing must occur at the carboxy-terminal end, and also that these carboxy-terminal modifications are functional. We tested this idea by placing stop codons at positions 159 and 151 of short M–CSF, and 222 of long M–CSF. In essence, these stop codons create deletion mutants of 66 (mutant 17), 74 (mutant 18) and 301 amino acids (mutant 19), respectively. As shown (Table 2), all three mutants are fully active. They actually express higher levels of M–CSF, presumably because removal of the transmembrane domain allows for more efficient secretion. Similar biologically active deletion mutants for large M–CSF have also been reported (17). These structure-function studies indicate that the actual receptor-binding portion of M–CSF resides in the amino-terminal one-half or one-third of the molecule, whereas the function of the carboxy-terminal portion is for membrane attachment or some other as yet unknown function.

VIII. SUMMARY

In this chapter we have described one of the more complex hemopoietic factors, M–CSF. The single-copy M–CSF gene is almost 21 kb in length and is arranged into 10 exons and 9 introns. Expression of the gene at the RNA level is heterogeneous, and several species of M–CSF mRNA have been found in human and murine cells and tissues. In human cells the different mRNAs arise from alternative splicing of the nuclear RNA precursor in both coding and noncoding regions. This results in mRNAs encoding two distinct M–CSF proteins, 256 and 554 amino acids in length. In murine cells only a 552-amino-acid form has been found thus far.

All forms of M–CSF have a 32-amino-acid signal peptide and a 23-amino-acid hydrophobic region near the carboxy-terminus, which resembles a transmembrane domain. A large portion of the carboxy-terminal end, including the hydrophobic region, is not found in the mature protein. Thus, the primary translation product of M–CSF is a prepropolypeptide, with processing occurring at both amino- and carboxy-terminal ends. The exact size of the mature protein is still somewhat in doubt, but deletion mutagenesis from the carboxy-terminal end indicates that the protein may be as small as 150 amino acids and still be functional. Site-directed mutagenesis has also shown that the first seven cysteines in the mature molecule are probably necessary for biological activity, whereas the next two cysteine residues are not. In spite of the heavy glycosylation found in the native protein, removal of the *N*-linked glycosylation signals does not seem to affect activity to any great degree.

The M–CSF gene and its receptor, C–FMS, are tightly linked on the long arm of chromosome 5, a unique finding in the ligand/receptor field. This region also contains the genes for GM–CSF, IL-3, ECGF, and the receptor for PDGF. A similar situation may exist on chromosome 11 of the mouse. The close linkage of these factors and receptors is the probable cause for the disorders of hemopoiesis that arise when deletions occur in this area.

The preceding discussion has shown how quickly the area of M–CSF molecular biology has advanced in the past 2-3 years. A great deal of effort is now being directed toward expressing M–CSF at high levels in a variety of prokaryotic and eukaryotic systems. Large amounts of purified recombinant M–CSF will be useful in unraveling the complex interactions of CSFs and for use in clinical situations. Finally, the availability of pure M–CSF DNA probes and protein will help in discovering and defining new functions and roles for this growth factor.

ADDENDUM

After this manuscript was completed, a third form of biologically active M–CSF was described (57). A cDNA clone was isolated encoding 438 amino acids. This clone shares the same 149 amino acid amino-terminus and 75 common carboxy-terminal residues of the 256 and 554 amino acid forms of M–CSF.

ACKNOWLEDGMENT

We thank the Departments of Molecular Biology, Cell Biology, and Protein Chemistry, and the DNA Synthesis Group at Cetus Corporation, without which, of course, we would not have beeen able to write this review. We would also like to acknowledge Cetus' management for their support and encouragement throughout the M–CSF project. Finally, a thanks to our Graphics Department for cheerfully providing us with figures.

REFERENCES

1. Das, S. K., and E. R. Stanley. Structure-function studies of a colony stimulating factor (CSF-1). *J. Biol. Chem. 257*:13679-13684 (1982).
2. Motoyoshi, K., T. Suda, K. Kusumoto, F. Takaku, and Y. Miura. Granulocyte macrophage-colony stimulating and binding activities of purified human urinary colony-stimulating factor to murine and human bone marrow cells. *Blood 60*:1378-1386 (1982).
3. Csejtey, J., and a. Boosman. Purification of human macrophage colony stimulating factor (CSF-1) from medium conditioned by pancreatic carcinoma cells. *Biochem. Biophys. Res. Commun. 138*:238-245 (1986).
4. Das, S. K., E. R. Stanley, L. J. Guilbert, and L. W. Forman. Human colony stimulating factor (CSF-1) radio-immunoassay: resolution of three subclasses of human colony-stimulating factors. *Blood 58*:630-641 (1981).
5. Stanley, E. R., L. J. Guilbert, R. J. Tushinski, and S. H. Bartelmez. CSF-1– a mononuclear phagocyte lineage specific haemopoietic growth factor. *J. Cell. Biochem. 21*:151-159 (1983).
6. Ralph, P., M. K. Warren, I. Nakoinz, M-T. Lee, L. Brindley, A. Sampson-Johannes, E. S. Kawasaki, M. B. Ladner, J. E. Strickler, A. Boosman, J.

Csejtey, and T. J. White. Biological properties and molecular biology of human macrophage growth factor, CSF-1. *Immunobiology 172*:194-204 (1986).
7. Kawasaki, E. S., M. B. Ladner, A. M. Wang, J. Van Arsdell, M. K. Warren, M. Y. Coyne, V. L. Schweickart, M.-T. Lee, K. J. Wilson, A. Boosman, E. R. Stanley, P. Ralph, and D. F. Mark. Molecular cloning of a complementary DNA encoding human macrophage-specific colony-stimulating factor (CSF-1). *Science 230*:291-296 (1985).
8. Ladner, M. B., G. A. Martin, J. A. Noble, D. M. Nikoloff, R. Tal, E. S. Kawasaki, and T. J. White. Human CSF-1: gene structure and alternative splicing of mRNA precursors. *EMBO J. 6*:2693-2698 (1987).
9. Wong, G. G., P. A. Temple, A. C. Leary, J. S. Witek-Giannotti, Y-C. Yang, A. B. Ciarletta, M. Chung, D. Murtha, R. Kriz, R. J. Kaufman, C. R. Ferenz, B. S. Sibley, K. J. Turner, R. M. Hewick, S. C. Clark, N. Yanai, H. Yokota, M. Yamada, M. Saito, K. Motoyoshi, and F. Takaku. Human CSF-1: molecular cloning and expression of 4 kb cDNA encoding the human urinary protein. *Science 235*:1504-1508 (1987).
10. DeLamarter, J. F., C. Hession, D. Semon, N. M. Gough, R. Rothenbuhler, and J-J. Mermod. Nucleotide sequence of a cDNA encoding murine CSF-1 (macrophage-CSF). *Nucleic Acids Res. 15*:2389-2390 (1987).
11. Rajavashisth, T. B., R. Eng, R. K. Shadduck, A. Waheed, C. M. Ben-Avram, J. E. Shively, and A. J. Lusis. Cloning and tissue-specific expression of mouse macrophge colony-stimulating factor mRNA. *Proc. Natl. Acad. Sci. USA 84*:1157-1161 (1987).
12. Ladner, M. B., G. A. Martin, J. A. Noble, V. P. Wittman, M. K. Warren, M. McGrogan, and E. R. Stanley. cDNA cloning and expression of murine CSF-1 from L929 cells. *Proc. Natl. Acad. Sci. USA 85*:6706-6710 (1988).
13. Maniatis, T., R. C. Hardison, E. Lacy, J. Lauer, C. O'Connell, D. Quon, G. K. Sim, and A. Efstratiadis. The isolation of structure genes from libraries of eucaryotic DNA. *Cell 15*:687-701 (1978).
14. Naora, H., and N. J. Deacon. Relationship between the total size of exons and introns in protein-coding genes of higher eukaryotes. *Proc. Natl. Acad. Sci. USA 79*:6196-6200 (1982).
15. Reynolds, G. A., S. K. Basu, T. F. Osborne, D. J. Chin, G. Gil, M. S. Brown, J. L. Goldstein, and K. L. Luskey. HMG CoA reductase: a negatively regulated gene with unusual promoter and 5′ untranslated regions. *Cell 38*: 275-285 (1984).
16. Mount, S. M. A catalogue of splice junction sequences. *Nucleic Acids Res. 10*:459-472 (1982).
17. Clark, S. C., and R. Kamen. The human haematopoietic colony-stimulating factors. *Science 236*:1229-1237 (1987).
18. Nevins, J. R. The pathway of eukaryotic mRNA formation. *Annu. Rev. Biochem. 52*:441-466 (1983).
19. Rich, A., A. Nordheim, and A. Wang. The chemistry and biology of left-handed Z-DNA. *Annu. Rev. Biochem. 53*:792-846 (1984).

20. Dynan, W. S., S. Sazer, R. Tjian, and R. T. Schimke. Transcription factor Sp1 recognizes a DNA sequence in the mouse dihydrofolate reductase promoter. *Nature 349*:246-248 (1986).
21. Ishii, S., G. T. Merlino, and I. Pastan. Promoter region of the human Harvey-ras proto-oncogene: similarity to the EGF receptor proto-oncogene promoter. *Science 230*:1378-1381 (1985).
22. Ishii, S., Y-H. Xu, R. H. Stratton, B. A. Roe, G. T. Merlino, and I. Pastan. Characterization and sequence of the promoter region of the human epidermal growth factor receptor gene. *Proc. Natl. Acad. Sci. USA 82*:4920-4924 (1985).
23. Valerio, D., M. Duyvesteyn, B. Dekker, G. Weeda, T. Berkrens, L. van der Voorn, H. van Ormondt, and A. van der Eb. Adenosine deaminase: characterization and expression of a gene with a remarkable promoter. *EMBO J. 4*:437-443 (1985).
24. Kim, S. H., J. C. Moores, D. David, J. G. Respess, D. J. Jolly, and T. Friedmann. The organization of the human HPRT gene. *Nucleic Acids Res. 14*:3103-3118 (1986).
25. Laimins, L. A., M. Kessel, N. Rosenthal, and G. Khoury. Viral and cellular enhancer elements. In *Communications in Molecular Biology, Enhancers and Eukaryotic Gene Expression*, Y. Gluzman and T. Shenk (Eds.). Cold Spring Harbor Laboratory, New York, pp. 28-87 (1983).
26. Wu, M-C., J. K. Cini, and A. A. Yunis. Purification of a colony-stimulating factor from cultured pancreatic carcinoma cells. *J. Biol. Chem. 254*:6226-6228 (1979).
27. Ralph, P., M. K. Warren, M-T. Lee, J. Csejtey, J. F. Weaver, H. E. Broxmeyer, D. E. Williams, E. R. Stanley, and E. S. Kawasaki. Inducible production of human macrophage growth factor, CSF-1. *Blood 68*:633-639 (1986).
28. Ralph, P., M. B. Ladner, A. M. Wang, E. S. Kawasaki, L. McConlogue, J. F. Weaver, S. A. Weiss, P. Shalde, K. Koths, and M. K. Warren. The molecular and biological properties of the human and murine members of the CSF-1 family. In *Molecular Basis of Lymphokine Action*, D. R. Webb, C. W. Pierce and S. Cohen (Eds.). Humana Press, City, New Jersey, pp. 295-311 (1987).
29. von Heijne, G. Signal sequences—the limits of variation. *J. Mol. Biol. 184*:99-105 (1985).
30. Kawasaki, E. S. Structure-function studies of human M–CSF. (In preparation.)
31. Sabatini, D. D., G. Kreibich, T. Morimoto, and M. Adesnik. Mechanisms for the incorporation of proteins in membranes and organelles. *J. Cell. Biol. 91*:1-22 (1982).
32. Stanley, E. R., M. Cifone, P. M. Heard, and V. Defendi. Factors regulating macrophage production and growth: identity of colony-stimulating factor and macrophage growth factor. *J. Exp. Med. 143*:631-647 (1976).

33. Cifoni, M., and V. Defendi. Cyclic expresion of a growth conditioning factor (MGF) on the cell surface. *Nature 252*:151-153 (1974).
34. Rettenmier, C. W., M. F. Roussel, R. A. Ashmun, P. Ralph, K. Price, and C. Sherr. Synthesis of membrane-bound colony-stimulating factor 1 (CSF-1) and down modulation of CSF-1 receptors in NIH 3T3 cells transformed by cotransfection of the human CSF-1 and c-fms (CSF-1 receptor) genes. *Mol. Cell. Biol. 7*:2378-2387 (1987).
35. Stanley, E. R., and P. M. Heard. Factors regulating macrophage production and growth. Purification and some properties of colony stimulating factor from medium conditioned by L-cells. *J. Biol. Chem. 252*:4305-4312 (1977).
36. Arya, S. K., F. Wong-Staal, and R. C. Gallo. T-cell growth factor gene: lack of expression in human T-cell leukemia-lymphoma virus-infected cells. *Science 223*:1086-1087 (1984).
37. Tsai, S., C. A. Sieff, and D. G. Nathan. Stromal cell-associated erythropoiesis. *Blood 67*:1418-1426 (1986).
38. Haskill, S., M. K. Warren, S. Becker, M. B. Ladner, C. Johnson, D. Eierman, P. Ralph, and D. F. Mark. Adherence induces CSF-1 gene expression in monocytes. *J. Leukocyte Biol. 41*:359 (Abstract 91) (1987).
39. Horiguchi, J., M. K. Warren, P. Ralph, and D. Kufe. Expression of the macrophage specific colony-stimulating factor (CSF-1) during monocytic differentiation. *Biochem. Biophys. Res. Commun. 141*:924-930 (1986).
40. Rambaldi, A., D. C. Young, and J. D. Griffin. Expression of the M–CSF (CSF-1) gene by monocytes. *Blood 69*:1409-1413 (1987).
41. Bartocci, A., J. W. Pollard, and E. R. Stanley. Regulation of colony-stimulating factor 1 during pregnancy. *J. Exp. Med. 164*:956-961 (1986).
42. Pollard, J. W., A. Bartocci, R. Arceci, A. Orlofsky, M. B. Ladner, and E. R. Stanley. Apparent role of the macrophage growth factor, CSF-1, in placental development. *Nature 330*:484-486 (1987).
43. Seelentag, W. K., J-J. Mermod, R. Montesano, and P. Vassalli. Additive effects of interleukin 1 and tumor necrosis factor-alpha on the accumulation of the three granulocyte and macrophage colony-stimulating mRNAs in human endothelial cells. *EMBO J. 6*:2261-2265 (1987).
44. Hunt, P., D. Robertson, D. Weiss, D. Rennick, R. Lee, and O. N. Witte. A single bone marrow-derived stromal cell type supports the in vitro growth of early lymphoid and myeloid cells. *Cell 48*:997-1007 (1987).
45. Pettenati, M. J., M. M. LeBeau, R. S. Lemons, E. A. Shima, E. S. Kawasaki, R. A. Larson, C. J. Sherr, M. O. Diaz, and J. D. Rowley. Assignment of CSF-1 to 5q33.1: evidence for clustering of genes regulating haemotopoiesis and for their involvement in the deletion of the long arm of chromosome 5 in myeloid disorders. *Proc. Natl. Acad. Sci. USA 84*:2970-2974 (1987).
46. Wisniewski, L. P., and K. Hirschorn. Acquired partial deletions of the long arm of chromosome 5 in haemotological disorders. *Am. J. Hematol. 15*: 295-310 91983).

47. Sherr, C. J., C. W. Rettenmier, R. Sacca, M. F. Roussel, A. T. Look, and E. R. Stanley. The c-fms proto-oncogene product is related to the receptor for the mononuclear phagocyte growth factor, CSF-1. *Cell 41*:665-676 (1985).
48. LeBeau, M. M., C. A. Westbrook, M. O. Diaz, R. A. Larson, J. D. Rowley, J. C. Gasson, D. W. Golde, and C. J. Sherr. Evidence for involvement of GM–CSF and FMS in the deletion (5q) in myeloid disorders. *Science 231*: 984-987 (1986).
49. LeBeau, M. M., N. D. Epstein, S. J. O'Brien, A. W. Nienhuis, Y-C. Yang, S. C. Clark, and J. D. Rowley. The interleukin-3 gene is located on human chromosome 5 and is deleted in myeloid leukemias with a deletion of 5q. *Proc. Natl. Acad. Sci. USA 84*:5913-5917 (1987).
50. Jaye, M., R. Howk, W. Burgess, G. A. Ricca, I-M. Chiu, M. W. Ravera, S. J. O'Brien, W. S. Modi, T. Maciag, and W. N. Drohan. Human endothelial cell growth factor: cloning, nucleotide sequence, and chromosomal localization. *Science 233*:541-545 (1986).
51. Yarden, Y., J. A. Escobedo, W. J. Kuang, T. L. Yang-Feng, T. O. Daniel, P. M. Tremble, E. Y. Chan, M. E. Ando, R. N. Harkins, U. Francke, V. A. Fried, A. Ullrich, and L. T. Williams. Structure of the receptor for platelet derived growth factor helps define a family of closely related growth factor receptors. *Nature 323*:226-232 (1986).
52. Barlow, D. P., M. Bucan, H. Lehrach, B. L. M. Hogan, and N. M. Gough. Close genetic and physical linkage between the murine haemopoietic growth factor genes GM–CSF and multi-CSF (IL3). *EMBO J. 6*:617-623 (1987).
53. Smith, M. In vitro mutagenesis. *Annu. Rev. Genet. 19*:423-463 (1985).
54. Stanley, E. R. Colony stimulating factors. In *The Lymphokines*, W. E. Stewart and J. W. Hadden (Eds.). Humana Press, City, New Jersey, pp. 102-132 (1981).
55. Shaw, G., and R. Kamen. A conserved AU sequence from the 3′ untranslated region of GM–CSF mRNA mediates selective degradation. *Cell 46*: 659-667 (1986).
56. Caput, D., B. Beutler, K. Hartog, R. Thayer, S. Brown-Shimer, and A. Cerami. Identification of a common nucleotide sequence in the 3′-untranslated region of mRNA molecules specifying inflammatory mediators. *Proc. Natl. Acad. Sci. USA 83*:1670-1674 (1986).
57. Cerretti, D. P., J. Wignall, D. Anderson, R. J. Tushinski, B. M. Gallis, M. Stya, S. Gillis, D. L. Urdal, and D. Cosman. Human macrophage-colony stimulating factor: alternative RNA and protein processing from a single gene. *Mol. Immun. 25*:761–770 (1988).

6
Molecular and Cellular Biology of Interleukin-3

CAROLYN F. MORRIS, IAN G. YOUNG, and ANDREW J. HAPEL
John Curtin School of Medical Research, Australian National University, Canberra, Australia

I. INTRODUCTION

The cellular components of blood comprise several distinct, specialized cell types. These include erythrocytes involved in oxygen transport, platelets that initiate clotting, and white blood cells (lymphocytes, granulocytes, and monocytes) that are involved in immune and inflammatory responses. The mature blood cells develop from pluripotent stem cells capable of reconstituting the entire blood cell system. These cells are found in embryonic yolk sac, fetal liver, and adult bone marrow. The functionally specialized mature cells are generally short lived and are constantly replaced. For example, approximately 1.2 million erythrocytes and 0.5 million white cells are replaced per second in the steady state and these rates are increased after bleeding or during an inflammatory or immune response.

The process of blood cell formation is, in part, regulated by a family of glycoproteins called colony-stimulating factors (CSFs). These factors appear to act in a sequential manner on successive generations of progeny from pluripotent stem cells. As cycles of cell division occur, daughter cells become committed to a particular cell lineage. Blood progenitor cells therefore generally contain a continuum representing the different stages of stem-cell development.

Progenitor cells were first identified during reconstitution experiments in which bone marrow cells were transplanted into irradiated recipients (1,2). Colonies, derived from a small proportion of injected cells, developed in the spleens of recipient mice, and these colonies differed in their content of transplantable stem cells. Colonies that contained predominantly granulocytes or macrophages were generally far less capable of reconstituting secondary recipients than were colonies containing immature blast cells (3). In in vitro experiments, Johnson and Metcalf (4) demonstrated that, as in vivo, single cells could give rise to colonies that contained cells with the morphology of granulocytes, macrophages, erythroblasts, or megakaryocytes, as well as mixed colonies containing many cell types. The advantage of the in vitro assays, where cells were plated at low density suspended in semisolid medium containing agar, was that the effect of "conditioning factors" on colony formation could be studied. These in vitro and in vivo systems complemented each other in demonstrating that the hemopoietic system contained cells that differed in their capacity for proliferation and differentiation.

The differences in colony type revealed in in vitro colony-forming assays were affected by two parameters: the nature of the "stem" cells used and the type of "conditioning" factors that were added. Culture supernatants from L929 fibroblasts promoted growth primarily of macrophage (M) colonies, and gave rise to the notion of a specific M–CSF (5). Culture supernatants from organ culture of whole lung gave rise to an increased proportion of granulocyte or granulocyte-macrophage colonies (6) while spleen cell conditioned medium also induced eosinophil, megakaryocyte, and erythroid colonies (7). Biochemical fractionation of these complex factor sources revealed the presence of several distinct (glyco)protein CSFs, whose activities were apparently restricted to inducing proliferation or differentiation in a particular subclass of hemopoietic cells.

A number of CSF activities found in spleen-cell-conditioned medium were refractory to biochemical separation, and eventually were shown to have great similarity with the CSF activities produced by the myelomonocytic leukemia WEHI-3B. These activities were designated erythroid burst (colony)-promoting activity (BPA) (8), mast-cell growth factor (9), Thy-1-inducing activity (10), hemopoietin-2 (11), 20α-hydroxysteroid dehydrogenase (20αSDH)-inducing activity, interleukin-3 (IL-3) (12), and so on. Some workers suggested that many of these activities were, in fact, the properties of a single CSF (13,14) and named the factor accordingly (e.g., multi-CSF). The latter interpretation was ultimately shown to be correct. The factor involved is now commonly referred to as a multi-CSF or IL-3 and is believed to play a central role in regulating the production of all of the major blood cell types including macrophages, megakaryocytes, mast cells, eosinophils, neutrophils, erythroblasts, and probably the precursors of T and B cells (15,16).

The primary source of IL-3 appears to be the antigen-reactive T lymphocyte that is present only in animals with a mature immune system. This creates a problem in ascribing a central role for IL-3 in embryonic or fetal blood cell development. Many investigators therefore consider IL-3 to be solely a mediator of the inflammatory events that follow T-cell stimulation. Our own view, in part because we have demonstrated constitutive production of IL-3 by endothelial cell cultures, is that IL-3 may be present and active from the earliest development of blood islets in the embryonic yolk sac, and may continue to play a role in steady-state adult hematopoiesis. Some of our findings that support this notion are presented here, along with a review of the cellular and molecular biology of this multifaceted hemopoietic cytokine.

II. BIOLOGICAL ACTIVITIES OF IL-3 IN VITRO

A. Biological Activities of Murine IL-3

Interleukin-3 was initially defined by its ability to induce the enzyme 20αSDH in spleen cells from athymic nu/nu mice (17). At the time 20αSDH was regarded as a unique marker for T-cell maturation and IL-3 was therefore proposed to mediate early events in T-cell differentiation (12). Using a highly sensitive and specific assay for 20αSDH we found that this enzyme is present in a variety of hemopoietic cells including bone-marrow-derived macrophages and the nonlymphoid cell lines, L929 (fibroblastlike) and WEHI-3B (myelomonocytic leukemia). We also found that, although 20αSDH is present in very low levels in spleens of nude mice, the enzyme occurs in high levels in the bone marrow of these animals as well as in the bone marrow and fetal liver of normal mice. Furthermore, the induction of 20αSDH is not an activity unique to IL-3; both GM–CSF and M–CSF maintain high levels of 20αSDH expression in normal macrophages and GM–CSF induces 20αSDH in nu/nu splenocytes and fetal liver cells (18). Thus, our results indicate that although the induction of 20αSDH is a highly sensitive assay for hemopoietic growth-factor activity, it is not specific for IL-3 and does not define IL-3 as a regulator of T-cell differentiation.

Purified or recombinant IL-3 induced the formation of colonies in soft agar cultures of bone marrow cells (15,19,20) and IL-3 has therefore been classified as a CSF. In these semisolid cultures, IL-3 supports the development of mixed myeloid/erythroid, macrophage, neutrophil, eosinophil, and basophil colonies as well as early growth of stem-cell colonies containing progenitors of the erythroid and megakaryocyte lineages. Direct effects of IL-3 on progenitors of mast cells, megakaryocytes, neutrophils, and macrophages have been established by experimentation with single isolated cells (21).

IL-3 is unable to support full megakaryocyte development to the platelet-shedding stage in the absence of thromobopoietin (22). Similarly, optimal production of mature erythrocytes requires erythropoietin (23), although under certain conditions IL-3 can apparently drive the complete sequence of events leading to the formation of mature red blood cells (24).

The involvement of IL-3 in the regulation of early multipotential stem cells has been studied using blast-cell colony assays (25,26). These experiments revealed that, in contrast to more mature progenitors, the progenitors for blast cell colonies are not in cell cycle and do not require IL-3 for survival (27). Furthermore, IL-3 does not stimulate stem cells in the Go state to begin active proliferation but rather provides a permissive milieu for the continued proliferation of early hemopoietic progenitors (28).

Interleukin-3 also supports the proliferation of a number of continuous cell lines, for example FDC-P1 and 32D cl-23. FDC-P1 cells, originally derived from "Dexter" long-term cultures of DBA/2 bone marrow (29), express the macrophage specific marker F4/80 and require IL-3 or GM–CSF for growth (30). 32D cl-23 cells were derived from cultures of C3H bone marrow (31), are alcian blue-positive and can give rise to mast cell/basophil colonies in soft agar. Proliferation of 32D cl-23 cells is supported by either IL-3 or IL-2 (30).

We routinely use these cell lines in assays to detect and measure levels of IL-3. The assays are based on measurements of the incorporation of tritiated thymidine by proliferating cells (30) or on spectrophotometric detection of dyes that stain living cells (32).

Experiments using continuous cell lines have highlighted the requirement of IL-3 for the survival of hemopoietic cells. Thus, culture of 32D cl-23 or FDC-P1 cells in the absence of IL-3 leads to rapid cell death (33,34). Whetton and associates have proposed that the effect of IL-3 on cell survival is due to the maintenance of adenosine triphosphate levels and activation of hexose transport and glycolysis (35).

In addition to its activity on continuous cell lines, IL-3 can support the proliferation of mucosal type mast cells derived from normal bone marrow or spleen cell cultures. In early work, before the availability of purified or recombinant IL-3, a number of groups reported that when bone marrow cells or splenocytes were cultured in the presence of T-cell conditioned medium, the cultures soon consisted almost entirely of cells phenotypically similar to mucosal mast cells (36-40). The continued growth of these mast cells depended on a factor present in the T-cell-conditioned medium. This factor was originally referred to as mast-cell growth factor (MCGF) or persisting cell stimulating factor (PCSF). It has since been shown that MCGF and PCSF are the same as IL-3 (15,19,20). In our laboratory, mucosal type mast cells are routinely derived from NIH Swiss and BALB/c mice by culture of spleen or bone marrow cells in IL-3. Mast cells generated in this way are dependent on IL-3 for their growth and can be used as an assay for IL-3 activity. Similar effects of IL-3 on proliferation of mucosal mast cells have been noted in the rat (41).

In addition to mucosal type mast cells, there is a phenotypically distinct subpopulation of mast cells in rodents, known as connective-tissue-type mast cells (CTMCs). Recent studies have suggested that IL-3 has only poor proliferative activity on CTMCs although this can be significantly enhanced by the presence of IL-4 (42).

B. Synergism Between Murine IL-3 and Other Factors

A particularly interesting aspect of IL-3 biology is its apparent synergism with other growth factors. Early evidence that IL-3 had synergistic activity came from the realization that a factor with BPA was probably identical to IL-3. Conceptually, BPA was a "lineage-indifferent" factor needed to expand a population of pluripotent hemopoietic precursors and their early committed progeny (13). Lineage specific factors such as M–CSF or erythropoietin acted only on later cells that had acquired the appropriate receptors as part of their differentiation programs. In support of this proposal, it was found that a preparation of BPA, extensively purified from WEHI-3B-conditioned medium, increased the numbers of erythroid bursts when added to bone marrow colony assays containing erythropoietin and also augmented macrophage colony growth in the presence of purified macrophage colony stimulating factor (M–CSF, CSF-1) (13).

Later studies confirmed that IL-3 synergizes with M–CSF when assayed on bone marrow enriched for early progenitor cells. The combination of IL-3 and M–CSF enhances M–CSF receptor levels synergistically and generates very large macrophage colonies in soft agar cultures (43,44). The IL-3 + M–CSF responsive cell that produces these distinctively large colonies is called the high proliferative potential colony forming cell and is proposed to be a primitive progenitor cell (45).

C. Strains of Mice with Abnormal Responsiveness to IL-3

A further development in the study of synergism between IL-3 and M–CSF came from our experiments on bone marrow cells from A/J mice (46). There is a large strain-dependent variation in the response of murine bone marrow cells to IL-3. Cells from A/J, AKR, and the congenic strains ATH and ATL are particularly unresponsive, rarely producing any colonies in soft agar supplemented with IL-3 alone. Kincade and colleagues (47) have observed that the NZB strain is also a poor responder to IL-3. We found that even though IL-3 apparently has no proliferative activity on bone marrow cells from A/J and AKR mice, it will nevertheless synergize with M–CSF to generate giant colonies. Interleukin-3 also enhances the expression of the M–CSF receptor (c-fms) mRNA in A/J bone marrow cells. This synergism between IL-3 and M–CSF, in the absence of direct IL-3 induced proliferation, provides strong evidence that the classical BPA concept cannot be applied to all of the synergistic activities of IL-3.

We have observed a similar phenomenon in embryonic hemopoietic cells. In colony assays using cells isolated from the yolk sac blood islets of 8-day mouse embryos we find that in BALB/c and NIH/Swiss as well as A/J embryos the predominant growth-factor-responsive colony-forming cell requires both IL-3 and M–CSF. These cells therefore seem to be key elements of embryonic blood cell development but are apparently superseded in the adult bone marrow by cells responsive to IL-3 alone. It is not clear why this transition does not occur in A/J mice. Perhaps a subpopulation of stem cells from which IL-3 responsive colony-forming cells derive is absent in the A/J strain. Alternatively, there may be a defect in the IL-3 receptor, or the proliferative signal from this receptor in the A/J system. As a corollary to this theory, the synergistic and mitogenic responses to IL-3 would have to be controlled by separate signalling pathways. Whatever the underlying mechanism may be, the separation of the synergistic from the mitogenic effects of IL-3 in the A/J system provides us with an ideal model for studying the role of the synergistic activities of IL-3 in hemopoiesis.

The synergistic activities of IL-3 extend beyond its interactions with M–CSF and erythropoietin. Interleukin-3 and human recombinant IL-6 act synergistically to support multilineage colony formation (48); the combination of IL-1 and IL-3 has a similar effect (49,50) but this may result from IL-1-induced production of IL-6 in culture (48). In our own experiments using the A/J model we have found that IL-3 and IL-1 synergize to produce large numbers of granulocyte colonies. This synergism, however, is only observed at high bone marrow cell densities, suggesting that the synergistic activity of IL-1 may be mediated by induction of secondary cytokines in culture. Interleukin-3 and IL-4 synergize to support the proliferation of CTMCs and mast-cell lines (42,51) although IL-4 is slightly inhibitory for IL-3-dependent colony formation by multipotential cells (52). In the A/J system, IL-3 and IL-4 synergize in the presence of erythropoietin, which may provide a pathway for erythroid differentiation in this strain.

D. Biological Properties of Rat, Human, and Gibbon IL-3

Colony assays done using recombinant rat IL-3 and rat bone marrow cells produce the full spectrum of colonies found using mouse IL-3 in the murine system. Rat and mouse IL-3s, however, show little cross-reactivity between species. Rat IL-3 has minimal activity on the murine-factor-dependent FDC-P1 and 32D cl-23 cell lines and on mouse bone marrow cells, and mouse IL-3 is completely inactive on rat bone marrow cells (53). Further studies using recombinant rat IL-3 have shown that it supports the growth of mucosal mast cells (MMCs) from rat multipotential progenitor cells (41). These MMCs are present in the majority of multilineage colonies in semisolid cultures, indicating that the MMC lineage originates from primitive multipotential progenitors.

We have compared the biological activities of human recombinant IL-3 to those of GM–CSF and G–CSF using colony assays and liquid proliferation assays established in our laboratory. To prepare target cells for these assays, mononuclear cells, obtained from human bone marrow by centrifugation onto a Ficoll cushion, are washed and plated at 1-2 × 10^7/60 mm Petri dish in Iscove's medium supplemented with 20% fetal calf serum and 10^{-5} *M* hydrocortisone. After 3 weeks, the population of nonadherent early progenitor cells is enriched using Percoll gradients. The resulting "stem" cell population is used as if it were a factor-dependent cell line in sensitive [^{3}H] thymidine incorporation assays and in standard colony assays.

In colony assays using fresh human bone marrow, or cultured "stem" cells, human IL-3, and GM–CSF at saturating levels both promote the formation of G, GM, M, mixed, and erythroid colonies. Interleukin-3 is superior to GM–CSF only in promoting growth of megakaryocyte-containing colonies from progenitor cells in fresh bone marrow. In liquid culture both IL-3 and GM–CSF maintain factor-responsive progenitor populations for short periods but IL-3 becomes more effective in this role as the period of culture increases. We compared the ability of IL-3, GM–CSF, and G–CSF to maintain multipotent progenitor cells in liquid culture using colony assays of cultured cells in the presence of G, CM–CSF, or IL-3. The total yield of nonadherent cells after 21 days of culture in IL-3 was 2½ times greater than that in GM–CSF and 10 times greater than that in G–CSF. A larger proportion of the nonadherent cells precultured in IL-3 subsequently produced colonies in IL-3 than in G or GM–CSF, and 17% of the colonies that grew in IL-3 were stem-cell colonies based on morphology and efficiency in replating experiments. Thus, culture of human bone marrow cells in IL-3 selects for IL-3 responsiveness and stem-cell survival, and may be a useful way of maintaining an enriched stem cell population in vitro for the purpose of autologous grafting. Interleukin-3 may also ultimately prove to be a useful lymphokine for treating cancer patients with myelosuppression after chemotherapy or radiotherapy.

The biological properties of gibbon IL-3 (which is 93% homologous to the human factor) are largely in agreement with those of human IL-3 described above (54). In addition, Lopez and colleagues (54) found that gibbon IL-3 strongly stimulates some functions of mature human eosinophils but not of neutrophils whereas GM–CSF stimulates both cell types.

Recently, gibbon IL-3 was found to be more effective than GM–CSF at supporting 21-day colony formation from cultures of purified bone marrow progenitors. These IL-3-derived colonies could be replated at higher efficiency than GM–CSF derived colonies to yield secondary colonies, many of which contained multiple cell lineages (55). Such results are consistent with our own work described above and with a murine study in which it was concluded that the multipotential progenitors responding to GM–CSF represents a subpopulation of the multipotential progenitors responding to IL-3 (56).

III. BIOLOGICAL ACTIVITIES OF IL-3 IN VIVO

A. Effects of Administration of Recombinant IL-3

In one of the earliest in vivo studies, continuous infusion of large amounts of recombinant IL-3 via an implanted osmotic pump resulted in significant increases in the numbers of progenitor cells and myeloid committed precursors in the spleen but not in the bone marrow (57). Broadly similar effects were observed after a 6-day regime involving regular intraperitoneal injection of 100,000 U of IL-3 (58). The most striking overall population changes in the latter study were in the peritoneal cavity (the site of IL-3 injection) where large increases in numbers of macrophages, eosinophils, and neutrophils were observed, probably due at least in part to an influx of cells from the circulation. In addition, there was evidence of increased peritoneal macrophage phagocytic activity.

Later Broxmeyer and co-workers (59) carried out a series of in vivo experiments using mice pretreated with purified iron-saturated lactoferrin, a material that suppresses release of some hemopoietic growth factors and inhibitors in vitro and also suppresses myelopoiesis in vivo. This pretreatment was designed to facilitate the detection of effects of relatively small quantities of exogenously administered CSFs by reducing both the background of normal hemopoiesis and the masking effects of CSF-induced release of suppressor molecules. Using this model, an infusion of 2,000 U of IL-3 was sufficient to enhance the cycling status and absolute numbers of all progenitors in the bone marrow and to a lesser extent in the spleen (59). Moreover, when administered with low doses of GM–CSF or M–CSF, which themselves have no observable effect in vivo, the concentration of IL-3 required to produce a stimulatory effect could be further reduced (60), thus demonstrating that combinations of CSFs can synergize in vivo.

B. Expression of IL-3 by Different Cell Types

The predominant cellular source of IL-3 appears to be antigen- or mitogen-activated T cells, particularly the $Lyt2^-$ $L3T4^+$ subpopulation, which is responsible for the majority of lymphokine production (61,62).

Because some stimulated $Lyt2^-$ $L3T4^+$ T cell clones produce IL-3, IL-2, GM–CSF, and IFN-γ, a number of groups have speculated that T-cell lymphokines may comprise a family of genes under common regulatory control. For example, Herold and associates (63) found that the T-cell lines AD9.1 and L.2 expressed the IL-3, IL-2, and IFN-γ genes coordinately, suggesting that the initial stimulatory event led to the concomitant expression of all three lymphokine genes. However, several observations suggest differential regulation of IL-3 and other lymphokine genes. For example, we have found that phorbol myristic acetate-stimulated EL4 cells express IL-2 at least 8 hr before IL-3 (Morris et al., unpublished data) and studies on stimulation of LB3 cells with either *Concanavalin A* or IL-2 show differences between the expression of the IL-3 and GM–CSF genes (64).

A further interesting aspect of the regulation of lymphokine gene expression is the demonstration that T-cell clones show a stable pattern of lymphokine production, both in terms of which lymphokines are expressed and their relative levels of production (64). This imples that T lymphocytes must acquire during maturation a stable clonotypic pattern of lymphokine gene expression (65). Mossmann and associates (66) have proposed that T-helper clones can be divided into two major groups: type 1 T-helper cells produce IL-2, IFNγ, GM–CSF, and IL-3, whereas type 2 T-helper cells produce GM–CSF, IL-3, IL-4, and IL-5. In the human, T cells that express the $\alpha\beta$ configuration of T-cell receptor produce IL-2, GM–CSF, and IL-3, whereas those expressing the $\gamma\delta$ configuration produce GM–CSF but not IL-2 (Warren et al., unpublished data). Further work is required to determine the biological significance of the proposed classes of helper T cells in both the murine and the human systems.

Probably the most well known non-T cell source of IL-3 is the WEHI-3B (D-) myelomonocytic cell line. This cell line, which displays constitutive rather than inducible IL-3 production, contains a rearranged gene for IL-3 that has an IAP retroviral genome inserted upstream of the promoter (67). The synthesis of IL-3 by WEHI-3B cells is thought to be a direct result of this rearrangement and therefore does not indicate that IL-3 production is a characteristic of normal cells from the myeloid or monocytic lineages.

Other non-T-cell sources reported to produce an IL-3-like activity include epidermal cells (68), yolk sac (69), and astrocytes (70), although the data presented in these cases is inconclusive because the factor was either not purified and/or the assays used were not specific for IL-3. In our laboratory we have isolated a CSF with the chromatographic properties of IL-3 in conditioned medium from cultures of human umbilical vein endothelial cells and have detected IL-3 mRNA by Northern blot analysis of RNA from these cells. Because these cultures are not clonally derived, the true source of IL-3 may, however, be contaminating subendothelial cells such as fibroblasts or myoblasts.

Attempts to detect IL-3 in vivo are beset by technical difficulties. Using the WEHI-3B cells as a model, we have found that the expression of IL-3 can be detected with greater sensitivity using biological assays than at the mRNA level using Northern analysis. The success of detecting IL-3 in vivo using biological assays is frustrated, however, not only by the lack of specificity of the assays but also by the possible consumption of the factor by IL-3 responsive cells.

Although there have been reports of an IL-3-like activity in the serum of mice with severe graft-versus-host disease (71) and of *Nippostrongylus brasiliensis*-infected mice after challenge with an intravenous injection of adult worm antigen (72), there has been no evidence of IL-3 in normal serum. Similarly, we have detected IL-3 mRNA in the mesenteric lymph nodes of mice infected with *Nematospiroides dubius* but have been unable to find the mRNA in normal mice despite an extensive study of adult and embryonic tissues (Morris et al., unpublished data).

C. In Vivo Role of IL-3

Although the activities of IL-3 in vitro have for the most part been established as direct effects on responding hemopoietic populations, the results of exogenous administration of IL-3 to whole animals are difficult to interpret because of the number of complex interactions in vivo. In addition, because it is difficult to detect IL-3 in normal serum, the massive doses applied in early, in vivo studies do not appear to imitate a normal physiological situation. The relevance of this work has, however, been defended on the basis of the short half-life of IL-3 in serum (73). The synergistic studies carried out using lactoferrin-treated mice, however, were successful in detecting the effects of relatively small single injections of IL-3 in combination with other factors and point to the probable significance of combinations of CSFs in the physiological regulation of myeloid blood cell production in vivo.

The absence of detectable levels of IL-3 in vivo in all but extreme immunological situations, and the predominantly T-cell source of IL-3 suggests that IL-3 may play a role in the immune response rather than in steady-state hemopoiesis. The massive increase in cellularity observed by Metcalf and co-workers (58) at the site of IL-3 injection suggests that localized IL-3 production by T cells may serve to recruit and activate cells at the site of infection.

Alternatively, IL-3 may be present at sites of normal hemopoiesis in vivo but at undetectably low levels. Indeed, the amount of IL-3 we observed produced by endothelium-associated cells in vitro was only just within our limit of detection and would therefore probably be impossible to detect in vivo using the assays currently available. Furthermore, IL-3 may be involved in direct cell-cell interactions in the hemopoietic microenvironment. In this context, an extracellular matrix that incorporates glycosaminoglycans has been shown to bind growth factors, such as GM–CSF, and present these growth factors to appropriate hemopoietic target cells (74). This provides a possible explanation for the earlier observation that, although growth factors are not detectable in the conditioned medium of marrow stromal cells, multipotential cells allowed to attach to the stroma undergo proliferation and differentiation to produce mature myeloid cells (75). Storage of IL-3 by binding to particular cells or to the extracellular matrix of the stroma could explain the difficulties encountered in detecting IL-3 in vivo. Future studies involving both in-situ hybridization to identify the location of IL-3 mRNA production and immunohistochemistry to locate the IL-3 protein are required to investigate this possibility.

Our observation that cells from the yolk sac are particularly responsive to IL-3 in combination with other growth factors suggests the availability of a non-T-cell source of IL-3 at this site. A similar conclusion was reached by Labastie and co-workers (69). Because we have evidence that IL-3 may be produced by endothelial cells, we propose that endothelial cells that surround blood islets in the yolk sac may be responsible for IL-3 production required for embyronic blood cell development. Furthermore, endothelial cells in the bone marrow stroma and lining blood vessel walls may produce IL-3, serving both to maintain steady-state hemopoiesis and the viability of circulating hemopoietic cells.

IV. CLONING AND EXPRESSION OF MURINE IL-3

A. Isolation of cDNA Clones for Murine IL-3

cDNA clones encoding murine IL-3 were isolated in our laboratory early in 1984 (76). The strategy used was based on the translation of mRNA in *Xenopus laevis* oocytes followed by assaying the translates for biologically active IL-3 using the IL-3 dependent cell line, 32D cl-23 (30,77). The source of IL-3 mRNA used in this study was the BALB/c-derived myelomonocytic cell line WEHI-3B, which produces IL-3 constitutively.

Poly(A) + RNA from WEHI-3B was fractionated by sucrose density gradient centrifugation, and the partially purified IL-3 mRNA was used as a template to synthesize double-stranded cDNA (78). This cDNA was then cloned into the plasmid pAT153 by G—C tailing (79). The resulting cDNA library was screened by the "hybrid release translation" method (80). This approach yielded a cDNA clone carrying the entire coding sequence for IL-3. The IL-3 cDNA sequence, together with the deduced primary structure of murine IL-3, is shown in Fig. 1.

At about the same time, Yokota and colleagues reported the isolation of a cDNA clone encoding the murine MCGF (81). In this study the source of mRNA was a cloned T-cell line, derived from C57BL/6 mice, which was known to produce MCGF after *Concanavalin A* stimulation (9). An unfractionated cDNA library was screened using the hybrid release method and *Xenopus* oocytes for translation. The activity was detected using a factor-dependent mast cell line. A feature of this cloning strategy was the use of a pCD cloning vector (82), which allows expression of cDNA inserts after transfection into monkey COS cells. The presence of MCGF activity directly expressed by transfected COS cells was used to verify that clones containing the complete MCGF coding sequence had been isolated.

```
                                                      10                            20
                           M  V  L  A  S  S  T  T  S  I  H  T  M  L  L  L  L  L  M  L  F
GAACCCCTTGGAGGACCAGAACGAGACAATGGTTCTTGCCAGCTCTACCACCAGCATCCACACCATGCTGCTCCTGCTCCTGATGCTCTT
        10        20        30        40        50        60        70        80        90
               ↓         30                            40                            50
 H  L  G  L  Q  A  S  I  S  G  R  D  T  H  R  L  T  R  T  L  N  C  S  S  I  V  K  E  I  I
CCACCTGGGACTCCAAGCTTCAATCAGTGGCCGGGATACCCACCGTTTAACCAGAACGTTGAATTGCAGCTCTATTGTCAAGGAGATTAT
       100       110       120       130       140       150       160       170       180
                         60                            70                            80
 G  K  L  P  E  P  E  L  K  T  D  D  E  G  P  S  L  R  N  K  S  F  R  R  V  N  L  S  K  F
AGGGAAGCTCCCAGAACCTGAACTCAAAACTGATGATGAAGGACCCTCTCTGAGGAATAAGAGCTTTCGGAGAGTAAACCTGTCCAAATT
       190       200       210       220       230       240       250       260       270
                         90                            100                           110
 V  E  S  Q  G  E  V  D  P  E  D  R  Y  V  I  K  S  N  L  Q  K  L  N  C  C  L  P  T  S  A
CGTGGAAAGCCAAGGAGAAGTGGATCCTGAGGACAGATACGTTATCAAGTCCAATCTTCAGAAACTTAACTGTTGCCTGCCTACATCTGC
       280       290       300       310       320       330       340       350       360
                         120                           130                           140
 N  D  S  A  L  P  G  V  F  I  R  D  L  D  D  F  R  K  K  L  R  F  Y  M  V  H  L  N  D  L
GAATGACTCTGCGCTGCCAGGGGTCTTCATTCGAGATCTGGATGACTTTCGGAAGAAACTGAGATTCTACATGGTCCACCTTAACGATCT
       370       380       390       400       410       420       430       440       450
                         150                           160
 E  T  V  L  T  S  R  P  P  Q  P  A  S  G  S  V  S  P  N  R  G  T  V  E  C  *
GGAGACAGTGCTAACCTCTAGACCACCTCAGCCCGCATCTGGCTCCGTCTCTCCTAACCGTGGAACCGTGGAATGTTAAAACAGCAGGCA
       460       470       480       490       500       510       520       530       540

GAGCACCTAAAGTCTGAATGTTCCTCATGGCCCATGGTCAAAAGGATTTTACATTCCTTTATGCCATCAAATGTCTTATCAATTTATCTA
       550       560       570       580       590       600       610       620       630

CTTTCTGAAATTTACAACTCTCCTTTGGCTTTACCTAATTATGTTCCTATTTTATTCCATTAAGGCTATTTATTTATGTATTTATGTATT
       640       650       660       670       680       690       700       710       720

TATTTATTTATTGCCTTCTGTGATGTGAGTATATCTGTTTTAGCTGAGGAGGAGTTTCTCCAAAGAAAATTCCAAGGAAGACTGGGGCCA
       730       740       750       760       770       780       790       800       810

TGTTCATTTGTCCCTTGTGGAAATAAATAACTTTGAACAAA
       820       830       840       850
```

Figure 1 Nucleotide sequence of cDNA for murine IL-3 and predicted amino acid sequence of the protein. The sequence was determined from cDNA clones derived from the WEHI-3B cell line (76,144) and is in complete agreement with the IL-3 gene sequence from BALB/c mice (85). The poly(A) tail is not shown. The arrow indicates the likely site of cleavage of the signal peptide (95), and potential *N*-glycosylation sites are underlined.

The nucleotide sequence of the MCGF cDNA is identical with that of the IL-3 cDNA except for a single nucleotide difference. This difference results in the substitution of Ala for Thr at residue 146 (Fig. 1). This difference in sequence probably represents allelic variation between the mouse strains BALB/c and C57BL/6 from which the cell lines used to isolate the IL-3 cDNA and MCGF cDNA, respectively, were derived.

Southern hybridization (83) using these cDNA clones, showed that there is a single copy of the IL-3 gene in the haploid mouse genome (84,85).

B. Characterization of Murine IL-3 Gene

Using Southern hybridization we found that the IL-3 gene was present on a single 8.5 kilobase (kb) *Eco*RI fragment in genomic DNA from BALB/c mice. This *Eco*RI fragment was cloned from BALB/c liver DNA (85) using the vector λ gt WES.λB (86). Further hybridization analysis localized the IL-3 gene to a 3.7 kb *Hin*c II/*Eco*RI fragment of the original 8.5-kb clone, and the nucleotide sequence of 3490 base pairs (bp) encompassing the IL-3 gene was determined on both strands (75) by the chain-termination method (87-89).

Miyatake and co-workers also cloned the mouse IL-3 gene from a BALB/c mouse sperm library in λ Charon 4A and sequenced a 3140-bp region covering the IL-3 gene (74). The only differences in the sequences obtained are four bases present in the sequence we reported but not in the sequence reported by Miyatake and co-workers (85). These differences are in noncoding regions. Both sequences show complete agreement with the sequence for IL-3 cDNA derived from the WEHI-3 cell line.

C. Expression of Recombinant Murine IL-3

The expression of recombinant IL-3 in both eukaryotic and prokaryotic systems has provided access to large quantities of the growth factor, thus facilitating biological experiments and allowing confirmation of the activities ascribed to IL-3.

To verify the biological properties of IL-3, we initially recloned the IL-3 cDNA into pSV2*neo* (90) and transfected monkey COS1 cells with the resulting construct (15). This provided a convenient source of recombinant IL-3 that was partially purified for biological studies. Similar studies were carried out by Rennick and co-workers (20).

The recombinant IL-3 was found to have the chromatographic properties of natural IL-3 from WEHI-3B conditioned medium, to induce 20αSDH in nu/nu splenic lymphocytes and to act as a CSF for granulocyte/macrophage, eosinophil, megakaryocyte, erythroid, and multipotential colony-forming cells from murine adult bone marrow and fetal liver (15,20). These studies provided verification that the multilineage activities claimed for IL-3 purified from natural sources were the properties of a single polypeptide.

Although the COS cell expression system produces recombinant IL-3 very similar in structure and glycosylation to natural IL-3, the amounts produced are limited. Also, controls need to be performed to guard against the possibility that low levels of other CSFs produced by the COS cells are influencing the results obtained with the recombinant material. A useful alternative is provided by the use of microbial expression systems. High-level expression of IL-3 in *Escherichia coli* [in collaboration with Biotechnology Australia Pty. Ltd. (Clark et al., unpublished results)] has provided large amounts of pure IL-3 that are free from contaminating eukaryotic proteins. Other independent studies have also achieved high-level expression of IL-3 in *E. coli* and yeast (57,91) The IL-3 expressed in *E. coli* appears to have all of the biological activities of natural IL-3 purified from WEHI-3B conditioned medium even though it is nonglycosylated and in some cases differs slightly from natural IL-3 in its *N*-terminal sequence.

D. Cloning and Expression of the Rat IL-3 Gene

As part of a study of the structure and biological role of IL-3 in other mammals we used the coding region of the mouse IL-3 cDNA as a probe, and carried out Southern hybridizations on DNA from a range of mammalian species including humans. The only species showing the presence of a significantly homologous sequence was the rat. The low conservation of mammalian IL-3 genes was unexpected and contrasts with the genes for other lymphokines such as IL-2, GM–CSF, and γ-interferon, which cross-hybridize between mouse and humans. As a result of this study, however, we were able to clone and sequence the rat IL-3 gene (53). The gene was then subcloned into pSV2*neo* and expressed in monkey COS1 cells to enable us to study the biological activities of the rat recombinant IL-3 both in the rat and murine systems (as discussed in previous section).

E. Cloning and Expression of Gibbon and Human IL-3 Genes

For some time the human IL-3 gene proved elusive but eventually gibbon and human IL-3 were identified by expression cloning (92). Conditioned media from a number of different cell lines were tested for activity in a routine proliferation assay for GM–CSF. This assay was based on the observation that primary leukemic blast cells from patients with chronic myelogenous leukemia (CML) proliferated in vitro when provided with GM–CSF (93). In the course of these experiments the conditioned medium from one gibbon T-cell line was found to be much more effective than human recombinant GM–CSF in supporting the proliferation of the CML blast cells. In addition, the activity from this conditioned medium was only partially inhibited by preincubation with a neutralizing antibody known to inhibit gibbon GM–CSF.

mRNA extracted from this gibbon cell line was fractionated by sucrose gradient centrifugation. The fractions containing the mRNA of interest were identified using translation in *X. laevis* oocytes, followed by assay of the activity of the secreted proteins on CML blast cells. RNAs from the peak fractions were then used as templates to prepare a 30,000 clone cDNA library. The clones were transfected initially as pools and then individually into monkey COS1 cells and the resulting conditioned medium was assayed against CML blast cells in the presence of the GM–CSF neutralizing antibody. This approach yielded a positive clone whose cDNA insert had low but significant homology with mouse IL-3. Using this gibbon cDNA as a probe that human IL-3 gene was isolated from a human genomic library.

To enable us to study the biological activities of human IL-3 we have also cloned the IL-3 gene and expressed it in monkey COS1 cells (van Leeuwen et al., unpublished data). Interleukin-3 activity in the stem-cell proliferation asaay reached between 64 and 128 ED50 U/ml by day 3 posttransfection. The biological properties of the recombinant protein are discussed in an earlier section. One interesting aspect of our human IL-3 clone is that residue 27 in the unprocessed polypeptide is Pro rather than Ser as reported by Yang and associates (92) (see Fig. 2). The significance of this difference in terms of biological activity remains to be established.

F. IL-3 Sequence and Gene Structure

The sequence of murine IL-3 cDNA is shown in Fig. 1. It predicts a mRNA of 851 bases excluding the poly(A) tail and encodes a polypeptide of 166 amino acids including a hydrophobic signal peptide. Analogous data on mRNA and protein sizes for rat and human IL-3 are given in Table 1. Initial studies by Edman degradation of purified murine IL-3 (94) suggested that the *N*-terminal residue of mature murine IL-3 is aspartic acid (residue 33). However, two later studies indicate that the amino terminus of murine IL-3 starts at alanine (residue 27) (95). Assuming that mature mouse IL-3 begins at residue 27, it would be 140 amino acid residues in length with an expected molecular weight (MW) of 15,674. The apparent MW of murine IL-3 determined by sodium dodecyl sulfate polyacrylamide gel electrophoresis, however, is 28,000 (94). Glycosylation is probably responsible for this discrepancy because there are four potential *N*-glycosylation sites in the predicted murine IL-3 sequence and considerable charge heterogeneity was revealed on examining IL-3 by isoelectric focusing (12). The observation that bacterial recombinant murine IL-3 has properties similar to "natural" IL-3, however, suggests that glycosylation is not necessary for biological activity.

We have generally been unable to detect significant overall sequence homology between mouse IL-3 and any other proteins including other hemopoietic

Table 1 Predicted Sizes of IL-3 mRNAs and Proteins

	Length of mRNA[a] (nucleotides)	Length of protein (amino acids)	
		Unprocessed	Mature
Mouse	851	166	140 (MW 15,674)[b]
Rat	872	166	140 (MW 15,787)
Human	910	152	133 (MW 15,082)[b]

[a]Excluding the poly(A) tail.
[b]The molecular weights (MWs) shown are calculated for the BALB/c IL-3 protein (73) and for the human IL-3 protein in which residue 27 in the unprocessed protein is pro (144; van Leeuwen and Young, unpublished data).

growth factors. This includes GM–CSF, which exhibits considerable overlap in biological activity with IL-3. However, by searching for chemical and structural similarity between amino acids over segments using the DIAGON program (96), we have located an area of apparent relatedness close to the C-terminal end of IL-3, IL-4, IL-5, IL-6, and GM–CSF (97). This may represent a conserved protein domain common to each lymphokine.

The degree of amino acid homology between the IL-3 proteins from different species is surprisingly low when compared with other lymphokines. For example, rat and mouse IL-3 have only 54% amino acid homology between the mature proteins (53). The homology between human and mouse IL-3 is also very low, with an amino acid homology of only 29% (92). In comparison, IL-2 and GM–CSF show mouse:human amino acid homologies of 60 and 54%, respectively (98,99). An alignment of the amino acid sequences encoded by the exons of mouse, rat, and human IL-3 is shown in Fig. 2. The IL-3 sequences in Fig. 2 are derived from the DNA sequences we determined for the BALB/c murine cDNA and gene (76,85), the PVG × DA rat gene (53), and a human IL-3 gene (van Leeuwen and Young, unpublished data). There are two cysteine residues (Cys 43 and 106) conserved between the three species, suggesting the possibility of a functional disulphide bridge in mammalian IL-3 species. Evidence supporting this hypothesis has been obtained by Clark-Lewis and associates (100,101) by chemical synthesis of IL-3 and modified forms.

Although there is a low level of amino acid homology between mouse and human IL-3, comparison of the two sequences using the program DIAGON (96) that employs a similarity matrix shows a clear relationship with four conserved regions in addition to that of the signal peptide (Fig. 3). This, taken together

```
            10        20      ↓ 30         40        50
(M) MVLASSTTSIHTMLLLLLMLFHLGLQASISGRDTHRLTRTLNCSSIVKEIIGKLP-------
(R)           LC    P      Q   I DR S A H L   D RT AL   LV
(H)  ------- RLPV    QL VRP    PMT-QT PLK SWV   NMID   TH KQPPLPLL
                   10        20         30        50         50
```

```
       60
(M) EPELKTDDEGPSLR
(R) VSG NNS DKAN
(H) DFNNLNGEDQDI M
        60
```

```
   70       80        90        100
(M) NKSFRRVNLSKFVESQGEVDPEDRYVIKSNLQ
(R)  STL     DE LK  E F SQ TTD   K
(H) ENNL  P  EA NRAVKSL--QNASA E I K
    70        80           90
```

```
          110
(M) KLNCCLPTSANDSA
(R)   K  I AA S  V
(H) N LP   LATAAPT
   100       110
```

```
      120         130       140       150       160
(M) LPGVF--IRDLDDFRKKLRFYMVHLNDLETVLTSRPPQPASGSVSPNRGTVEC
(R)     Y  NK      K      VI  K  QP SV       T S DNFRPM
(H) RHPIHIKDG WNE  R  T   LKT ENAQAQQ TLSLAIF-------------
         120       130       140       150
```

Figure 2 Alignment of the coding segments of the mouse (M), rat (R), and human (H) IL-3 genes. Identical residues are not shown. The arrow indicates the likely site of signal peptide cleavage for mouse IL-3.

with the similarities in gene structure and biological activity, leaves no doubt of their relationship.

Although there is only low homology between the coding regions of the mammalian IL-3 genes studied to date, there is considerably more homology evident in their gene structure and organization. Each gene has five relatively small exons separated by three small and one relatively large intron (intron 2). In each case, the introns interrupt the exons exactly between codons. The murine gene organization is summarized in Fig. 4. It has four introns of 96, 993, 135, and 122 bp in order from the 5′ end of the gene (84,85). The intron/exon boundaries of the murine gene are fully conserved in ther at IL-3 gene, giving coding segments of identical size. The flanking regions (200 nucleotides either side of the gene) and the introns show 90 and 80% homology, respectively, between the mouse and rat genes. The coding regions of the human gene are very similar in size to those in the mouse and rat genes (Fig. 2).

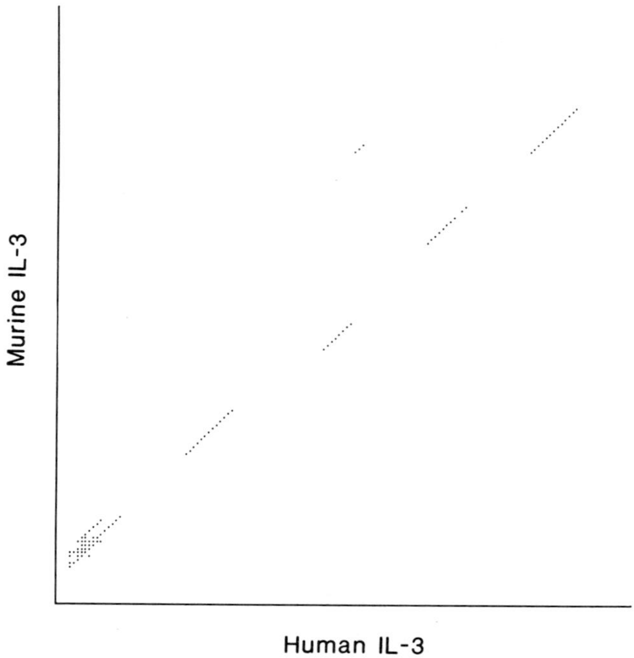

Figure 3 Comparison of the human and mouse IL-3 protein sequences using the program DIAGON (96). The dots indicate those spans of 11 residues that give a similarity score over 132 using the mutation data matrix.

Intron 2 in the murine gene contains a tandem repeating sequence made up of 12 similar 14-bp units with a consensus sequence AGCTCTCACCTCCC (85). In the corresponding position in the rat gene there is a similar region composed of ~40 14-bp units that have an identical consensus sequence to that of the murine repeats (53). These 14-bp repeats share strong homology with a series of 20-bp repeats in the human genome that show some enhancerlike activity. Therefore, it is speculated that this unusual sequence element in the IL-3 intron may play an important role in the regulated expression of the gene in activated T cells. Intron 2 of the human IL-3 gene does not possess analogous repeats although it is interesting to note that an Alu repeat is present at a similar position to the repeats in the mouse and rat genes (van Leeuwen and Young, unpublished data).

There is also a TATT sequence that is repeated nine times between the translational termination codon and the polyadenylation signal of the mouse IL-3

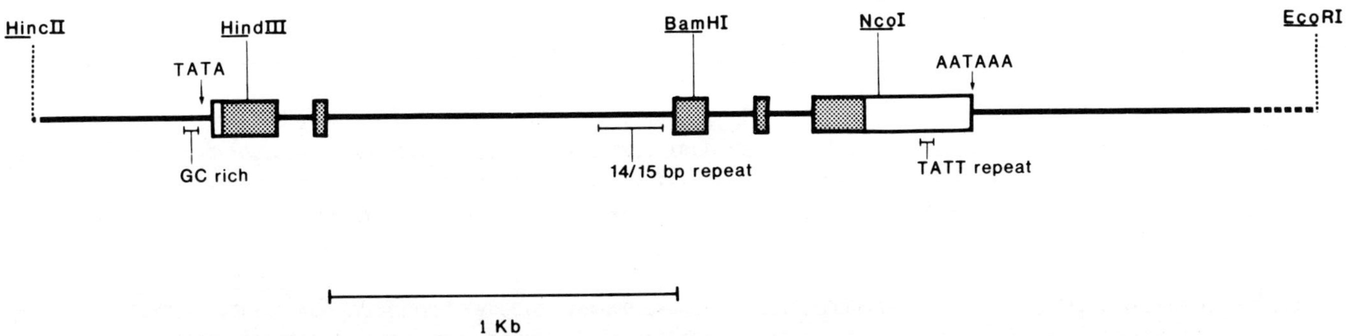

Figure 4 Gene organization of the murine IL-3 gene (85). The boxes on the lines indicate the exons. Shaded portions denote coding sequences whereas unshaded regions indicate the untranslated regions of the IL-3 mRNA. Unique *Hin*dIII, *Bam*HI, and *Nco*I sites in the 3.7-kb fragment that was sequenced are shown. The locations of the TATA box, the AATAAA polyadenylation signal, and the GC-rich region upstream of the TATA box are indicated. Areas of tandem repeats, one based on a 14/15-bp repeating unit, the other on a TATT unity, are also shown.

```
(H) AGGAGAAGCTGCCCGAAGCCCATGGGACAAACCACTGGGGACTGGAACACCAGTAATTCTGTATTGGGAAGCGGCACCAAGAGATGTGCTTCTCAGAGCCTGAGGCTGAACGTGGA
     :  :::  :  :  :::  :           :  ::::  :  ::::                                  :::::             ::::    :: :   :: ::::
(M) ATAAGAGTCAACGGGAAAGCTGCCTAGAGATACACTTAGCTCTGGCTAC-----------------------CACCAGC-----------CAGACGACTAGCCCACACATGGA

TGTTTAGCAGCGTGACCGGCTACCAGACAAACTCTCATCTGTTCCAGTGGCCTCCTGGCCACCCACCAGG---------------ACCAAGCAGGGCGGGCAGCAGAGGGCCAGGGTA
 ::::: :   ::: : :  : ::   :::  ::   :::: ::::::: :       : :  :: : :                  ::::   :::  :  :::    :: : ::  :
AGTTTAACCATGTGCCAGAATGCCTACCAAGTTCAAGTCTGCTCCAGTGACTCTGGTGACTACCTCTATCAGGTAGGGTGTACAGTCACCACATAGGCTGTACAGTCAGGGTCAAGTTTG

GTCCAGGTGATGGCAGATGAGATCCCACTGGGCAGGAGGCCTCAGTGAGCT--GAGTCAGGCTTCCCCTTCCTGCCACAGGGGTCCTCTCACCTGCTGCCATGCTTCCCATCTCTCATCC
  : :::    :    :::::::: ::::::   :: : :::  ::   :::  ::: ::::::    :::::: :::::            ::::: : :::  :   ::::::::
TGCAAGGGATGGTAGGATGAGATTCCACTGCATAGAAAGCCCAAGCTGGCTCAGAGCCAGGCTA---CTTCCTCCCACA-----------ACCTGTTTCCACTCCGTCCATCTCTA----
                  10-mer

TCCTTGACAAGATGAAGTGATACCGTTTAAGTAA----TCTTTTTTCTTGTTTCACTGATCTTGAGTACTAGAAAGTCATGGATGAATAATTACGTCTGTGGTTTTCTATGGAG------
    ::::::    :   :::  : ::: : :::    ::::: :::   : ::  : ::::::::::::::::::   ::::::::: :     :::::::: ::::::::
----TGACAAAGGAAGAAGATGGCCTTTGAATAAGCAGTCTTTCTTCCCATGTCGATAATCTTGAGTACTAGAAAGT----GATGAATAAGTT----TGTGGTTTGCTATGGAGGTTCCA
                                                                                                                10-mer

--TCAGATAAAGATCCTTCCGACGCCTGCCCCACACCACCACCTCCCCCGCCTTGCCCGGGGTTGTGGGCACCTTGCTGCTGC-ACATATAAGGCGGGAGGCTGTTGCCAACTCTTCAG
  :::::::::: : :::: :: ::::::::              ::::::: :    :::::       ::::   ::      ::::::::: :  :::::  ::      :::::::
TGTCAGATAAAGCTGCTTCTGATGCCTGCCC------------TCCCCCCTGCCCCGCCGGGCCCCGCCCCACCCCTCTCTGAATACATATAAGGTGA-AGGCTCCTGTGGCTTCTTCAG
                                   GC-rich region                                 TATA Box               CAP Site
```

Figure 5 Alignment of the 5′-flanking regions of the mouse (M) IL-3 gene (75), and the human (H) IL-3 gene (van Leeuwen and Young, unpublished data). The putative TATA box, GC-rich region, and conserved 10-mers found in different lymphokine genes (103) are shown.

gene (85). This tandem repetitive element is conserved in both the rat and human genes. Similar AT-rich regions are present in the genes encoding mouse GM–CSF, mouse and human IL-2, human G–CSF, and members of the human interferon family. Experimental evidence suggests that these sequences regulate mRNA stability (102) and may, therefore, play a critical role in the regulation of hemopoietic growth-factor mRNA levels.

In view of the analogous expression of lymphokine genes in activated T cells, the 5′ untranslated regions of the IL-3 genes and other sequenced lymphokine genes have been examined for common nucleotide sequences. An alignment of the 5′ flanking regions of the human and mouse IL-3 genes is given in Fig. 5. Each of the mammalian IL-3 genes has a very GC-rich region just upstream of the putative TATA box. Also shown in Fig. 5 are two decanucleotide sequences that share homology with sequences present in the 5′ flanking regions of the mouse GM–CSF gene and the mouse and human IL-2 genes (103). One of these 10-mer sequences is also present in the human IL-3 gene we have isolated (Fig. 5). Whether these regions play a role in the coordinate regulation of lymphokine genes remains to be established.

The fact that the two most conserved regions of the mammalian IL-3 genes are the promoter regions and the untranslated region of exon 5 suggests conservation of sequences regulating gene expression. In contrast, the relatively low amino acid sequence homology could suggest that significant evolution of the IL-3 proteins has occurred in concert with their specific receptors. This is supported to some extent by the observation that mouse IL-3 shows essentially no activity in rat or human assays (53,104).

G. Chromosomal Location of IL-3 Genes

In mice, the IL-3 gene has been shown to be located on chromosome 11 using mouse/Chinese hamster hybrids (105,106). In addition, genetic analysis of a *Mus musculus* × *Mus spretus* interspecies back-cross indicated linkage with the GM–CSF and SPARC genes (105), the latter having been mapped to subband B1 of mouse chromosome 11 by in situ hybridization (107,108). Using pulsed-field gel electrophoresis, Barlow and co-workers (105) reported that the IL-3 and GM–CSF genes were not more than 230 kb from each other but that they were at least 400–500 kb distant from the SPARC locus.

In recent work from our laboratory (109) we have mapped the IL-3 gene directly by in situ hybridization to subband B1 of mouse chromosome 11 and used pulsed-field gel electrophoresis and genomic λ clones to show that the IL-3 and GM–CSF genes are much closer to each other than previously realized. They are 14 kb apart, in the same orientation, and with the IL-3 gene located 5′ to the GM–CSF gene.

In humans, the IL-3 gene is located on the long arm of chromosome 5 in the region of 5q23.3 and 5q32 (110), close to the genes for IL-5, M–CSF, GM–CSF, M–CSF receptor, and platelet derived growth factor receptor (111-116). These genes are commonly deleted in the 5q- syndrome, which leads to proliferative disorders such as macrocytic anaemia and secondary acute nonlymphocytic leukemia. The involvement of the deletion of the IL-3 gene in the pathogenesis of these disorders has not yet been established.

More detailed studies have shown that the IL-3 and GM–CSF genes are also adjacent in humans, being 8 kb apart and in the same orientation (117).

The close linkage of the IL-3 and GM–CSF genes suggests that they may have arisen from an ancient gene deplication and subsequently diverged, performing slightly different roles in the hemopoietic system and being regulated differentially. The exon structure of the two genes is quite similar, particularly if one assumes that exon 3 of the GM–CSF gene has been split into two exons in the case of the IL-3 gene (85,103). In both cases the exons are interrupted exactly between codons.

V. IL-3 AND LEUKEMIA

One proposed mechanism of oncogenesis is abnormal expression of a growth factor gene by a cell that can also respond to that factor (118-123). This "autocrine" hypothesis predicts that a cell already committed to a particular self-renewing lineage would be susceptible to this type of transformation. An uncommitted stem cell, on the other hand, should retain the ability to differentiate normally in the presence of lineage-specific molecules even though it may abnormally produce large amounts of stem-cell growth factor. Additional genetic changes that result in uncoupling of differentiation from proliferation would be required for such "stem" cells to become tumorigenic.

Interleukin-3 appears to be primarily a promoter of cell division. It acts on uncommitted stem cells and partially committed stem cells of most hemopoietic lineages. Final maturation events are regulated by other CSFs such as IL-5, M–CSF, and erythropoietin. Thus, the hemopoietic system permits the testing of some of the concepts outlined above.

A. IL-3 Producing Autocrine Leukemias

There are several factor-dependent hemopoietic cell lines that proliferate in the presence of IL-3 but do not differentiate. FDC-P1 is myelomonocytic and will also proliferate in response to GM–CSF and IL-4, whereas 32D cl-23 cells are related to basophils and proliferate in response to IL-2 and IL-4 as well as to IL-3. These differentiation-defective cell lines provided an opportunity to construct experimental IL-3 producing autocrine leukemias.

The approach we took to produce autocrine leukemias relied on infection of the factor-dependent cells with a defective retroviral genome containing the IL-3 gene and the selectable marker *neo*, which confers on eukaryotic cells resistance to the antibiotic G418. A novel aspect of our approach was the use of HIX and 4070A helper viruses, which do not themselves cause transformation, to pseudotype the replication-incompetent viral vector (124). Such an approach permits incorporation of the retroviral expression vector into a large proportion of treated cells by active infection of the culture with helper virus.

The vector itself, fpGV-1, contains the bacterial Col E1 origin of replication, the SV40 origin of replication, the neomycin (G418) resistance gene from transposon Tn5, and the long-terminal repeats (LTRs) from HT-1 MSV (125). The *Hin*cII-*Nco*I fragment of the murine IL-3 gene (85) was inserted into an M13-derived polylinker in the viral vector using *Eco*RI linkers, to give fpGV-IL3. To produce infective virus, fpGV-IL-3 was transfected into COS1 cells that had previously been infected with HIX or 4070A viruses. Supernatants from transfected COS cells were used to infect NIH3T3 cells that were selected in G418. Cell lines derived by this process produced relatively high levels of IL-3 and served as a source of virus for infection of FDC-P1, 32D cl-23, and normal bone marrow cells.

Pseudotyping fpGV-IL-3 with either HIX or 4070A viruses allowed efficient infection of FDC-P1 cells, but 32D cl-23 cells were resistant to HIX pseudotyped virus. This suggests that 32D cl-23 cells may have previously been infected by a virus of the HIX interference group (126). Alternatively, there may be a degree of cell lineage specificity with the HIX virus. When infected FDC-P1 or 32 D cl-23 cells were selected in G418, IL-3-producing cell lines were generated that did not require added growth factors to maintain proliferation. All cell lines produced in this way were leukemic in mice (124). Normal bone marrow cells from NIH Swiss mice could also be infected with either virus. In this instance bone marrow cultures selected in G418 produced IL-3, but no leukemic cells were generated even after two years in culture. The growth of autocrine FDC-P1-IL-3 cells, but not of WEHI-3B cells (see later) could be prevented by addition of a neutralizing anti-IL-3 serum in keeping with the autocrine nature of this leukemia, but surprisingly, proliferation of 32D cl-23-IL-3 cells was not inhibited by the same antibody.

Cell surface molecules expressed by FDC-P1 and 32D cl-23 cells were affected differently after leukemic transformation. Levels of IL-2 receptor and Thy1 were increased on FDC-P1-IL3 cells while the level of LFA1 and Mel14 decreased. This was associated with preferential growth of FDC-P1-IL-3 clones that expressed particularly high levels of IL-2 receptor in lymph node (127). Thy1 and Mel14 were not present on normal or infected 32D cl-23 cells but the level of phagocytic glycoprotein (PGP) increased after infection. Although

32D cl-23 cells express a functional IL-2 receptor (128), they did not grow in lymph node presumably because lack of Mel14 antigen prevented them from homing to that site (129). FDC-P1 cells do not proliferate in response to IL-2, although they express considerably higher levels of IL-2 receptor than 32D cl-23 cells, which can be maintained in IL-2 alone. The expression of high levels of IL-2 receptor on FDC-P1-IL3 cells that grow in lymph node correlates with improved cloning efficiency in IL-2 containing soft agar cultures indicating that the IL-2 receptor on FDC-P1 cells is not completely inert. The apparent ability of IL-2 to synergize with IL-3 in promoting growth of FDC-P1-IL3 cells has some parallels with the synergistic effects of IL-3 and M–CSF in growth of bone marrow colony-forming cells from A/J mice (46).

Bone marrow cells, cultured for almost 2 years after infection with fpGV-IL-3 virus, did not produce leukemias in mice. The cultures did, however, contain G418-resistant stem cells capable of reconstituting sublethally irradiated mice, from which G481*R* colony-forming cells could later be recovered. The major distinguishing feature of infected cultures was that, in contrast to uninfected controls, they produced nonadherent colony-forming cells for many months and could be amplified by splitting the cultures once each week like many continuous cell lines. The nonadherent cells formed normal G, GM, and M colonies in the presence of IL-3, M–CSF, G–CSF, or GM–CSF and, if left at high density, matured into granulocytes and macrophages in liquid "Dexter"-type cultures. These observations demonstrate that IL-3-producing stem cells can self-renew more effectively than can noninfected cultures, but that these stem cells remain capable of normal differentiation in vivo and in vitro. This implies, as discussed earlier, that additional genetic changes are required for the development of tumorigenicity.

B. Constitutive IL-3 Production by the WEHI-3B Leukemia

WEHI-3B is a myelomonocytic cell line that is leukemic (130). It is morphologically similar to FDC-P1 but constitutively produces IL-3 (131) and was used by us as the source of IL-3-specific mRNA for the cloning of murine IL-3 cDNA (76). Normal myelomonocytic cells and FDC-P1 cells proliferate in response to IL-3 and some investigators have proposed that abnormal expression of IL-3 in WEHI-3B cells may have played a role in development of this leukemia (132-135).

Using Southern analysis, a rearrangement was detected in the vicinity of the IL-3 gene in DNA from WEHI-3B cells (67). We studied this rearrangement in detail and found that it was due to insertion of a 5.1-kb intracisternal A particle (IAP) genome just upstream of the IL-3 gene (Fig. 6). Intracisternal A particle genomes are analogous to the proviral forms of retroviruses (136). There are up to 1000 IAP genomes per copy of the haploid mouse genome,

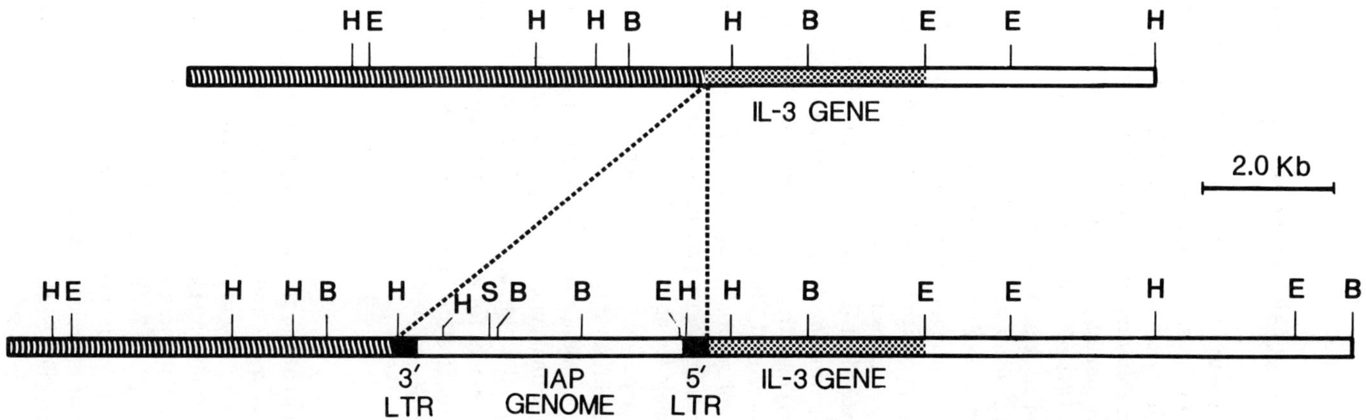

Figure 6 Restriction map of the normal and rearranged IL-3 genes from WEHI-3B. The inserted intracisternal A-particle gene and its LTRs are indicated. H, *Hin*dIII; B, *Bam*HI; E, *Eco*RI; S, *Sal*I.

and the level of transcription of IAP elements is increased in some murine tumors (137).

Insertional mutations due to IAP genomes have resulted in inactivation of k light-chain gene function (138), and conversely, in activation of the cellular oncogene c-mos (139). In the case of the rearranged IL-3 gene in WEHI-3B cells, the IAP genome results in abnormal expression. This was demonstrated by transfection of λ clones containing copies of either the normal or rearranged IL-3 gene, into monkey COS1 cells and murine 3T3 cells. Only λ clones containing the IAP genome in association with the IL-3 gene expressed IL-3 in this system, in keeping with the silent nature of the normal IL-3 gene in fibroblast cell lines. The mechanism of IAP-induced activation of the IL-3 gene in WEHI-3B cells is not certain but is probably due to an enhancer in the IAP LTR. The IAP element is inserted head to head with the IL-3 gene: That is, the 5′ LTR is close to the 5′ end of the IL-3 gene. The insertion is 215 bp upstream from the IL-3 gene's TATA box. Most IL-3 gene transcripts in WEHI-3B cells appear to come from the gene's normal promoter because all three cDNA clones obtained from WEHI-3B (76) have the same 5′ end as the full-length cDNA clones from a T-cell line (81). Because LTRs of IAP elements contain sequences homologous to the SV40 core enhancer sequence (140) and LTR enhancer activity has been demonstrated in expression systems (141-143), this type of enhancer activity might be responsible for abnormal IL-3 production in WEHI-3B cells.

A possible evolutionary pathway for the generation of WEHI-3B leukemia is that it arose from an IL-3 responsive granulocyte-macrophage progenitor, which gained the ability to produce IL-3 due to IAP insertion and which subsequently underwent additional genetic changes that resulted in uncoupling of differentiation from proliferation as has occurred in FDC-P1 and 32D cl-23 cells. However, the requirement for an autocrine loop appears to have since been bypassed because proliferation of WEHI-3B cells cannot be inhibited by neutralizing anti-IL-3 antibodies.

VI. SUMMARY

The cloning and expression of the genes for rat, mouse, and human IL-3 have resulted in major advances in our knowledge of this lymphokine and the biological processes in which it is involved. Such advances include the primary structures of the IL-3 proteins, the nucleotide sequences of the IL-3 genes, and the elucidation of the retroviral insertion that resulted in constitutive synthesis of IL-3 by the leukemic cell line WEHI-3B. These advances have facilitated studies of the control of IL-3 gene expression in normal and malignant cells, which is an ongoing area of interest. The use of the cloned IL-3 gene in a retroviral expression vector has allowed the generation of experimental autocrine leukemias from IL-3-dependent cell lines.

The use of recombinant rat, mouse, and human IL-3 has verified that IL-3 is a multilineage hemopoietic regulator in each of these species. Although it is clear that in each of the above species IL-3 promotes the growth of a wide variety of myeloid progenitors, the regulation of stem-cell division and maturation by IL-3 and its synergism with other regulators requires clarification and remains an active area of research. The discovery that hemopoietic stem cells from certain murine strains do not proliferate in response to IL-3 alone has provided a model system in which to analyze the synergistic activities of IL-3. It will be interesting to see whether analogous differences in the responsiveness of stem cells to IL-3 also exist in humans.

Our work on the ontogeny of blood cell development in embryonic and fetal mice indicates that in every strain, early hemopoietic progenitor cells respond to IL-3 plus M–CSF, or to IL-3 plus IL-4, rather than to IL-3, M–CSF, or IL-4 alone. The difficulties encountered in acquiring sufficient cells to work with from embryonic sources could possibly be overcome by developing a fetal sheep model in which access to the embryo and fetus can be achieved at several sequential points in development. It may then be possible to determine the molecular events that regulate primitive stem-cell responsiveness to IL-3 as well as other hemopoietic cytokines.

Recombinant IL-3 proteins should continue to prove valuable in further biological studies of IL-3, in the generation of monoclonal antibodies against IL-3, and for further studies of the IL-3 receptor. The molecular description of the IL-3 receptor and a detailed understanding of the mechanism of IL-3-induced cell division and maturation awaits further research. Administration of recombinant IL-3, alone and in combination with other CSFs in both animal experiments and clinical trials, should clarify whether IL-3 will be of value in the treatment of myelosuppressed patients or in autologous bone marrow transplantation.

ACKNOWLEDGMENT

This work was supported in part by a grant to A. J. Hapel from the National Heart Foundation of Australia.

REFERENCES

1. Ford, C. E., J. L. Hammerton, D. W. H. Barnes, and J. F. Loutit. Cytological identification of radiation-chimaeras. *Nature 177*:452-454 (1956).
2. Till, J. E., and E. A. McCulloch. Early repair processes in marrow cells irradiated and proliferating in vivo. *Radiat. Res. 18*:96-105 (1963).
3. Becker, A. J., E A. McCullogh, and J. E. Till. Cytological demonstration of the clonal nature of spleen colonies derived from transplanted mouse marrow cells. *Nature 197*:452-454 (1963).

4. Johnson, G. R., and D. Metcalf. Pure and mixed erythroid colony formation in vitro stimulated by spleen conditioned medium with no detectable erythropoietin. *Proc. Natl. Acad. Sci. U.S.A. 74*: 3879-3882 (1985).
5. Stanley, E. R., and P. M. Heard. Factors regulating macrophage production and growth. Purification and some properties of the colony stimulating factor from a medium conditioned by mouse L cells. *J. Biol. Chem. 252*:4305-4312 (1977).
6. Metcalf, D., J. Parker, H. M. Chester, and P. W. Kincade. Formation of eosinophilic-like granulocytic colonies by mouse bone marrow cells in vitro. *J. Cell. Physiol. 84*:275-290 (1974).
7. Metcalf, D. The granulocyte-macrophage stimulating factors. *Science 229*: 16-22 (1985).
8. Iscove, N. N. Erythropoietin-independent stimulation of early erythropoiesis in adult marrow cultures by conditioned media from lectin-stimulated mouse spleen cells. In *Hematopoietic Cell Differentiation*, D. C. Golde, M. J. Cline, D. Metcalf, and C. F. Fox (Eds.). Academic Press, New York, pp. 37-52 (1978).
9. Nabel, G., S. J. Galli, A. M. Dvorak, H. F. Dvorak, and H. Cantor. Inducer T lymphocytes synthesize a factor that stimulates proliferation of clonal mast cells. *Nature 291*:332-334 (1981).
10. Schrader, J. W., F. Battye, and R. Scollay. Expression of Thy-1 antigen is not limited to T cells in cultures of mouse hemopoietic cells. *Proc. Natl. Acad. Sci. USA 79*:4161-4165 (1982).
11. Bartelmez, S. H., R. Sacca, and E. R. Stanley. Lineage specific receptors used to identify a growth factor for developmentally early haemopoietic cells: assay of hemopoietin-2. *J. Cell. Physiol. 122*:362-369 (1985).
12. Ihle, J. N., L. Rebar, J. Keller, J. C. Lee, and A. J. Hapel. Interleukin 3: possible roles in the regulation of lymphocyte differentiation and growth. *Immunol. Rev. 63*:5-32 (1982).
13. Iscove, N. N., C A. Roitsch, N. Williams, and L. J. Guilbert. Molecules stimulating early red cell, granulocyte, macrophage and megakaryocyte precursors in culture: similarity in size, hydrophobicity, and charge. *J. Cell Physiol* [*Suppl. 1*] *65-78 (1982).*
14. Burgess, A. W., D. Metcalf, S. H. Russell, and N. A. Nicola. Granulocyte/macrophage-, megakaryocyte-, eosinophil- and erythroid-colony-stimulating factors produced by mouse spleen cells. *Biochem. J. 185*:301-314 (1980).
15. Hapel, A. J., M.-C. Fung, I. G. Young, G. Johnson, and D. Metcalf. Biological properties of molecularly cloned and expressed murine interleukin-3. *Blood 65*:1453-1459 (1985).
16. Palacios, R., and M. Steinmetz. IL-3-dependent mouse clones that express B-220 surface antigen, contain Ig genes in germ-line configuration and generate B lymphocytes in vivo. *Cell 41*:727-734 (1985).
17. Ihle, J. N., L. Pepersack, and L. Rebar. Regulation of T-cell differentiation: in vitro inuction of 20α-hydroxysteroid dehydrogenase in splenic lymphocytes from athymic mice by a unique lymphokine. *J. Immunol. 126*:2184-2189 (1981).

18. Hapel, A. J., J. M. Osborne, M.-C. Fung, I. G. Young, W. Allen, and D. A. Hume. Expression of 20α hydroxysteroid dehydrogenase in mouse macrophages, haemopoietic cells, and cell lines. *J. Immunol. 134*:2492-2497(1985).
19. Ihle, J. N., J Keller, S. Oroszlan, L. E. Henderson, T. D. Copeland, F. Fitch, M. B. Prystowsky, E. Goldwasser, J. W. Schrader, E. Palaszynski, M. Dy, and B. Lebel. Biologic properties of homogeneous interleukin-3. 1. Demonstration of WEHI-3 growth factor activity, mast cell growth factor activity, P cell-stimulating factor activity, colony-stimulating factor activity and histamine producing cell-stimulating factor activity, *J. Immunol. 131*:282–287 (1983).
20. Rennick, D. M., F. D. Lee, T. Yokota, K.-I. Arai, H. Cantor, and G. J. Nabel. A cloned MCGF cDNA encodes a multilineage hematopoietic growth factor: multiple activities of interleukin 3. *J. Immunol. 134*:910-914 (1985).
21. Clark-Lewis, I., and J. W. Schrader. Molecular structure and biological activities of P cell-stimulating factor (Interleukin 3). In *Lymphokines*, Vol. 15, J. W. Schrader (Ed.). Academic Press, San Diego, pp. 1-37 (1988).
22. Williams, N., H. Jackson, N. N. Iscove, and P. P. Dukes. The role of erythropoietic, thrombopoietic stimulating factor, and myeloid colony-stimulating factors on murine megakaryocyte colony formation. *Exp. Hematol. 12*:734-740 (1984).
23. Goldwasser, E. Erythropoietin and the differentiation of red blood cells. *Fed. Proc. 34*:2285-2292 (1975).
24. Moore, M. A. S. Interleukin 3: an overview. In *Lymphokines*, Vol. 15, J. W. Schrader (Ed.). Academic Press, San Diego, pp. 219-280 (1988).
25. Nakahata, T., and M. Ogawa. Identification in culture of a class of hemopoietic colony-forming units with extensive capability to self-renew and generate multipotential colonies. *Proc. Natl. Acad. Sci. USA 79*:3843-3847 (1982).
26. Suda, T., J. Suda, and M. Ogawa. Proliferation kinetics and differentiation of murine blast cells in culture: evidence for variable Go periods and constant doubling rates of early pluripotent haemopoietic progenitors. *J. Cell Physiol. 117*:308-318 (1983).
27. Suda, T., J. Suda, M. Ogawa, and J. N. Ihle. Permissive role of interleukin 3 (IL-3) in proliferation and differentiation of multipotential haemopoietic progenitors in culture. *J. Cell. Physiol. 124*:182-190 (1985).
28. Koike, K., J. N. Ihle, and M. Ogawa. Declining sensitivity to interleukin 3 of murine multipotential hemapoietic progenitors during their development. Application to a culture system that favors blast cell colony formation. *J. Clin. Invest. 77*:894-899 (1986).
29. Dexter, T. M., J. Garland, D. Scott, E. Scolnick, and D. Metcalf. Growth of factor-dependent haemopoietic cell lines. J. Exp. Med. 152:1036-1047 (1980).
30. Hapel, A. J., H. W. Warren, and D. A. Hume. Different colony-stimulating factors are detected by the "interleukin-3" dependent cell lines FDC-P1 and 32Dcl-23. *Blood 64*:786-790 (1984).

31. Greenberger, J. S., R. J. Eckner, M. Sakakeeny, P. Marks, D. Reid, G. Nabel, A. J. Hapel, J. N. Ihle, and K. C. Humphries. Interleukin-3-dependent haemopoietic progenitor cell lines. *Fed. Proc. 42*:2762-2771 (1983).
32. Mosmann, T. Rapid colorimetric assay for cellular growth and survival: application to proliferation and cytotoxicity assays. *J. Immunol. Methods 65*:55-63 (1983).
33. Metcalf, D. Multi-CSF-dependent colony formation by cells of a murine hemapoietic cell line: specificity and action of multi-CSF. *Blood 65*:357-362 (1985).
34. Whetton, A. D., and T. M. Dexter. Effect of haematopoietic cell growth factor on intracellular ATP levels. *Nature 303*:629-631 (1983).
35. Whetton, A. D., G. W. Bazill, and T. M. Dexter. Haemopoietic cell growth factor mediates cell survival via its action on glucose transport. *EMBO J. 3*:409-413 (1984).
36. Schrader, J. W., S. J. Lewis, I. Clark-Lewis, and J. G. Culvenor. The persisting (P) cell: histamine content, regultion by a T cell-derived factor, origin from a bone marrow precursor, and relationship to mast cells. *Proc. Natl. Acad. Sci. USA 78*:323-327 (1981).
37. Nagao, K., K. Yokoro, and S. A. Aaronson. Continuous lines of basophil/mast cells derived from normal mouse bone marrow. *Science 212*:333-335 (1981).
38. Razin, E., C. Cordon-Cardo, and R. A. Good. Growth of a pure population of mouse mast cells in vitro with conditioned medium derived from concanavalin A-stimulated splenocytes. *Proc. Natl. Acad. Sci. USA 78*:2559-2561 (1981).
39. Seldin, D. C., and K. F. Austen. Mast cell subclasses and their growth dependence in mice, rats, and humans. In *Lymphokines*, Vol. 15, J. W. Schrader (Ed.). Academic Press, San Diego, pp. 313-340 (1988).
40. Tertian, G., Y.-P. Yung, D. Guy-Grand, and M. A. S. Moore. Long-term in vitro culture of murine mast cells. 1. Description of a growth factor-dependent culture technique. *J. Immunol. 127*:788-794 (1981).
41. Haig, D. McK., C. McMenamin, J. Redmond, D. Brown, I. G. Young, D. Cohen, and A. Hapel. Rat interleukin-3 stimulates the growth of rat mucosal mast cells in culture. *J. Immunol (in press) (1988).*
42. Hamaguchi, Y., Y. Kanabura, J. Fujita, S.-I. Takeda, T. Nakano, S. Tarui, T. Honju, and Y. Kitamura. Interleukin-4 as an essential factor for in vitro clonal growth of murine connective tissue-type mast cells. *J. Exp. Med. 165*:268-273 (1987).
43. Bartelmez, S. H., R. Sacca, and E. R. Stanley. Lineage specific receptors used to identify a growth factor for developmentally early hemopoietic cells: assay of hemopoietin-2. *J. Cell. Physiol. 122*:362-369 (1985).
44. McNiece, I. K., A. B. Kriegler, A. J. Hapel, M.-C. Fung, I. G. Young, T. R. Bradley, and G. S. Hodgson. Recombinant interleukin-3 exhibits synergistic factor activity. *Cell Biol. Int. Rep. 8*:812 (1984).
45. McNiece, I. K., N. T. Williams, G. R. Johnson, A. B. Kriegler, T. R. Bradley,

and G. S. Hodgson. Generation of murine hematopoietic precursor cells from macrophage high-proliferative-potential colony-forming cells. *Exp. Hematol. 15*:972-977 (1987).
46. Morris, C. F., J. Salisbury, M. Kobayashi, P. V. Townsend, and A. J. Hapel. Absence of interleukin-3 induced proliferation in A/J bone marrow cell cultures: a novel system for studying the synergistic activities of IL-3. *J. Cell. Physiol.*: (submitted) (1989).
47. Kincade, P. W., G. Lee, G. Fernandes, M. A. S. Moore, N. Williams, and R. A. Good. Abnormalities in clonable B lymphocytes and myeloid progenitors in autoimmune NZB mice. *Proc. Natl. Acad. Sci. USA 76*:3464-3468 (1979).
48. Ikebuchi, K., G. C. Wong, S. C. Clark, J. N. Ihle, Y. Hirai, and M. Ogawa. Interleukin 6 enhancement of interleukin 3-dependent proliferation of multipotential haemopoietic progenitors. *Proc. Natl. Acad. Sci. USA 84*: 9035-9039 (1987).
49. Mochizuki, D. Y., J. R. Eisenman, P. J. Conlon, A. D. Larsen, and R. J. Tuchinski. Interleukin 1 regulates hematopoietic activity, a role previously ascribed to hemopoietic 1. *Proc. Natl. Acad. Sci. USA 84*:5267-5271 (1987).
50. Dexter, T. M. Growth factors involved in haemopoiesis. *J. Cell. Sci. 88*: 1-6 (1987).
51. Smith, C. A., and D. M. Rennick. Characterisation of a murine lymphokine distinct from interleukin 2 and interleukin 3 (IL-3) possessing a T-cell growth factor activity and a mast-cell growth factory activity that synergises with IL-3. *Proc. Natl. Acad. Sci. USA 83*:1857-1861 (1986).
52. Rennick, D., G. Yang, C. Muller-Sieburg, C. Smith, N. Arai, Y. Takabe, and L. Gemmell. Interleukin 4 (B cell stimulating factor 1) can enhance or antagonize the factor-dependent growth of hemopoietic progenitor cells. *Proc. Natl. Acad. Sci. USA 84*:6889-6893 (1987).
53. Cohen, D. R., A. J. Hapel, and I. G. Young. Cloning and expression of the rat interleukin-3 gene. *Nucleic Acid Res. 14*:3641-3658 (1986).
54. Lopez, A. F., L. B. To, Y.-C. Yang, J. R. Gamble, M. F. Shannon, G. F. Burns, P. G. Dyson, C. A. Juttner, S. Clark, and M. A. Vadas. Stimulation of proliferation, differentiation, and function of human cells by primate interleukin 3. *Proc. Natl. Acad. Sci. USA 84*:2761-2765 (1987).
55. Clark, S. C., and R. Kamen. The human hematopoietic colony-stimulating factors. *Science 236*:1229-1237 (1987).
56. Koike, K., M. Ogawa, J. N. Ihle, T Miyake, T. Shimizu, A. Miyajima, T. Yokora, and K.-I. Arai. Recombinant murine granulocyte-monocyte (GM) colony-stimulating factor supports formation of GM and multipotential blast cell colonies in culture: comparison with the effects of IL-3. *J. Cell. Physiol. 131*:458-464 (1987).
57. Kindler, V., B Thorens, S. De Kossodo, B. Allet, J. F. Eliason, D. Thatcher, N. Farber, and P. Vassali. Stimulation of haemopoiesis in vivo by recombinant bacterial murine interleukin-3. *Proc. Natl. Acad. Sci. USA 83*:1001-1005 (1986).

58. Metcalf, D., C. G. Begley, G. R. Johnson, N. A. Nicola, A. F. Lopez, and D. J. Williamson. Effects of purified bacterially synthesized murine multi-CSF (IL-3) on hematopoiesis in normal adult mice. *Blood 68*:46-57 (1986).
59. Broxmeyer, H. E., D. E. Williams, S. Cooper, R. K. Shadduck, S. Gillis, A. Waheed, D. L. Urdal, and D. C. Bicknell. Comparative effects in vivo of recombinant murine-stimulating factor-1, and recombinant murine granulocyte-macrophage colony-stimulating factor on myelopoiesis in mice. *J. Clin. Invest. 79*:721-730 (1987).
60. Broxmeyer, H. E., D. E. Williams, G. Hangoc, S. Cooper, S. Gillis, R. K. Shadduck, and D. C. Bicknell. Synergistic myelopoietic actions in vivo after administration to mice of combinations of purified natural murine colony-stimulating factor 1, recombinant murine interleukin-3, and recombinant murine granulocyte-macrophage colony-stimulating factor. *Proc. Natl. Acad. Sci. USA 84*:3871-3875 (1987).
61. Miller, R. A., and O. Stutman. Enumeration of IL2-secreting helper T cells by limiting dilution analysis, and demonstration of unexpectedly high levels of IL-2 production per responding cells. *J. Immunol. 128*:2258-2264 (1982).
62. Kelso, A., and D. Metcalf. Clonal heterogeneity in colony stimulating factor production by murine T-lymphocytes. *J. Cell. Physiol. 123*:101-110 (1985).
63. Herold, K. C., D. W. Lancki, D E. Dunn, K. Arai, and F. W. Fitch. Activation of lymphokine genes during stimulation of cloned T cells. *Eur. J. Immunol. 16*:1533-1538 (1986).
64. Kelso, A., and N. Gough. Expression of hemopoietic growth factor genes in murine T lymphocytes. In *Lymphokines*, Vol. 13, D. R. Webb, and D. V. Goeddel (Eds.). Academic Press, Orlando, pp. 209-238 (1987).
65. Miller, R. A., and O. Stutman. Limiting dilution analysis of T helper cell heterogeneity: a single class of T cell makes both IL-2 and IL-3. *J. Immunol. 130*:1749-1753 (1983).
66. Mosmann, T. R., H. Cherwinski, M. W. Bond, M. A. Gieldin, and R. L. Coffman. Two types of murine helper T cell clone. 1. Definition according to profiles of lymphokine activities and secreted proteins. *J. Immunol. 136*:2348-2357 (1986).
67. Ymer, S., W Q. Tucker, C. J. Sanderson, A. J. Hapel, H. D. Campbell, and I. G. Young. Constitutive synthesis of interleukin-3 by leukaemia cell line WEHI-3B is due to retroviral insertion near the gene. *Nature 317*:255-258 (1985).
68. Luger, T. A., U. Wirth, and A. Köck. Epidermal cells synthesize a cytokine with interleukin 3-like properties. *J. Immunol. 134*:915-919 (1985).
69. Labastie, M. C., J. P. Thiery, and N. M. Le Douarin. Mouse yolk sac and intra embryonic tissues produce factors able to eleicit differentiation of erythroid burst-forming units and colony-forming units respectively. *Proc. Natl. Acad. Sci. USA 81*:1453-1456 (1984).

70. Frei, K., S. Bodmer, C. Schwerdel, and A. Fontana. Astrocyte-derived interleukin-3 as a growth factor for microglia cells and peritoneal macrophages. *J. Immunol. 137*:3521-3527 (1986).
71. Crapper, R. M., and J. W. Schrader. Evidence for the in vivo production and release into the serum of a T-cell lymphokine, persisting-cell stimulating factor (PSF), during graft-versus-host reactions. *Immunology 57*:553-558 (1986).
72. Filho, M. A., M. Dy, R. Lebel, G. Luffau, and J. Hamburger. In vitro and in vivo histamine-producing cell-stimulating factor (or IL-3) production during *Nippostrongylus brasiliensis* injection: coincidence with self-cure phenomenon. *Eur. J. Immunol. 13*:841-845 (1983).
73. Metcalf, D. The role of the colony-stimulating factors in resistance to acute infections. *Immunol. Cell. Biol. 65*:35-43 (1987).
74. Gordon, M. Y., G. P. Riley, S. M. Watt, and M. F. Greaves. Compartmentalization of a haemopoietic growth factor (GM–CSF) by glycosaminoglycans in the bone marrow microenvironment. *Nature 326*:403-405 (1987).
75. Spooncer, E., C. M. Heyworth, A. Dunn, and T. M. Dexter. Self-renewal and differentiation of interleukin-3-dependent multipotent stem cells are modulated by stromal cells and serum factors. *Differentiation 31*:111-118 (1986).
76. Fung, M.-C., A. J. Hapel, S. Ymer, D. R. Cohen, R. M. Johnson, H. D. Campbell, and I. G. Young. Molecular cloning of cDNA for murine interleukin-3. *Nature 307*:233-237 (1984).
77. Greenberger, J. S., M. A. Sakakeeny, R. K. Humphries, C. J. Eaves, and R. J. Eckner. Demonstration of permanent factor-dependent multipotential (erythroid/neutrophil/basophil) hematopoietic progenitor cell lines. *Proc. Natl. Acad. Sci. USA 80*:2931-2935 (1983).
78. Land, H., M. Grez, H. Hauser, W. Lindenmaier, and G. Schütz. 5′-Terminal sequences of eukaryotic mRNA can be cloned with high efficiency. *Nucleic Acids. Res. 9*:2251-2266 (1981).
79. Michelson, A. M., and S. H. Orkin. Characterisation of the homopolymer tailing reaction catalysed by terminal deoxynucleotidyl transferase. Implications for the cloning of cDNA. *J. Biol. Chem. 257*:14773-14782 (1982).
80. Parnes, J. R., B. Velan, A. Felsenfeld, L. Ramanathan, U. Ferrini, E. Appella, and J. G. Seidman. Mouse beta 2-microglobulin cDNA clones: a screening procedure for cDNA clones corresponding to rare mRNAs. *Proc. Natl. Acad. Sci. USA 78*:2253-2257 (1981).
81. Yokota, T., F. Lee, D. Rennick, C. Hall, N. Arai, T. Mosmann, G. Nabel, H. Cantor, and K. Arai. Isolation and characterisation of a mouse cDNA clone that expresses mast cell growth factor activity in monkey cells. *Proc. Natl. Acad. Sci. USA 81*:1070-1074 (1984).
82. Okayama, H., and P. Berg. A cDNA cloning vector that permits expression of cDNA inserts in mammalian cells. *Mol. Cell. Biol. 3*:280-289 (1983).
83. Southern, E. M. Detection of specific sequences among DNA fragments separated by gel electrophoresis. *J. Mol. Biol. 98*:503-517 (1975).

84. Miyatake, S., T. Yokota, F. Lee, and K. Arai. Structure of the chromosomal gene for murine interleukin 3. *Proc. Natl. Acad. Sci. USA 82*:316-320 (1985).
85. Campbell, H. D., S. Ymer, M.-C. Fung, and I. G. Young. Cloning and nucleotide sequence of the murine interleukin-3 gene. *Eur. J. Biochem. 150*:297-304 (1985).
86. Leder, P., D. Tiemeier, and L. Enquist. Ek2 derivatives of bacteriophage lambda useful in the cloning of DNA from higher organisms: the λgt WES system. *Science 196*:175-177 (1977).
87. Sanger, F., S. Nicklen, and A. R. Coulson. DNA sequencing with chain-terminating inhibitors. *Proc. Natl. Acad. Sci. USA 74*:5463-5467 (1977).
88. Deininger, P. L. Random subcloning of sonicated DNA: application to shotgun DNA sequence analysis. *Anal. Biochem. 129*:216-223 (1983).
89. Sanger, F., A. R. Coulson, G. F. Hong, D. F. Hill, and G. B. Peterson. Nucleotide sequence of bacteriophage λ DNA. *J. Mol. Biol. 162*:729-773 (1982).
90. Southern, P. J., and P. Berg. Transformation of mammalian cells to antibiotic resistance with a bacterial gene under control of the SV40 early region promoter. *J. Mol. Appl. Gen. 1*:327-341 (1982).
91. Lee, F., J. Abrams, K. Arai, A. Miyajima, S. Miyatake, T. Mosmann, D. Rennick, J. Schreurs, C. Smith, Y. Takebe, T. Yokota, G. Zurawski, and S. Zurawski. The expression and characterization of recombinant mouse IL-3. In *Lymphokines*, Vol. 15, J. W. Schrader (Ed.). Academic Press, San Diego, pp. 163-182 (1988).
92. Yang, Y.-C., A. B. Ciarletta, P. A. Temple, M. P. Chung, S. Kovacic, J. S. Witek-Giannotti, A. C. Leary, R. Kriz, R. E. Donahue, G. G. Wong, and S. C. Clark. Human IL-3 (multi-CSF): identification by expression cloning of a novel hematopoietic growth factor related to murine IL-3. *Cell 47*:3-10 (1986).
93. Griffin, J. D., R. Sullivan, R. P. Beveridge, P. Larcom, and S. F. Schlossman. Induction of proliferation of purified human myeloid progenitor cells: a rapid assay for granulocyte colony-stimulating factors. *Blood 63*: 904-911 (1984).
94. Ihle, J. N. Biochemical and biological properties of interleukin-3: a lymphokine mediating the differentiation of a lineage of cells that includes prothymocytes and mastlike cells. In *Contemporary Topics in Molecular Immunology*., Vol. 10, S. Gillis and F. P. Inman (Eds.). Plenum Press, New York, pp. 93-119 (1984).
95. Clark-Lewis, I., S. B. Ketn, and J. W. Schrader. Purification to apparent homogeneity of a factor stimulating the growth of multiple lineages of hemopoietic cells. *J. Biol. Chem. 259*:7488-7494 (1984).
96. Staden, R. An interactive program for comparing and aligning nucleic acid and amino acid sequences. *Nucl. Acids Res. 10*:2951-2961 (1982).
97. Sanderson, C. J., H. D. Campbell, and I. G. Young. Molecular and cellular biology of eosinophil differentiation factor (interleukin-5) and its effects on human and mouse B cells. *Immunol. Rev. 102*:29-50 (1988).

98. Fiers, W., W. Degrave, R. Devros, G. Plaetinck, H. Cheroutre, J. Tavernier, G. Simons, and E. Remaut. Cloning, characterization and expression of human and mouse lymphokine genes (interferon-γ and interleukin-2). In *Cellular and Molecular Biology of Lymphokines*, C. Sorg and A. Schimple (Eds.). Academic Press, Orlando, pp. 595-603 (1985).
99. Cantrell, M. A., D. Anderson, D. P. Cerretti, V. Price, K. McKereghan, R. J. Tuchinski, D. Y. Mochizuki, A. Larsen, K. Grabstein, S. Gillis, and D. Cosman. Cloning, sequence, and expression of a human granulocyte/macrophage colony-stimulating factor. *Proc. Natl. Acad. Sci. USA 8*: 6250-6254 (1985).
100. Clark-Lewis, I., R. Aebersold, H. Ziltener, J. W. Schrader, L. E. Hood, and S. B. H. Kent. Automated chemical synthesis of a protein growth factor for hemopoietic cells, interleukin-3. *Science 231*:134-139 (1986).
101. Clark-Lewis, I., and J. W. Schrader. Molecular structure and biological activities of P cell-stimulating factor (interleukin 3). In *Lymphokines*, Vol. 15, J. W. Schrader (Ed.). Academic Press, San Diego, pp. 1-37 (1988).
102. Shaw, G., and R. Kamen. A conserved AU sequence from the 3′ untranslated region of GM–CSF mRNA mediates selective mRNA degradation. *Cell 46*:659-667 (1986).
103. Stanley, E., D. Metcalf, P. Sobieszczuk, N. M. Gough, and A. R. Dunn. The structure and expression of the murine gene encoding granulocyte-macrophage colony stimulating factor: evidence for utilization of alternative promoters. *EMBO J. 4*:2569-2573 (1985).
104. Hapel, A. J., and I. G. Young. Molecular biology of interleukin 3: a multilineage hemopoietic growth regulator. In *Lymphokines*, Vol. 15, J. W. Schrader (Ed.). Academic Press, San Diego, pp. 91-126 (1988).
105. Barlow, D. P., M. Bucan, H. Lehrach, B. L. M. Hogan, and N. M. Gough. Close genetic and physical linkage between the murine haemopoietic growth factor genes GM–CSF and multi-CSF (IL-3). *EMBO J. 6*:617-623 (1987).
106. Roberts, B., P. Barton, A. Minty, P. Doubas, A. Weydert, F. Bonhomme, J. Catalan, D. Chazottes, J.-L. Guenet, and M. Buckingham. Investigation of genetic linkage between myosin and actin genes using an interspecific mouse back-cross. *Nature 314*:181-183 (1985).
107. Mason, I. J., A. Taylor, J. G. Williams, H. Sage, and B. L. M. Hogan. Evidence from molecular cloning that SPARC, a major product of mouse embryo parietal endoderm is related to an endothelial cell "culture shock" glycoprotein of M 43000. *EMBO J. 5*:1465-1472 (1986).
108. Mason, I. J., D. Murphy, M. Münke, U. Francke, R. W. Elliot, and B. L. M. Hogan. Developmental and transformation-sensitive expression of the SPARC gene on mouse chromosome 11. *EMBO J. 5*:1831-1837 (1986).
109. Lee, J. S., G. C. Webb, and I. G. Young. Fine-structure mapping of the murine IL-3 and GM–CSF genes by pulse-field gel electrophoresis and molecular cloning. *Genomics 5*, in press (1989).

110. Le Beau, M. M., N. D Epstein, S. J. O'Brien, A. W. Nienhuis, Y.-C. Yang, S. C. Clark, and J. D. Rowley. The interleukin 3 gene is located on human chromosome 5 and is deleted in myeloid leukemias with a deletion of 5q. *Proc. Natl. Acad. Sci. USA 84*:5913-5917 (1987).
111. Sutherland, G. R., E. Baker, D. F. Callen, H. D. Campbell, I. G. Young, C. J. Sanderson, O. M. Garsen, A. F. Lopez, and M. A. Vadas. Interleukin-6 is at 5q31 and is deleted in the 5q-syndrome. *Blood 71*:1150-1152 (1988).
112. Neinhuis, A. W., H. F. Bunn, P. H. Turner, T. V. Gopal, W. G. Nash, S. J. O'Brien, and C. J. Sherr. Expression of the human c-*fms* proto-oncogene in hematopoietic cells and its deletion in the 5q- syndrome. *Cell 42*:421-428 (1985).
113. Le Beau, M. M., C. A. Westbrook, M. O. Diaz, R. A. Larson, J. D. Rowley, J. C. Gasson, D. W. Golde, and C. J. Sherr. Evidence for the involvement of GM–CSF and FMS in the deletion (5q) in meyloid disorders. *Science 231*:984-987 (1986).
114. Pettenati, M. J., M. M. Le Beau, R. S. Lemons, E. A. Shima, E. S. Kawasaki, R. A. Larson, C. J. Sherr, M. O. Diaz, and J. D. Rowley. Assignment of CSF-1 to 5q33.1: evidence for clustering of genes regulating hematopoiesis and for their involvement in the deletion of the long arm of chromosome 5 in myeloid disorders. *Proc. Natl. Acad. Sci. USA 84*:2970-2974 (1987).
115. Huebner, K., M. Isobe, C. M. Croce, D. W. Golde, S. E. Kaufman, and J. C. Gasson. The human gene encoding GM–CSF is at 5q21-q32, the chromosome region deleted in the 5q- anomaly. *Science 230*:1282-1285 (1985).
116. Yarden, Y., J. A. Escobedo, W.-J. Kuang, T. L. Yang-Feng, T. O. Daniel, P. M. Tremble, E. Y. Chen, M. E. Ando, R. N. Harkins, U. Francke, V. A. Fried, A. Ullrich, and L. T. Williams. Structure of the receptor for platelet-derived growth helps define a family of closely related growth factor receptors. *Nature 323*:226-232 (1986).
117. Yang, Y.-C., S. Kovacic, R. Kriz, S. Wolf, S. C. Clark, T. E. Wellems, A. Nienhuis, and N. Epstein. The human genes for GM–CSF and IL3 are closely linked in tandem on chromosome 5. *Blood 71*:958-961 (1988).
118. Todaro, G. J., and J. E. De Larco. Growth factors produced by sarcoma virus-transformed cells. *Cancer Res. 38*:4147-4154 (1978).
119. Heldrin, C. H., and B. Westermark. Growth factors: mechanism of action and relation to oncogenes. *Cell 37*:9-20 (1984).
120. Sporn, M. B., and A. B. Roberts. Autocrine factors and cancer. *Nature 313*:745-747 (1985).
121. Waterfield, M. D., G. T. Scrace, N. Whittle, P. Stroobant, A. Johnsson, A. Wasteson, B. Westermark, C.-H. Heldrin, J. S. Huang, and T. F. Deuel. Platelet-derived growth factor is structurally related to the putative transforming protein $P28^{sis}$ of simian sarcoma virus. *Nature 304*:35-39 (1983).

122. Doolittle, R. F., M. W. Hunkapiller, L. E. Hood, S. G. Devare, K. C. Robbins, S. A. Aaronson, and H. N. Antoniades. Simian sarcoma virus *onc* gene, v-sis, is derived for the gene (or genes) encoding a platelet-derived growth factor. *Science 221*:275-277 (1983).
123. Hunter, T. Oncogenes and growth control. *Trends. Biochem. Sci. 10*: 275-280 (1985).
124. Hapel, A. J., and I. G. Young. Abnormal expression of interleukin-3 and leukaemia. *Anticancer Res. 7*:661-668 (1987).
125. Jhappan, C., G. F. Vande Woude, and T. S. Robins. Transduction of host cellular sequences by a retroviral shuttle vector. *J. Virol. 60*:750-753 (1986).
126. Rein, A., and A. Shultz. Different recombinant murine leukaemia viruses use different cell surface receptors. *Virology 136*:144-152 (1984).
127. Ceredig, R., T. Robins, H. D. Campbell, I. G. Young, and A. J. Hapel. Correlation of interleukin-2 receptor expression with tissue specific growth of an interleukin-3-dependent autocrine leukaemia. *J. Natl. Cancer Inst. 80*:165-170 (1988).
128. Warren, H. S., J. Hargreaves, and A. J. Hapel. Some interleukin-3 dependent mast cell lines also respond to interleukin-2. *Lymphokine Res. 4*: 195-204 (1985).
129. Gallatin, M., T. P. St. John, M. Siegelman, R. Reichert, E. C. Butcher, and I. L. Weissman. Lymphocyte homing receptors. *Cell 44*:673-680 (1986).
130. Warner, N. L., M. A. S. Moore, and D. Metcalf. A transplantable myelomonocytic leukemia in BALB/c mice: cytology, karyotype, and muramidase content. *J. Natl. Cancer Inst. 43*:963-982 (1969).
131. Lee, J. C., A. J. Hapel, and J. N. Ihle Constitutive production of a unique lymphokine (IL-3) by the WEHI-3 cell line. *J. Immunol. 128*:2393-2398 (1982).
132. Schrader, J. W., and R. M. Crapper. Autogenous production of a hemopoietic growth factor, persisting-cell-stimulating factor, as a mechnism for transformation of bone marrow-derived cells. *Proc. Natl. Acad. Sci. USA 80*:6892-6896 (1983).
133. Iscove, N. N., and C. Roitsch. The multilineage hemopoietic growth factor (Multi HGF/BPA/MCGF/IL-3). In *Cellular and Molecular Biology of Lymphokines*, C. Sorg and A. Schimpl (Eds.). Academic Press, Orlando, pp. 397-424 (1985).
134. Garland, J. M. Involvement of interleukin 3 in lymphocyte biology and leukemogenesis. In *Lymphokines*, Vol. 9, E. Pick (Ed.). Academic Press, Orlando, pp. 153-200 (1984).
135. Dexter, T. M., and T. D. Allen. The regulation of growth and development of normal and leukaemic cells. *J. Pathol. 141*:415-433 (1983).
136. Finnegan, D. J. Transposable elements in eukaryotes. *Int. Rev. Cytol. 93*:281-326 (1985).
137. Augenlicht, L. H., D. Kobrin, A. Pavlovec, and M. E. Royston. Elevated expression of an endogenous retroviral long terminal repeat in a mouse colon tumor. *J. Biol. Chem. 259*:1842-1847 (1984).

138. Kuff, E. L., A. Feenstra, K. Leuders, L. Smith, R. Hawley, N. Hozumi, and M. Shulman. Intracisternal A-particle genes as movable elements in the mouse genome. *Proc. Natl. Acad. Sci. USA 80*:1992-1996 (1983).
139. Canaani, E., O. Dreazen, A. Klar, G. Rechavi, D. Ram, J. B. Cohen, and D. Givol. Activation of the c-mos oncogene in a mouse plasmacytoma by insertion of an endogenous intracisternal A-particle genome. *Proc. Natl. Acad. Sci. USA 80*:7118-7122 (1983).
140. Chen, H. R., and W. C. Barker. Nucleotide sequences of the retroviral long terminal repeats and their adjacent regions. *Nucleic Acids Res. 12*: 1767-1778 (1984).
141. Christy, R. J., A. R. Brown, B. B. Gourlie, and R. C. Huang. Nncleotide sequences of murine intracisternal A-particle gene LTRs have extensive variability within the R region. *Nucleic Acids. Res. 13*:289-302 (1985).
142. Koury, G., and P. Gruss. Enhancer elements. *Cell 33*:313-314 (1983).
143. Gruss, P. Magic enhancers? *DNA 3*:1-5 (1984).
144. Dorssers, L., H. Burger, F. Bot, R. Delwel, G. Van Kessel, R. Lowenberg, and G. Wagemaker. Characterization of a human multilineage-colony-stimulating factor cDNA clone identified by a conserved noncoding sequence in mouse interleukin-3. *Gene 55*:115-124 (1987).

7
Megakaryocyte Growth Factors

NEIL WILLIAMS
University of Melbourne, Parkville, Victoria, Australia

I. INTRODUCTION

Megakaryocytopoiesis is the formation of new platelets in megakaryocytes in hemopoietic tissue, predominantly the bone marrow. The production of megakaryocytes from precursors, and their development and formation into platelets involves the commitment of precursors to the cell lineage, cell expansion followed by endomitosis, and cytoplasmic maturation of megakaryocytes before release of platelets into the circulation. The mechanisms controlling these various differentation events are poorly understood. This chapter provides an overview of the interaction of the known stimulatory factors.

It is quite clear from a large body of literature that megakaryocytopoiesis is under some form of feedback control. Changes in platelet mass have been shown to be associated with changes in the number, ploidy, and size of megakaryocytes (1-11). The simplest interpretation of data up to the 1970s was that megakaryocytopoiesis and platelet production was controlled by a plasma hormone, thrombopoietin, which functioned as a regulatory molecule stimulating new platelet production in response to thrombocytopenia (Fig. 1).

Early transfusion experiments (12-14) supported such a notion, and later a thrombopoietin assay was developed in which radioisotope incorporation into newly released platelets gave a bioassay of this material (15-17). Such an assay is relatively insensitive and inaccurate, but remains the only "classical" approach to monitoring the true potential of a substance to increase platelet counts in experimental animals.

Other potential methods have not been successful; the development of a radioimmunoassay for thrombopoietin has been hampered by lack of discrete

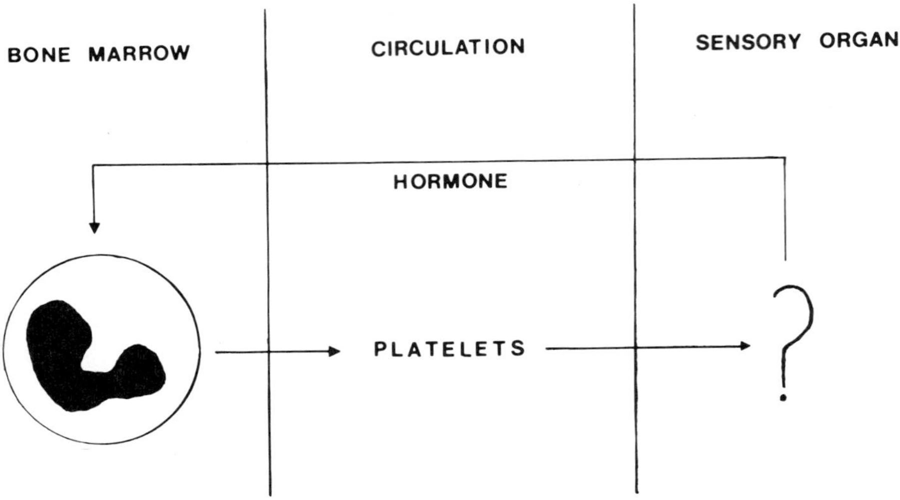

Figure 1 A simple feedback loop controlling megakaryocytopoiesis and thrombopoiesis by a single hormone, thrombopoietin. Hormone production occurs in response to altered platelet demand.

purified factors, and attempts to form platelets in vitro have been unrewarding. Recent approaches of quantitating thrombopoietic substances have been the measurement of GpIIb/IIIa complex and α-granule synthesis (18,19).

In an attempt to find a suitable alternative source of thrombopoietin to plasma, McDonald and colleagues have biochemically and biologically characterized an activity termed thrombopoietic stimulatory factor (TSF) from human embryonic kidney cells (20-24). This activity shows all the characteristics of crude plasma thrombopoietin (25). However, if it is a single molecule, the activity displays a markedly wide range of activities, from stimulating an increase in acetylcholinesterase-positive cells in animals through to the release of "young" platelets. It could even be envisaged that the activity acts on stem cells directing all of megakaryocytopoiesis (24). An alternative explanation is that TSF manifests fewer activities, one being the stimulation of acetylcholinesterase levels. Increase in the level of this enzyme is the endpoint of several assays and may help explain the wide range of activities attributed to TSF.

Thrombopoietic stimulatory factor is distinguishable from thrombopoietin in that it is not obtained from the plasma of experimental animals (20). A criticism of all approaches to the analysis of thrombopoietic stimuli is that platelet production can be nonspecifically stimulated in vivo (26). Specificity of action of a factor with a direct relationship to the circulating platelet mass remains central to the concept of a thrombopoietin-stimulated process. It might be argued that TSF acts via an indirect and nonspecific manner (i.e., independent of platelet mass); however, the nonspecific stimuli tested have been found to be required in high-dose amounts (27). On balance it appears likely that TSF when purified will be found to induce an authentic thrombopoietic response.

II. IN VITRO STUDIES

A. Thrombopoietic and Megakaryocyte Colony-Stimulating Factor

A separate approach to analyzing megakaryocyte and thrombopoietic stimuli became available with the in vitro growth of murine megakaryocytes (28) and cloning of their precursors in cell culture (29,30). It was clear from the earliest studies that megakaryocyte colonies were usually of a single cell type, although a proportion of erythroid and mixed colony types were observed (31,32). The activity responsible for the growth of these precursors was termed "megakaryocyte colony stimulating factor" (Mk-CSF). The term has conjured up images of lineage specificity back to the stem cell, thus recalling some of the earlier notions of stem-cell competition by hormones such as erythropoietin. Plasma thrombopoietin and thrombopoietic-stimulating factors were (with a single exception) found to be routinely ineffective when added to bone marrow colony assay (33-37), even though the factors could alter megakaryocyte number size and ploidy in experimental animals (33). Thus, these in-vivo stimulators did not share biological activity with megakaryocyte CSF.

It is clear now that the predominant murine megakaryocyte-CSF is interleukin-3 (IL-3), a multilineage stimulator of hemopoiesis (38-40). This observation comes from studies with IL-3 purified against a thymidine incorporation assay and the megakaryocyte colony assay (40-43). Moreover, an antibody to IL-3 completely neutralized all megakaryocyte-CSF activity from both WEHI-3 cell and mitogen-stimulated spleen-cell-conditioned medium (44). These data support the concept that IL-3 may be the primary stimulus of early events of megakaryocytopoiesis in cell culture and that part of the differentiation pathway may be nonspecifically and constitutively stimulated. Recent data suggest that recombinant granulocyte macrophage (GM)-CSF can also act as a megakaryocyte-CSF (45). This finding is in apparent contradiction with earlier studies in which purified native GM–CSF was found to be active in colony assays only at very high concentrations (46,47). The differences between the data remain unresolved but cross-binding of receptors at such high concentrations, or cross-binding of a differently folded recombinant molecule remains possible.

Erythropoietin has also been suggested as a modifier of platelet production, and its role in thrombopoiesis has remained controversial. Unfortunately its effects are in vitro equally enigmatic. Erythropoietin clearly can stimulate megakaryocytopoiesis in plasma clot assays (48,49), but not in methylcellulose or agar cultures (46). Mizogushi et al. (50) made a pertinent study using separate assays with these stimulators, confirming earlier studies that erythropoietin was effective as a postendomitotic stimulator of megakaryocytopoiesis in agar cultures, but it can stimulate both cell amplification and developmental aspects in plasma clot assays. We are no closer to understanding the effect of different growth factors on megakaryocyte colony formation in the different support matrices. Will methylcellulose and agar not support erythropoietin binding and stimulation? Clearly the BFU-e and CFU-e assays show excellent erythropoietin-dependent responses. Alternatively, stimulation of megakaryocyte colony formation by erythropoietin in the plasma clot assay may be related to the cascade of proteolytic enzymes released in the clot process.

Stimulators of human megakaryocyte colony formation is not nearly as well established, with plasma from patients with aplastic anemia being the superior stimulus (36). Other sources include conditioned media from PHA-stimulated leukocytes, Mo cell line, 5637-stimulated cell line, and erythropoietin (51-57). Granulocyte macrophage colony-stimulating factor is the only purified and cloned CSF active thus far in the megakaryocyte colony assay, but this molecule does not stimulate maximal colony numbers. Also, megakaryocyte development may not be optimal, with few colony cells realizing full maturation relative to normal megakaryocytes (57). It is likely that human IL-3 will be the maximally active component in the other preparations. It is not clear that the culture system is optimal and other factors may be required as in the mouse assay.

B. Second Growth Factors or Potentiators

The contention that multiple growth factors are required for full megakaryocyte development has been suggested by two sets of data: (a) the ability of thrombopoietin and thrombopoietic factors to stimulate in-vivo megakaryocytopoiesis but not in-vitro colony formation (33,34); (b) the in-vivo observation that megakaryocyte precursors were little affected by large and abrupt changes in platelet mass, yet megakaryocyte development was markedly changed (58,59).

The inability of thrombopoietic stimuli to promote colony formation and the potentiation of megakaryocyte colonies by bone-marrow-conditioned medium from cultures with active megakaryocytopoiesis (34,60) prompted studies to investigate whether those activities were second factors acting on the more mature cells, i.e., postmitotic megakaryocytes. Such activity was readily obtained from media conditioned by lung, peritoneal exudate cells, and bone marrow (61). The activity was termed megakaryocyte potentiator to distinguish it from thrombopoietic stimuli, defined in vivo.

Preparations of megakaryocyte potentiator influence megakaryocyte development in a similar manner to TSF by increasing megakaryocyte size and DNA content, while having no effect on progenitor cells (62-64). In addition, the activities titrate in a similar fashion, and are additive with each other at suboptimal levels but not at maximal levels of stimulation (65,66). Furthermore, both activities will synergize with low doses of IL-3, producing a nonadditive increase in the number of megakaryocyte colonies (43,67). Synergy is not restricted to precursors, however, and good synergy between TSF and IL-3 or megakaryocyte potentiator and IL-3 (but not TSF and megacaryocyte potentiator) when megakaryocyte growth is assessed (Table 1).

All the data observed thus far have not shown any biological distinction between TSF and sources of megakaryocyte potentiator. Specificity of these activities to megakaryocytopoiesis in cell culture is inferred because highly purified TSF (68) and a 6000-fold purified megakaryocyte potentiator preparation (Oon and Williams, unpublished data) does not stimulate myeloid colony growth or enhance the growth of myeloid colonies stimulated by suboptimal levels of IL-3 (Table 2).

A summary of the properties of these two classes of activities is shown in Table 3. The data suggest strongly that they act in a similar manner and may be

Table 1 The Interaction of Megakaryocyte Potentiator or Thrombopoietic Stimulatory Factor with Recombinant Interleukin-3

Megakaryocyte growth factors added	Number of megakaryocytes per culture (± SD)
Experiment 1	
TSF	22 ± 4
Mk-potentiator	26 ± 4
r IL-3	12 ± 4
TSF + TSF	35 ± 3
TSF + Mk-potentiator	48 ± 4
r IL-3 + TSF	70 ± 3
Experiment 2	
Mk-potentiator	4 ± 2
r IL-3	17 ± 5
r IL-3 + Mk-potentiator	47 ± 5

TSF, thrombopoietic stimulatory factor; Mk, megakaryocyte.
All factors were tested suboptimally in the single megakaryocyte growth assay at doses that fall within the linear portion of their portion of their individual growth response curves. Results (mean ± SD) are expressed as the mean number of megakaryocytes stimulated above cultures containing only medium. TSF was used at 25% of an optimal dose, stage 5 purified Mk-potentiator at 40%, and rIL-3 at 20% of a maximal response. The results of IL-3 and Mk-potentiator and IL-3 and TSF were significantly different from an additive response as determined by chi-square test ($p > 0.05$).

Table 2 The Effect of Megakaryocyte Potentiator and Thrombopoietic Stimulatory Factor on Myeloid Colony Formation

Factor added to each culture	Number of myeloid colonies/5 × 10^4 cells
medium	0
120 U rIL-3	69 ± 3
0.3 U TSF	0
2 U Mk-potentiator	0
rIL-3 + Mk-potentiator	57 ± 4

Abbreviations as in Table 1.
TSF was purified as described (68) and Mk-potentiator was purified 6000-fold from serum-free P388D1 CM. Colonies were counted after 7 days of culture. The results are the mean ± SD from triplicate cultures.

derived from the same molecule. Less is known about the molecular characteristics of these activities and a comparison of several properties is shown in Table 4.

Despite in-vitro colony assays being established in 1979 for human megakaryocytes (48), no human system has been developed that allows detection of any thrombopoietic substances. The murine systems clearly document that this activity is only effective as a cofactor in the colony assay, but is active alone in a megakaryocyte growth assay from immature megakaryocyte populations

Table 3 Culture Properties of Thrombopoietic Stimulators and Megakaryocyte Potentiators

All factors are inactive in colony assays
Factors enhance megakaryocyte colony numbers in the presence of titrated doses of IL-3
At high doses they stimulate enzyme synthesis (acetylcholinesterase), DNA content (but not cell division), and cell size of developing megakaryocytes
At low doses they synergize with IL-3 increasing acetylcholinesterase levels, DNA content, and cell size
They do not synergize but are additive with each other in culture assays
They do not enhance megakaryocyte numbers when added to maximum concentrations of the other
They have identical dose-response curves

Table 4 Comparison of the Biochemical and Biological Properties Between Megakaryocyte Growth Factors

Properties	IL-3	TPO	TSF	Mk-Pot
Est. MW (kilodaltons)	28-32		32	30
$(NH_4)_2SO_4$ solubility (%)	50-80	60-80	40-80	60-70
Binds to DEAE exchange resin pH 7.2	No	Yes	Yes	Yes
Binds to WGA		Yes	Yes	Yes
Mk-CSF activity	Yes	No	No	No
Stimulate immature megakaryocyte growth	Yes	Yes	Yes	Yes
Stimulate 32D cells	Yes		No	No
GM–CSF activity	Yes		No	No
Neutralized by α IL-3	Yes		No	No

TSF, thrombopoietic stimulatory factor; Mk-pot, megakaryocyte potentiator; WGA, wheat germ agglutinin, TPO, ; DEAE, ; CSF, colony-stimulating factor.
Data are taken from Refs. 41, 42, 67-72 and unpublished observation from Oon and Williams.
Highly purified megakaryocyte potentiator from P388D1 CM was obtained using sequential steps involving DEAE-sepharone pH 7.2, binding to wheat-germ agglutinin, gel filtration, and reversed-phase liquid chromatography. The material retains full biological activity and stimulates half maximal growth at 1×10^9 M.

(64,65). Comparably defined systems with human cells still have to be developed. It is of note that thrombopoietic stimuli appears to act across species. Thus it is more like erythropoietin, an end-cell stimulator, rather than behaving like some of the CSFs, which are more species specific and act on precursor-cell populations as well. Human embryonic kidney TSF is quite active in mouse assays, which are presently the only broadly defined culture procedures for assessing human thrombopoietic stimuli. New procedures measuring PF4 and GpIIb/IIIa should provide better quantitative data. Caution should be placed on using only a single parameter for measuring megakaryocyte development. For example, rat IL-3 was not active in the mouse colony assay, but markedly enhanced the detection of single acetylcholinesterase-positive cells in the megakaryocyte growth assay (Fig. 2). A notable feature of these studies was that no change in size or morphological appearance of the megakaryocytes was observed.

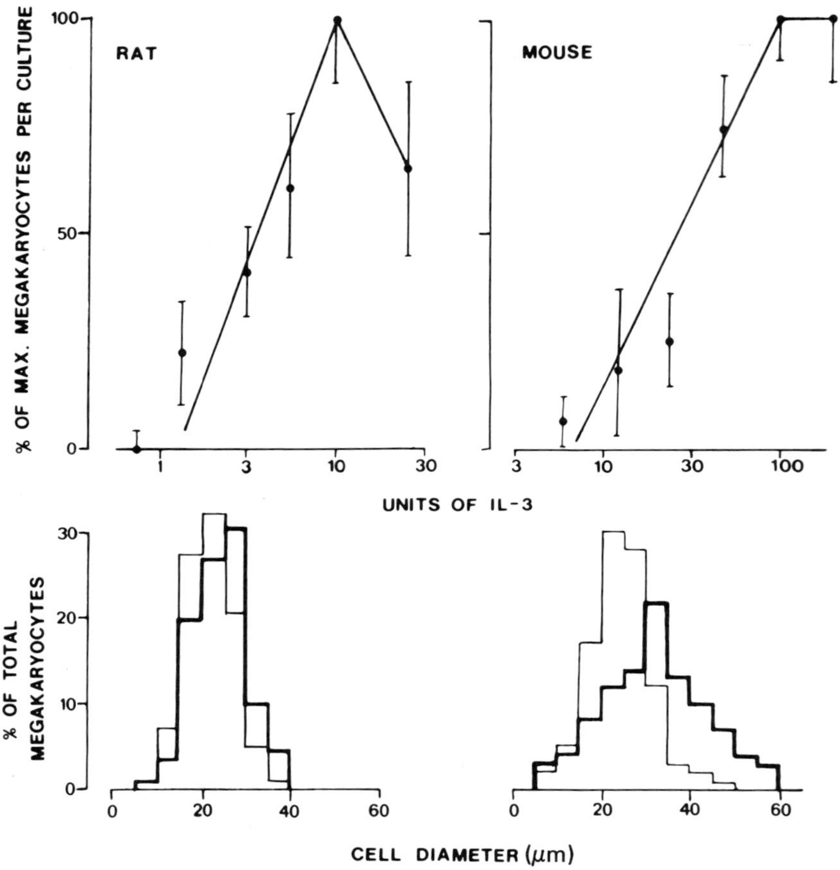

Figure 2 The response of murine megakaryocytes to mouse and rat IL-3. The upper response curves show the increase in detectable acetylcholinesterase-positive cells with increase in factor concentration. Single megakaryocytes were grown from fractionated bone marrow cell populations (Oon and Williams, unpublished data). The lower figures show the effect of both mouse (left) and rat (right) IL-3 on megakaryocyte diameter (µm). Cultures containing maximal levels of IL-3 (——) are compared with serum controls (**——**). Size of cells was measured on at least 200 cells from each group. The profiles of mouse IL-3 and serum-stimulated cultures were significantly different ($p < 0.001$), whereas no difference was found for the rat IL-3 profile and control.

It is concluded that rat IL-3 stimulated protein synthesis in the mouse megakaryocytes but without any concomitant stimulation of maturation events proceeding to platelet formation.

III. SUMMARY

A substantial body of data now indicate that megakaryocytopoiesis is regulated, like erythropoiesis, on at least two levels—a constitutive maintenance of progenitor cells and a megakaryocyte pool that respond to thrombopoietic stimuli in response to altered platelet demand.

The above model is assembled on the following observations: (a) Megakaryocyte progenitors are not regulated by platelet mass and therefore are not controlled via thrombopoietin; (b) IL-3 is the most potent stimulator of megakaryocyte progenitor cells described thus far, and therefore is a likely candidate for in-vivo maintenance of progenitor cell numbers; (c) thrombopoietin is the major stimulus for megakaryocyte development into platelets with their subsequent release into the circulation.

Can such a model meet physiological requirements? It is not clear that IL-3 can be produced in the bone marrow. At best it can be speculated that megakaryocyte progenitors are sustained by an "IL-3-like" or GM–CSF mechanism but the triggering mchanism may be membrane bound or via a separate molecule that activates a similar mechanism. It is quite possible that IL-3 has little to do with normal megakaryocytopoiesis (72,73), and that IL-3 levels are elevated only in immune responses that stimulate T cells to release IL-3. It is now known that IL-3 will stimulate all levels of megakaryocyte development, and this system may be a thrombopoietin-independent nonspecific mechanism of platelet formation (72). Such a concept demands that IL-3 can stimulate platelet release in the in-vivo thrombopoietin assay, as should appropriate stimulation in vivo by a strong and specific T-cell mitogen.

A second hypothesis is that IL-3 may be involved in a synergistic pathway with either TSF or thrombopoietin, based on the synergy between these two types of activities in cell culture. This concept requires that IL-3 be produced in very low but active amounts in the bone marrow or by an IL-3-like mechanism. Amplification of the IL-3-type-induced signal would then depend on the second factor (74). This mode requires only low levels of IL-3 (may be subthreshold by Northern blot analysis or biological assays). Synergy with thrombopoietin with IL-3 or an IL-3-like mechanism would mean that only low levels of thrombopoietin would also be required to sustain normal megakaryocytopoiesis. This concept fits earlier data where thrombopoiesis persists at reduced rates in transfusion-induced thombocytotic mice (75). Thus, in thrombocytopenic conditions, when thrombopoietin levels become elevated, an increased rate of megakaryocytopoiesis would be observed. Therefore, a synergistic mechanism would allow a rapid response from low background levels, increasing responsiveness of megakaryocytes to growth factors.

The above models assume no distinction between the thrombopoietic stimulators, thrombopoietin, and megakaryocyte potentiator. They further assume that megakaryocyte protein synthesis (especially membrane synthesis and demarcation membrane synthesis) correlates with endomitosis. However, induction of platelet-specific proteins, membrane and α granule synthesis may require coordinately controlled cofactors. Recently, Rosenberg and colleagues (18,76) have purified a factor (megakaryocyte stimulating factor) that fulfills those requirements, it elevates platelet factor 4 synthesis in the α granules of developing megakaryocytes without influencing general protein synthesis, cell size, DNA content, or demarcation membrane development: all the latter characteristics are hallmarks of a response to a thrombocytopenic condition.

Determination of conditions and factors controlling megakaryocytopoiesis remains unresolved. The purification of thrombopoietic factors to homogeneity and the assignment of biological activity to a single molecular species remains a single important step. The cloned factors are mouse and human IL-3, and human GM–CSF do not show lineage specificity required for the basic biological finding that a specific response of new platelet formation occurs in response to platelet demand. The most plausible conclusion to date is that specificity is obtained by the thrombopoietic stimuli; however, fidelity of all the various factors already described remains to be established.

ACKNOWLEDGMENTS

This work was supported by the National Health and Medical Research Council of Australia and the Anti-Cancer Council of Victoria. Rat IL-3 was a gift from A. Hapel, Canberra, Australia.

REFERENCES

1. Harker, L. A. Control of platelet production. *Annu. Rev. Med. 25*:383-399 (1974).
2. Ebbe, S. Biology of megakaryocytes. *Prog. Hemost. Thromb. 3*:211–229 (1976).
3. Ebbe, S. Megakaryocytopoiesis in vivo. In *Megakaryocyte Biology and Precursors: In Vitro Cloning and Cellular Properties*, B. L. Evatt, R. F. Levine, N. T. Williams (Eds.). Elsevier/North Holland, New York, pp. 1-13 (1981).
4. Paulus, J. M. DNA metabolism and development of organelles in guinea-pig megakaryocytes; a combined ultrastructural, autoradiographic and cytophotometric study. *Blood 35*:298-311 (1970).

5. Odell, T. T. Jr., C. W. Jackson, and R. S. Reiter. Depression of the megakaryocyte-platelet system in rats by transfusion of platelets. *Acta Haematol. 38*:34-42 (1967).
6. Odell, T. T. Jr., C. W. Jackson, and T. J. Friday. Megakaryocytopoiesis in rats with special reference to ploidy. *Blood 35*:775-782 (1970).
7. Odell, T. T. Jr., and C. Shelton. Increasing stimulation of megakaryocytopoiesis with decreasing platelet count. *Proc. Soc. Exp. Biol. Med. 161*:531-533 (1979).
8. Odell, T. T. Jr., C. W. Jackson, T. J. Friday, and D. E. Charsha. Effects of thrombocytopenia on megakaryocytopoiesis. *Br. J. Haematol. 17*:91-101 (1969).
9. Odell, T. T. Jr., and D. A. Boran. The mitotic index of megakaryocytes of mice after thrombocytopenia. *Proc. Soc. Exp. Biol. Med. 155*:149-151 (1977).
10. Penington, D. G., and T. E. Olsen. Megakaryocytes in states of altered platelet production: cell numbers, size and DNA content. *Br. J. Haematol. 18*:447-463 (1970).
11. Goldberg, J., E. Phalen, D. Howard, S. Ebbe Jr., and F. Stohlman. Thrombocytotoc suppression of megakaryocyte production from stem cells. *Blood 49*:59-69 (1977).
12. Keleman, E., D. Lehoczky, J. Perkedy, I. Cserhati, and K. Rak. Serum thrombocytopoietic activity in idiopathic thrombocytopenic purpura. *Lancet 1*:1134 (1960).
13. Spector, B. In vivo transfer of a thrombopoietic factor. *Proc. Soc. Exp. Biol. Med. 108*:146-149 (1961).
14. Odell, T. T. Jr., T. P. McDonald, and T. C. Detwiler. Stimulation of platelet-depleted rats. *Proc. Soc. Exp. Biol. Med. 107*:428-431 (1961).
15. Penington, D. G. Assessment of platelet production with ^{75}Se selenomethionine. *Br. Med. J. 4*:782-784 (1969).
16. Evatt, B. L., and J. Levin. Measurement of thrombopoiesis in rabbits using ^{75}seleno-methionine. *J. Clin. Invest. 48*:1615-1626 (1967).
17. Cooper, G. W., B. Cooper, and C. Chang. Demonstration of a circulating factor regulating blood platelet production using ^{35}S-sulfate in rats and mice. *Proc. Soc. Exp. Biol. Med. 134*:1123–1127 (1970).
18. Tayrien, G., and R. D. Rosenberg. Purification and properties of a megakaryocyte stimulating factor present in both the serum-free conditioned medium of human embryonic kidney cells and in thrombocytopenic plasma. *J. Biol. Chem. 262*:3262-3268 (1987).
19. Grant, B. W., N. L. Nichols, L. A. Solberg, D. J. Yachiniak, and K. G. Mann. Quantitiation of human in vitro megakaryocytopoiesis by radioimmunoassay. *Blood 69*:1334-1339 (1987).
20. McDonald, T. P., R. Clift, R. D. Lange, D. Nolan, I. I. E. Tribby, and G. H. Barlow. Thrombopoietin production by human embryonic kidney cells in culture. *J. Lab. Clin. Med. 85*:59-66 (1975).
21. McDonald, T. P. Assay and site of production of thrombopoietin. *Br. J. Haematol. 49*:493-499 (1981).

22. McDonald, T. P., and G. D. Kalmaz. Effects of thrombopoietin and erythropoietin on the number and diameter of marrow megakaryocytes of mice. *Exp. Hematol. 11*:91-97 (1983).
23. McDonald, T. P. Effect of thrombopoietin on platelet size of mice. *Exp. Hematol. 8*:527-532.
24. McDonald, T. P., M. Cottrell, C. C. Congdon, O. Walasek, and G. H. Barlow. Stimulation of megakaryocyte spleen colonies on mice by thrombopoietin. *Life Sci. 22*:1853-1858 (1978).
25. McDonald, T. P. Thrombopoietin and its control of thrombopoiesis and megakaryocytopoiesis. In *Megakaryocyte Biology and Precursors*, B. L. Evatt, R. F. Levine, and N. T. Williams (Eds.). Elsevier/North-Holland, New York, pp. 39-58 (1981).
26. Odell, T. T. Jr., T. P. McDonald, and F. L. Howsden. Native and foreign stimulators of platelet production. *J. Lab. Clin. Med. 6*:418-424 (1964).
27. McDonald, T. P., and R. Clift. Effects of thrombopoietin and erythropoietin on platelet production in rebound-thrombocytotic and normal mice. *Am. J. Hematol. 6*:219-228 (1979).
28. Nakeff, A., and K. A. Dicke. Stem cell differentiation into megakaryocytes from mouse bone marrow cultured with the thin layer again technique. *Exp. Hematol. 22*:58-60 (1972).
29. Metcalf, D., H. R. MacDonald, N. Odartchenko, and B. Sordat. Growth of mouse megakaryocytes colonies in vitro. *Proc. Natl. Acad. Sci. USA 72*:1744-1748 (1975).
30. Nakeff, A., and S. D. McQueen. In vitro colony assay for a new class of megakaryocyte precursor: colony-forming unit–megakaryocyte (CFU–M). *Proc. Soc. Exp. Med. Sci. 151*:587-590 (1976).
31. McLeod, D. L., M. M. Shreeve, and A. A. Axelrad. Chromosome marker evidence for the bipotentiality of BFU-3. *Blood 56*:318-322 (1980).
32. Ogawa, M., P. N. Porter, and T. Nakahata. Renewal and commitment to differentiation of hemopoietic stem cells (an interpretive review). *Blood 61*:823-829 (1983).
33. Levin, J., F. C. Levin, D. F. Hull III, and D. G. Penington. The effects of thrombopoietin on megakaryocyte-CFC, megakaryocytes and thrombopoiesis: with studies of ploidy and platelet size. *Blood 60*:989-998 (1982).
34. Williams, N., T. P. McDonald, and E. M. Rabellino. Maturation and regulation of megakaryocytopoiesis. *Blood Cells 5*:43-53 (1979).
35. Freedman, M. H., T. P. McDonald, and E. F. Saunders. Differentiation of murine marrow megakaryocytes progenitors: humoral control in vitro. *Cell Tissue Kinet. 14*:53-58 (1981).
36. Mazur, E. M., R. Hoffman, and E. Bruno. Regulation of human megakaryocytopoiesis. *J. Clin. Invest. 68*:733-741 (1981).
37. Keller, K. L., Z. Rolovic, B. L. Evatt, E. T. Sewell, and R. B. Ramsey. Thrombopoietin stimulation of megakaryocytopoiesis in agar culture. *Prog. Clin. Biol. Res. 215*:209-213 (1986).
38. Ihle, J. N., J. Kelller, S. Oroszlar, L. E. Henderson, T. D. Copeland, F.

Fitch, M. B. Prystowsky, E. Godwasser, J. W. Schrader, E. Plasaszynski, M. Dy, and B. Label. Biological properties of homologous interleukin-3. *J. Immunol. 131*:282-286 (1982).

39. Hapel, A. J., M. C. Fung, R. M. Johnson, I. G. Young, G. Johnson, and D. Metcalf. Biologic properties of molecularly cloned and expressed murine interleukin-3. *Blood 65*:1433-1459 (1985).
40. Metcalf, D., C. G. Begley, G. R. Johnson, N. A. Nicola, A. F. Lopez, and D. G. Williamson. Effects of purified bacterially-synthesized murine multi-CSF (IL-3) on hemopoiesis in normal adult mice. *Blood 68*:46-57 (1986).
41. Ihle, J. N., J. Keller, L. Henderson, F. Klein, and E. Palaszynski. Procedures for the purification of interleukin-3 to homogeneity. *J. Immunol. 129*: 2431-2436 (1982).
42. Quesenberry, P. J., J. N. Ihle, and E. McGrath. The effect of interluekin-3 and GM–CSA-2 on megakaryocyte and myeloid colony formation. *Blood 65*:214-217 (1985).
43. Williams, N., R. Sparrow, K. Gill, D. Yasmeen, and I. McNiece. Murine megakaryocyte colony stimulating factor: its relationship to interleukin-3. *Leuk. Res. 9*:1487-1495 (1985).
44. Sparrow, R. L., S. H. Oon, and N. Williams. Haemopoietic factors stimulating murine megakaryocytopoiesis: interleukin-3 is immunologically distinct from megakaryocyte potentiator. *Leuk. Res. 11*:31-36 (1987).
45. Robinson, B., E. McGrath, and P. Quesenberry. Megakaryocyte stimulating and interleukin-3 potentiating effects of recombinant granulocyte-macrophage colony stimulating factor. *J. Clin. Invest. 79*:1648-1652 (1987).
46. Williams, N., H. Jackson, N. N. Iscove, and P. P. Dukes. The role of erythropoietin, thrombopoietic stimulating factor and myeloid colony-stimulating factors on murine megakaryocyte colony formation. *Exp. Hematol. 12*: 734-740 (1984).
47. Metcalf, D., A. W. Burgess, G. R. Johnson, N. A. Nicola, E. C. Nice, J. De Lamarter, D R. Thatcher, and J. J. Mermod. In vitro actions on hemopoietic cells of recombinant murine GM–CSF purified after production in *E. coli*: comparison with purified native GM–CSF. *J. Cell. Physiol. 128*: 421-431 (1986).
48. Vainchenker, W., J. Bouguet, J Guichard, and J. Breton-Gorius. Megakaryocyte colony formation from human bone marrow precursors. *Blood 54*: 940-945 (1979).
49. Dukes, P. P., J. C. Egrie, T. W. Strickland, J. K. Browna, and F.-K. Lin. Megakaryocyte colony stimulating activity of recombinant human and monkey erythropoietin. *Prog. Clin. Biol. Res. 215*:105-110 (1986).
50. Mizoguchi, H., Y. Fujiwara, R. Sasaki, and H. Chiba. The effect of interleukin-3 and erythropoietin on murine megakaryocyte colony formation. *Prog. Clin. Biol. Res. 215*:111-116 (1986).
51. Kawakita, M., T. Miyake, S. Kishimoto, and M. Ogawa. Apparent heterogeneity of human megakarocyte colony- and thrombopoiesis-stimulating

factors: studies on urinary extracts from patients with aplastic anaemia and idioplastic thrombocytopenic patients. *Br. J. Haematol. 52*:429-438 (1982).

52. Hoffman, R., H. H. Yang, E. Bruno, and J. E. Stravena. Purification and partial characterization of a megakaryocyte colony-stimulating factor from human plasma. *J. Clin. Invest. 75*:1174-1182 (1985).
53. Kimura, H., S. A. Burstein, D. Thorning, J. S. Powell, L. A. Harker, P. J. Fialkow, and J. W. Adamson. Human megakaryocytic progenitors assayed in methylcellulose. *J. Cell. Physiol. 118*:87-96 (1984).
54. Solberg, L. A., N. Jamal, and H. A. Messner. Characterization of human megakaryocytic colony formation in human plasma. *J. Cell. Physiol. 124*: 67-74 (1985).
55. Myers, C. D., F. E. Katz, G. Joshi, and J. L. Miller. A cell line secreting stimulating factors for CFU–GEMM culture. *Blood 64*:152-155 (1984).
56. Levene, R. B., N. Williams, J-M. D. Lamaziere, and E. M. Rabellino. Human megakaryocytes IV. Growth and characterization of clonable megakaryocytes in agar. *Exp. Hematol. 15*:181-189 (1987).
57. Arriaga, M., K. South, J. L. Cohen, and E. M. Mazur. Interrelationship between mitosis and endomitosis in cultures of human megakaryocyte progenitor cells. *Blood 69*:486-492 (1987).
58. Levin, J., F. C. Levin, and D. Metcalf. The effects of acute thrombocytopenia on megakaryocyte CFC and granulocyte-macrophage CFC in mice: studies of bone marrow and spleen. *Blood 56*:274-283 (1980).
59. Burstein, S. A., J. W. Adamson, S. K. Erb, and L. A. Harker. Megakaryocytopoiesis in the mouse: response to varying platelet demand. *J. Cell. Physiol. 109*:333-341 (1981).
60. Williams, N., H. Jackson, A. P. C. Sheridan, J. J. Murphy Jr., A. Elste, and M. A. S. Moore. Regulation of megakaryocytopoiesis in long-term murine bone marrow cultures. *Blood 51*:245-255 (1978).
61. Williams, N., H. Jackson, P. Ralph, and I. Nakoinz. Cell interactions influencing murine marrow megakaryopoiesis: nature of the potentiating cell in bone marrow. *Blood 57*:157-163 (1981).
62. Long, M. W., N. Williams, and S. Ebbe. Immature megakaryocytes in the mouse. Physical characteristics, cell cycle status, and in vitro responsiveness to thrombopoietic stimulating factor. *Blood 59*:569-575 (1982).
63. Long, M. W., N. Williams, and T P. McDonald. Immature megakaryocytes in the mouse: the in vitro relationship to mature megakaryocytes and clonable megakaryocyte progenitor cells. *J. Cell. Physiol. 112*:339-344 (1982).
64. Williams, N., and H. Jackson. Kinetic analysis of megakaryocyte numbers and ploidy levels in developing colonies from mouse bone marrow cells. *Cell. Tissue Kinet. 15*:483-494 (1982).
65. Williams, N., S. H. Oon, H. Jackson, and R. Lim. Studies on megakaryocyte potentiator: its production and some biochemical characteristics. *Prog. Clin. Biol. Res. 215*:91-103 (1986).

66. Oon, S.-H., and N. Williams. Immature megakaryocytes in the mouse: synergistic response to megakaryocyte potentiator, thrombopoietic stimulatory factor with interleukin 3. *Leukemia 1*:772-776 (1987).
67. Sparrow, R. L., and N. Williams. Megakaryocyte colony stimulating factor: its identity to interleukin-3. *Prog. Clin. Biol. Res. 215*:123-128 (1986).
68. McDonald, T. P., M. Cottrell, R. Clift, J. A. Khouri, and M. D. Long. Studies on the purification of thrombopoietin from kidney cell culture medium. *J. Lab. Clin. Med. 106*:162-174 (1985).
69. Kellar, K. L., B. L. Evatt, C. R. McGrath, and R. B. Ramsey. Stimulation of DNA synthesis in megakaryocytes by thrombopoietin in vitro. In *Megakaryocyte Biology and Precursors*, B. L. Evatt, R. F. Levine, and N. T. Williams (Eds.). Elsevier/North Holland, New York, pp. 21-38 (1981).
70. Oon, S.-H., and N. Williams. Biochemical characterization of an in vitro megakaryocyte growth activity: megakaryocyte potentiator. *Leuk. Res. 10*:403-412 (1986).
71. Hill, R., and J. Levin. Partial purification of thrombopoietin using lectin chromatography. *Exp. Hematol. 14*:752-759 (1986).
72. Ishibashi, T., and S. A. Burstein. Interleukin 3 promotes the differentiation of isolated single megakaryocytes. *Blood 67*:1512-1514 (1986).
73. Alberico, T. A., J. N. Ihle, C.-M. Liang, H. E. McGrath, and P. J. Quesenberry. Stromal growth factor production in irradiated lectin exposed long-term murine bone marrow cultures. *Blood 69*:1121-1128 (1987).
74. Williams, N., R R. Eger, H. Jackson, and D. J. Nelson. Two factor requirement for megakaryocyte colony formation. *J. Cell. Physiol. 110*:101-104 (1982).
75. Shreiner, D. P., and J. Levin. Detection of thrombopoietic activity in plasma by stimulation of suppressed thrombopoiesis. *J. Clin. Invest. 49*: 1709-1713 (1970).
76. Greenberg, S. M., D. J. Kuter, and R. D. Rosenberg. In vitro stimulation of megakaryocyte maturation by megakaryocyte stimulatory factor. *J. Biol. Chem. 262*:3269-3277 (1987).

8
Eosinophil Differentiation Factor (Interleukin-5)

COLIN J. SANDERSON
National Institute for Medical Research, London, England

I. INTRODUCTION

The nomenclature of the cytokines is confusing and still evolving. The name eosinophil differentiation factor (EDF) was first used for this factor because in the mouse it stimulated the production of eosinophils in liquid cultures but produced relatively few eosinophil colonies in semisolid cultures (1). Subsequent work, however, particularly with human cultures, has shown that it is indeed an eosinophil colony-stimulating factor (Eo-CSF) and so this name could have been used. Because of its activity on B cells in the mouse where it has attracted a number of names based on different bioassays [for example, B-cell growth factor II (BCGFII) and T-cell replacing factor (TRF)], the term interleukin-5 (IL-5) has also become accepted.

Eosinophils are present at such low frequency in normal animals as to be almost undetectable in the blood and tissues. They are characteristic of allergic responses and infection by parasitic helminths, where they can become the predominant leukocyte in the inflammatory response. This fluctuation from low to very high levels provides a fascinating model for the control of hemopoiesis, and coupled with their distinctive morphological characteristics, has led to a large number of reports of their association with different pathological processes. These established an association between eosinophilia and high levels of IgE antibody in parasitic infection and allergy. It has become clear that eosinophils are characteristic of helminth infections and allergies, and are also found in a variety of other pathological responses. There are, however, many situations in which a profound cellular inflammatory response occurs in the complete absence of eosinophils, suggesting some control mechanism that is specific for the production and localization of eosinophils.

Although descriptive reports go back into the last century, it is only relatively recently that the biological activity of eosinophils has begun to be unraveled. This began with the observation that they are capable of killing large metazoan parasites in vitro (reviewed in Ref. 2), and was followed by the demonstration of cytotoxic activity against the protozoan *Trypanosoma cruzi* (3-5), erythrocytes (6,7), and tumor cells (8). In each case the cytotoxicity was antibody dependent and enhanced by complement. While these experiments in vitro demonstrate the cytotoxic potential of the eosinophil, medical epidemiology is providing important insights into the biological importance of these cells. Thus, what was until recently little more than a hematological curiosity has become a major focus in a number of important human diseases.

For example, the ability of eosinophils to kill helminths in vitro suggests an important role in the expulsion of these parasites from the infected host. Evidence supporting this view comes from two important surveys of reinfection of children with *Schistosoma mansoni* (9) and *S. haematobium* (10). In both cases high blood eosinophil counts correlated with increased resistance to reinfection.

Similarly, the cytotoxic activity of eosinophils against tumor cells in vitro suggests a potential role in cancer and graft rejection. Indeed, the presence of eosinophils in association with tumors has been noted for many years, but it is only recently that follow-up clinical studies have suggested an important biological role. For example, in a study of lung cancer it was found that local infiltration of eosinophils was associated with a very good prognosis (11). Similarly, eosinophil infiltration of colonic carcinoma (12) and gastric carcinoma (13) was found to have a positive prognostic significance. In Hodgkin's disease a selective blood eosinophilia was found to be a good prognostic indicator (14), and correlated with an extensive deposition of eosinophil peroxidase in a proportion of biopsies, suggesting eosinophil degranulation in these lymphomas (15). In renal transplantation the presence of eosinophils in the intragraft cellular infiltrate of patients with allografts was found to be a specific indicator of irreversible rejection (16), although it is not known whether the eosinophils were directly involved in the rejection.

Despite the presence of large numbers of eosinophils in allergic diseases such as asthma, little attention was given to their potential as agents of tissue destruction. However, there is an increasing awareness that the eosinophil is involved in these pathological processes, and indeed could be the predominant cause of tissue damage. For example, there is increasing evidence that eosinophils could be a major cause of respiratory malfunction in asthma. This conclusion is based on the ability of the products of eosinophil granules such as the major basic protein, eosinophil cationic protein and eosinophil peroxidase, to cause damage to lung epithelium in vitro, and the demonstration of these agents in areas of tissue damage in lung biopsies from patients with asthma (reviewed in Ref. 17).

Furthermore, in a number of other clinical syndromes characterized by eosinophilia, compelling evidence for an eosinophil involvement in the pathological process has been obtained. Thus, allergic granulomatosis (Churg-Strauss syndrome) is characterized by large numbers of eosinophils in the blood and tissues, and granulomatous lesions containing eosinophils occur in the heart and other organs. The demonstration of secreted eosinophil cationic proteins in cardiac lesions suggests that the eosinophils play a role in the development of these lesions (18). A similar deposition observed in renal biopsies suggests that eosinophils also play a part in the development of glomerulonephritis in this syndrome (E. J. Clutterbuck, unpublished observations). Similarly, in idiopathic hypereosinophilia, deposits of eosinophil granule proteins in the cardiac tissue of patients with endomyocardial disease suggests that the eosinophils cause the heart lesions (19).

Although much remains to be understood, it has become clear that eosinophils can deliver both positive and negative effects. Thus, understanding the control of the production, distribution, and activation of eosinophils could provide a rational approach to the treatment of some of these important diseases. In this chapter our work on the identification of a lymphokine controlling eosinophil production is reviewed. The possibility that a lymphokine was involved in this process came from the observations that the development of eosinophilia in parasitic infections is T-lymphocyte dependent (20), and that supernatants of activated spleen cells (and therefore a putative T-lymphocyte product) stimulated the production of eosinophil colonies (21).

II. MODEL SYSTEMS FOR STUDYING EOSINOPHILIA

Laboratory models for the production of eosinophils have included peritoneal lavage in guinea pigs and injections of foreign proteins in guinea pigs and mice. In addition, certain strains of rat have increased numbers of eosinophils in the peritoneal exudate. Although the mechanism of this eosinophilia is unknown, it provides a useful source of eosinophils in this species (3). The most widely used laboratory models have been experimental parasitic infections. In preliminary experiments we compared *Trichinella spiralis, Toxocara canis, Taenia crassiceps*, and *Mesocestoides corti* for their ability to induce eosinophilia in mice. In practice we have found that *M. corti* induces a strong eosinophilia and is simple and relatively safe to maintain.

III. ASSAYS FOR EDF

Early studies mainly from Metcalf's laboratory had indicated tht mouse eosinophil colonies were produced in semisolid agar cultures using medium conditioned by spleen cells stimulated with pokeweed mitogen (21). This had given rise to the concept of an eosinophil colony-forming unit as a precursor to the eosinophil, and the concept of an Eo-CSF. In our preliminary experiments in mice we found this colony assay insensitive and unreliable for detecting factors stimulating eosinophil differentiation. Although small numbers of eosinophil colonies were obtained, this did not appear to reflect the ability of the bone marrow cells to produce eosinophils. For example, bone marrow taken from *M. corti*-infected mice, which were undergoing an impressive eosinophilia and in which the bone marrow contained more than 50% of cells containing eosinophil granules, gave less than five eosinophil colonies/10^5 marrow cells, which was not significantly different from the number obtained with normal bone marrow. In these studies there were 50-100 colonies/10^5 marrow cells of other cell types in the same cultures, indicating that the assay system was satisfactory for other cell lineages (22).

In parallel studies with liquid cultures (23), bone marrow from eosinophilic mice were cultured in the absence of added growth factors. Surprisingly, the eosinophils disappeared within a few days whereas the neutrophils remained. The cultures continued to produce neutrophils for many weeks, as had been demonstrated by Dexter (24) with normal bone marrow, but no eosinophils were observed. This suggested that eosinophils required additional "factors" to those required for neutrophil differentiation. This was found to be the case, because the additon of spleen-cell supernatants induced the production of eosinophils, and there was a marked increase in production from bone marrow cells from eosinophilic mice compared with normal mice. This difference in the ability of eosinophils to proliferate and differentiate in liquid culture but apparently not as well in semisolid medium is not understood, and does not apply to human cells (see Section XI C). From a practical point of view it was clear that liquid cultures provided a better assay system than the colony assay for factors controlling eosinophil production. It has also meant that studies of eosinophil differentiation have been limited because many experiments that are possible with the colony assay have not yet been done in the mouse system.

To increase the versatility of the assay a microplate modification of this culture system was developed as an assay for eosinophil differentiation factors. In this assay, bone marrow cells in 0.1 ml of medium are cultured in the presence of factors for 5-7 days. The number of eosinophils in each well can be determined by counting the total cells with a Coulter counter, and carrying out a differential count on a stained preparation (1). A simple alternative is to assay for eosinophil peroxidase in the wells of the microplate, and determine the eosinophils indirectly by measuring absorbance in a multiplate reader (25). This was found to be specific for eosinophils in the mouse, apparently because of the relatively low concentration of myeloperoxidase in the mouse neutrophil. Although no detailed kinetic analysis has been carried out, this short-term assay system is probably measuring survival of inoculum eosinophils as well as eosinophil production. Therefore, it detects any factor that prolongs their survival or increases their production. In addition to EDF the assay detects IL-3 and granulocyte-macrophage CSF (GM–CSF).

In humans either the colony assay or a liquid culture system can be used as an assay for EDF (26,27). The mononuclear fraction is obtained from human bone marrow by centrifugation over Ficoll-Hypaque. To conserve reagents and simplify analysis we carry out the colony assay in the wells of a migration inhibition plate, using 0.4 ml of semisolid medium. After 14 days these are removed to a microscope slide, dried, and stained to identify the colony types. The liquid culture system is carried out in microplates. Eosinophil production is detectable at 14 days but is best assayed at 21 days. Eosinophils are quantified by a combination of total counting on a Coulter counter and differential counting of stained preparations. Assay for eosinophil peroxidase is less satisfactory than in the mouse assay because it is complicated by the presence of myeloperoxidase in residual neutrophils. However, provided that no factors stimulating neutrophil production are present, the peroxidase assay can be used after 21 days when the number of neutrophils is relatively small. In this liquid culture system the cells require feeding after 12 days. Because the assay time is 14 days in the colony assay and 21 days in the liquid culture system, it must be concluded that these bioassays for human EDF are less than ideal.

IV. IDENTIFICATION OF MURINE EDF

After the observation that supernatants of activated spleen cells induced eosinophil production in bone marrow cultures (23), a panel of T-cell clones was tested for their ability to induce eosinophil differentiation. Supernatants from a high proportion of both parasite antigen and alloantigen reactive T-cell clones were positive in the EDF assay. A coordinate analysis of the production of different lymphokines by these T-cell clones indicated that this activity was distinct from IL-2 and IL-3. The activity from one of these clones was found to have an M_r of 45,000 and thus was clearly distinct from IL3 and GM–CSF (1). Subsequent work indicated that a high proportion of T-cell clones raised from antigens not normally associated with eosinophilia (*Bacillus Calmette-Guerin* and sheep red blood cells) also produced EDF (C. J. Sanderson, unpublished observations).

Examination of several sublines of the lymphoma EL-4 indicated that some of these also produce EDF as well as a number of other lymphokines, particularly IL-2. In an attempt to provide a better source of the lymphokine, a series of T-cell hybrids were produced by fusing spleen cells to the lymphoma BW5147. These were induced with phorbol myristate acetate (PMA) and supernatants tested for EDF production. One of the lines was carried through six reclonings, and the subclone producing the highest levels of EDF selected at each stage. The resulting hybrid (NIMP–TH1) had a modal chromosome number of 102, suggesting that it was the product of the fusion of more than two cells. It produced high levels of EDF but no IL-2, IL-3, interferon, or GM–CSF (22). This hybrid has provided a source of this factor for biological and biochemical analysis.

V. BIOCHEMICAL CHARACTERIZATION

Eosinophil differentiation factor from a T-cell clone (NIMP-T2), the T-cell hybrid (NIMP–TH1), and the EL-4 cell line fractionates on gel filtration as a heterogeneous molecule over a M_r range of 35,000-65,000, with a peak at 45,000 (1). The protein binds to lentil lectin and can be eluted with α-methyl mannoside, indicating the presence of carbohydrate. When cells are grown in the presence of tunicamycin, which blocks *N*-linked glycosylation, biological activity is recovered at M_r of approximately 30,000 (D. J. Warren, unpublished observations). This indicates a maximum size for the protein moiety, and demonstrates that *N*-linked carbohydrate is not necessary for biological activity. The biological activity is stable for months at 4°C in the presence of carrier proteins, but is partially destroyed by heating at 70°C for 30 min. It is stable over the pH range 2-10, and in the presence of 2% sodium dodecyl sulfate.

The demonstration that murine EDF was identical to BCGFII and TRF (which is discussed below) provided additional biochemical information because TRF has been quite extensively characterized (28). After purification, the glycosylated native form runs at approximately M_r 18,000 on polyacrylamide gels under reducing conditions. This is compatible with the structure predicted by the cDNA sequence (see Section X) indicating a mature protein with a molecular weight of ~13,000. Although there are some conflicting reports in the literature, in our experience reduction destroys the biological activity of EDF (S. Barry, unpublished observations).

VI. ISOLATION OF cDNA CLONES

The T-cell hybrid NIMP–TH1 after stimulation with PMA for 8 hr provided a source of mRNA. Using enriched mRNA a cDNA library was prepared and screened by direct expression in COS cells using diethylaminoethyl dextran and chloroquine transfection (29). One pool of 48 clones was dectected that expressed EDF activity, and a single expressing clone was then isolated from this pool (pEDFM-4). The nucleotide sequence was identical to the sequence reported for BCGFII (30). The sequence of the insert in pEDFM-4 terminated at nucleotide 653, with a run of approximately 85 As. A longer clone (pEDFM-5) was also isolated, which contained an insert of 1534 base pairs (bp) followed by 51 As. The shorter clone indicates the presence of an alternative polyadelylation signal. A third clone was isolated that was missing exon 2, but expressed material from this clone had no biological activity (29).

VII. STRUCTURE OF THE CHROMOSOMAL GENES

A mouse EDF cDNA coding region fragment was used as a probe for Southern blotting and isolation of the mouse and human genes. A 3.2-kilobase (kb)-fragment encompassing the human gene and a 6.7-kb fragment encompassing the mouse gene were isolated and sequenced (29,31). The coding region is divided between four exons in both species. The three introns of the mouse gene are 829, 1875, an 79 bp in length and in the human gene the introns are 208, 950, and 105 bp, respectively (Fig. 1). The relative positions of the introns within the coding regions are identical in the two genes and all introns begin with GT and end in AG. A potential TATA box is located 29 bp (human) and 30 bp (mouse) upstream from the start of transcription, and there is at least one potential CAAT box ~80 bp upstream of the start of transcription of each gene. The 3′ untranslated region of the mouse gene carries an additional segment of 738 bases. Short inverted repeats at the boundaries of this insert flanked by possible duplications of host sequence suggest it could have resulted from the insertion of a mobile genetic element.

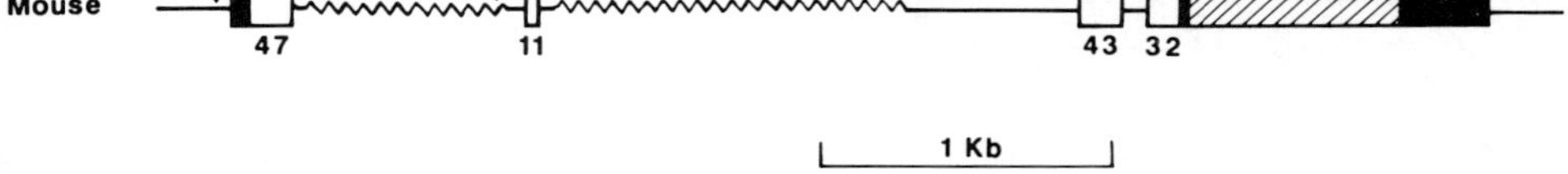

Figure 1 Diagram indicating the structure of the mouse and human IL-5 genes. Exons are shown by boxes, and the open boxes show the coding regions. The number of amino acids encoded by each exon is indicated. Shaded areas indicate untranslated regions of the mRNA. The hatched area represents the *Alu*-like insert in the mouse gene. ▾, putative TATA box, ——, parts of introns showing significant homology; ∿∿, parts of introns not showing significant homology. (From Ref. 46.)

VIII. CHROMOSOMAL LOCATION

The human IL-5 gene is located at 5q31 and is deleted in patients with the 5q- syndrome (32). Both IL-3 (33) and GM–CSF (34) genes have been located to the same region of this chromosome, and are separated by only 9 kb (35). The mouse gene is located on chromosome 11 (H. D. Campbell, unpublished observation). It has been established that the murine IL-3 and GM–CSF genes lie close together on mouse chromosome 11 (36), but the exact location of IL-5 in relation to these genes is not yet known in either the human or mouse genome.

IX. COMPARISON WITH OTHER LYMPHOKINE GENES

An identical 10-base sequence occurs upstream of the TATA box of the EDF and GM–CSF genes in both the mouse and human genes (31). Similar sequences are present in some other cytokine genes including the human G–CSF gene (37) and the mouse IL-4 gene (28), suggesting that they may function as regulatory elements. Apart from this there is no significant nucleotide homology between IL-5 and any of the genes encoding other cytokines. However, there is an overall structural similarity between the genes encoding the lymphokines IL-3 (39, 40), IL-5, and GM–CSF (41). The introns begin with GT and end in AG, and

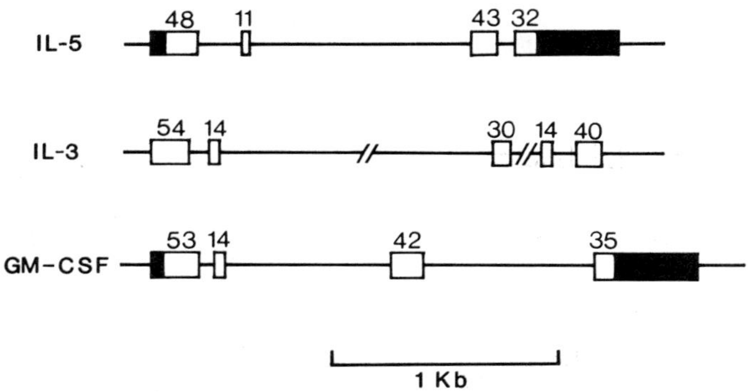

Figure 2 Diagram of the structures of human IL-3, IL-5, and GM–CSF genes illustrates similarities in intron/exon sizes. Coding regions are shown as open boxes, with the number of amino acids encoded indicated in each case. Shaded areas indicate untranslated areas of the mRNA. The full sequence of the IL-3 gene has not been published, so the length of introns 2 and 3 is not known accurately. (Data from Refs. 31,40,41.)

each exon contains a complete set of codons. Apart from an additional intron in IL-3, so that exons 4 and 5 correspond to exon 4 in the other two genes, the similarity is clear (Fig. 2).

X. POLYPEPTIDE STRUCTURES OF MURINE AND HUMAN EDF

The genes encode EDF precursors of 133 amino acids in the mouse and 134 in humans (Fig. 3). The predicted site of cleavage of the signal peptide using the von Heijne rules (42) is after amino acid 18 in the mouse and 19 in humans to give a processed protein of 115 amino acids in each case. Two independent reports indicate that native mouse IL-5 has an *N*-terminal methionine, indicating cleavage after amino acid 20, giving an M_r of 13,111 (43,44). Our attempts to obtain a sequence have been unsuccessful because the *N*-terminus is blocked. No sequence information is yet available for the native human protein but if the von Heijne prediction is correct the mature protein would have a M_r of 13,149.

As discussed above, the native mouse protein appears to have an M_r of approximately 30,000 in deglycosylated form, and so is likely to be a dimer. An alignment of the two amino acid sequences as encoded by the exons of the genes is shown in Fig. 3. There are three potential *N*-glycosylation sites in the mouse EDF, the first and last of which are conserved in the human protein. Two cyteine residues are present in each mature protein at corresponding positions (Fig. 4). Thus, at least one of these cysteines must be involved in dimerization, so it is unlikely that there is internal disulphide bridging. Our work with

```
         (H)   MR.MLLHLSLLALGAAYVYAIPTEIPTSALVKETLALLSTHRTLLIANE
Exon 1:        ** ****** * *    * *   *** *  *****  ** ** **  **
         (M)   MRRMLLHLSVLTLS..CVWATAMEIPMSTVVKETLTQLSAHRALLTSNE

         (H)   TLRIPVPVHKN
Exon 2:        * * *** ***
         (M)   TMRLPVPTHKN

         (H)   HQLCTEETFQGIGTLESQTVQGGTVERLFKNLSLIKKYIDGQK
Exon 3:        ****  ******   *  ***  ****** ** *********** **
         (M)   HQLCIGEIFQGLDILKNQTVRGGTVEMLFQNLSLIKKYIDRQK

         (H)   KKCGEERRRVNQFLDYLQEFLGVMNTEWIIES
Exon 4:         *********  ************** ***   *
         (M)   EKCGEERRRTRQFLDYLQEFLGVMSTEWAMEG
```

Figure 3 Alignment of the amino acid sequences of human (H) and mouse (M) IL-5 as coded by each of the four exons. Asterisks indicate identical residues. (Adapted from Ref. 31.)

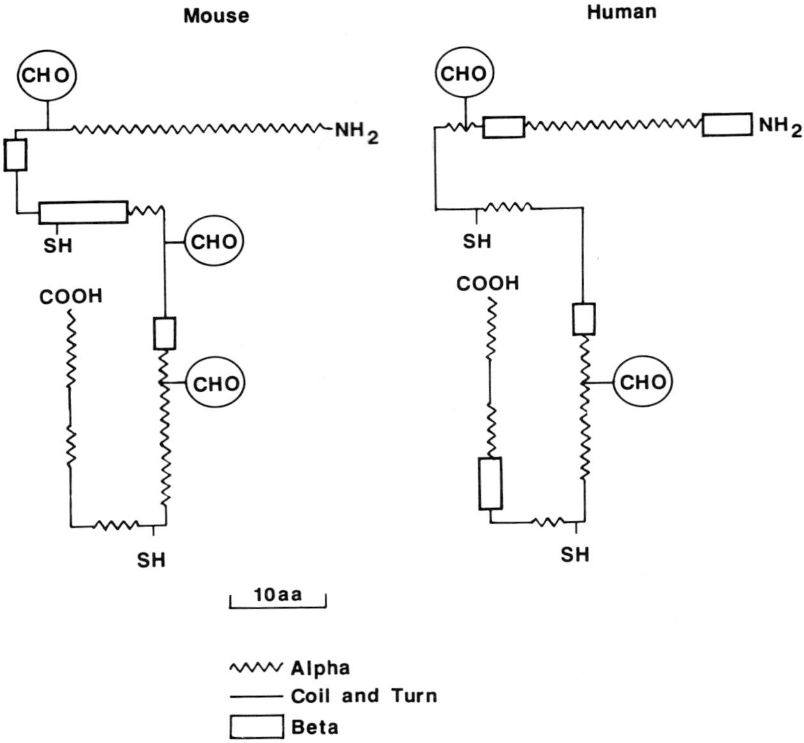

Figure 4 Diagrammatic representation of the mouse and human IL-5 mature polypeptides, based on Robson/Garnier prediction (72). CHO, potential *N*-linked glycosylation site; SH, position of cyteine residues. For diagrammatic purposes each predicted turn is shown as a right angle. It should be noted that the biologically active native form of both human and mouse IL-5 is a covalently linked dimer of these polypeptides.

recombinant human IL-5 from COS cells indicates a molecule of M_r ~45,000 by gel permeation, suggesting that this is also a glycosylated dimer (C. J. Sanderson, unpublished observation).

There is 67% amino acid sequence identity between mouse and human EDF. This compares with 29% with IL-3 (39,40) and 52% with GM–CSF (41). There is no significant homology between EDF and any other proteins. However, by searching for similarity rather than homology to the IL-5 sequence, using a matrix derived from evolutionary mutation (45) a small area of similarity is detected near the C-terminal end of IL-5 and IL-3, GM–CSF, IL-4, and IL-6 (46).

Although the gene structure of IL-3, IL-4, IL-5, and GM–CSF show similarities that might suggest a common origin, IL-6 does not fit into this group and so the significance of this area of similarity is not clear.

XI. PRODUCTION OF EOSINOPHILS IN VITRO

A. Mouse Cultures

The addition of native or recombinant IL-5 to flasks containing murine bone marrow, by refeeding at weekly intervals with medium containing IL-5, induces the production of eosinophils, with no increase in any other cell type (1,23). The kinetics of this production shows a number of interesting features, the most obvious of which is the higher production by bone marrow from mice infected with *M. corti* compared with normal mice. Because these mice are undergoing a profound eosinophilia and the bone marrow contains large numbers of eosinophils at all stages of differentiation, this difference is interpreted as due to an increase in eosinophil precursors in the infected mice. In the absence of added IL-5, resident marrow eosinophils rapidly decrease in number, to become virtually undetectable at the end of the first week.

While IL-5 is added at the time of culture initiation, the number of eosinophils is maintained for several weeks, but then slowly declines to disappear by 8-10 weeks. Although no kinetic analysis has been carried out it seems likely that IL-5 is increasing the survival time of resident eosinophils, as well as causing the generation of eosinophils from earlier differentiation stages. The liquid assay system is carried out under similar conditions for a 5-day period, and is therefore measuring both increased survival and production of eosinophils.

If the first addition of IL-5 is delayed for a week, so that the resident eosinophils disappear, there is a production of eosinophils that reaches a peak 10-14 days after the factor was first added. Again the numbers of eosinophils decline to very low levels by 8-10 weeks. If the first addition is delayed for longer periods, the total number of eosinophils recovered progressively decreases, and the difference between replicate cultures increases, suggesting that production is derived from a relatively small and decreasing number of precursors. This interpretation is supported by the observation that hemopoiesis is taking place in a small number of islands in the flasks (typically less than five in $25cm^2$ flasks initiated with 10^7 bone marrow cells).

In contrast to the transient production of eosinophils, as shown previously (24), neutrophil production is maintained in the absence of added growth factors for many weeks. This suggests that the cessation of eosinophil production is not due to trivial causes such as exhaustion of the culture medium but is more likely due to failure of the cultures to generate more eosinophil progenitors. It is also interesting to note that there is a decreased production of neutrophils during the peak production of eosinophils, suggesting some kind of interaction between the production of these two lineages. This means that ~90% of the nonadherent cells are eosinophils (1).

Addition of IL-3 or GM–CSF under similar conditions to those described for IL-5 causes increased production of neutrophils, macrophages, and eosinophils. The number of eosinophils produced is considerably smaller with both these factors than the number produced with IL-5 (M. Strath, unpublished observations). Although both IL-3 and GM–CSF induce eosinophil differentiation, only IL-5 is specific for the eosinophil lineage.

In the absence of added growth factors, these types of long-term bone marrow culture maintain a population of cells capable of forming colonies of mixed cells in the spleens of irradiated mice, usually interpreted as hemopoietic stem cells or cells closely related to stem cells (24). Furthermore, there is a continuous production of progenitor cells capable of forming granulocyte-macrophage colonies in semisolid medium. The types of experiment described above suggest that IL-5 acts only on precommitted eosinophil progenitors, and that these precursors are not being maintained by the cultures, even in the presence of IL-5. Thus, there appears to be no production of eosinophil progenitor cells from stem cells in the cultures, so that normal marrow gives rise to very few eosinophils, and bone marrow from eosinophilic mice produce eosinophils until the progenitors are exhausted and then production stops.

To study possible mechanisms for producing eosinophil progenitors a number of experiments have been carried out. The experimental approach has been to preincubate normal bone marrow with different factors. Any changes in progenitor numbers are then estimated in a secondary incubation with IL-5. Interleukin-3 induces a massive proliferation in the bone marrow cultures so that by 7 days the cultures contain large numbers of neutrophils and macrophages. When IL-5 is added at this time there is an increased production of eosinophils compared with cultures preincubated with IL-5 or no added factor. However, the effect was variable, with some experiments showing little effect (M. Strath, unpublished observations).

Using either bone marrow or spleen cells from 5-fluorouracil-treated mice, in which only primitive stem cells survive, Yamaguchi et al. (47) confirmed that IL-5 was only active at a late stage of eosinophil differentiation, and showed that either IL-3 or G–CSF but not GM–CSF were able to induce IL-5-responsive eosinophil progenitors. However, in comparison with the total number of colonies induced by IL-3 or G–CSF alone, the number of eosinophil colonies was small.

In contrast to these results Warren and Moore (48) reported that preincubation of normal bone marrow with IL-3 and GM–CSF, but not G–CSF or M–CSF, induced a spectacular increase in IL-5 responsive eosinophil progenitors.

Again in contrast to Yamaguchi et al., these authors reported that IL-3 had no effect in a preincubation experiment with cells from mice treated with 5-fluorouracil. A combination of IL-1 and IL-3 was active in this system. In our experience IL-1 has no significant production of eosinophils by normal bone marrow in the presence of IL-5 (M. Strath, unpublished observations).

The biological significance of these preincubation experiments remains unclear for two reasons. First, the contrasting results obtained in different laboratories raises doubts about their general validity. Second, and more importantly, all of these factors (IL-3, GM–CSF, and G–CSF) are potent stimulants of the production of neutrophils. Because eosinophilia is frequently a specific cell response, with a massive increase in eosinophil numbers and little or no increase in neutrophils, it is not clear how any of these factors could play a major role in the control of eosinophilia, without also increasing the number of neutrophils and macrophages.

B. Cross-Species Activity of IL-5

The observation that mouse IL-5 was active in inducing human eosinophil differentiation (see below) was interesting because few of the lymphokines show activity between these species. By comparison of the dilution giving a 50% endpoint in eosinophil differentiation assays, mouse IL-5 is approximately 20-fold less active on human bone marrow than on homologous bone marrow. However, human IL-5 is approximately 100-fold less active on mouse than on human bone marrow (49). Both human and mouse IL-5 are active on rat, hamster, guinea pig, gerbil, cotton rat (*Sigmodem hispidus*), multimammate rat (*Mastomys natalensis*), bank vole (*Clethrionomys glareolus*), and ferret bone marrow. These experiments were carried out with highly active recombinant preparations, but no information is available on the relative activities in these species (M. Strath, unpublished observations). These cross-reactivities are an illustration of the conservation of IL-5 and its specific receptor, and indicate that experiments are probably possible with bone marrow from most mammals, with highly active preparations of either human or mouse IL-5.

C. Human Cultures

No native human IL-5 has yet been characterized. The first observation with human bone marrow that can be attributed to IL-5 was the observation that crude murine-spleen-conditioned medium stimulated the specific production of eosinophil colonies (50). The failure of either mouse GM–CSF or IL-3 to react with human cells makes it clear that IL-5 was responsible for that observation. Subsequent work with purified mEDF confirmed that it is an Eo-CSF for human cells (51). Furthermore, it increased the survival time of peripheral

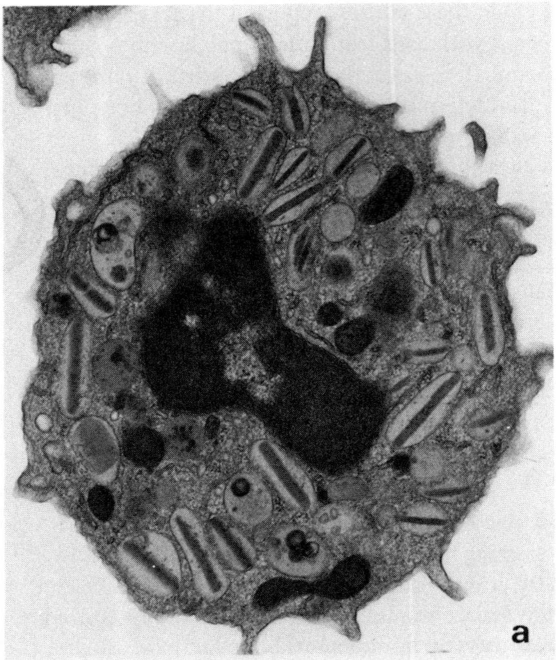

Figure 5 Electron micrographs of eosinophils produced in bone marrow cultures with IL-5. (a) Mouse eosinophil showing the typical granules containing a rod-shaped crystal of major basic protein. (Reproduced from Ref. 1, with permission.) (b) Human eosinophil showing typical lobed nucleus. The granules contain an amorphous electron-dense area, probably representing major basic protein, but the crystalline structure seen in peripheral blood eosinophils is not present. (Reproduced from Ref. 26, with permission.)

blood eosinophils in vitro (52). It was surprising that mouse IL-5 would stimulate eosinophil human colony formation when it is a poor CSF with mouse bone marrow. However, this has been a consistent finding in this laboratory, and it is clear that using identical reagents and culture conditions, either mouse or human IL-5 will give rise to between 10 and 50 eosinophil colonies/10^5 normal human marrow cells, but less than 5 from bone marrow taken from eosinophilic mice.

Human bone marrow cultured in the presence of mIL-5 shows a progressive selective increase in eosinophils (26). They are first detectable at approximately day 14 and reach a peak between days 21 and 35. There is a progressive decrease in other cell types, which parallels the effect in cultures without added

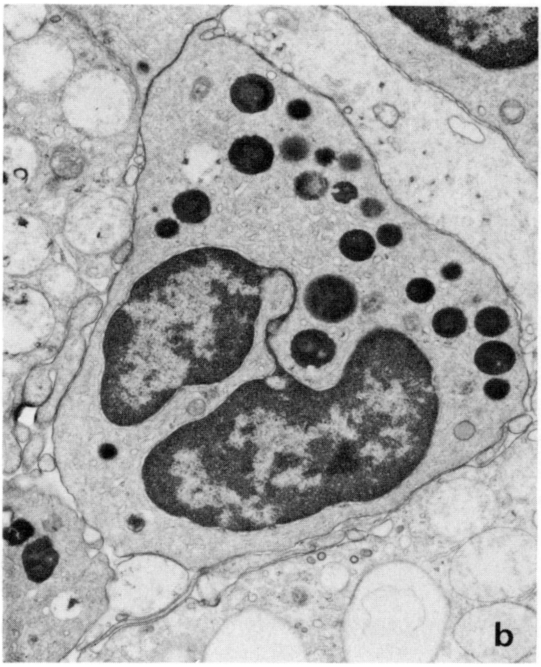

Figure 5 (continued)

factor. Day 14 eosinophils were typical immature myelocytes with characteristic eosinophilic granules, which under the electron microscope appeared homogeneously electron dense. After 3 weeks one-fifth of the eosinophils had the typical lobulated nucleus of the mature cell (Fig. 5b). After 7 weeks almost all the eosinophils were of mature nuclear morphology. The granules had electron-dense cores that were clearly demarcated from the less dense matrix, but the sharply cuboidal crystalline cores typical of peripheral blood eosinophils were not seen. Similar morphological observations were reported with eosinophils produced as colonies in semisolid medium (53,54).

As in the mouse it appears that IL-5 stimulates committed eosinophil progenitor cells to proliferate and differentiate into mature eosinophils, but has no activity in the production of these progenitors. To investigate the possibility that other cytokines are involved in this process, preincubation experiments have been carried out in which human bone marrow is first preincubated with various cytokines, and then the number of colony-forming cells is assayed in semisolid medium containing rhIL-5. Both rhGM–CSF and rhIL-3 increase the number of IL-5-responsive colony-forming cells, but IL-1 and IL-6 have no activity in this system (55). As discussed above, the relevance of this observation to the control of eosinophilia is not clear.

XII. FUNCTIONAL PROPERTIES OF EOSINOPHILS PRODUCED IN VITRO

Eosinophils taken at the peak of production from mouse bone marrow in vitro can be further purified by density gradient centrifugation. These cells have normal morphology (Fig. 5a), express normal levels of eosinophil peroxidase, and have a cytotoxic capacity similar to cells recovered in the peritoneal exudate of mice infected with *M. corti*. In a comparison of the ability of culture derived and peritoneal exudate eosinophils, it was demonstrated that cells from both sources could lyse chick red blood cells coated with monoclonal antibody of IgG1, IgG2a and IgG2b isotypes, but had no activity with IgM, IgA, or IgE (1, 56). Furthermore, the culture derived eosinophils were capable of killing antibody-coated schistosomula (1).

The cytotoxic activity of human eosinophils produced in vitro was assessed on antibody-coated mouse tumor cell. Both immature myelocytes and mature cells were cytotoxic for the first 6-7 weeks of culture. After this their cytotoxic activity declined, although they remained viable and morphologically normal (26).

XIII. ACTIVATION OF EOSINOPHILS BY IL-5

The concept of activation (57) refers to the ability of certain agents to increase the functional capacity of eosinophils when incorporated into short-term assay systems. The mechanism of this phenomenon remains uncertain, and it is not clear whether it is a reversible or irreversible phenomenon. In the latter case it could probably be regarded as part of the differentiation pathway, rather than a separate phenomenon.

Murine IL-5 was shown to increase the ability of human peripheral blood eosinophils to lyse antibody-coated tumor cells (51). In the absence of mIL-5 and at limiting dilutions of antibody, eosinophils had only low cytotoxic activity. However, when IL-5 was added to the 3-hr assay system there was a marked enhancement of cytotoxicity. This work has been extended with rhIL-5, which was found to induce changes in eosinophil shape ("polarization") possibly associated with increased cytokinesis and to increase the production of superoxide anion production by eosinophils, as well as increasing the antibody-dependent cytotoxic activity of eosinophils (58).

In an attempt to study the possible reversibility of activation, human bone marrow cultures that had been maintained in IL-5 were washed free of IL-5, and eosinophil cytotoxic activity was assessed during a 7-9 day period (26). Cultures containing mostly immature myelocytes were cytotoxic, and this activity was increased by including IL-5 in the assay. Surprisingly there was little change in either the direct cytotoxic activity or on the activation response of the eosinophils during this period, when it was expected at the very least that eosinophil numbers would decrease in the absence of IL-5. This makes it difficult to interpret these experiments. However, the fact that immature myelocytes could be activated by IL-5 suggests that this phenomenon might not be simply part of the differentiation pathway. The fact that washing the cultures had no effect on the eosinophils might suggest that IL-5, although removed from the culture medium, remained available in some insoluble form. For example, GM–CSF is known to bind to components of the extracellular matrix (59). If the same applies to IL-5, then eosinophils may have been maintained in an activated state by the bound IL-5.

A number of attempts have been made in this laboratory to demonstrate activation of mouse eosinophils, by incorporating mIL-5 into cytotoxicity assay systems (Strath and Sanderson, unpublished work). However, these experiments have not shown increased cytotoxicity, and it is possible that the eosinophils are fully activated either by IL-5 in the culture system or by IL-5 and possibly other activating agents in the parasitized mice used as a source of eosinophils (60). In confirmation of experiments with human eosinophils (58), IL-5 has recently been shown to increase superoxide anion production by eosinophils from mice infected with *Toxicara canis* (47).

XIV. OTHER BIOLOGICAL ACTIVITIES OF IL-5

A. Mouse B-Cell Growth Factor Activity

Before EDF had been identified in this laboratory, a number of factors active on mouse B cells had been identified and partially characterized. The common factor between these factors was a molecular weight of about 50,000. It was probably not surprising when purification of one of them, known as TRF, indicated that it was identical with another known as BCGFII (28). It was, however, surprising when EDF was found to be identical to BCGFII. This was established when these two activities were found to be coordinately expressed by a panel of T-cell clones (61), and when the two activities were demonstrated to copurify (62).

It is not necessary to review the literature on the activities of IL-5 on mouse B dells, because many reviews are available (63,64). A summary of these activities is appropriate as a basis for discussing the differences seen with IL-5 in humans. The murine B-cell tumor line, BCL1, responds to IL-5 with an increased rate of proliferation, and this gives a simple assay for the lymphokine

(65). On primary B cells IL-5 is a late-acting factor in differentiation (66), requiring a primary stimulus to make resting B cells responsive. This can be either polyclonal stimulants such as bacterial lipopolysaccharide and antiimmunoglobulin, or specific antigen. Whether it acts specifically to increase particular antibody isotypes or nonspecifically to increase secretion from all immunoglobulin-producing cells remains unclear. Large splenic B cells, presumed to have been activated in vivo, when cultured with IL-5 for 7 days show markedly enhanced numbers of IgM- and IgG-producing cells (66). When added to similar cultures in the presence of lipopolysaccharide, which itself is a potent polyclonal stimulator of murine B cells, a significant increase in IgM, IgG1, and IgA immunoglobulins was observed (67,68). However, the effect on the latter two was small in comparison to the total levels of IgM produced. In a study of B cells from gut associated lymphoid tissue (Peyer's patches), IL-5 increased production of IgA from B cells selected for IgA surface expression, but did not induce IgA secretion from cell populations depleted of IgA surface-positive cells (69). This suggests that IL-5 does not induce switching of IgA production, but acts nonspecifically to enhance the production from IgA committed precursors.

B. Apparent Lack of B-Cell Activity in Humans

Recombinant hIL-5 has been tested in a wide range of assay systems with human B cells, and has been found not to have activities analogous to the activity of IL-5 in the mouse B-cell systems (49). It is interesting to note that rhIL-5 is active on mouse B cells, but at very low specific activity. The human assays included: (a) proliferation assays with tonsillary or splenic B cells in the presence of the costimulators anti-μ or PMA. This is the assay system most commonly used to identify human BCGF, and is analogous to assays used to identify BCGFII in mice, (b) a restimulation assay in which tonsillar or splenic B cells are first activated with a primary stimulant and then tested with rhIL-5. This is a sensitive assay for hIL-4, (c) assay for immunoglobulin production in restimulation assays or assay with a human B lymphoblastoid cell line; (d) assay for increased proliferation with primary explanted chronic lymphocytic leukemia cells (B–CLL) or with two human B-cell lines that are known to detect human B-cell differentiation factors, (e) assay for the ability of rhIL-5 to replace T cells in specific antibody responses. This is precisely analogous to the TRF assay used to identify murine IL-5.

These results indicate that IL-5 in humans has no activities analogous to the BCGF or TRF activities of IL-5 in the mouse. Furthermore, although hIL-5 may be shown to be active in some other assay system with human B cells, it is clearly not responsible for any of the human BCGF activities that have been described but not yet fully characterized, because these have used assays included in the above panel. These negative results appear to be in conflict with two reports claiming that rhIL-5 is active on human B cells (67,70), but in each of these cases a significant proportion of accessor cells including T cells were present in the cultures. It is now accepted that hIL5 has no effect on purified B cells.

C. Activity on Mouse T Cells

A single report indicates that rmIL-5, in the form of crude oocyte-conditioned medium, increases the cytotoxic activity of peanut-agglutinin-positive splenic cells (71). The cells were incubated in the presence of both IL-2 and IL-5, with mitomycin-C-treated syngeneic spleen cells that had been coupled with trinitrophenol. As yet there is no reported activity of hIL-5 on human T cells, but the possibility that this activity exists makes it imperative that human B-cell assays be done in the complete absence of T cells.

XV. SUMMARY

The development of eosinophilia has an interesting biological specificity, in that the numbers of eosinophils can increase from a normal paucity to very high levels, without a concomitant increase in other leukocytes. This suggests a unique controlling mechanism. Although IL-3, GM–CSF, and IL-5 all induce eosinophil differentiation, IL-5 is the only identified cytokine specific for the eosinophil lineage. This together with the fact that IL-5 can be detected in the serum of mice with eosinophilia (60) suggests that it may be a critical controlling factor.

The mechanism of this control is not yet clear. It appears that a high proportion of T-cell clones produce IL-5, even when isolated using antigens that do not induce eosinophilia. If IL-5 is produced in response to all antigens, what is the basis for the biological specificity of eosinophilia? It is possible that the production of lymphokines by isolated T-cell clones does not reflect the true biological situation, because there may be a selection process in the isolation of T cells that is not yet understood. In addition, IL-5 appears not to stimulate the production of precursors committed to the eosinophil lineage, at least in vitro. Thus, it would seem likely that other controlling mechanisms exist to generate the precursors. Attempts to study this stage of eosinophil production are difficult to interpret. The experiments are technically difficult, and conflicting results have been reported. Interleukin-3 and GM–CSF appear to have this activity in vitro, but it is not clear how these factors could be active in the prodicution of eosinophil precursors without also inducing increases in other leukocyte types, particularly neutrophils.

Another intriguing aspect of the biology of IL-5 is the activity on other cell types, particularly B cells in the mouse. This activity as a murine BCGF precedes the identification of IL-5 as an eosinophil differentiation factor, and has been studied in considerable detail for many years. However, if IL-5 is important in the production of antibody, why are eosinophils not seen in all immune responses? An important insight into this problem has come from the observation that human IL-5 has no analogous activity on human B cells. If IL-5 is not active on human B cells, it raises questions about the biological role of IL-5 as a BCGF in the mouse. In view of the large amount of information on this activity in vitro, it must be admitted that at the present time there are no clear answers to these questions.

In addition to these biological aspects of the control of eosinophilia and its relationship to the control of the production of other blood cells, the molecular biology of these factors is beginning to provide information that raises some fascinating questions about the evolution of the hemopoietic system. In particular, the similarities in gene structure and chromosomal location of IL-3, IL-5, and GM–CSF suggest the possibility that gene duplication and divergent evolution has provided the basis for the different hemopoietic lymphokines. Despite this possible common origin it is interesting to note the differences in rates of change of the three factors IL-3, GM–CSF, and IL-5 between mouse and human. Interleukin-3 has undergone quite remarkable mutation, IL-5 is highly conserved, and GM–CSF is intermediate. Although this might suggest that eosinophils have been particularly important in mammalian evolution, it is difficult to reconcile with the obvious fact that macrophages are more primitive, and that IL-3, because of its broad spectrum of activity on different cell lineages, is a candidate for the precursor of these other genes.

ACKNOWLEDGMENT

I thank my colleagues Malcolm Strath, Elaine Clutterbuck, Simon Barry, Hugh Campbell, and Ian Young for their contributions to this work, particularly work that is not yet published.

REFERENCES

1. Sanderson, C. J., D. J. Warren, and M. Strath. Identification of a lymphokine that stimulates eosinophil differentiation in vitro. Its relationship to IL3, and functional properties of eosinophils produced in cultures. *J. Exp. Med. 162*: 60-74 (1985).
2. Butterworth, A. E. Cell-mediated damage to helminths. *Adv. Parasitol. 23*: 144-235 (1984).
3. Sanderson, C. J., A F. Lopez, and M. Bunn Moreno. Eosinophils and not lymphoid K cells kill *Trypanosoma cruzi* epimastigotes. *Nature 268*:340-341 (1977).

4. Lopez, A. F., M. M. Bunn Moreno, and C. J. Sanderson. The lysis of *Trypanosoma cruzi* epimastigotes by eosinophils and neutrophils. *Int. J. Parasitol.* *8*:485-489 (1978).
5. Sanderson, C. J., and W. De Souza. An electron microscopic study of the killing of *Trypanosoma cruzi* epimastigotes and trypomastigotes by eosinophils and neutrophils. *J. Cell Sci.* *37*:275-286 (1979).
6. Sanderson, C. J., and J. A. Thomas. A comparison between the cytotoxic activity of eosinophils and other cells by ^{51}Cr release and time lapse microcinematography. *Immunology 34*:771-780 (1978).
7. Lopez, A. F., M. Strath, and C. J. Sanderson. IgG and complement receptors on purified mouse eosinophils and neutrophils. *Immunology 43*:779-786 (1981).
8. Lopez, A. F., and C. J. Sanderson. Antibody-dependent, cell-mediated cytotoxicity of nucleated mammalian cells by rat eosinophils and neutrophils. *Int. Arch Allergy Appl. Immunol.* *67*:200-205 (1981).
9. Sturrock, R. F., R. Kimani, B. J. Cottrell, A. E. Butterworth, H. M. Seitz, T. K. Siongok, and V. Houba. Observations on possible immunity to reinfection amoung Kenyan schoolchildren after treatment for *Schistosoma mansoni.* *Trans. R. Soc. Trop. Med. Hyg.* *77*:363-371 (1983).
10. Hagan, P., H. A. Wilkins, U. B. Blumenthal, R. J. Hayes, and B. M. Greenwood. Eosinophilia and resitance to *Schistosoma haemotobium* in man. *Parasite Immunology* *7*:625-632 (1985).
11. Kolb, E., and E. Muller. Local responses in primary and secondary human lung cancers. II. Clinical correlations. *Br. J. Cancer 40*:410-416 (1979).
12. Pretlow, T. P., E. F. Kieth, A. K. Cryar, A. A. Bartolucci, A. M. Pitts, T. G. Pretlow, P M. Kimball, and E. A. Boohaker. Eosinophil infiltration of human colonic carcinomas as a prognostic indicator. *Cancer Res. 43*:2997-3000 (1983).
13. Iwasaki, K., M. Torisu, and T. Fujimura. Malignant tumor and eosinophils. I. Prognostic significance in gastric cancer. *Cancer 58*:1321-1327 (1986).
14. Vaughan Hudson, B., D. C. Linch, E A. MacIntyre, M. H. Bennett, K. A. MacLennan, G. Vaughan Hudson, and A. M. Jelliffe. Selective peripheral blood eosinophilia associated with survival advantage in Hodgkin's disease (BNLI Report No. 31). *J. Clin. Pathol. 40*:247-250 (1987).
15. Samozuk, M. K., B. N. Nathwani, and R. J. Lukes. Extensive deposition of eosinophil peroxidase in Hodgkin's and Non-Hodgkin's lymphomas. *Am. J. Pathol. 125*:426-429 (1986).
16. Kormendi, F., and W. J. C. Amend. The importance of eosinophil cells in kidney allograft rejection. *Transplantation 45*:537-539 (1988).
17. Gleich, G. J., and C. R. Adolphson. The eosinophilic leukocyte: Structure and function. *Adv. Immunol. 39*:177-253 (1986).
18. Tai P-C., M. E. Holt, P. Denny, A. Gibbs, B. D. Williams, and C. J F. Spry. Deposition of eosinophil cationic protein in granulomas in allergic granulomatosis and vasculitis: the Churg-Strauss syndrome. *Br. Med. J. 289*:400-402 (1984).

19. Tai, P-C., S. J. Ackerman, C. J. F. Spry, S. Dunnette, E. G. F. Olsen, and G. J. Gleich. Deposits of eosinophil granule proteins in cardiac tissues of patients with eosinophilic endoymocaridal disease. *Lancet 1*:643-647 (1987).
20. Basten, A., and P. B. Beeson. Mechanism of eosiniphilia. II. Role of the lymphocyte. *J. Exp. Med. 131*:1288-1307 (1970).
21. Metcalf, D., J. W. Parker, H. M. Chester, and P. W. Kincade. Formation of eosinophilic-like granulocytic colonies by mouse bone marrow cells in vitro. *J. Cell. Physiol. 84*:275-290 (1974).
22. Warren, D. J., and C. J. Sanderson. Production of a T cell hybrid producing a lymphokine stimulating eosinophil differentiation. *Immunology 54*:615-623 (1985).
23. Strath, M., and C. J. Sanderson. The production and functional properties of eosinophils from bone marrow cultures. *J. Cell Sci. 74*:207-217 (1985).
24. Dexter, T. M., T. D. Allen, and L. G. Lajtha. Conditions controlling the proliferation of haemopoietic stem cells in vitro. *J. Cell. Physiol. 91*:335-344 (1977).
25. Strath, M., D. J. Warren, and C. J. Sanderson. Detection of eosinophils using a eosinophil peroxidase assay. Its use as an assay for eosinophil differentiation factors. *J. Immunol. Methods 83*:209-215 (1985).
26. Clutterbuck, E. J., and C. J. Sanderson. Human eosinophil production in vitro studied by means of murine eosinophil differentiation factor (IL5): production of functionally active eosiniphils from normal human bone marrow. *Blood 71*:646-651 (1988).
27. Strath, M., E. J. Clutterbuck, and C. J. Sanderson. Production of human and murine eosinophils in vitro and assay for eosinophil differentiation factors. In *Methods in Molecular Biology, Vol. 5. Tissue Culture*. J. Walker and J. W. Pollard (Eds.). Humana Press, Clifton, New Jersey (in press).
28. Harada, N., Y. Kikuchi, A. Tominaga, S. Takaki, and K. Takatsu. BCGFII activity on activated B cells of purified murine T cell replacing factor (TRF) from a T cell hybridoma (B151K12). *J. Immunol. 134*:3944-3951 (1985).
29. Campbell, H. D., C. J. Sanderson, Y. Wang, Y. Hort, M. E. Martinson, W. Q. J. Tucker, A. Stellwagen, M. Strath, and I. G. Young. Isolation, structure and expression of cDNA and genomic clones for murine eosinophil differentiation factor. Comparison with other eosinophilopoietic lymphokines and identity with interleukin-5. *Eur. J. Biochem. 174*:345-352 (1988).
30. Kinashi, T., N. Harada, E. Severinson, T. Tanabe, P. Sideras, M. Konishi, C. Azuma, A. Tominaga, S. Bergstedt-Lindqvist, M. Takahashi, F. Matsuda, Y. Yaoita, K. TAkatsu, and T. Honjo. Cloning of a complementary DNA encoding T-cell replacing factor and identity with B-cell growth factor II. *Nature 324*:70-73 (1986).
31. Campbell, H. D., W. Q. J. Tucker, Y. Hort, M. E. Martinson, G. Mayo, E. J. Clutterbuck, C. J. Sanderson, and I. G. Young. Molecular cloning, nucleotide sequence,and expression of the gene encoding human eosino-

phil differentiation factor (interleukin 5). *Proc. Natl. Acad. Sci. USA 84*: 6629-6633 (1987).

32. Sutherland, G. R., E. Baker, D. F. Callen, H. D. Campbell, I. G. Young, C. J. Sanderson, O. M. Garson, A. F. Lopez, and M. A. Vadas. Interleukin-5 is at 5q31 and is deleted in the 5q- syndrome. Blood 71:1150-1152 (1988).
33. Le Beau, M. M., N. D. Epstein, S. J. O'Brien, A. W. Nienhuis, Y-C. Yang, S. C. Clark, and J. D. Rowley. The interleukin-3 gene is located on human chromosome 5 and is deleted in myeloid leukemias with a deletion of 5q. *Proc. Natl. Acad. Sci. USA 84*:5913-5917 (1987).
34. Huebner, K., M. Isobe, C. M. Croce, D. W. Golde, S. E. Kaufman, and J. C. Gasson. The human gene encoding GM–CSF is at 5q21-q32, the chromosome region deleted in the 5q- anomaly. *Science 230*:1282-1285 (1985).
35. Yang, Y-C., S. Kovacic, R. Kriz, S. Wolf, S. C. Clark, T. E. Wellems, A. Nienhuis, and N. Epstein. The human genes for GM–CSF and IL3 are closely linked in tandem on chromosome 5. *Blood 71*:958-961 (1988).
36. Brolow, D. P., M. Bucan, H. Lehrach, B. L. M. Hogan, and N. M. Gough. Close genetic and physical linkage between the murine haemopoietic growth factor genes GM–CSF and multi-CSF (IL3). *EMBO J. 6*:617-623 (1987).
37. Nagata, S., M. Tsuchiya, S. Asano, O. Yamamoto, Y. Hirata, N. Kubota, M. Oheda, H. Nomura, and T. Yamazaki. The chromosomal gene structure and two mRNA's for human granulocyte colony-stimulating factor. *EMBO J. 5*:575-581 (1986).
38. Otsuka, T., D. Villaret, T. Yokota, Y. Takebe, F. Lee, N. Aria, and K. Aria. Structural analysis of the mouse chromosomal gene encoding interleukin 4 which expresses B cell, T cell and mast cell stimulating activity. *Nucleic Acids. Res. 15*:333-344 (1987).
39. Campbell, H. D., S. Ymer, M-C. Fung, and I. G. Young. Cloning and nucleic acid sequence of the murine interleukin-3 gene. *Eur. J. Biochem. 150*:297-304 (1985).
40. Yang, Y.-C., A. B. Ciarletta, P. A. Temple, M. P. Chung, S. Kovacic, J. S. Witek-Giannotti, A. C. Leary, R. Kriz, R. E. Donahue, G. G. Wong, and S. C. Clark. Human IL-3 (multi-CSF): identification by expression cloning of a novel hematopoietic growth factor related to murine IL–3. *Cell 47*: 3-10 (1986).
41. Miyatake, S., T. Otsuka, T. Yokota, F. Lee, and K. Aria. Structure of the chromosomal gene for granulocyte-macrophage colony stimulating factor: comparison of the mouse and human genes. *EMBO J. 4*:2561-2568 (1985).
42. von Heijne, G. Patterns of amino acids near signal-sequence cleavage sites. *Eur. J. Biochem. 133*:17-21 (1983).
43. Bond, M. W., B. Shrader, T. R. Mosmann, and R. L. Coffman. A mouse T cell product that preferentially enhances IgA production. II. Physicochemical characterisation. *J. Immunol. 139*:3691-3696 (1987).
44. McKenzie, D. T., H. I. Filutowitcz, S. L. Swain, and R. W. Dutton. Purification and partial sequence analysis of murine B cell growth factor II (interleukin 5). *J. Immunol. 139*:2661-2668 (1987).

45. Dayhoff, M. O., R. V. Eck, and C. M. Park. A model of evolutionary change in proteins. In *An Atlas of Protein Sequence and Structure*, Vol. 5, M. O. Dayhoff (Ed.). Biomedical Research Foundation, Silver Springs, Maryland, pp. 89-99 (1972).
46. Sanderson, C. J., H. D. Campbell, and I. G. Young. Molecular and cellular biology of eosinophil differentiation factor (interleukin-5) and its effects on human and mouse B cells. *Immunol. Rev. 102*:29-50 (1988).
47. Yamaguchi, Y., Y. Hayashi, Y. Sugama, Y. Miura, T. Kasahara, S. Kitamura, M. Torisu, S. Mita, A. Tominaga, K. Takatsu, and T. Suda. Highly purified murine interleukin 5 (IL-5) stimulates eosinophil function and prolongs in vitro survival. IL-5 as an eosiniphil chemotactic factor. *J. Exp. Med. 167*: 1737-1742 (1988).
48. Warren, D. J., and M. A. S. Moore. Synergism among interleukin 1, interleukin 3, and interleukin 5 in the production of eosinophils from primitive hemopoietic stem cells. *J. Immunol. 140*:94-99 (1988).
49. Clutterbuck, E. J., J. Shields, J. Gordon, H. D. Campbell, I. G. Young, and C. J. Sanderson. Recombinant human interleukin-5 is an eosinophil differentiation factor but has no activity in standard human B cell growth factor assays. *Eur. J. Immunol. 17*:1743-1750 (1987).
50. Metcalf, D., R. L. Cutler, and N. A. Nicola. Selective stimulation by mouse spleen conditioned medium of human eosinophil colony formation. *Blood 61*:999-1005 (1983).
51. Lopez, A. F., C. G. Begley, D. J. Williamson, D. J. Warren, M. A. Vadas, and C. J. Sanderson. Murine eosinophil differentiation factor: an eosinophil-specific colony-stimulating factor with activity for human cells. *J. Exp. Med. 163*:1085-1099 (1986).
52. Begley, C. G., A. F. Lopez, N. A. Nicola, D. J. Warren, and C. J. Sanderson. Purified colony stimulating factors enhance the survival of human neutrophils and eosinophils in vitro: a rapid and sensitive microassay for colony stimulating factors. *Blood 68*:162-166 (1986).
53. Zucker-Franklin, D., G. Grusky, and P. L'Esperance. Granulocyte colonies derived from lymphocyte fractions of normal human peripheral blood. *Proc. Natl. Acad. Sci. USA 71*:2711-2714 (1974).
54. Butterfield, J. H., A. B. Eisenbrey, and G. J. Gleich. Membrane marker characterisation of th eosinophil colony-forming cell. *Br. J. Haematol. 51*: 209-216 (1982).
55. Clutterbuck, E. J., and C. J. Sanderson. Regulation of the production of eosinophil colony forming cells in human bone marrow cultures by cytokines. (Submitted for publication).
56. Lopez, A. F., M. Strath, and C. J. Sanderson. Mouse immunoglobulin isotypes mediating cytotoxicity of target cells by eosinophils and neutrophils. *Immunology 48*:503-509 (1983).
57. Silberstein, D. S., and J. R. David. The regulation of human eosinophil function by cytokines. *Immunology Today 8*:380-385 (1987).
58. Lopez, A. F., C. J. Sanderson, J. R. Gamble, H. D. Campbell, I. G. Young,

and M. A. Vadas. Recombinant human interleukin-5 is a selective activator of human eosinophil function. *J. Exp. Med. 167*:219-224 (1988).

59. Gordon, M. Y., G. P. Riley, S. M. Watt, and M. F. Greaves. Compartmentalization of a haemopoietic growth factor (GM–CSF) by glycoaminoglycans in the bone marrow microenvironment. *Nature 326*:403-405 (1987).
60. Strath, M., and C. J. Sanderson. Detection of eosinophil differentiation factor and its relationship to eosinophilia in *Mesocestoides corti*-infected mice. *Exp. Hematol. 14*:16-20 (1986).
61. Sanderson, C. J., M. Strath, D. J. Warren, A. O'Garra, and T. B. L. Kirkwood. The production of lymphokines by primary alloreactive T cell clones. A co-ordinate analysis of 233 clones in seven lymphokine assays. *Immunology 56*:575-584 (1985).
62. Sanderson, C. J., A. O'Garra, D. J. Warren, and G. G. B. Klaus. Eosinophil differentiation factor also has B cell growth factor activity. Proposed name interleukin 4. *Proc. Natl. Acad. Sci. USA 83*:437-440 (1986).
63. Takatsu, K., A. Tominaga, N. Harada, S. Mita, M. Matsumota, T. Takahashi, Y. Kikuchi, and N. Yamaguchi. T cell replacing factor (TRF)/interleukin 5 (IL5): molecular and functional properties. *Immunol. Rev. 102*:107-135 (1988).
64. Swain, S. L., D. T. McKenzie, R. W. Dutton, S. L. Tonkonogy, and M. English. The role of IL4 and IL5: characterisation of a distinct helper T cell subset that makes IL4 and IL5 (Th2) and requires priming before induction of lymphokine secretion. *Immunol. Rev. 102*:77-105 (1988).
65. Swain, S., and R. W. Dutton. Production of a B cell growth promoting activity, (DL)BCGF, from a cloned T cell line and its assay on the BCL_1 B cell tumor. *J. Exp. Med. 156*:1821-1834 (1982).
66. O'Garra, A., D. J. Warren, M. Holman, A. M. Popham, C. J. Sanderson, and G. G. B. Klaus. Interleukin 4 (B cell growth factor II/eosinophil differentiation factor) is a mitogen and differentiation factor for preactivated murine B lymphocytes. *Proc. Natl. Acad. Sci. USA 83*:5228-5232 (1986).
67. Yokota, T., R. L. Coffman, H. Hagiwara, D. M. Rennick, Y. Takebe, K. Yokota, L. Gemmell, B. Shrader, G. Yang, P. Meyerson, J. Luh, P. Hoy, J. Pene, F. Briere, H. Spits, J. Banchereau, J. de Vries, F. D. Lee, N. Aria, and K. Arai. Isolation and characterisation of lymphokine cDNA clones encoding mouse andhuman IgA enhancing factor and eosinophil colony stimulating factor activities: relationship to interleukin 5. *Proc. Natl. Acad. Sci. USA 84*:7388-7392 (1987).
68. Coffman, R. L., B. Shrader, J. Carty, T. R. Mosmann, and M. W. Bond. A mouse T cell product that preferentially enhances IgA production. I. Biologic characterisation. *J. Immunol. 139*:3685-3690 (1987).
69. Murray, P. D., D. T. McKenzie, S. L. Swain, and M. F. Kagnoff. Interleukin 5 and interleukin 4 produced by Peyer's patch T cells selectively enhance immunoglobulin A expression. *J. Immunol. 139*:2669-2674 (1987).

70. Azuma, C., T. Tanabe, M. Konishi, T. Noma, F. Matsuda, Y. Yaoita, K. Takatsu, L. Hammerstrom, C. I. E. Smith, E. Severinson, and T. Honjo. Cloning of cDNA for human T cell replacing factor (interleukin-5) and comparison with the murine homologue. *Nucleic Acids. Res. 14*:9149-9158 (1986).
71. Takatsu, K., Y. Kikuchi, T. Takahashi, T. Honjo, M. Matsumoto, N. Harada, N. Yamaguchi, and A. Tominaga. Interleukin 5, a T cell derived B cell differentiation factor also induces T lymphocytes. *Proc. Natl. Acad. Sci. USA 84*:4234-4238 (1987).
72. Garnier, J., D. J. Osguthorpe, and B. Robson. Analysis of the accuracy and implications of simple methods for predicting the secondary structure of globular proteins. *J. Mol. Biol. 120*:97-120 (1978).

9
Erythropoietin

EUGENE GOLDWASSER, NEGA BERU, and DAVID SMITH
The University of Chicago, Chicago, Illinois

I. INTRODUCTION

Although the existence of a circulating substance that regulates the rate of red blood cell formation had been postulated since 1906 (1), verification of this concept was delayed for more than 30 years (2,3). This factor, termed erythropoietin (epo), has been the subject of several reviews and symposium volumes (4-13) during the past 2-3 decades and its role in erythropoiesis is firmly established. This review will, therefore, not be an exhaustive one but will be selective and concerned for the most part with advances made in the more recent past as research and accumulation of information have accelerated.

II. CHEMISTRY OF EPO

The advance in our understanding of the structure of human epo followed the first purification of it from urine of patients with aplastic anemia. This preparation, shown to be homogeneous by sodium dodecyl sulfate gel electrophoresis, consisted of two forms with essentially the same potency (70-120,000 U/mg protein) (14). It was later found that the difference between these two reflected somewhat different carbohydrate compositions (15). It is possible that this difference was brought about by either the excretion in the urine, i.e., some deglycosylation in the kidney or bladder, or loss of carbohydrate during the multistep purification process. In any case it would appear that the form with the lower carbohydrate content may possibly represent a trivial artefact.

The role of carbohydrate in the biological activity of epo is still not completely settled but it appears that it may be required for in vivo stability rather than for ability to bind to appropriate receptors on hemopoietic precursor cells. The small amount of data available suggest that deglycosylated epo whether of urinary (u-epo) or recombinant (r-epo) origin had signficant biological activity when assayed in cell culture but none when assayed in vivo (15,16). The enzymically deglycosylated u-epo has a great tendency to aggregate to a multimeric form without biological activity, but the small amount of monomeric u-epo remaining after removal of the oligosaccharide chains has enhanced activity on bone marrow cells in culture (15).

Purified human u-epo was used to show that there are two internal disulfide bonds and that they are required for appropriate folding of denatured u-epo to be renatured to the active conformation (17). It was also used to determine the primary structure of human u-epo by direct amino acid sequencing (18). The sequence so determined was the same as that prediced from the nucleotide sequence of the coding regions of the human epo gene (19,20). Figure 1 shows the amino acid sequences of epo from three species. These are the only ones available at the time of this writing. The sequence reveals three potential sites for *N*-glycosylation (asparagines 24, 38, and 83). Results of selective *N*-glycanase hydrolysis (16) show that all three are glycosylated. Serine 126 is a potential site of *O*-glycosylation. Similar enzymic hydrolysis with *O*-glycanase (after sialidase treatment) verified the presence of an *O*-linked oligosaccharide, and the fact that the automated amino acid sequencing showed no residue at position 126 (18), presumably because the amino acid had the sugar attached, permitted the assignment of the *O*-linked carbohydrate to the serine at position 126.

In a very recent article (21) Recny et al. found that both r-epo and u-epo lack the *C*-terminal arginine predicted from the DNA sequence and found earlier by direct amino acid sequencing in a different preparation of u-epo (18). They suggest that desArg166 epo may be the naturally circulating form of the hormone, although it has not been determined what the structure of circulating plasma epo is.

Comparison of the amino acid sequences of human, monkey, and mouse epos shows a high degree of identity (Fig. 1) and conservation of predicted secondary structure (22). At the amino acid level, the human and monkey epos are 92% identical. The corresponding figures for the human/mouse and monkey/mouse comparisons are 80 and 82%, respectively. Some differences exist, however: in the Human and monkey epo molecules, the serine at position 126 is a site of *O*-glycosylation; in the mouse, this serine is replaced by a proline Also, in the monkey, the amino acid corresponding to the lysine at position 116 in the human epo is missing. An evolutionary analysis of the nucleotide sequences of the three cloned epos was presented in Ref. 22.

Recombinant human epo expressed in Chinese hamster ovary cells and purified from the culture medium was compared with u-epo by circular dichro-

```
-27
MET GLY VAL HIS GLU CYS PRO ALA TRP LEU TRP LEU LEU LEU SER LEU LEU
--- --- --- --- --- --- --- --- --- --- --- --- --- --- --- --- VAL
--- --- --- PRO --- ARG --- THR     --- LEU --- --- --- --- --- ---
                                        +1
SER LEU PRO LEU GLY LEU PRO VAL LEU GLY ALA PRO PRO ARG LEU ILE CYS
--- --- --- --- --- --- --- --- PRO --- --- --- --- --- --- --- ---
LEU ILE --- --- --- --- --- --- --- CYS --- --- --- --- --- --- ---
        10                                      20                *
ASP SER ARG VAL LEU GLU ARG TRY LEU LEU GLU ALA LYS GLU ALA GLU ASN
--- --- --- --- --- --- --- --- --- --- --- --- --- --- --- --- ---
--- --- --- --- --- --- --- --- ILE --- --- --- --- --- --- --- ---
                    30                                *       40
ILE THR THR GLY CYS ALA GLU HIS CYS SER LEU ASN GLU ASN ILE THR VAL
VAL --- MET --- --- SER --- SER --- --- --- ASN --- --- --- --- ---
VAL --- MET --- --- --- --- GLY PRO ARG --- SER --- --- --- --- ---
                                50
PRO ASP THR LYS VAL ASN PHE TYR ALA TRP LYS ARG MET GLU VAL GLY GLN
--- --- --- --- --- --- --- --- --- --- --- --- --- --- --- --- ---
--- --- --- --- --- --- --- --- --- --- --- --- --- --- --- GLU GLU
    60                                      70
GLN ALA VAL GLU VAL TRP GLN GLY LEU ALA LEU LEU SER GLU ALA VAL LEU
--- --- --- --- --- --- --- --- --- --- --- --- --- --- --- --- ---
--- --- ILE --- --- --- --- --- --- SER --- --- --- --- --- ILE ---
                80              *                           90
ARG GLY GLN ALA LEU LEU VAL ASN SER SER GLN PRO TRP GLU PRO LEU GLN
--- --- --- --- VAL --- ALA --- --- --- --- --- PHE --- --- --- ---
GLN ALA --- --- --- --- ALA --- --- --- --- --- PRO --- THR --- ---
                            100
LEU HIS VAL ASP LYS ALA VAL SER GLY LEU ARG SER LEU THR THR LEU LEU
--- --- MET --- --- --- ILE --- --- --- --- --- ILE --- --- --- ---
--- --- ILE--- ---- --- ILE --- --- --- --- --- --- --- SER --- ---
110                                     120
ARG ALA LEU GLY ALA GLN LYS GLU ALA ILE SER PRO PRO ASP ALA ALA SER
--- --- --- --- --- ---     --- --- --- --- LEU --- --- --- --- ---
--- VAL --- --- --- --- --- --- LEU MET --- --- --- --- THR THR PRO
            130                                     140
ALA ALA PRO LEU ARG THR ILE THR ALA ASP THR PHE ARG LYS LEU PHE ARG
--- --- --- --- --- --- --- --- --- --- --- --- CYS --- --- --- ---
PRO --- --- --- --- --- LEU --- VAL --- --- --- CYS --- --- --- ---
                        150                                     160
VAL TYR SER ASN PHE LEU ARG GLY LYS LEU LYS LEU TYR THR GLY GLU ALA
--- --- --- --- --- --- --- --- --- --- --- --- --- --- --- --- ---
--- --- ALA --- --- --- --- --- --- --- --- --- --- --- --- --- VAL
                    166
CYS ARG THR GLY ASP ARG
--- --- ARG --- --- ---
--- --- ARG --- --- ---
```

Figure 1 Comparison of the amino acid sequences of human (top), monkey (middle) and mouse (bottom) erythropoietins. Sequences of the leader peptide as well as the mature proteins are shown. Numbering is from the amino terminus of the human protein. Differences from the human sequence are indicated. Asterisks indicate the sites of potential *N*-linked glycosylation.

ism, ultraviolet absorbance, and fluorescence spectroscopy with no differences found between the two preparations (23). In the same article, the authors used the sedimentation equilibrium method to find a molecular weight of 30,400 g mol^{-1}. With the peptide molecular weight (from the amino acid sequence including the *C*-terminal arginine) of 18,398 the carbohydrate content was determined to be 39%, which is significantly higher than the previous estimate of 31%, which was based on a holomolecular weight of 34,000 (15). A corrected value for carbohydrate content based on M_r = 30,400 would be 35% for u-epo, which is in fair agreement with the 39% calculated for r-epo (23).

The structures of the *O*-linked and *N*-linked oligosaccharides of u-epo were determined, after chromatographic isolation of the individual chains, by fast atom bombardment–mass spectrometry and methylation analysis (24) and by exoglycosidase digestion, Smith degradation, and methylation analysis (25). The major component of the oligosaccharide linked to Ser-126 is *N*-acetyl neuraminic acid 2 → 3 galactose β → 3 (*N*-acetyl neuraminic acid 2 → 6) *N*-acetyl galactosamine–serine. The *N*-linked oligosaccharides consist of the following (expressed as percent of total carbohydrate): biantennary 1.4%, triantennary 10%, triantennary with one *N*-acetyl lactosaminyl repeat 3.5%, tetraantennary 31.8%, tetraantennary with one such repeat 32.1%, with two repeats 16.5% or with three repeats 4.7%. The tetraantennary chains for the most part are di- or trisialated with α2 → 3-linked sialic acid. An example of the structure of the predominant species (31.8%) of tetraantennary *N*-linked oligosaccharide with one sialic acid per penultimate galactose residue is seen in Fig. 1. The same authors found that there is only a slight difference in sialic acid content between u-epo and r-epo, with no other difference in carbohydrate.

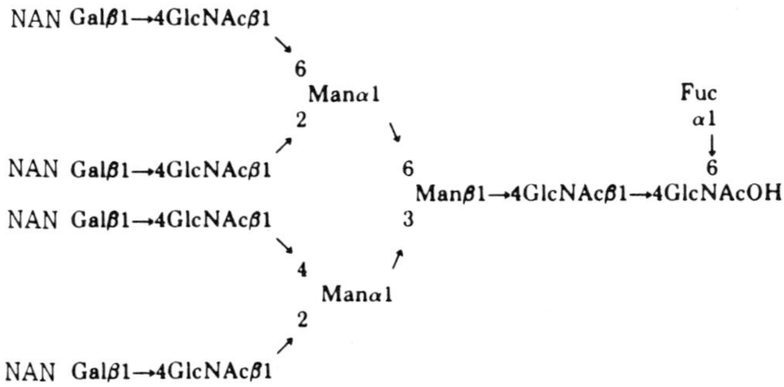

Figure 2 An example of the structure of the predominant species of tetraantennary *N*-linked oligosaccharide with one sialic acid per penultimate galactose residue.

No definitive study of the active site (or sites) of epo has been published although older evidence had indicated that tryptophan (26), tryosine (27), and α or ϵ amino groups (27) may be required for biological activity. In addition, the disulfide bonds, or the conformation imposed on epo by the disulfide bridging, are important for activity (17). Studies using antibodies made to synthetic peptides indicate that an antibody made to the *N*-terminal 26 residues does not impair biological function (28). Of five other antibodies to small peptides, only those directed against resides 99-118 and 111-129 had an effect on epo action. These regions of the epo protein do not have either a tryptophan or tryosine although there is one lysine at residue 116. The current state of knowledge does not permit an unequivocal interpretation of these seemingly inconsistent observations; it is clear that much more data are needed before the structure responsible for the biological activity of epo can be identified.

Some important information has been derived from experiments with site-directed mutagenesis experiments using the ratio of activity by an in vitro assay (29) to activity by radioimmunoassay (30) as the indicator. The substitution of leucine at residue 2 for proline has no effect on activity, nor did phenylalanine at 49 for tyrosine, or glutamic acid at 52 for lysine. The replacement of cysteine-33 with proline [as it is in the mouse (22)] resulted in decreased biological activity (31). This suggests that the loop formed by the 29-33 disulfide bond is not exactly replaced by putting a turn in the peptide chain with a proline. It might also suggest that pure mouse epo would have a lower potency than human epo.

III. SPECIFIC EPO RECEPTORS

The first demonstration that receptors for epo were expressed on erythropoietic precursor cells was an indirect one, showing that the effect of epo on transcription by rat bone marrow cells in culture was abolished by trypsinization of the cells (32). Confirmation had to wait until pure epo and a homogeneous epo-responsive cell population were available. Homogeneous u-epo labeled with ^{3}H in modified sialic acid residues (33) was used with a uniform preparation of cells from the spleens of mice infected with the anemic variant of the Friend erythroleukemic virus (FVA cells) (34). The use of tritium at this stage of experimentation was dictated by the evidence that iodinated u-epo was devoid of biological activity (27) whereas the ^{3}H-labeled material was fully active (33). Because of the low specific radioactivity of the epo, high cell concentrations had to be used to obtain reasonable count rates and the results may have been distorted by the high cell number. Re-examination of the question with ^{125}I-r-epo was feasible once it was found that specific binding of ^{125}I-epo occurred even though there might be a question about biological activity. The later experiments (35) showed that FVA cells have two classes of receptors with k_D values of 0.09 nM and 0.6 nM, with ~300 of the former and 500-700 of the latter. These FVA cells show maximal response to epo when ~180 receptors are occupied, indicating that it is the higher-affinity class that is biologically effective.

Binding of epo to cell surface receptors is followed by internalization of the ligand-receptor complex and the intracellular, iodinated epo is then broken down and released from the cells into the medium (35). Whether the internalization step is necessary for the biological action of epo, or whether the binding of epo to the external domain of the receptor is sufficient for biological action on appropriate target cells, is still not known.

Another binding study (36) used r-epo biolabeled with ^{35}S by Chinese hamster ovary cells containing an amplified human epo gene. The cells used to bind epo in these experiments were enriched from the spleens of mice made anemic with phenylhydrazine. The k_D found for the 100-200 receptors on these cells was found to be 0.75 nM although the data at the lower levels of bound epo (<0.02 fmol/10^6 cells) may have been too sparse to permit detection of a possible second class of higher-affinity receptors. The authors also showed that asialo r-epo was a better competitor for receptors than native epo with these cells. Our own recent data (37) indicate that with FVA cells the k_D value for asialo r-epo is significantly lower than that of native epo. In the same article Mufson and Gesner (36) found that ^{125}I r-epo was essentially not internalized whereas the ^{35}S r-epo was almost completely internalized under the same conditions. The difference between these findings and those reported by Sawyer et al. (35) using FVA cells may reside in the cells, or, perhaps, in the degree of iodination (I atoms/molecule), which was not specified by Mufson and Gesner, although they reported that the iodinated r-epo had <10% of the original activity. Sawyer et al. showed that at 0.6 I/epo molecule they found 95% of the original biological activity but at 4 I/epo molecule binding activity was decreased to 10%.

Sakaguchi et al. (38) have reported that the cell line DA1 (species was not indicated), which is interleukin-3 dependent, can be converted to epo dependence by growth in the presence of 1 U/ml of r-epo. These cells [termed DA1 (cl.14)] express ~130 epo receptors per cell with a k_D of 0.5 nM and only a single class. Similarly, Boussios and Bertles (39) reported that hamster yolk sac erythroid cells have a mean of ~700 epo receptors per cell with a k_D of 0.2 nM and again with only a single class. Cells infected with the nonanemic, polycythemic variant of Friend virus also have been reported to bind epo with a k_D of 0.5 nM and 300-600/cell (40). The same authors (40) showed that rat fetal liver cells have ~490 receptors with a k_D of 0.17 nM. Similarly, Todokoro et al. (41) showed that an erythroleukemic cell line responsive to epo has ~470 receptors per cell with a k_D of 0.15 n*M*. In still another study of epo receptors (42), murine erythroleukemic cells were reported to change both k_D and number of receptors when induced with dimethylsulfoxide.

A number of publications have reported on the results of crosslinking studies to determine the apparent molecular weight of epo receptors. In general, two labeled components were found with the following sizes: 78 kDa, 94 kDa (43), 63 kDa, 94 kDa, 119 kDa (41), 120 kDa, 140 kDa (42), 85 kDa, 100 kDa (44) and 95 kDa, 110 kDA (45). In this laboratory we have also found two bands with sizes of 90 kDa and 105 kDa (47). Despite an unusually large number of publications (for this hormone) it still appears that much work needs to be done before epo receptors are well characterized. The common thread in all of these studies is that whether cells are responsive to epo or not, they express a small number of surface receptors.

IV. MOLECULAR BIOLOGY OF EPO

A. Cloning and Analysis of the Human, Monkey, and Mouse Erythropoietin Genes

The human epo gene was cloned recently (19,20) using a mixture of oligonucleotide probes based on the sequences of selected tryptic peptide fragments of pure human u-epo. The old-world monkey epo gene was also cloned from a cDNA library using the same approach (46) as was the mouse epo gene from a genomic library using the monkey epo cDNA probe (22) and a human epo cDNA probe (47).

Comparison of the human and mouse epo genes shows that the overall organization of the gene is conserved (Fig. 3). The number of exons and introns is the same (22,47). In addition, exon length and sequence is well conserved, the cod-

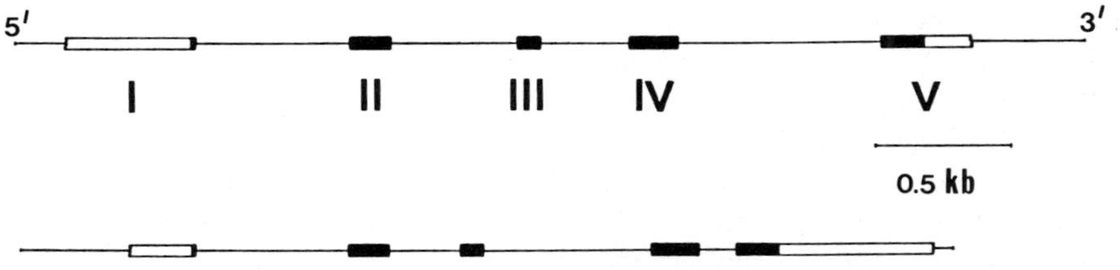

Figure 3 Organization of the mouse (top) and human (bottom) genes. Open boxes indicate exons; closed boxes indicate the coding regions. The uncertainty of the exact transcript ends is indicated.

ing sequence identity being 79%. Intron length and sequence, on the other hand, is not conserved (beyond the first 10-20 nucleotides flanking the exon boundaries) with the exception of the first intron, which exhibits moderate sequence conservation. The 5′- and 3′-untranslated sequences also showed substantial conservation (47). The monkey coding sequence is 94% identical to the human gene (46).

A 600-base pair (bp) region upstream of the protein initiation codon of the human epo gene did not contain any promoterlike sequences such as "TATA" or "CCAAT" boxes (20,31). This appears to be true of the promoter region of mouse epo gene also, although there are sequences with some homology to these elements. Similarly, in both the human and monkey epo genes, the normal polyadenylation sequence AAUAAA is not present (31). The site of polyadenylation of the mouse transcript has not been determined (22,47). There are other peculiarities of these genes. For example, in all three species, the 5′ untranslated region of the transcripts appears to be longer than most mRNAs studied (48). In addition, translation starts at the second AUG in the mRNA instead of the first AUG as is the case in most mRNAs (48).

There was no disagreement between the amino acid sequence of pure human epo (18) and that derived from the nucleotide sequence of the cloned gene (19, 20). Comparison of the human and mouse epo genes has made it possible to delineate the mouse epo coding sequences. The human gene codes for a 27 amino acid signal peptide and a 166 amino acid secreted protein (19,20). The mouse gene codes for the same size mature epo but the length of the leader peptide is 26 amino acid residues (22,47). The amino acid sequence of the monkey epo derived from the cDNA clone (31) shows that the monkey gene codes for a 192 amino acid protein. Expression of monkey epo in CHO cells and *N*-terminal analysis has shown that monkey epo has three more amino acids than does human epo (31,46). Thus, the monkey gene codes for a 168 amino acid secreted protein and a 24 amino acid signal peptide.

B. Expression of the Epo Gene

The availability of the cloned mouse gene has finally made it possible to confirm a finding made in 1957 (49) that the kidney is the major source of epo in the adult animal. Using two PstI fragments of the mouse gene (one containing exons II, III, and part of IV and the other containing the remainder of exon IV and exon V) as probes, epo mRNA was demonstrated in the kidney of rats, 10 hr after bleeding, by Northern blotting. No epo message was detected in the kidney of control (unbled) rats nor in the salivary glands, thymuses, spleens, or livers from bled or control rats (50). The epo mRNA comigrated with the 18S ribosomal RNA on denaturing gels, indicating that it is ~1.85 kb in size. When cobalt chloride was used to stimulate epo production (50), epo mRNA was always detected in the kidney of treated animals and to a much lesser extent in the liver in some experiments but not others. As was the case in the bleeding experiments, no epo message was detected in the salivary gland or spleen with or without cobalt chloride (50). Similar results have been obtained in response to anemia induced by bleeding (51) as well as in response to hypoxia induced by exposure of the experimental animals to a hypobaric chamber (52).

Time course studies showed that epo mRNA in the kidney first appears sometime between 3 and 6 hr after cobalt injection a result consistent with serum epo levels which at 3 hr are the same as at zero time, but rise by sevenfold at 6 hr after injection (50). In mice, a similar time course study showed that the earliest time epo mRNA could be detected was at 10 hr after cobalt injection. Again no epo mRNA was detected in liver, spleen, salivary gland, or thymus.

That no increase in epo levels is seen before the appearance of message argues against the notion that the first response to hypoxia may be a release of stored epo. Thus, the regulation of epo gene expression would appear to be transcriptional or related to the stability of message.

C. Cellular Localization of the Site of Synthesis of Erythropoietin

Cloned mouse epo gene probes have also been instrumental in the attempt to identify the cell type that synthesizes epo. In situ hybridization studies on kidney sections from mice made profoundly anemic showed that epo is made in peritubular cells of the cortex and, to a lesser extent, in the outer medulla (53,54). These cells were nonglomerular and nontubular and were tentatively identified as capillary endothelial cells. On the other hand, in situ hybridization studies done in our laboratory point to the cells of the proximal tubule as the source of epo (55).

D. Chromosomal Localization of the Epo Gene and Identification of Restriction Fragment Length Polymorphism

Using a human epo cDNA probe and Southern blot analysis on DNA isolated from a number of human-Chinese hamster (56) and human-mouse somatic cell hybrids (57) containing reduced numbers and different combinations of human chromosomes, the epo gene has been localized to chromosome 7 as hybridization of the probe correlated with the presence of human chromosome 7 in the cell hybrids. Southern analysis of DNA from a cell hybrid containing a translocated derivative of chromosome 7 localized the epo gene to 7pter-q22 (57). Localization of the epo gene to the region q11-q22 of chromosome 7 was also accomplished by in situ hybridization to metaphase chromosomes (56). In the mouse, the epo gene has been localized to chromosome 5 (58).

When the cDNA probe was used to screen genomic DNA from a number of individuals digested with a variety of restriction enzymes, restriction fragment length polymorphism (RFLP) was observed in those digested with HindIII (56,57) as well as HinfI (57). Each enzyme revealed two alleles. Family studies have shown that these RFLPs display segregation patterns consistent with Mendelian inheritance making them useful genetic markers for this region of the human genome.

E. Studies of Epo Gene Expression in Transformed Cells

The IW32 and NN10 cell lines are erythroleukemic cell lines, isolated from spleens of mice infected with two different biologically cloned viruses, the former with a virus derived from a helper-independent ecotropic Friend murine leukemia virus and the latter with a virus derived from a complete Friend virus (anemia strain) (59-61). Both cell lines constitutively secrete epo and contain epo mRNA of normal size (50,62). Southern blot analysis using several restriction enzymes has shown that the IW32 cell line has a normal as well as a rearranged and amplified epo gene (63). The amplification may be related to the polyploidy of these transformed cells. The NN10 cell line, on the other hand, contains only the normal epo gene (62,64). The mechanism or constitutive expression is being studied in these cells because it may lead to an understanding of the normal regulation of epo gene expression. Although both cell lines are being studied, the initial focus has been on the IW32 cell line, which has been studied in more detail.

The restriction maps of both the normal and rearranged epo loci of IW32 cells are depicted in Fig. 4. The rearranged locus contains an apparently normal epo gene and downstream region but an altered upstream region that contains a PstI, BamHI, and XbaI sites not present upstream of the normal epo gene. Moreover, it does not contain one PstI site present upstream of the normal gene. The entire upstream regions of both genes shown in Fig. 4 have been sequenced and the rearrangement breakpoint has been localized to 530 bp upstream of the XbaI site found immediately 5′ to the gene denoted by the thick line (65).

Studies of chromatin structure by DNase hypersensitivity methods (63,66) have shown that the normal epo locus is relatively resistant to DNase I, whereas the rearranged epo locus is hypersensitive. The DNase I hypersensitive site maps as shown in Fig. 4. If the correlation between nuclease hypersensitivity and gene expression holds, it would appear that IW32 cells express the rearranged but not the normal gene.

There was no integration of the FMuLV provirus within the region 4.5 kb upstream and 9.5 kb downstream of the epo gene at the rearranged locus. Similarly, no transduction of the epo gene into the retorviral genome has been found. Thus, these two events had to be discounted as possible mechanisms by which the epo gene at the rearranged locus was activated (63). Sequence analysis of the upstream region of the rearranged locus encompassing the rearrangement

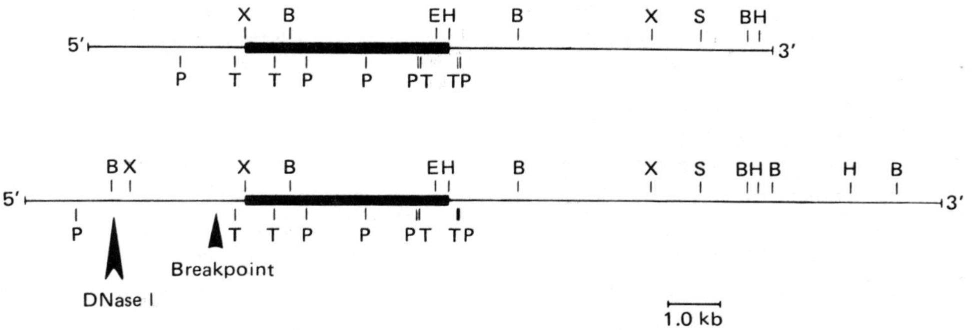

Figure 4 Restriction maps of the two clones containing the normal (top) and the rearranged (bottom) epo gene loci. Fine-structure PstI and BstXI analysis was limited to the immediate region of the epo gene that is contained within the thick line. The rearrangement breakpoint deduced from sequencing as well as the DNase I hypersensitive site are indicated. X, XbaI; B, BamHI; E, EcoRI; H, HindIII; S, SalI; P, PstI; T, BstXI.

breakpoint indicates that the epo gene at this locus has been activated because the rearrangement event has brought a novel active gene into close proximity and opposite transcriptional orientation to the epo gene (65) and activation appears to be due to bidirectional transcription. The mechanism by which the epo gene in NN10 cells has been turned on is still under study.

V. MODE OF ACTION OF EPO

The mechanism of action of epo on its target cells within the bone marrow has been studied extensively by a number of methods, yet the complex course of biochemical events leading to the expression of the erythroid phenotype by these cells remains unclear. The actions of epo may be viewed in terms of its effects on pluripotent stem cells toward lineage commitment, differentiation, and proliferation. Erythropoietin is thought to be a humoral agent committing stem cells to the erythroid phenotype (67,68). Supporting this concept is evidence that epo can compete in vitro with other colony-stimulating factors in committing stem cells towards their respective lineages (69). Nonetheless, the mechanism of hematopoietic commitment, erythroid or otherwise, is unknown; and until this mechanism is fully understood, the effect of epo on erythroid commitment will remain controversial. In contrast, there is some limited knowledge of the events of erythroid differentiation and proliferation, a process where the regulatory action of epo is well established and a process that appears to proceed in an orderly and predictable manner. Therefore, this discussion will address what is known of the role that epo plays in differentiation and proliferation.

In earlier studies, crude epo preparations containing endotoxins and/or colony stimulating factors that have been shown to influence the effects of epo, made interpretations uncertain. In addition, difficulties arose because of the use of bioassays in which the cellular components were heterogeneous with respect to their responsiveness to epo (e.g., bone marrow cells), or capable of generating epo inherently (e.g., fetal liver cells). These problems have been reviewed elsewhere (70). More recent studies of the events after epo exposure have used mouse spleen cells, either virally transformed or erythropoietic by virtue of the induction of anemia, as a source of relatively homogeneous erythroid precursor cells for study. The cost of the improved specificity that these models provide is the caution required in extrapolating events within virally transformed cells and spleen cells to the phenomena characteristic of normal adult bone marrow.

The earliest known event after the exposure of target cells to epo is an accumulation of cytosolic Ca^{2+}. This was demonstrated in studies of FVA cells within 1 min of addition of a high concentration of epo (2 U/ml) at 4°C (71). In support for the potential role of calcium in erythroid differentiation is its effects on colony-forming units (CFU)-E. Colony formation has been shown by Misti and Spivak (72) to increase when cultured with carboxylic ionophores, which increases cytosolic Ca^{2+}, and to decrease under the influence of ethyleneglycol tetraacetic acid (EGTA) chelation of Ca^{2+}, an effect that may reflect an action of calcium on epo-receptor binding (72). These results are not unchallenged because Bonnanou-Tzedaki et al. (73) have found no evidence of calcium flux with enriched bone marrow erythroblast fractions under physiologic temperatures and epo concentrations. Nonetheless, Bonnanou-Tzedaki et al. found a calcium-associated effect when intracellular calcium levels are manipulated by pharmacologic means. An epo response (measured by enhanced DNA synthesis, a somewhat late effect of epo action) was blunted by calcium chelation with EGTA, by calcium channel blockade with verapamil, and by the intracellular calcium antagonist TMB-8. Artificially induced changes in cytosolic calcium levels within effector cells, without exposure to epo, cannot induce erythroid differentiation (72,73). This suggests its role is accessory to the differentiation and proliferation process. It is unknown whether alterations in intracellular Ca^{2+} might involve a calmodulin- or protein kinase C-mediated process (73).

The possible role of sodium ions in erythroid differentiation and proliferation is even less convincing. Exposure of CFU-E to pharmacologic doses of Na^+/K^+ adenosine triphosphatase inhibitors have had variable and conflicting influences on erythroid colony formation depending on the concentrations of the inhibitors used (74,75).

There has been no general agreement within the literature regarding the role that cyclic nucleotides might play in mediating the action of epo. The weight of the evidence, however, outlined in depth by Spivak (10), persuasively argues against cyclic adenosine monophosphate as a second messenger within this process. Similarly, a role for cycline guanosine 3′5′-monophosphate has not been found (76).

Although the earliest events within stem cells remain speculative, later occurrences leading to the erythroid phenotype have been characterized more clearly. Using a posthypoxic, erythropoietic spleen assay, investigators have determined the timing of developmental processes after in vivo exposure to exogenous epo [reviewed by Spivak (10)]. Increased RNA polymerase II activity occurs within 0.5-2.0 hr, followed by increases in RNA polymerase I and nonhistone protein synthesis (3-12 hr). Both a second increase in RNA polymerase II activity and the earliest events leading to DNA replication (DNA polymerase alpha and histone synthesis) occur between 12 and 24 hr after epo injection. Maximal DNA synthesis occurs after 48 hr.

The synthesis of hemoglobin, the principal protein of mature erythrocytes, might be expected to be influenced by the action of epo. In fact, investigators have demonstrated that the differences in the expression of globin mRNA, between suppressed and stimulated cells within rat bone marrow, could be seen with 2 hr of the addition of epo (77). With FVA cells, the earliest β-globin mRNA was seen 6 hr after exposure to epo (78). Nijhof et al. presented data (79) indicating that globin mRNA transcription in purified, normal erythroid precursor cells can be induced by epo in less than 1 hr. This suggests that globin expression and protein synthesis are one of the earliest, if not primary, effects of epo. The effect of epo on globin gene transcription probably precedes the effect of epo on heme synthesis (see below). Both α- and β-globin gene expressions appear to be coordinately regulated, but the mechanism of epo action on the expression of these genes is unknown.

The enzymes responsible for heme synthesis appear to be present within normal, suppressed bone marrow cells before stimulation with epo and in sufficient quantities to provide heme for the initial days of hemoglobin synthesis (80). These same authors have shown that epo stimulates the synthesis of porphobilinogen deaminase without discernible effects on the other enzymes of the heme synthetic pathway (80). The synthesis of this apparent rate-limiting enzyme occurred between 20 and 40 hr after the addition of epo.

Sawyer and Krantz (81) have demonstrated that FVA cells on exposure to epo increase transferrin binding sites within 6 hr, and that the number of transferrin receptors doubled within 24 hr. Study of suppressed rat marrow cells in culture showed that epo had a stimulatory effect on several components of the red blood cell membrane and on some components that were only transiently expressed during differentiation (82). Koury et al. have studied the effects of epo on the synthesis of other proteins on the mature red blood cell (83). In FVA cells, de novo synthesis of bands 3 (12 hr) and 4.1 (48 hr) occurred as a result of epo stimulation. Spectrin, constitutively produced before hormone exposure, accumulated sharply within these cells 12 hr after the addition of epo.

VI. CLINICAL USE OF EPO

From the time it was first understood that epo was made in the kidney (49), the inference has been made that the anemia of chronic renal disease might be alleviated by administration of exogenous epo. This was supported by the findings that patients with this type of anemia did not respond to it with elevated serum epo level, but that their levels were either lower than normal or normal (30). The desirability of treatment of the anemia with epo was heightened by the realization that if transfusions could be replaced by epo administration, problems such as iron overload, sensitization, hepatitis, and now acquired immune deficiency syndrome, could be eliminated. Production of epo by recombinant methods has now made all of this a realization. In two published articles (84,85) the data are quite clear that the anemia of chronic renal disease of patients on dialysis can be corrected and that the requirement for transfusion can be either eliminated or very greatly reduced.

Another possible therapeutic use of epo derives from the observations made by Al-Khatti et al. (86) showing that fetal hemoglobin appearance in the circulation can be stimulated in adult baboons by exogenous epo. This could form a potential therapy for patients with sickle cell anemia in whom fetal hemoglobin has been known to diminish the occurrence of sickling episodes.

REFERENCES

1. Carnot, P. Sur le mechanisme d'hyperglobulie provoquee par le serum d'animaux en renovation sanguine. *C. R. Acad. Sci. (Paris) 111*:344-346 (1906).
2. Krumdieck, N. Erythropoietic substance in the serum of anemic animals. *Proc. Soc. Exp. Biol. Med. 54*:14-17 (1943).
3. Erslev, A. J., P. H. Lavietes, and G. Van Wagenen. Erythropoietic stimulation induced by "anemic" serum. *Proc. Soc. Exp. Biol. Med. 83*:548-550 (1953).
4. Krantz, S. B., and L. O. Jacobson. *Erythropoietin and the Regulation of Erythropoiesis*. The University of Chicago Press, Chicago (1970).
5. Gordon, A. S., M. Condorelli, and C. Peschle, eds. *Regulation of Erythropoiesis*. Il Ponte, Milan (1971).
6. Nakao, K., J. W. Fisher, and F. Takaku. *Erythropoiesis*. The University of Tokyo Press, Tokyo (1975).

7. Graber, S. E., and S. B. Krantz. Erythropoietin and the control of red cell production. *Annu. Rev. Med. 29*:51-66 (1978).
8. Sherwood, J. B. The chemistry and physiology of erythropoietin. *Vitam. Horm. 41*:161-211 (1984).
9. Anagnostou, A. Erythropoietin: a hematopoietic hormone produced by the kidney. *Semin. Nephrol. 5*:104-114 (1985).
10. Spivak, J. The mechanism of action of erythropoietin. *Int. J. Cell Cloning 4*:139-166 (1986).
11. Orlic, D., and J. LoBue (Eds.). Controls of hematopoiesis. In *Blood Cells*, Vol. 10. Springer Verlag, New York (1984).
12. Rich, I. N. (Ed.). *Molecular and Cellular Aspects of Erythropoietin and Erythropoiesis*. Springer Verlag, Berlin (1987).
13. Jelkmann, W. Renal erythropoietin: properties and production. *Rev. Physiol. Biochem. Pharmacol. 104*;140-215 (1986).
14. Miyake, T., C. K.-H. Kung, and E. Goldwasser. Purification of human erythropoietin. *J. Biol. Chem. 252*:5558-5564 (1977).
15. Dordal, M. S., F. F. Wang, and E. Goldwasser. The role of carbohydrate in erythropoietin action. *Endocrinology 116*:2293-2299 (1985).
16. Egrie, J., T. W. Strickland, J. Lane, A. Aoki, A. M. Cohen, R. Smalling, G. Trail, F. K. Lin, J. K. Browne, and D. K. Hines. Characterization and biological effects of recombinant human erythropoietin. *Immobiol. 172*: 213-224 (1986).
17. Wang, F. F., C. K.-H. Kung, and E. Goldwasser. Some chemical properties of human erythropoietin. *Endocrinology 116*:2286-2292 (1985).
18. Lai, P. H., R. Everett, F. F. Wang, T. Arakawa, and E. Goldwasser. Structural characterization of human erythropoietin. *J. Biol. Chem. 261*:3116-3121 (1986).
19. Jacobs, K., C. Shoemaker, R. Rudersdorf, S. D. Neill, R. J. Kaufman, A. Mufson, J. Seehra, S. S. Jones, R. Hewick, E. F. Fritsch, M. Kawakita, T. Shimizu, and T. Miyake. Isolation and characterization of genomic and cDNA clones of human erythropoietin. *Nature 313*:806-810 (1985).
20. Lin, F. K., S. Suggs, C. H. Lin, J. Browne, R. Smalling, J. Egrie, K. Chen, G. Fox, F. Martin, Z. Stabinsky, S. Badrawi, P. H. Lai, and E. Goldwasser. Cloning and expression of the human erythropoietin gene. *Proc. Natl. Acad. Sci. USA 82*:7580-7584 (1985).
21. Recny, M. A., H. A. Scoble, and Y. Kim. Structural characterization of natural human urinary and recombinant DNA-derived erythropoietin. *J. Biol. Chem. 262*:17156-17163 (1987).
22. McDonald, J., F. K. Lin, and E. Goldwasser. Cloning, sequencing and evolutionary analysis of the mouse erythropoietin gene. *Mol. Cell. Biol. 6*:842-848 (1986).
23. Davis, J. M., T. Arakawa, T. W. Strickland, and D. A. Yphantis. Characterization of recombinant human erythropoietin produced in Chinese hamster ovary cells. *Biochemistry 26*:2633-2638 (1987).

24. Sasaki, H., B. Bothner, A. Dell, and M. Fukuda. Carbohydrate structure of erythropoietin expressed in Chinese hamster ovary cells by a human erythropoietin cDNA. *J. Biol. Chem. 262*:12059-12076 (1987).
25. Takeuchi, M., S. Takasaki, H. Miyazaki, T. Kato, S. Hoshi, N. Kochibe, and A. Kobata. Comparative study of the asparagine-linked sugar chains of human erythropoietin purified from urine and the culture medium of recombinant Chinese hamster ovary cells. *J. Biol. Chem. 263*:3657-3663 (1988).
26. Lowy, P., and G. Keighly. Inactivation of erythropoietin by Koshland's tryptophan reagent and by membrane filtration. *Biochim. Biophys. Acta 160*:413-419 (1968).
27. Goldwasser, E. Erythropoietin and red cell differentiation. In *Control of Cellular Division and Development: Part A*, D. Cunningham, E. Goldwasser, J. Watson, and C. F. Fox (Eds.). Alan R. Liss, New York, pp. 487-494 (1981).
28. Sytkowski, A. J., and K. A. Donahue. Immunochemical studies of human erythropoietin using site-specific anti-peptide antibodies. *J. Biol. Chem. 262*.1161-1165 (1987).
29. Goldwasser, E., J. F. Eliason, and D. Sikkema. An assay for erythropoietin in vitro at the milliunit level. *Endocrinology 97*:315-323 (1975).
30. Sherwood, J. B., and E. Goldwasser. A radioimmunoassay for erythropoietin. *Blood 53*:935-946 (1979).
31. Lin, F. K. The molecular biology of erythropoietin. In *Molecular and Cellular Aspects of Erythropoietin and Erythropoiesis*, I. N. Rich (Ed.). Springer Verlag, Berlin, pp. 23-36 (1987).
32. Chang, C. S., D. Sikkema, and E. Goldwasser. Evidence for an erythropoietin receptor protein on rat bone marrow cells. *Biochem. Biophys. Res. Commun. 57*:399–405 (1974).
33. Krantz, S. B., and E. Goldwasser. Specific binding of erythropoietin to spleen cells infected with the anemia strain of Friend virus. *Proc. Natl. Acad. Sci. USA 81*:7574-7578 (1984).
34. Koury, M., S. T. Sawyer, and M. C. Bondurant. Splenic erythroblasts in anemia-inducing Friend disease: a source of cells for studies of erythropoietin-mediated differentiation. *J. Cell. Physiol. 121*:526-532 (1984).
35. Sawyer, S. T., S. B. Krantz, and E. Goldwasser. Binding and receptor-mediated endocytosis of erythropoietin in Friend virus-infected erythroid cells. *J. Biol. Chem. 262*:5554-5562 (1987).
36. Mufson, R. A., and T. G. Gesner. Binding and internalization of recombinant erythropoietin in murine erythroid precursor cells. *Blood 69*:1485-1490 (1987).
37. Dong, Y.-J., and E. Goldwasser. Unpublished data.
38. Sakaguchi, M., Y. Koishihara, H. Tsuda, K. Fujimoto, K. Shibuya, M. Kawakita, and K. Takatsuki. The expression of functional erythropoietin receptors on an interleukin-3 dependent cell line. *Biochem. Biophys. Res. Commun. 146*:7-12 (1987).

39. Boussios, T., and J. F. Bertles. Receptors specific for erythropoietin on yolk-sac erythroid cells. In *Developmental Control of Globin Gene Expression*, G. Stammatoyannopoulos and A. W. Nienhuis (Eds.). Alan R. Liss, New York, pp. 35-41 (1987).
40. Mayeux, P., C. Billat, and R. Jacquot. Murine erythroleukaemia cells (Friend cells) possess high-affinity binding sites for erythropoietin. *FEBS Lett. 211*: 229-233 (1987).
41. Todokoro, K., S. Kanazawa, H. Amanuma, and Y. Ikawa. Specific binding of erythropoietin to its receptor on responsive mouse erythroleukemia cells. *Proc. Natl. Acad. Sci. USA 84*:4126-4130 (1987).
42. Sasaki, R., S. Yanagawa, K. Hitomi, and H. Chiba. Characterization of erythropoietin receptors of murine erythroid cells. *Eur. J. Biochem. 168*: 43-48 (1987).
43. Mayeux, P., C. Billat, and R. Jacquot. The erythropoietin receptor of rat erythroid progenitor cells. *J. Biol. Chem. 262*:13985-13990 (1987).
44. Sawyer, S. T., S. B. Krantz, and J. Luna. Identification of the receptor for erythropoietin by cross-linking to Friend virus-infected erythroid cells. *Proc. Natl. Acad. Sci. USA 84*:3690-3694 (1987).
45. Tojo, A., H. Fukamachi, M. Kasuga, A. Urabe, and F. Takaku. Identification of erythropoietin receptors on fetal liver erythroid cells. *Biochem. Biophys. Res. Commun. 148*:443-448 (1987).
46. Lin, F.-K., C.-H. Lin, P.-H. Lai, J. K. Browne, J. C. Egrie, R. Smalling, G. M. Fox, K. K. Chen, M. Castro, and S. Suggs. Monkey erythropoietin gene: cloning, expression and comparison with the human erythropoietin gene. *Gene 44*:201-209 (1986).
47. Shoemaker, C. B., and L. D. Mistock. Murine erythropoietin gene: cloning, expression, and human gene homology. *Mol. Cell. Biol. 6*:849-858 (1986).
48. Kozak, M. Compilation and analysis of sequences upstream from the translational start site on eukaryotic mRNA's. *Nucleic Acids Res. 12*: 857-872 (1984).
49. Jacobsen, L. O., E. Goldwasser, W. Fried, and L. F. Plzak. Role of the kidney in erythropoiesis. *Nature 179*:633-634 (1957).
50. Beru, N., J. McDonald, C. Lacombe, and E. Goldwasser. Expression of the erythropoietin gene. *Mol. Cell. Biol. 6*:2571-2575 (1986).
51. Bondurant, M. C., and M. J. Koury. Anemia induces accumulation of erythropoitin mRNA in the kidney and liver. *Mol. Cell. Biol. 6*:2731-2733 (1986).
52. Nishida, J., H. Hirai, M. Kubota, F.-K. Lin, T. Okabe, A. Urabe, and F. Takaku. Detection of erythropoietin message in hypoxic mouse kidney. *Jpn. J. Exp. Med. 56*:321-323 (1986).
53. Lacombe, C., J. L. Da Silva, P. Bruneval, J.-G. Fournier, F. Wendling, N. Casadevall, J.-P. Camilleri, J. Bariety, B. Varet, and P. Tambourin. Peritubular cells are the site of erythropoietin synthesis in the murine hypoxic kidney. *J. Clin. Invest. 81*:620-623 (1987).

54. Koury, S. T., M. C. Bondurant, and M. J. Koury. Localization of erythropoietin synthesizing cells in murine kidneys by in situ hybridization. *Blood 71*:524-527 (1987).
55. Smith, D. H., and E. Goldwasser. Unpublished data.
56. Law, M. L., G.-Y. Cai, F.-K. Lin, Q. Wei, S.-Z. Huang, J. H. Hartz, H. Morse, C.-H. Lin, C. Jones, and F.-T. Kao. Chromosomal assignment of the human erythropoietin gene and its DNA polymorphism. *Proc. Natl. Acad. Sci. USA 83*:6920-6924 (1986).
57. Watkins, P. C., R. Eddy, N. Hoffman, P. Stanislovitis, A. K. Beck, J. Galli, V. Vellucci, J. F. Gusella, and T. B. Shows. Regional assignment of the erythropoietin gene to human chromosome region 7pter → q22. *Cell Genet. 42*:214-218 (1986).
58. Lacombe, C., M. G. Mattei, D. Simon, J. L. Guenet, and P. Tambourin. The murine erythropoietin gene is localized on chromosome 5. *Blood 72*: 1440-1442 (1988).
59. Tambourin, P., N. Casadevall, J. Choppin, C. Lacombe, J. M. Heard, S. Fichelson, F. Wendling, and B. Varet. Prodcution of erythropoietin-like activity by a murine erythroleukemia cell line. *Proc. Natl. Acad. Sci. USA 80*:6269-6273 (1983).
60. Choppin, J. C., C. Lacombe, N. Casadevall, O. Muller, P. Tambourin, and B. Varet. Characterization of erythropoietin produced by IW32 murine erythroleukemia cells. *Blood 64*:341-347 (1984).
61. Choppin, J. C., N. Casadevall, C. Lacombe, F. Wendling, E. Goldwasser, R. Berger, P. Tambourin, and B. Varet. Production of erythropoietin by cloned malignant murine erythroid cells. *Exp. Hematol. 13*:610-615 (1985).
62. Lacombe, C., N. Casadevall, J. Choppin, O. Muller, E. Goldwasser, B. Varet, and P. Tambourin. Erythropoietin production and erythropoietin receptors on murine erythroleukemia cell lines. In *Molecular and Cellular Aspects of Erythropoietin and Erythropoiesis*, I. N. Rich (Ed.). Springer-Verlag, Berlin, pp. 61-72 (1987).
63. McDonald, J., N. Beru, and E. Goldwasser. Rearrangement and expression of erythropoietin genes in transformed mouse cells. *Mol. Cell. Biol. 7*: 365-370 (1987).
64. Beru, N., J. McDonald, and E. Goldwasser. Studies of the constitutive expression of the mouse erythropoietin gene. *Ann. N.Y. Acad. Sci. 554*: 29-35 (1989).
65. Beru, N., J. McDonald, and E. Goldwasser. Activation of the erythropoietin gene due to the proximity of an expressed gene. *DNA 8*:253-259 (1989).
66. Goldwasser, E., J. McDonald, and N. Beru. The molecular biology of erythropoietin and the expression of its gene. In *Molecular and Cellular Aspects of Erythropoietin and Erythropoiesis*, I. N. Rich (Ed.). Springer-Verlag, Berlin, pp. 849-858 (1987).
67. Goldwasser, E. The action of erythropoietin as the inducer of erythroid differentiation. In *Hematopoietic Stem Cell Physiology*, E. P. Cronkite,

N. Daniak, J. Palek, R. P. McCaffrey, and P. Quesenberry (Eds.). Alan R. Liss, New York, pp. 77-84 (1984).
68. Goldwasser, E. Erythropoietin and its mode of action. *Blood Cells 10*:147-162 (1984).
69. Van Zant, G., and E. Goldwasser. Competition between erythropoietin and colony-stimulating factor for target cells in mouse marrow. *Blood 53*:946-965 (1979).
70. Van Zant, G., and E. Goldwasser. Erythropoietin and its target cells. In *Growth and Maturation Factors*, G. Guroff (Ed.). John Wiley, New York, pp. 3-36 (1984).
71. Sawyer, S. T., and S. B. Krantz. Erythropoietin stimulates $^{45}Ca^{2+}$ uptake in Friend virus-infected erythroid cells. *J. Biol. Chem. 259*:2769-2774 (1984).
72. Misti, J., and J. L. Spivak. Erythropoiesis in vitro role of calcium. *J. Clin. Invest. 64*:1573-1579 (1979).
73. Bonnanou-Tzedaki, S. A., M. K. Sohi, and H. Arnstein. The role of cAMP and calcium in the stimulation of proliferation of immature erythroblasts by erythropoietin. *Exp. Cell Res. 170*:276-289 (1987).
74. Spivak, J. L., J. Misti, R. Stuart, and S. J. Sharkis. Suppression and potentiation of mouse hematopoietic progenitor cell proliferation by ouabain. *Blood 56*:315-317 (1980).
75. Gallichio, V. S., M. G. Chen, and M. J. Murphy. Modulation of murine in vitro erythroid and granulocytic colony formation by ouabain, digoxin, and theophylline. *Exp. Hematol. 10*:682-688 (1982).
76. Graber, S. E., S. E. Bomboy, W. D. Salmon, and S. B. Krantz. Evidence that endotoxin is the cyclic 3′: 5′-cGMP-promoting factor in erythropoietin preparations. *J. Lab. Clin. Med. 93*:25-31 (1979).
77. Sahr, K., and E. Goldwasser. The effects of erythropoietin on the biosynthesis of translatable globin mRNA. In *Regulation of Hemoglobin Biosynthesis*, E. Goldwasser (Ed.). Elsevier, New York, pp. 153-161 (1983).
78. Bondurant, M. C., R. N. Lind, M. J. Koury, and M. E. Ferguson. Control of globin gene transcription by erythropoietin in erythroblasts from Friend virus-infected mice. *Mol. Cell. Biol. 5*:675-683 (1985).
79. Nijhof, W., P. Wierenga, K. Sahr, N. Beru, and E. Goldwasser. Induction of globin mRNA transcription in differentiating precursor cells. *Exp. Hematol. 15*;779-784 (1987).
80. Beru, N., and E. Goldwasser. The regulation of heme biosynthesis during erythropoietin induced erythroid differentiation. *J. Biol. Chem. 260*: 9251-9257 (1985).
81. Sawyer, S. T., and S. B. Krantz. Transferrin receptor number, synthesis and endocytosis during erythropoietin-induced maturation of Friend virus-infected erythroid cells. *J. Biol. Chem. 261*:9187-9195 (1986).
82. Tong, B. D., and E. Goldwasser. The formation of erythrocyte membrane proteins during erythropoietin-induced differentiation. *J. Biol. Chem. 256*: 12666-12672 (1981).

83. Koury, M. J., M. C. Bondurant, and T. J. Mueller. The role of erythropoietin in the production of principal erythrocyte proteins other than hemoglobin during terminal differentiation. *J. Cell. Biol. 126*:259-265 (1986).
84. Winearls, C. G., D. O. Oliver, M. J. Pippard, C. Reid, M. R. Downing, and P. M. Cotes. Effect of human erythropoietin derived from recombinant DNA on the anaemia of patients maintained by chronic haemodialysis. *Lancet 2*:1175-1177 (1986).
85. Eschbach, J. W., J. C. Egrie, M. R. Downing, J. K. Browne, and J. W. Adamson. Correction of the anemia of end-stage renal disease with recombinant human erythropoietin. *N. Engl. J. Med. 316*:73-77 (1987).
86. Al-Khatti, A., R. Veith, T. Papayannopoulou, E. Fritsch, E. Goldwasser, and G. Stamatoyannopoulos. Stimulation of fetal hemoglobin synthesis by erythropoietin in baboons. *N. Engl. J. Med. 317*:415-420 (1987).

10

Hematopoietic Growth Factors and Progenitor Cells in Human Acute Leukemia

BOB LÖWENBERG, RUUD DELWEL, and IVO TOUW
Dr. Daniel den Hoed Cancer Center, Rotterdam, The Netherlands

I. HEMATOPOIETIC STEM CELLS: IN VITRO ASSAYS AND REGULATORY FACTORS

A complex network of positive and negative regulatory factors and cellular interactions governs hematopoiesis and permits a continuous high production rate of mature blood cells as well as the adaptation of blood cell production to the variable needs of particular blood cell compartments (e.g., granulocytes during infection).

In vitro assays for primitive multilineage hematopoietic precursor cells and assays for unipotent progenitors have been advanced during the last 2 decades. Thus, precursors with the capacity to form colonies of granulocytic cells, macrophages, erythrocytes, megakaryocytes, collectively designated as CFU–GEMM (1-3), and more mature progenitor cells that are committed to one hematopoietic lineage, can all be assayed in vitro: erythroid burst-forming units (BFU–E) (4,5), erythroid colony-forming units (CFU–E) (6), granulocyte/monocyte CFU (CFU–GM) (7,8), granulocyte CFU (CFU–G), monocyte CFU (CFU–M), eosinophil colony-forming units (CFU–Eo), or megakaryocyte CFU (CFU–Mega) (9). These methods are also useful to define the role of hematopoietic growth factors in normal hematopoiesis. Several regulatory molecules that induce colony formation from those precursor cells have been isolated after molecular cloning.

The same technology is currently being applied for elucidating growth-factor responses of neoplastic clonogenic cells in acute myeloid and lymphoid leukemia. In acute leukemia, a transformed malignant precursor cell arises that is unable to generate terminally differentiated cells. This neoplastic cell population apparently escapes normal maturation impulses and expands dispropor-

Table 1 Human Hematopoietic Growth Factors

Acronym	Protein size (kDa)	Hematopoietic progenitor cell targets			
		Normal	Ref.	Leukemia	Ref.
IL-3	14-28	CFU-GEMM BFU-E CFU-Eo	10-12	AML-CFU	13-17
GM-CSF	14-35	CFU-GEMM BFU-E CFU-Eo	18-24	AML-CFU	14-17,25-28
G-CSF	18-22	CFU-G	29-31	AML-CFU	14-17,28
M-CSF	35-45 (x2) 18-26	CFU-M	32,33	AML-CFU (rare)	14,16,34
EPO	36	BFU-E CFU-E	35,36	AML-CFU (M-6 type)	14,37

Table 2 Human Lymphoid Growth and Differentiation Factors

Acronym	Protein size (kDa)	Lymphoid target cells: Normal	Lymphoid target cells: Leukemic	Ref.
IL-1	17		Unknown	88
IL-2	15	B lymphocytes, T lymphocytes thymocytes	T ALL T NHL ATL B CLL	82,89-92
IL-4 (BSF-1)	15		Unknown	93
IL-5 (TRF)	13	B cells(?)[a]	Unknown	94
IL-6 (BSF-2,IFN β2)	26	B lymphocytes	Myeloma	95-98
TNF	17	B lymphocytes T lymphocytes	B CLL	99-101

BSF, B- cell stimulatory factor; TRF, T-cell replacing factor; IFN, interferon; TNF, tumor necrosis factor.
[a]Activity on lymphoid cells not yet defined in humans.

tionally in marrow and blood. In this review we shall discuss our understanding of the role of hematopoietic growth factors in leukemic blood cell formation in human acute leukemia mainly based on the study of in-vitro growth.

II. HEMATOPOIETIC COLONY-STIMULATING FACTORS AND INTERLEUKINS

The hematopoietic colony-stimulating factors (CSFs) and interleukins (ILs) and their target precursor cells are listed in Tables 1 and 2. The CSFs regulate the proliferation and outgrowth of different hematopoietic progenitors and frequently also have a role in stimulating the metabolic and functional activities of the mature cell types.

An extending series of human growth factors acting on lymphoid cells has become available through molecular cloning as well (Table 2). The spectrum of activities of these factors is not restricted to lymphoid cells. In mice, for instance, IL-1, IL-4, and IL-6 promote proliferation of a variety of hematopoietic progenitor cells. Thus these factors have multiple roles in the regulatory network controlling blood cell proliferation and maturation.

III. HEMATOPOIESIS IN ACUTE MYELOBLASTIC LEUKEMIA

The malignantly transformed cell population expands progressively when clinical acute myeloblastic leukemia (AML) evolves. In the marrow of patients with AML immature cells, i.e., mainly myeloblasts, predominate. Although AML is the result of a clonal expansion of a single cell (38), the resulting cell population is biologically heterogeneous according to morphological, [^{3}H] thymidine labeling (39), immunological, cytogenetic, and colony culture parameters. For instance, a small population of proliferating cells expressing self-renewal capacities can be identified among the AML cell mass in colony assays (40).

Since the availability of pure growth factors it has become possible to understand the exact role of these hormones, either as individual or as combined activities in controlling the proliferation and differentiation of human AML cells and to identify possible abnormalities of AML growth in vitro. In-vitro studies are now undertaken to elucidate the complex system of growth-factor regulation of human AML precursors. In this regard the recently designed serum-free cultur method for human AML cells appears to be a useful addition to the experimental arsenal (41).

A. Cellular Heterogeneity

The heterogeneous cellular composition of human AML may be regarded as a hierarchy of subpopulations of cells of different maturation stages (40,41). This concept is supported by immunological data that indicate that among AML blasts the in-vitro colony-forming cells (AML–CFU) carry immature surface phenotypes, whereas the immunophenotypically more mature cells among the leukemia lack proliferative (i.e., colony-forming) ability (43). On proliferation in vitro, colony-forming cells give rise to maturing progeny but maturation is incomplete (44,45). Apparently the dynamics and negative linkage of proliferation and differentiation events characteristic of the normal hematopoietic system have to a certain extent been retained in human AML. It is difficult to determine the in-vivo significance of AML–CFU. In-vitro observations support the concept that they maintain and expand the malignancy and indeed act as progenitor cells: (a) they have capacities for self renewal (46,47); (b) like normal hematopoietic precursors, they have the capacity not only to proliferate but also to undergo some further differentiation in vitro (43,44,48,49).

Immunologic analysis has shown that the phenotypes of AML–CFU vary markedly among patients with AML. In certain cases AML–CFU share immunologic features with CFU–GEMM cells, whereas in other instances AML–CFU express surface-antigen combinations typical of committed granulocyte/macrophage progenitors in normal bone marrow (CFU–GM) (40). Thus, it has been appreciated that the immunological phenotypes of AML progenitors and particular types of normal hematopoietic precursors are largely similar. However, detailed surface marker studies have revealed that discrepancies between the immunologic surface phenotypes of AML–CFU and the majority normal marrow progenitors may exist. In studies that considered not only the presence or absence of antigenic determinants but also antigen density at the cell surface, differences between the immunophenotypes of AML–CFU from most of the normal hematopoietic progenitor cells were recognized (50-52). Crucial in these studies was the fact that a combination of surface markers was selected that identified a constant surface profile for CFU–GEMM, BFU–E as well as two subpopulations of CFU–GM. This permitted the recognition of phenotypes of AML–CFU at variance with these normal progenitors. The progenitor phenotypes that predominate in AML probably reflect those of infrequent, although essentially normal, hematopoietic precursors that have expanded selectively in AML as a consequence of the abnormality of maturation.

B. The Role of CSFs

Acute myeloblastic leukemia cells have not become independent of CSFs. Interleukin-3 and GM–CSF each as individual factors induce colony formation in vitro (15,27,28), stimulate DNA synthesis (14,16) and enhance the survival of AML progenitors in vitro (15). Granulocyte-macrophage colony-stimulating factor has also been shown to support the selfrenewal of AML precursors (26). In addition to IL-3 and GM–CSF, G–CSF is successful in inducing AML cell growth in culture, although somewhat less frequently. Often, the same leukemias are susceptible to multiple CSFs. Each of these three CSFs as single factors is able to generate significant proliferation in vitro. Infrequently, M-CSF induces AML growth in culture and erythropoietin has occasionally been reported to act as a growth factor in human AML, i.e., in cases of erythroleukemia (M6) (14,37). The fact that in certain cases, proliferation can only be maximally stimulated in the presence of an impure phytohemagglutinin (PHA)-leukocyte conditioned medium leaves the possibility open that yet other soluble growth factors operate in AML.

The CSFs affect overlapping stages of the hematopoietic pathways and thus, to a significant extent, exert stimulating effects on the same target cells. Hence, the cooperative interactions between these molecules in controlling human AML growth are of particular interest.

In tritiated thymidine uptake assays it appeared that the responsiveness of AML cells to IL-3 parallels the response to GM–CSF (14,16). This suggests that IL-3 and GM–CSF responsive subpopulations of AML precursors to a certain extent are identical. Also, it has become evident that IL-3 and GM–CSF usually do not produce additive proliferative effects, again suggesting that the two CSFs act mainly on the same AML precursor cell subpopulation. In contrast, G–CSF often produces additive growth-potentiating effects in AML, when supplemented to cultures with IL-3 or GM–CSF. In colony culture, IL-3 has been shown to elevate colony numbers in a majority of AML cases when supplemented along with G–CSF or GM–CSF to the cultures (15), but the total numbers of colonies formed in response to the combinations of CSFs are less than the sum of the individual effects (28). Therefore, most likely the AML cell mass contains subsets of precursors with variable growth-factor reactivities that can be recruited into cell proliferation, depending on the exact combination of growth factors supplied.

From dose-effect studies with CSFs a marked patient-to-patient variation with regard to growth-factor responsiveness of AML cell proliferation has become apparent (Table 3). Thus, different patterns of response can be distinguished among clinical cases of AML. Generally, growth-factor reactivity did not correlate with morphological classification of human AML, except in the rare cases of erythroleukemia that appeared to be especially sensitive to erythropoietin. Acute myeloblastic leukemia progenitor (AML–CFU) immunophenotypes did not show a relationship with in-vitro growth factor dependence either (14).

The biological basis of the heterogeneity of growth-factor responsiveness in clinical AML is not clear. The abilities of the AML blasts to react with cer-

Table 3 Frequency of Recombinant Hematopoietic Growth Factors Acting as Stimuli of Human AML Growth In Vitro

Growth (assay)	IL–3	GM–CSF	G–CSF	M–CSF	Refs.
Colony formation	–	32/48	28/48	0/48	28,83
Colony formation	–	16/20	–	–	27
Colony formation	8/9	9/9	5/9	–	15
DNA synthesis (serum-free)	19/25	16/25	13/25	4/25	14,16
Colony formation (serum-free)	7/20	9/20	6/20	2/20	14,16

Ratios indicate quotients of positive responses and total number of cases examined.

tain CSFs and not with others could reflect the physiological diversity of normal hematopoietic progenitors as regards their susceptibility to the hematopoietic growth factors. In this sense, the diversity of response could indicate the selective increase of certain precursors that have accumulated as a result of the differentiation arrest. An alternative hypothesis to explain these phenomena of growth-factor responses is that variations of cell growth are expressions of the transformed status of the cells and that asynchronous maturation is associated with abnormalities (e.g., a loss of responsiveness) of growth-factor regulation. Because our understanding of specific differences of growth-factor regulation between discrete precursor cell stages in normal hematopoiesis is only beginning to emerge, it is impossible to distinguish between these two explanations at the present time.

It is also unclear whether and how the features of AML growth could determine the imbalance of growth between AML and the normal hematopoietic tissue. Direct evidence indicating that the overgrowth of AML in normal marrow is caused by regulatory abnormalities has not been put forward. Preferential growth of AML cells could result from an enhanced susceptibility to one of the hematopoietic growth factors, e.g., due to an overexpression of receptors or their permanent activation by structural receptor abnormalities. Generally the in-vitro susceptibility of human AML to IL-3, GM–CSF, or G–CSF (22,24, 26) and the density of GM–CSF receptors on AML blasts are in the same order as those of normal cells (53,54). For instance, GM–CSF acts optimally at concentrations of 1-5 ng/ml (40-200 U/ml, 60-300 p*M*) (22,24). Acute myeloblastic leukemia blasts express comparatively low levels (8-100 sites/cell) of GM–CSF receptors (53,54).

Another possible way of conferring a growth advantage on AML cells would be the creation of an autocrine loop of stimulation. RNA message of GM–CSF has been demonstrated in the blasts of certain cases of AML (54-56). Colony-stimulating activity was demonstrated in supernatants conditioned by AML cells. This activity could be neutralized by an anti-GM–CSF antibody (56). These data provided evidence for the autocrine production and secretion of GM–CSF in human AML. Indeed, in certain cases, AML cells will form colonies in vitro in the absence of any exogenously supplemented growth stimulus. It remains questionable whether GM–CSF production and GM–CSF response truly reside in one and the same cell of the leukemic clone.

Release of CSFs from monocytes (58,59), fibroblasts (60), and endothelial cells (61-63) is normally induced by specific stimuli, e.g., tumor necrosis factor (63-65), or IL-1 (59-61). A question that has to be addressed is whether the synthesis and release of CSFs by AML cells is controlled by physiological regulators, or whether it is constitutive. It is uncertain how the proposed CSF-mediated autocrine process of stimulation of AML is related to the transformed status of the malignant cells. Autocrine-mediated growth in hematopoiesis is not unusual and does not necessarily represent neoplastic alteration. For example, the production of IL-2 by T lymphocytes and the stimulability of the same cells by this factor, or the synthesis of M–CSF by monocytes, are physiological phenomena (68).

Taken together, no experimental evidence has been obtained to explain why AML growth gains a selective advantage over normal hematopoiesis. Aspects of cell-cell interactions and the susceptibility of progenitor cells to recombinant inhibiting factors have not yet been critically evaluated in experimental contexts of the recombinant hematopoietic growth factors.

C. Cell Maturation

Although the hematopoietic growth factors are highly active in triggering cell proliferation, they usually appear to be incapable of inducing maturation in human AML blasts. Surface markers indicative of progressive myeloid maturation did not appear as the result of incubation of the cells with IL-3, GM–CSF, or G–CSF (15,16,26). Only in rare instances do IL-3, GM–CSF, G–CSF, or M–CSF bring about morphological alterations that are indicative of monocytic or granulocytic maturation (16,34). Thus, although the cells express surface receptors for these factors and their cell cycle can be activated by these molecules, they are generally unable to proceed along the maturation pathway. Why are AML cells incompetent to mature to terminally differentiated granulocytic cells when exposed to the appropriate growth factors? Obviously the answer to this question is fundamental as regards our understanding of neoplastic growth. Because the cells are intact at the level of ligand receptor binding, one may assume that the defect preventing proper maturation resides further downstream in the cell and involves the signal transduction route. Why a normal hematopoietic progenitor cell that responds to growth-factor activation is capable of mounting a sequential series of maturation events and why the leukemic cell is incapable of eliciting a similar response is unresolved and remains a main target of future research.

IV. HEMATOPOIESIS IN ACUTE LYMPHOBLASTIC LEUKEMIA

Acute lymphoblastic leukemia (ALL), like AML, comprises a group of diseases expressing a marked heterogeneity with regard to the stage of maturation of the neoplastic blast cells. Immunophenotypic analysis has revealed two major classes of ALL, i.e., precursor B ALL and T ALL, which can be further subdivided according to additional distinctive immunological features (66,67). To a major extent the immunophenotypes of ALL cells of these subgroups reflect those of the early normal lymphoid developmental stages, although certain discrepancies may exist (67-70). Fresh T ALL and pre-B ALL cells contain a significantly lower proliferative fraction than do their normal counterparts as measured with 5-bromo-2′-deoxyuridine incorporation (71).

A. Growth In Vitro

The development of in-vitro culture methods for ALL has encountered considerably more difficulties than that for AML. Consequently, our insights into the mechanisms controlling the proliferation of ALL cells are still limited. Colonies of precursor B ALL cells can be induced in cultures supplemented with soluble factors produced by PHA-stimulated peripheral blood T lymphocytes and irradiated T lymphocytes as feeder cells (72). The formation of ALL colonies may be enhanced under hypoxic, i.e., 5% O^2, conditions (72). However, these culture conditions are not always beneficial for ALL cell growth (73).

Other methods derived from the colony culture techniques developed for AML, with PHA-activated leukocytes or their supernatants as sources of stimulatory factors, have been used but appeared less efficient for ALL colony formation with respect to the incidence of colony formation, cloning efficiencies, and reproducibility (74-77). The ill-defined sources of stimulatory activity supplemented to the cultures are clearly suboptimal for ALL clonogenic cells, either because they do not contain the proper stimuli at adequate concentrations or because contaminant inhibitor factors interfere. Obviously, the advancement of convenient in-vitro proliferation assays for ALL cells may greatly benefit from the production of pure growth-stimulatory compounds.

B. Lymphoid Growth and Differentiation Factors

1. Activity on Precursor B ALL

The role of growth factors in the proliferation of precursor B ALL cells has been investigated to some extent. Positive effects of IL-2 on precursor B ALL colony growth have been reported in certain cases but only when combined with feeder leukocyte stimulation (74). This would fit with the demonstration that the IL-2 receptors inducible on precursor B ALL cells were functional, i.e., expressing a high affinity for IL-2 (74,78). Nevertheless, a general role of IL-2 as a proliferation inducer of precursor B ALL cells has remained controversial and it has been suggested that other cytokines may have stimulatory influences as well (74,78, 79).

In a recent study the effects of nine recombinant growth factors on the activation of DNA synthesis in precursor B ALL cells in suspension culture have been investigated (80). These compounds included the lymphoid factors mentioned in Table 2 (except for IL-5) as well as the hematopoietic growth factors [except for erythropoietin (EPO)] listed in Table 1. None of these molecules, alone or in various combinations, induced a proliferative response in the ALL cells. The apparent unresponsiveness of the cells to these factors does not rule out that some of them may have profound effects on precursor B ALL cell growth, synergistically with other as yet unknown stimuli. Recently, a new murine bone marrow stroma-derived factor termed IL-7 has been shown to stimulate the proliferation of murine B-cell progenitors (85). The murine IL-7 preparation did not induce proliferation of human precursor B ALL cells (Touw and Löwenberg, unpublished data, 1988), but this lack of activity may be explained by interspecies incompatibility. It will be of great interest to analyze the effects of the human homologue of IL-7 on the proliferation and maturation of ALL cells.

In this regard it is also of note that proliferation of precursor B ALL cells has been induced with a nonrecombinant preparation of B-cell growth factor(s) BCGF (78,79). The activity in this T-cell-derived material is attributed to a 12-kDa protein referred to as low-molecular-weight BCGF. However, this partially purified material also contains other factors (81,100). A more detailed study founded on cytogenetic analyses of cultured ALL samples indicated that the majority of cells proliferating in suspension culture in response to BCGF expressed normal karyotypes, and thus were most likely nonleukemic contaminants (80). When highly purified (by fluorescence cell sorting of CD10 + CD19 stained cells) ALL samples were cultured, only a moderate proliferative response of the cytogenetically abnormal malignant cells was observed (80). These results not only raise doubts as regards an important stimulatory role of BCGF but also stress the importance of separating the ALL cells from the normal cell admixture prior to culture assay, so that artifacts related to interfering normal cell growth are avoided.

2. Action on T ALL Cells

Interleukin-2 may stimulate proliferation of T ALL cells in culture either in combination with leukocyte-conditioned media or with the phorbol ester 12-O-tetradecanoylphorbol-13-acetate (TPA). Apparently, apart from IL-2, other stimuli are needed for proliferation (82,83). TPA was found to induce IL-2 receptors on the membrane of T ALL cells. On the other hand, PHA, a potent activator of mature T lymphocytes and inducer of IL-2 receptors on these cells, exerted no effect in T ALL. It has been suggested that this was characteristic of the immature status of these cells because mature T non-Hodgkin's lymphoma cells were found to be responsive to PHA (82). In agreement with the growth-promoting role of IL-2, inducibility of high-affinity binding of IL-2 to cell-surface receptors on the malignant T cells could be demonstrated (83; Touw and Löwenberg, unpublished data, 1987). On the other hand, T ALL cases not responsive to IL-2 have been reported as well (84), indicating that IL-2 is not a general growth factor for T ALL cells. Little is known of other factors that may control the proliferation of T ALL cells.

The current understanding of the regulation of proliferation in precursor B ALL and T ALL is only fragmentary. At the same time, it is becoming increasingly clear that as yet fully undefined components associated with the bone marrow and thymic stroma may play crucial roles in the regulation of normal and neoplastic B and T lymphoid precursor cell proliferation (86,87). Whether our insight into the mechanism of growth control of ALL will improve in the near future shall mainly depend on the production of growth factors that act on the early cell stages of lymphoid development.

V. SUMMARY

In acute leukemia a neoplastic cell population escapes normal maturation impulses and expands selectively at the expense of normal hematopoiesis. How human acute leukemia cells react in the interplay of regulatory factors that control hematopoiesis and whether growth-factor abnormalities determine abnormal growth is only beginning to become evident. In this chapter developments resulting from the cloning of hematopoietic growth factors and the characterization of hematopoietic progenitor cells have been discussed.

ACKNOWLEDGMENT

This work was supported by The Netherlands Cancer Foundation "Koningin Wilhelmina Fonds."

REFERENCES

1. Johnson, G. R., and D. Metcalf. Pure and mixed erythroid colony formation in vitro stimulated by spleen conditioned medium with no detectable erythropoietin. *Proc. Natl. Acad. Sci. USA 74*:3879-3882 (1977).
2. Fauser A. A., and H. A. Messner. Identification of megakaryocytes, macrophages, and eosinophils in colonies of human bone marrow containing neutrophilic granulocytes and erythroblasts. *Blood 53*:1023-1027 (1979).
3. Messner, A. H., C. A. Izaguiree, and N. Jamal. Identification of T lymphocytes in human mixed hemopoietic colonies. *Blood 58*:402-405 (1981).
4. Axelrad, A. A., D. L. McLeod, M. M. Shreeve, and D. S. Heath. Properties of cells that produce erythrocytic colonies in vitro. In *Hemopoiesis in Culture*, W. A. Robinson (Ed.). U.S. Government Printing Office, Washington, D.C., pp. 226-234 (1974).
5. Gregory, C. J. Erythropoietin sensitivity as a differentiation marker in the hemopoietic system: studies of three erythropoietic colony responses in culture. *J. Cell. Physiol. 89*:289-302 (1976).

6. Stephenson, J. R., A. A. Axelrad, D. L. McLeod, and M. M. Shreeve. Induction of colonies of hemoglobin-synthesizing cells by erythropoietin in vitro. *Proc. Natl. Acad. Sci. USA 68*:1542-1546 (1971).
7. Bradley, T. R., and D. Metcalf. The growth of mouse bone marrow cells in vitro. *Aust. J. Exp. Biol. Med. 44*:287-299 (1966).
8. Pluznick, D. H., and L. Sachs. The cloning of normal "mast" cells in tissue culture. *J. Cell. Physiol. 66*:319-324 (1965).
9. Metcalf, D., H. R. MacDonald, N. Odartchenko, and B. Sordat. Growth of mouse megakaryocyte colonies in vitro. *Proc. Natl. Acad. Sci. USA 72*: 1744-1748 (1975).
10. Dorssers, L., H. Burger, F. Bot, R. Delwel, A. H. M. Geurts van Kessel, B. Löwenberg, and G. Wagemaker. Characterization of a human miltilineage colony stimulating factor cDNA clone identified by a conserved noncoding sequence in mouse interleukin-3. *Gene 55*:115-124 (1987).
11. Yang, Y. C., A. B. Ciarletta, P. A. Temple, M. P. Chung, S. Kovacic, J. S. Witek-Gianotti, A. C. Leary, R. Kriz, R. E. Donahue, G. G. Wong, and S. C. Clark. Human IL-3 (multi-CSF): identification by expansion cloning of a novel hematopoietic growth factor related to murine IL-3. *Cell 47*:3-10 (1986).
12. Bot, F. J., L. Dorssers, G. Wagemaker, and B. Löwenberg. Stimulating spectrum of human recombinant multi-CSF (IL-3) on human marrow precursors: importance of accessory cells. *Blood 71*:1609-1614 (1988).
13. Delwel, R., L. Dorssers, I. Touw, G. Wagemaker, and B. Löwenberg. Human recombinant multilineage colony stimulating factor (interleukin-3): stimulator of acute myeloid progenitor cells in vitro. *Blood 70*:333-336 (1987).
14. Delwel, R., M. Salem, C. Pellens, L. Dorssers, G. Wagemaker, S. Clark, and B. Löweberg. Growth regulation of human acute myeloid leukemia: effects of five recombinant hematopoietic growth factors. *Blood 72*:1944-1949 (1988).
15. Vellenga, E., D. Ostapovics, B. O'Rourke, and J. D. Griffin. Effects of recombinant IL-3, GM–CSF, and G–CSF on proliferation of leukemic clonogenic cells in short term and long term cultures. *Leukemia 1*:584-589 (1987).
16. Löwenberg, B., M. Salem, and R. Delwel. Effects of recombinant multi-CSF, GM–CSF, G–CSF, and M–CSF on the proliferation and maturation of human AML in vitro. *Blood Cells 14*:539-549 (1988).
17. Miyauchi, J., C. A. Kelleher, Y. Yang, G. G. Wong, S. C. Clark, M. D. Minden, S. Minkin, and E. A. McCulloch. The effects of three recombinant growth factors, IL-3, GM–CSF, and G–CSF, on the blast cells of acute myeloid leukemia maintained in short term culture. *Blood 70*:657-663 (1987).
18. Lee, F., T. Yokota, T. Otsuka, L. Gemmell, N. Larson, J. Luh, K. I. Arai, and D. Rennick. Isolation of cDNA for a human granulocyte-macrophage colony stimulating factor by functional expression in mammalian cells. *Proc. Natl. Acad. Sci. USA 82*:4360-4364 (1985).

19. Wong, C. G., J. S. Witek, P. A. Temple, K. M. Wilkins, A. C. Leary, D. P. Luxenberg, S. S. Jones, E. L. Brown, R. M. Kay, E. C. Orr, C. Shoemaker, D. W. Golde, R. J. Kaufman, R. M. Hewick, E. A. Wang, and S. C. Clark. Human GM–CSF: molecular cloning of the complementary DNA and purification of the natural and recombinant proteins. *Science 228*:810-815 (1985).
20. Sieff, C. A., S. G. Emerson, R. E. Donahue, D. G. Nathan, E. A. Wang, G. G. Wong, and S. C. Clark. Human recombinant granulocyte-macrophage colony stimulating factor: a multi-lineage hematopoietin. *Science 230*:1171-1173 (1985).
21. Donahue, R. E., S. G. Emerson, E. A. Wang, G. G. Wong, S. C. Clark, and D. G. Nathan. Demonstration of burst-promoting activity of recombinant GM–CSF on circulating erythroid progenitors using an assay involving the delayed addition of erythropoietin. *Blood 66*:1479-1481 (1985).
22. Metcalf, D., C. G. Begley, G. R. Johnson, N. A. Nicola, M. A. Vadas, A. F. Lopez, D. J. Williamson, G. G. Wong, S. C. Clark, and E. A. Wang. Biologic properties in vitro of a recombinant human granulocyte-macrophage colony stimulating factor. *Blood 67*:37-45 (1986).
23. Burgess, A. W., C. G. Begley, G. R. Johnson, A. F. Lopez, D. J. Williamson, J. J. Mermod, R. J. Simpson, A. Schmitz, and J. F. DeLamarter. Purification and properties of bacterially synthesized human granulocyte-macrophage colony stimulating factor. *Blood 69*:43-53 (1987).
24. Tomonago, M., D. W. Golde, and J. C. Gasson. Biosynthetic (recombinant) human GM–CSF: effect on normal bone marrow and leukemia cell lines. *Blood 67*:31-36 (1986).
25. Griffin, J. D., D. Young, F. Herrmann, D. Wiper, K. Wagner, and K. D. Sabbath. Effects of recombinant human GM–CSF on proliferation of clonogenic cells in acute myeloblastic leukemia. *Blood 67*:1448-1453 (1986).
26. Hoang, T., N. Nara, G. Wong, S. Clark, M. D. Minden, and E. A. McCulloch. Effects of recombinant GM–CSF on the blast cells of acute myeloid leukemia. *Blood 68*:313-316 (1986).
27. Vellenga, E., H. R. Delwel, I. P. Touw, and B. Löwenberg. Patterns of acute myeloid leukemia colony growth in response to recombinant granulocyte-macrophage colony-stimulating factor (rGM–CSF). *Exp. Hematol. 15*: 652-656 (1987).
28. Vallenga, E., D. C. Young, K. Wagner, D. Wiper, D. Ostapovicz, and J. D. Griffin. The effects of GM–CSF and G–CSF in promoting growth of clonogenic cells in acute myeloblastic leukemia. *Blood 69*:1771-1776 (1987).
29. Souza, L., T. C. Boone, J. Gabrilove, P. H. Lai, K. M. Zsebo, D. C. Murdock, V. R. Chazin, J. Bruszewski, H. Lu, K. K. Chen, J. Barendt, E. Platzer, M. A. S. Moore, R. Mertelsmann, and K. Welte. Recombinant human granulocyte colony-stimulating factor: effects on normal and leukemic myeloid cells. *Science 232*:61-65 (1986).

30. Nagata, S., M. Tsuchiya, S. Asano, Y. Kaziro, T. Yamazaki, O. Yamamoto, Y. Hirata, N. Kubota, M. Obeda, H. Nomura, and M. Ono. Molecular cloning and expression of cDNA for human granulocyte colony-stimulating factor. *Nature 319*:415-418 (1986).
31. Bot, F. J., L. van Eijk, P. Schipper, and B. Löwenberg. Effects of interleukin-3 on granulocytic colony forming cells in human bone marrow. *Blood* (in press).
32. Kawasaki, E. S., M. B. Ladner, A. M. Wang, J. van Arsdell, M. K. Warren, M. Y. Coyne, V. L. Schweickart, M. T. Lee, K. J. Wilson, A. Boosman, E. R Stanley, P. Ralph, and D. F. Mark. Molecular cloning of a complementary DNA encoding human macrophage specific colony-stimulating factor (CSF-1). *Science 230*:291-296 (1985).
33. Wong, G. G., P. A. Temple, A. C. Leary, J. S. Witek-Giannotti, Y. C. Yang, A. B. Ciarletta, M. Chung, P. Murtha, R. Kriz, R. J. Kaufman, C. R. Ferenz, B. S. Sibley, K. J. Turner, R. M. Herwich, S. C. Clark, N. Yanai, H. Yokota, M. Yamata, M. Saito, K. Motoyoshi, and F. Takaku. Human CSF-1: molecular cloning and expression of 4 kb cDNA encoding the human urinary protein. *Science 235*:1504-1508 (1987).
34. Salem, M., R. Delwel, L. A. Mahmoud, S. Clark, E. M. Elbasousy, and B. Löwenberg. Maturation of human acute myeloid leukemia in vitro: the response to five recombinant hematopoietic growth factors in a serum-free system. *Br. J. Haematol* (in press).
35. Lin, F. K., S. Suggs, L. H. Liu, J. K. Browne, R. Smalling, J. L. Egrie, K. K. Chen, G. M. Fox, F. Martin, Z. Stabinsky, S. M. Badrawi, P. H. Lai, and E. Goldwasser. Cloning and expression of the human erythropoietin gene. *Proc. Natl. Acad. Sci. USA 82*:7580-7584 (1985).
36. Jacobs, K., C. Shoemaker, R. Rudersdorf, S. D. Neill, R. J. Kaufman, A. Mufson, J. Seehra, S. S. Jones, R. Hewick, E. F. Fritsch, M. Kawakita, T. Shimizu, and T. Miyake. Isolation and characterization of genomic and cDNA clones of human erythropoietin. *Nature 313*:806-810 (1985).
37. Tomonaga, M., I. Jinnai, M. Tagawa, T. Amenomori, K. Nishino, E. Yao, H. Nonaka, K. Kuriyama, Y. Yoshida, T. Matsuo, M. Ichimaru, and T. Suematsu. Leukemic blast cell colony formation in semisolid culture with erythropoietin: a case report of poorly differentiated erythroid leukemia. *Blood 69*:546-552 (1987).
38. Fialkow, P. J., J. W. Singer, J. W. Adamson, K. Vaidya, L. W. Dow, J. Ochs, and J. W. Moohr. Acute nonlymphocytic leukemia: heterogeneity of stem cell origin. *Blood 57*:1068-1073 (1981).
39. Clarkson, B., J. Fried, A. Strife, Y. Sakai, K. Ota, and T. Ohkita. Studies of cellular proliferation in human leukemia. III. Behavior of leukemic cells in three adults with acute leukemia given continuous infusions of ^{3}H-thymidine for 8 or 10 days. *Cancer 25*:1237-1260 (1970).
40. Griffin, J. D., and B. Löwenberg. Clonogenic cells in acute myeloblastic leukemia. *Blood 68*:1185-1195 (1986).
41. Salem, M., R. Delwel, I. Touw, L. Mahmoud, and B. Löwenberg. Human AML colony growth in serum-free culture. *Leuk. Res. 12*:157-165 (1988).

42. Wouters, R., and B. Löwenberg. On the maturation order of AML cells: a distinction on the basis of self-renewal properties and immunologic phenotypes. *Blood 63*:684-689 (1984).
43. Löwenberg, B., and J. J. G. Bauman. Further results in understanding the subpopulation structure in AML: clonogenic cells and their progeny identified by differentiation markers. *Blood 66*:1225-1232 (1985).
44. Marie, J. P., C. A. Izaguirre, C. I. Civin, J. H. Mirro, and E. A. McCulloch. Granulopoietic differentiation in AML blasts in culture. *Blood 58*:670-674 (1981).
45. Touw, I., and B. Löwenberg. Variable differentiation of human acute myeloid leukaemia during colony formation in vitro: a membrane marker analysis with monoclonal antibodies. *Br. J. Haematol. 59*:37-44 (1985).
46. Buick, R. N., M. D. Minden, and E. A. McCulloch. Self renewal in culture of proliferative blast progenitor cells in acute myeloblastic leukemia. *Blood 54*:95-104 (1979).
47. Chang, L. J. A., J. E. Till, and E. A. McCulloch. The cellular basis of self renewal in culture by human acute myeloblastic leukemic blast cell progenitors. *J. Cell. Physiol. 102*:217-222 (1980).
48. Ferrero, D., S. Pessano, G. L. Pagliardi, and G. Rovera. Induction of differentiation of human myeloid leukemias. *Blood 61*:171-179 (1983).
49. Moore, M. A. S., N. Williams, and D. Metcalf. In vitro colony formation by normal and leukemic human hematopoietic cells: characterization of the colony forming cells. *J. Natl. Cancer Inst. 50*:603-623 (1973).
50. Delwel, R., I. Touw, and B. Löwenberg. Towards detection of minimal disease: discrimination of AML precursors from normal myeloid precursors using a combination of surface markers. In *Minimal Residual Disease in Acute Leukemia: 1986*, A. Hagenbeek and B. Löwenberg (Eds.). Martinus Nijhoff Publishers, Dordrecht/Boston, pp. 68-76 (1986).
51. Löwenberg, B., and R. Delwel. Unfrequent phenotypes of clonogenic cells in AML: evidence for abnormal differentiation and possibilities for detecting minimal disease. In *Tumor Differentiation: 1987*. J. Aarbakke, P. K. Chiang, and H. P. Koefler (Eds.). Humana Press, Clifton, NJ, pp. 71-78 (1987).
52. Delwel, R., F. Bot, I. Touw, and B. Löwenberg. Phenotype of acute myelocytic leukemia (AML) progenitors: an approach for tracing minimal numbers of AML cells among normal bone marrow. *Leukemia 2*:814-820 (1988).
53. DiPersio, J. F., P. Billings, S. Kaufman, P. Echtesad, R. E. Williams, and J. Gasson. Characterisation of the human granulocyte-macrophage colony stimulating factor receptor. *J. Biol. Chem. 263*:1837-1839 (1989).
54. Kelleher, C. A., G. C. Wong, S. C. Clark, P. F. Schendel, M. D. Minden, and E. A. McCulloch. Binding of iodinated recombinant human GM–CSF to the blast cells of acute myeloblastic leukemia. *Leukemia 2*:211-216 (1988).
55. Young, D. C., and J. D. Griffin. Autocrine secretion of GM–CSF in acute myeloblastic leukemia. *Blood 68*:1178-1181 (1986).

56. Cheng, G. Y. M., C. A. Kelleher, J. Miyauchi, G. Wang, G. Wong, S. C. Clark, E. A. McCulloch, and M. D. Minden. Structure and expression of genes of GM–CSF and G–CSF in blast cells from patients with acute myeloblastic leukemia. *Blood 71*:204-208 (1988).
57. Young, D. C., K. Wagner, and J. D. Griffin. Constitutive expression of the granulocyte-macrophage colony stimulating factor gene in acute meyloblastic leukemia. *J. Clin. Invest. 79*:100-106 (1987).
58. Rambaldi, A., D. C. Young, and J. D. Griffin. Expression of the M–CSF (CSF-1) gene by human monocytes. *Blood 69*:1409-1413 (1987).
59. Fibbe, W. E., J. van Damme, A. Billiau, P. J. Voogt, N. Duinkerkern, P. M. C. Kluck, and J. H. F. Falkenburg. Interleukin-1 (22-k factor) induced release of granulocyte macrophage colony-stimulating activity from human mononuclear pharocytes. *Blood 68*:1316-1321 (1986).
60. Lee, M., G. M. Segal, and G. C. Bagby. Interleukin-1 induces human bone marrow derived fibroblasts to produce multilineage hematopoietic growth factors. *Exp. Hematol. 15*:983-988 (1987).
61. Zsebo, K. M., V. N. Yuschenkoff, S. Schiffer, D. Chang, E. McCall, C. A. Dinarello, M. A. Brown, B. Altrock, and G. C. Bagby. Vascular endothelial cells and granulopoiesis: interleukin-1 stimulates release of G–CSF and GM–CSF. *Blood 71*:99-103 (1988).
62. Seiff, C. A., S. Tsai, and D. V. Faller. Interleukin 1 induces cultural human endothelial cell production of granulocyte-macrophage colony stimulating factor. *J. Clin. Invest. 79*:48-51 (1987).
63. Broudy, V., K. Kaushansky, G. Segal, J. Harlan, and J. Adamson. Tumor necrosis factor type α stimulates human endothelial cells to produce granulocyte macrophage colony stimulating factor. *Proc. Natl. Acad. Sci. USA 83*:7467-7471 (1986).
64. Munker, R., J. Gasson, M. Ogawa, and H. P. Koeffler. Recombinant human TNF induces production of granulocyte-macrophage colony stimulating factor. *Nature 328*:79-82 (1986).
65. Koeffler, H. P., J. Gasson, J. Ranyard, L. Souza, M. Shepard, and R. Munker. Recombinant human TNFα stimulates production of granulocyte colony stimulating factors. *Blood 70*:55-59 (1987).
66. Foon, K. A., and R. F. Todd. Immunologic classification of leukemia and lymphoma. *Blood 68*:1-31 (1986).
67. Freedman, A. S., and L. M. Nadler. Cell surface markers in hematologic malignancies. *Semin. Oncol. 14*:193-212 (1987).
68. Loken, M. R., V. O. Shah, K. L. Dattilio, and C. I. Civin. Flow cytometric analysis of human bone marrow. II. Normal B lymphocyte development. *Blood 70*:1316-1324 (1987).
69. Anderson, K. C., M. P. Bates, B. L. Slaughenhoupt, G. S. Pinkus, S. F. Schlossman, and L. M. Nadler. Expression of human B cell-associated antigens on leukemias and lymphomas: a model of human B cell differentiation. *Blood 63*:1424-1433 (1984).
70. Greaves, M. F. "Target" cells, cellular phenotypes and lineage fidelity in human leukemia. *J. Cell Physiol. Suppl. 1*:113-125 (1982).

71. Campana, D., and C. Janossy. Proliferation of normal and malignant human immature lymphoid cells. *Blood 71*:1201-1210 (1988).
72. Izaguirre, C. A., J. Curtis, H. Messner, and E. A. McCulloch. A colony assay for blast cell progenitors in non-B non-T (common) acute lymphoblastic leukemia. *Blood 57*:823-829 (1981).
73. Stong, R. C., and J. H. Kersey. In vitro culture of leukemic cells in t (4;11) acute leukemia. *Blood 66*:439-433 (1985).
74. Touw, I., R. Delwel, R. Bolhuis, G. van Zanen, and B. Löwenberg. Common and pre-B acute lymphoblastic leukemia cells express interleukin 2 receptors, and interleukin 2 stimulates in vitro colony formation. *Blood 66*: 556-561 (1985).
75. Uckun, F. M., J. H. Kersey, K. J. Gajl-Peczalska, et al. Heterogeneity of cultured leukemic lymphoid progenitor cells from B cell precursor acute lymphoblastic leukemia (ALL) patients. *J. Clin. Invest. 80*:639-646 (1987).
76. Uckun, F. M., Gajl-Peczalska, J. H. Kersey, L. L. Houston, and D. L. Vallera. Use of a novel colony assay to evaluate the cytotoxicity of an immunotoxin containing pokeweed antiviral protein against blast progenitor cells freshly obtained from patients with common B-lineage acute lymphoblastic leukemia. *J. Exp. Med. 163*:347-368 (1986).
77. Uckun, F. M., K. J. Gajl-Peczalska, D. E. Meyers, et al. Marrow purging in autologous bone marrow transplantation for T-lineage acute lymphoblastic leukemia: efficacy of ex vivo treatment with immunotoxins and 4-hydroperoxycyclophosphamide against fresh leukemic marrow progenitor cells. *Blood 69*:361-366 (1987).
78. Wörmann, B., J. M. Anderson, Z.-D. Ling, and T. W. Le Bien. Structure/function analyses of IL-2 binding proteins on human B cell precursor acute lymphoblastic leukemias. *Leukemia 1*:660-666 (1987).
79. Wörmann, B., S. R. Mehta, A. L. Maizel, and T. W. Le Bien. Low molecular weight B cell growth factor induces proliferation of human B cell precursor acute lymphoblastic leukemias. *Blood 70*:132-138 (1987).
80. Touw, I., J. Groot-Loonen, L. Broeders, T. van Agthoven, K. Hählen, A. Hagemeijer, and B. Löwenberg. Nine recombinant hematopoietic growth factors fail to induce a proliferative response in precursor B acute lymphoblastic leukemia. *Leukemia* (in press).
81. Mehta, S. R., D. Conrad, R. Sandler, J. Morgan, R. Montagna, and A. L. Maizel. Purification of human B cell growth factor. *J. Immunol. 135*: 3298-3302 (1985).
82. Touw, I., R. Delwel, G. E. van Zanen, and B. Löwenberg. Acute lymphoblastic leukemia and non-Hodgkin's lymphoma of T lineage: colony forming cells retain growth factor (interleukin-2) dependence. *Blood 68*:1088-1094 (1986).
83. Uckun, F. M., D. E. Meyers, J. A. Ledbetter, S. E. Swaim, K. J. Gajl-Peczalska, and D. A. Vallera. Use of colony assays and anti-T cell immunotoxins to elucidate the immunobiologic features of leukemic progenitor cells in T-lineage acute lymphoblastic leukemia. *J. Immunol. 140*:2103-2111 (1988).

84. Morishima, Y., Y. Morishita, K. Adachi, M. Tanimoto, R. Ohno, and H. Saito. Phorbol ester induces interleukin-2 receptor on the cell surface of precursor thymocyte leukemia with no rearrangement of T cell receptor β and γ genes. *Blood 70*:1291-1296 (1987).
85. Namen, A. E., S. Lupton, K. Hjerrild, et al. Stimulation of B-cell progenitors by murine interleukin-7. *Nature 333*:571-573 (1988).
86. Kierney, P. C., and K. Dorshkind. B Lymphocyte precursors and myeloid progenitors survive in diffusion chamber cultures but B cell differentiation requires close association with stromal cells. *Blood 70*:1418-1424 (1987).
87. Denning, S. M., J. Kurtzberg, P. T. Le, D. T. Tuck, K. H. Singer, and B. F. Haynes. Human thymic epithelial cells directly induce activation of autologous immature thymocytes. *Proc. Natl. Acad. Sci. USA 85*:3125-3129 (1988).
88. Oppenheim, J. J., E. J. Kovacs, K. Matsushima, and S. K. Durum. There is more than one interleukin 1. *Immunol. Today 7*:45-56 (1986).
89. Smith, K. A. Interleukin 2: inception, impact and implications. *Science 240*:1169-1176 (1988).
90. Maeda, M., N. Arima, Y. Daitoku, et al. Evidence for the interleukin-2 dependent expansion of leukemic cells in adult T cell leukemia. *Blood 70*: 1407-1411 (1987).
91. Touw, I., L. Dorssers, B. Löwenberg. The proliferative response of B cell chronic lymphocytic leukemia to interleukin 2: functional characterization of the interleukin 2 membrane receptors. *Blood 69*:1667-1673 (1987).
92. Karray, S., H. Merle-Béral, A. Vazquer, J.-P. Gerard, P. Debre, and P. Galanaud. Functional heterogeneity of B-CLL lymphocytes: dissociated responsiveness to growth factors and distinct requirements for a first activation signal. *Blood 70*:1105-1110 (1987).
93. Yokota, T., T. Otsuka, T. Mosmann, J. Banchereau, et al. Isolation and characterization of a human interleukin cDNA clone, homologous to mouse B-cell stimualtory factor 1, that expresses B-cell and T-cell stimulating activities. *Proc. Natl. Acad. Sci. USA 83*:5894-5898 (1986).
94. Azuma, C., T. Tanabe, M. Konishi, et al. Cloning of cDNA for human T-cell replacing factor (interleukin-5) and comparison with the murine homologue. *Nucl. Acids Res. 14*:9149-9158 (1986).
95. Hirano, T., K. Yasukawa, H. Harada, et al. Complementary DNA for a noval human interleukin (BSF-2) that induces B lymphocytes to produce immunoglobulin. *Nature 324*:73-76 (1986).
96. Billiau, A. Interferon β2 as a promoter of growth and differentiation of B cells. *Immunol. Today 8*:84-87 (1987).
97. Brakenhoff, J. P. J., E. R. de Groot, R. F. Evers, H. Pannekoek, and L. A. Aarden. Molecular cloning and expression of hybridoma growth factor in Escherichia coli. *J. Immunol. 139*:4116-4121 (1987).
98. Kawano, M., T. Hirano, T. Matsuda, et al. Autocrine generation and requirement of BSF-2/IL-6 for human multiple myelomas. *Nature 332*: 83-85 (1988).

99. Pennica. D., G. E. Nedwin, J. S. Hayflck, et al. Human tumor necrosis factor: precursor structure, expression and homology to lymphotoxin. *Nature 312*:724-279 (1984).
100. Kehrl, J. H., M. Alvarez-Mon, G. A. Delsing, and A. S. Fauci. Lymphotoxin is an important T cell-derived growth factor for human B cells. *Science 238*:1144-1146 (1987).
101. Cordingley, F. T., A. V. Hoffbrand, H. E. Heslop, et al. Tumour necrosis factor as an autocrine tumour growth factor for chronic B-cell malignancies. *Lancet 1*:969-971 (1988).

11
Lymphocytes, Lymphokines, and Hematopoiesis

JOHN M. GARLAND
University of Manchester Medical School, Manchester, England

I. INTRODUCTION

A major thrust of current research into the control of hematopoiesis centers on the soluble growth factors described in other chapters. The evidence is overwhelming that they can indeed mediate both growth and differentiation, from the earliest stem cell through committed progenitors to the final mature progeny. Equally certain is that a number of key growth factors can be elaborated by cells of the immune system [interleukin- (IL)-1, IL-3, IL-4, IL-5, IL-6, granulocyte-macrophage colony-stimulating factor (GM–CSF)], and one (IL-3) appears to be exclusively restricted to T cells ouside of the central nervous system. Furthermore, the general rule is that these powerful proteins are synthesized by one major subset of immunocytes, the T_H cells, which require a mitogenic signal before any significant amounts are produced or the genes coding for them are transcribed (see Sect. IV). The biological implication is that cells of the immune system, particularly T cells, can modulate or control hematopoiesis through the production of hematopoietic growth factors (HPGFs). In addition, the hematopoietic system may itself suffer from a variety of pathologies in which immune responses play a greater or lesser part. Therefore, the evidence also appears to be overwhelming that the immune system, through its T cells and T-derived products, must exert a significant effect on normal hematopoiesis. However, this conclusion is not as evident as first appears, and is perhaps one of the most perplexing areas to examine particularly following the introduction of hematopoietic growth factors, usually of T-cell type, into clinical regimes with the objective of improving hematological status. The perplexity arises from numerous observations of in-vitro marrow culture and circumstantially from lymphocyte-deficient animals in which hematopoiesis appears to be entirely normal in the absence of defined T-cells (and thus their derived soluble factors). This is apparently in contradiction to the power of these products to control hematopoiesis, when used (almost universally) in experimental investigations. The implication is that two tiers of hematopoietic control exist, one exerted by cell-to-cell contact, the other mediated by soluble factors such as T-derived proteins. This is discussed later. This chapter attempts to summarize what is known about the control of hematopoiesis by lymphocytes generally, the specific effects of defined, purified factors on hematopoietic cells are fully discussed in other chapters.

II. GENERATION OF HPGFs BY LYMPHOCYTES

At present, there is little evidence that B cells produce significant amounts of HPGFs, principally because few studies have been performed. However, quite extensive data exist on the effects of T cells on hematopoiesis in vitro. The two-tier system referred to above is also reflected here; one operates through direct cell contact, the other through the soluble factors. Regardless of which tier is involved, activation of the effector function is in response to activation signals presented to the T cell. These may be cell-associated antigens, or peptides processed during the cooperative interaction of T cells with "antigen-presenting cells," usually macrophages. However, the process of signaling does not follow a universal pathway; this raises the question, do all routes to T-cell activation lead to the production of the same soluble factors for a given T cell? The relevance of this is that different cytokines may be produced depending on, for example, whether the cells are responding to processed antigen during infection, or to allogeneic cells during a marrow graft. Almost certainly, different cytokine axes are involved in each, and the end-products are different (activation of neutrophils and macrophages, B-cell help in infection; generation of anti-allogeneic cytotoxic T cells for a graft). It is not difficult to see that an appropriate marrow response in each situation will also be different; for infection, rapid replenishment of myeloid cells and expansion of the B-cell pool, relying heavily on T_H cells and their secreted cytokines; in the graft, generation of cytotoxic T cells with less involvement of T_H cells and minimal call on myeloid homeostasis. The ability of the marrow to differentially communicate with the systemic circulation, and the immune system in particular, is one of the most intriguing areas in immunohematology; its basis is the way in which T cells are induced to respond to antigens. Although signaling generally has been dealt with elsewhere, because T cells are activated by so many molecules and because they represent the focus for elaboration of soluble HPGF, the relationship of T-cell signaling to HPGF gene activation is discussed below.

A. T-Cell Signaling

The process of T-cell signaling is complex, the final event (usually cell division) taking place several days after the initial signal. Much happens during this period, and elaboration of HPGFs is just one event in many. The process of signaling can conveniently be divided into two phases. In the first, the primary recognition event leads to a restricted upregulation or induction of certain genes, in particular transcription of the gene for IL-2 through transactivation of IL-2 promoter regions (1); and the modification of the low-abundance, low-affinity IL-2 receptor (α chain, p55, or tac) to one of high affinity. This is brought about through formation of heterodimers between the α chain, which appears to be upregulated or induced, and the second chain (β, p70), which may be constitutively expressed (1-5). The second phase involves responses of the primed cells expressing high-affinity IL-r to IL-2. The relationship of these two phases to HPGF production is an area requiring further investigation. However, the overall control of the T-cell proliferation is briefly summarized below.

B. T-Cell Priming: Lymphokine Synthesis Versus Cell Division

It is generally accepted that most peripheral T cells are in a "resting" or G_0 phase. They express few IL-2 receptors, do not respond to IL-2, and require activating signals before any progression through the cell cycle is possible. Because the prime function of the T_H cell is to elaborate bioactive peptides or to mediate effects through cellular activities, cell division may be considered as incidental and directed to enlarging the overall capability through clonal expansion. It should also be noted that most of the peptides produced by T cells have extremely short half-lives in vivo, and the system is self-regulating; removal of the stimulus to T-cell activation results in a rapid closing-down of the secondary effectors. Of fundamental interest, therefore, is the relationship of elaboration of factors to the process of signal transduction, its control, and coupling to cell division; it may well be possible for T cells to go through initial activation stages and produce certain factors but without necessarily progressing through division. The apparent contribution of T cells to any inflammatory condition or immune response may not, therefore, be accurately reflected in T-cell numbers or overall degree of activation, as represented, for example, by expression of αIL-2 receptor chains (tac) whose functional activity belongs to the second phase of activation. Even here, the precise relationship of commitment to division, through occupancy of a critical number of IL-2 receptors (6), to secretory functions is not established.

The study of initial responses of T cells to activation signals is also complicated by two other factors: the existence of a number of membrane-bound molecules that can all be involved in transducing a signal, either singly or in cooperation; and the variety of cells used experimentally and the criteria of "activation" applied. Thus, induction of IL-2 and IL-2 receptors by phorbol esters in growth-factor-independent, malignant lines such as Jurkat (7,8) may not faithfully represent events occurring in resting peripheral T cells because they are autonomous for the signals leading from the binding of ligand to the IL-2r to nuclear proliferation. However, certain conclusions may be reached on the kinds of signaling processes used by T cells depending on the stimulus applied. A detailed review on cell signaling is given earlier in this book, and other detailed reviews are available (9,10). The ultimate questions to be addressed are whether all stimuli lead to the same lymphokine production, and if not, why not; and does the immune system select certain cells or stimuli to restrict production of HPGFs? The possibility must be entertained that the "trophic" function of T cells may be divorced from the processes of cell division and specific antigen recognition. Because lymphokine production is associated with cell "activation," the kinds of signal available to the T cell are important because they may determine which lymphokine genes are expressed. The following summary indicates areas to which attention relating lymphokine secretion to T-cell activation is being directed.

III. MOLECULES IN THE T-CELL MEMBRANE ASSOCIATED WITH CELL ACTIVATION

A wide variety of ligands will cause stimulation of T cells. They range from lectins, which interact with specific sugar residues on glycoproteins, thereby potentially crosslinking them, to specific antigen receptors. Table 1 lists the principal membrane molecules that have been seriously examined as candidates for signal transduction. As such, all produce a signal known to be associated with progression of the T cell from its resting G_0 phase into G_1. How far the cell progresses depends on what combination of activities are induced or on the presence of second stimuli, and many studies use antibodies to membrane molecules as "artificial" ligands. Because of the complexity of the signaling cascades, it is not surprising that differing conclusions are reached as to which events are the most important.

A. The T-Antigen Receptor

The most obvious molecular complex generating activation is the antigen receptor (Ti/T3). This consists of the antigen receptor proper, a heterodimer of α/β (or γ/δ) chains closely but noncovalently linked to the T3 complex consisting of a minimum of four chains (γ, δ, ϵ, and τ). The membrane expression of Ti requires T3 (11), a fifth chain (ω) being transiently associated; only T3 has the required length of cytoplasmic "tails" to mediate transduction to the cytoplasm.

Table 1 Membrane Molecules Transducing a Signal in T Cells

Molecule	Examples of activity on binding ligand or antibody	Ref.
IL-1r	Second signal for activation; IL-2 and IL-2r upregulation	174
Ti/T3	Signal transduction by antigens, LK synthesis	11
IL-2r	Activation of early genes, LK synthesis	10
T11 (SRBCr/CD2)	Calcium metabolism, proliferation	37
Fc-IgGr	LK synthesis	175
IL-4r	Activation of tyrosine kinase, proliferation	62
CD3 (LFA-1)	Suppression or enhancement of activation; increase in IL-2, IL-2r	176
CD4	Calcium metabolism, LK synthesis	34
CD5	Accessory function, calcium metabolism	177
CD8	Expression of IL-2, IL-2r	178
Thy1	Similar to T3	179
IL-6r	Second signal for IL-2 production	180
TGF-b	Suppression of T cytotoxic development	181
T44	Accessory to T3	42

LK, lymphokines; SRBC, sheep red blood cell.

However, the Ti/T3 complex has no kinase motif, although T3 is phosphorylated on both serine and tyrosine residues, involving activation of both protein kinase c and an endogenous tyrosine kinase (Tck) when the complex is stimulated (12, 13). Binding of ligand to Ti/T3 results in changes in free calcium, activation of protein kinase c, phosphoinositide metabolism, and upregulation of both IL-2 and IL-2 receptors. The upregulation of the IL-2 receptor occurs with either anti-Ti or anti-T3 antibodies, but IL-2 secretion requires crosslinking of the Ti/T3 complex, furthermore, resting (G_0) T cells also need an initial stimulation with IL-1 before cells respond to the Ti/T3 triggering event (14). Both Ti and T3 can cause alterations in intracellular calcium when probed by antibodies (15); however, with anti-Ti, only antibodies to the antigen binding site are effective, and the level of calcium change induced suggests that the effect is mediated through T3.

The critical stage that decides whether a T cell progresses through the cycle is the expression of high-affinity IL-2 receptors (IL-2r) and production of IL-2. How this is controlled and which of the above cellular changes is responsible is still uncertain. For example, in the factor-independent T-cell Jurkat line, commonly used to analyze T-cell signal transduction, phorbol ester induction of IL-2 production needs costimulation with anti-T3 or lectin, both of which cause a rise in intracellular calcium (16). However, mutants can be isolated in which anti-T3 still produces a transient rise in calcium without production of IL-2 (17). In this case, the maintenance of the raised intracellular calcium appears to be the critical factor, because only wild-type Jurkat cells maintain the higher levels. In fresh T cells, there also appear to be two phases of cell activation. The first, associated with intracellular calcium, is mediated by lectins, calcium ionophores, or anti-Ti/T3 in which sustained calcium rises are important (7,18). The second is induced by translocation and activation of protein kinase c associated with metabolism of inositol phosphate (18,19). The precise mechanisms by which these two essential changes are brought about is uncertain. Ti chains can activate phosphoinositide (PI) turnover (20-22), perhaps through interaction with G proteins (23-26), and T cells possess the correct subspecies of protein kinase C that respond sensitively to one of the products of phosphoinositide metabolism, diacylglycerol (27,28). The results of these two activations results in phosphorylation of various proteins (29,30) and the production of high-affinity IL-2 receptors and secretion of IL-2.

B. Alternative Receptors or Molecules Activating T Cells

The above discussion has dealt with cell stimulation by protein-kinase-c activators and the Ti/T3 complex, leading to IL-2/IL-2r signaling. Very little is known of the mechansm of signaling by other ligands. Particular interest has been focused on the CD4 molecule, because this is implicated as the receptor for human immune deficiency virus. It is related to the immunoglobulin gene family (31), appears to be complexed to a kinase related to sarc (32), and is phosphorylated when T cells are stimulated (33). Possibly, CD4 is a regulatory molecule for the recognition of class II antigens acting through Ti/T3 and can be either stimulatory or inhibitory, appearing to act through calcium regulation (34,35). Of the other receptors, most appear to influence PI or calcium metabolism.

Table 1 indicates that there are many potential activation signals to which T cells can respond. Although they provide evidence for alternative pathways of T-cell activation, few have been investigated in detail to examine whether they all induce the same cluster of lymphokines. The most studied is the signal delivered through the T11 complex [CD2, sheep red blood cell (SRBC) receptor] which consists of a 50 kDa peptide expressing T11-1, T11-2, and T11-3 epitopes. Its ligand is the LFA-3 antigen present both on erythrocytes and on T cells themselves (36). The T11 receptor is developmentally expressed before the Ti/T3 complex. The T11-3 epitope is not present on resting T cells but is induced by interaction of a ligand with T11-2 or by binding SRBC (37,38); the SRBC binding site is confined to the T11-1 epitope. Potent cell activation in the absence of IL-1 or accessory cells is mediated by combinations of antibodies to T11-2 + T11-3 epitopes, but no stimulation is produced by antibodies to T11-1. The stimulation by anti T11-2 + T11-3 is independent of exogenous IL-2, but results in IL-2 secretion that upregulates itself (15,38,39). T11 is not associated with the T3 complex, although in thymocytes crosslinking the Ti/T3 complex results in abrogation of T11 responses. Stimulation through the T11 receptor is accompanied by changes in transmembrane calcium flux, protein-kinase-c activation, and inositol metabolism (15), but the receptor is not itself a calcium channel as shown by studies in which the T11 molecule has been transfected into non-T cells (40); however, stimulation through T11 is accompanied by increases in cyclic adenosine monophosphate similar to those induced by other adenyl cyclase activators (41). Stimulation through T11 is thought mainly to be involved in natural killer cell (NK) activity, or where cell-cell contact is required for activation.

In addition to the T11 molecule itself, an associated 44 kDa protein (Tp44, CD28) that exists as a 90 kDa heterodimer appears to be capable of modulating responses of both the CD3 and T11 receptors (42-44). Its function is not clear, because although it can activate resting T cells and replace certain requirements for PMA, it does not translocate protein kinase c nor synergize with calcium ionophores, and produces only transient calcium changes with moderate changes in PI metabolism. Of the other molecules delivering a stimulatory signal, little is known concerning their stimulation of HPGF gene expression in T cells.

C. Proliferation Driven by IL-2

The precise way in which the IL-2 receptor signals growth is not known. The receptor is itself devoid of a recognizable kinase domain despite the induction of protein phosphorylation, furthermore, it does not activate PI metabolism (45). Although expression of both α and β receptor chains are normally required for maximal signaling, the β chains can deliver a signal on their own (46). Interleukin-2 can itself upregulate the IL-2r (47), which leads to further increased cellular response. Other proliferation-associated genes, e.g., the transferrin receptor, are upregulated by IL-1/IL-2r interaction but may be independent of the receptor itself (48). However, once the cell has expressed high-affinity IL-2r, it appears to be fully competent to respond to IL-2 by eventually proliferating, a process that involves sequential changes in metabolism, RNA synthesis, protein synthesis, and then entry into S phase (49-51). Again, responses may be classified into "early" or "late," the early including very rapid transcription of c-*fos*, c-*myc*, and heat-shock proteins (HSPs) (minutes to 3 hr) (52), whereas lymphokine transcription and autoregulation of the IL-2r gene are essentially late (3-6 hr induction). Proliferation after IL-2 signaling occurs only after many hours, and thus lymphokine gene activation and secretion is a preproliferative response.

IV. TRANSCRIPTION OF HEMATOPOIETIC GROWTH FACTOR GENES IN T CELLS

Of the current array of HPGFs, the following are elaborated by T cells: GM-CSF, IL-3, IL-4, IL-5, IL-6. M- and G-CSF are produced by a variety of other nonlymphoid cells, including macrophages and fibroblasts or endothelial cells. Granulocyte-macrophage colony-stimulating is unusual in being an almost ubiquitous growth factor, IL-3 in being restricted virtually to T cells; both have very similar target cells in the marrow, but are quite different structurally, are not co-regulated, and appear to function through different mechanisms on the target cells. Two aspects are central to T-cell control of HPGFs: transcriptional regulation itself, and the state of T-cell activation that initiates the change in regulation.

A. Transcriptional Regulation

Detailed reviews have been given elsewhere in this book concerning the individual control regions of HPGF genes. The most pertinent observations are that those capable of being synthesized by T cells possess regions of DNA homology 5′ to the coding regions, the CK1 and CK2 enhancer motifs (Chapter 4). These motifs are shared by several lymphokine genes, including IL-2 and interferon-γ (IFN-γ). Furthermore, although the coding sequences for the individual genes differ between mouse and humans, the CK enhancer sequences are remarkably conserved. Recently, it has been shown that the CK1 sequences from different growth-factor-enhancer regions can bind the same protein (53,54). Therefore, one might expect coordinate regulationof HPGFs possessing these sequences in cells in which one is activated. However, this does not appear to be the case in T cells.

B. Noncoordinate Expression

In the mouse, two subsets of T-helper cells have been described based on the array of lymphokines secreted by cell lines (55,56). Table 2 lists the division. Both kinds secrete IL-3. Th1 secrete high amounts of GM-CSF and no IL-4, and Th2 secrete low amounts of GM-CSF and high amounts of IL-4. Th1 cells respond to IL-2, and Th2 cells respond to IL-4. However, within this grouping there appears to be no obligatory coordinate regulation of the secretion of lymphokines, despite the close proximity of several of the genes to each

Table 2 Phenotypes of Murine T Helper Cells

Th1	Th2
IL-2	IL-4
GM–CSF	Low GM-CSF
TNF	Low TNF
IL-3	IL-3
No IL-5	IL-5
High IFN	Low IFN

Source: Data from Ref. 55 and Mosmann et al., *J. Immunol. 136*:2348 (1986).

other, as mentioned elsewhere. Thus, in bulk cultures even within a cloned population, the ratios of secreted GM-CSF to IL-3 vary widely (57). Other evidence suggests, however, that transcription may be linked, because different T-cell stimuli induce coexpression of several lymphokine genes (58,59). At the single cell level also, murine T cells may differentially express individual HPGF (60), which is a stable attribute within a single-cell-derived sibling clone. This suggests that the homologous enhancers shared between the various growth-factor genes are recessive to individual promoters, and that lymphokine secretion is genetically "imprinted" on a clone. Although most experiments have been done in vitro, there is evidence that the same distinction of lymphokine-producing T-cell phenotypes exist in vivo, as revealed by frequency of cells responding to IL-2 or IL-4 (61). This, therefore, has a bearing on the complexity of potential control of hematopoiesis by T-cell-derived factors, particularly because the Th1 and Th2 subclasses respond to different stimuli and may themselves be controlled differently, e.g., by steroids (62).

In human T cells, the distinction between the Th1 and Th2 classes of T cells is much less clear. A large survey of clones from different sources (peripheral blood, tonsil, spleen) showed that although there were preferences for the grouping of lymphokine (IL-2, IL-4, IFN-γ) secretion, it was not clear-cut (63). Furthermore, the human GM-CSF promoter does not seem to require the 5′ flanking enhancer sequences for expression in T cells (64); the promoter is inactive in B cells. One possible explanation for discrepancies between murine and human cells, or between clones, is the ability of T cells to self-modulate their activities through secretion of factors, for example as illustrated by IL-2. The decision as to which lymphokines are transcribed and secreted may therefore rest with which receptors are being used and which autostimulatory factors are produced as a result of the primary transduction signal.

C. Temporal Expression

Although the general model of T-cell proliferation centers around the IL-2/IL-2r concept, secretion of lymphokines is a function of prereplicative T cells. Furthermore, which lymphokines are secreted may depend on the stimulatory signal. First, the mere acquisition of the IL-2r and secretion of IL-2 may not be entirely sufficient to promote growth; in some cases interaction of ligand with the TI/T3 antigen receptor is also needed (65). In addition, not all clones are able to produce detectable IL-2 in culture supernates in response to activation signals (59,60); this, however, may reflect quantitative changes. Second, the activation signal itself may determine lymphokine secretion. In human T cells, PHA alone induces HSP70 mRNA; PHA or PMA induces c-*fos*, c-*myc*, and IL-2r; PHA with PMA further induces IL-2 and IFN-γ; and exogenous IL-2 has little effect on IL-2 or IFN-γ but increases c-*myc*, IL-2r, and HSP (66). Lu et al. found that both IFN-γ and tumor necrosis factor (TNF)-α enhanced production of G- and GM-CSF in PHA-stimulated human T cells; this enhancing effect occurred only in Th cells (67). In murine T-cell clones, the ratio of GM-CSF to IL-3 also depends on the stimulus applied (68).

Resting T cells have no detectable mRNA for hematopoietic lymphokines. Induction occurs within a few hours of a mitogenic stimulus, and is generally insensitive to protein synthesis inhibitors such as cycloheximide, suggesting that chemical modification of preexisting inactive inducers or active repressors via signal transducers is responsible for initiation of transcription (60). In murine T cells promoted by either IL-2 or antigen, IL-2, IL-3, and IFN-γ mRNAs were coinduced during the first 6 hr (69). Secretion of transcribed lymphokines follows expression of the mRNA very quickly and before cells enter the replicative cycle; secretion is independent of cell-cycle inhibitors such as hydroxyurea (60).

D. Relationship to Signal Transduction

The indication above that HPGFs are expressed early but are variously subject to the kind of stimulus applied raises several questions. For example, can lymphokines be secreted without the cell proceeding through the cycle? How tightly linked is the induction of HPGF gene expression to different receptors that may all provide a stimulatory signal? Are the same regulatory proteins eventually always used regardless of the receptors transducing a signal? The difficulty of providing precise answers is evident from the array of different signals transduced with a T-cell population as reviewed above, and from the difficulty of assessing lymphokines produced. Thus, for example, many studies use a bioassay of crude supernates for secreted GM-CSF or IL-3, which are prone to error and lack standardization. Although proliferation may not itself be required for lymphokine secretion, different stimuli are undoubtedly of different potency in stimulating proliferation, which needs to be taken into account, especially in experiments lasting several days, many stimuli generate self-perpetuating loops of factors that may synergize.

There is little precise information on the direct coupling of different receptors to activation of HPGF genes, or of their relationship to the cell cycle, insofar as inhibition of the cycle itself by exogenous reagents does not give much information on whether cycle-inducing or phase transition proteins (70,71) are still produced. These may be part of the activation signals required for HPGF expression; nor is it clear whether the same kinds of signals leading to "cell activation" as revealed by IL-2r/IL-2 mRNA expression are obligatory and the same for all HPGF expression. The length of time a signal is applied may be of importance, both in stimulating the cell and in maintaining the levels of expressed mRNA (48,50). Furthermore, once the realm of protein-kinase-c activation is entered, matters are complicated by whether it is activated with or without translocation (72) or is itself regulated by the exposure to the activator during the course of the experiment (73-75). Evidence has been presented that at least one HPGF gene may use alternative promoters (76), and coordinate induction may involve negative regulation; expression of the IL-2 gene involves binding of a protein induced by activation to enhancer regions of the DNA and loss of other DNA–binding proteins present in uninduced cells (77). Relative stability of mRNA is also important. In the case of murine Th clones, mRNA transcription for IL-3 and IL-2 was initiated some 2 hr after stimulation with *Concanavalin A* (78). However, the mRNA for IL-3 and IFN-γ were more stable than that for IL-2. The potential complexities of relating signaling to specific gene expression can be illustrated with HTLV infection of T cells. Human T-lymphotropic virus type I (HTLV-I) infection results in constitutive secretion of GM-CSF, which is nevertheless still susceptible to reguation by protein-kinase-c activation (51). The transactivating proteins of HTLV (p40x and p37xII) have no direct action on GM-CSF expression but do activate the IL-2r. Reporter genes containing the 5′ flanking regions of the IL-2 and IL-2r (α) can be activated by the p40x protein of HTLV similarly to the activation induced by mitogens or anti-Ti/T3 (79); however, the action of p40x here on the IL-2r gene appears to be indirect, possibly by modifying an existing DNA-binding factor (80) operating in a region with high homology to the Ig k-β enhancer. The existence of consensus sequences shared, for example, between GM-CSF and IL-2 thus is not sufficient in itself to decide on coordinate control even within the same signaling process. Because even constitutive expression of a gene does not necessarily remove regulation, there would appear to be potentially numerous mechanisms of normal induction of growth factors in T cells.

Overall, much needs to be established concerning which factors are induced by which signals in T cells. Taking account of the precision with which all cell processes are controlled, it appears unlikely that T cells deliver an uncontrolled and indiscriminate burst of lymphokines whenever they are activated. The fine-tuning of T-cell-derived HPGF elaboration is perhaps one area where a focus of attention can be expected.

V. LYMPHOCYTE CONTROL OF HEMATOPOIESIS

An influence of lymphocytes on in-vitro and in-vivo hematopoiesis has been known for some time. Classical studies of the W/W^v mutant mouse, which is defective in spleen colony-forming cells and has a macrocytic anemia, showed that T cells could be implicated in correcting the defect. Thus the inability of T-cell-depleted normal marrow to cure W/W^v is restored by the addition of small numbers of thymocytes or T cells (81). Incidentally, the replacement also cures an associated defect in mast cells. Because the W/W^v mouse appears to have normal T-cell function, the effect of donor T cells on donor marrow suggests that the T cells directly interact in the marrow engraftment process. Similarly, experience with human marrow transplantation indicates that removal of all marrow T cells ("stripping") is associated with significantly reduced engraftment. However, this apparent facilitation of hematopoiesis by T cells appears to be contradicted by other observations, thus, the well-known nu/nu mouse is defective in classical T cells but appears to have normal hematopoiesis, and marrow cultures in which T cells are absent can be maintained for many months. These aspects are briefly discussed below.

A. Hematopoiesis and T Cells in In-Vivo Models

Mention has already been made of the W/W^v mouse. Although thymocytes can successfully restore curative properties to T-cell-depleted marrow, processing by a thymus is not required (82). Furthermore, the marrow of defective mice still contains normal colony-forming unit (CFU)-c and multipotential stem cells revealed by a CFU-mix assay (83,84), although classical CFUs and "transient erythroid" spleen colonies are defective (85). This association of T cells with hematopoietic recovery in the W/W^v mouse is not yet explained; it seems unlikely to be due to lymphokine deficiency because spleen cells are quite capable of generating IL-3 and GM-CSF (J. M. Garland, unpublished data). In the thymus-deficient nu/nu mouse, hematopoiesis is apparently normal, but there is evidence that nu/nu mice do produce a small number of T-like cells (86,87). However, even in normal animals the thymus can be shown to exert a modulating effect on hematopoiesis; thymectomy is accompanied by changes in marrow stem cells in both mice and rats (88-93), and indeed the thymus is a hematopoietic organ in birds (94). Some studies suggest that thymic hormones are important hematopoietins (95), whereas others suggest that thymectomy reduces the capacity of

T cells for generating lymphokines even if their numbers are partially restored (96). In vivo activation of the T cell system also has a stimulatory effect on stem cell proliferation (97). Stimulation of stem cells in vivo might be expected to result in a panleukocytosis, but conversely stimulation of stem cells in vitro by a marrow non-T-cell-derived stimulator before transplantation inhibits B lymphocyte responses (as assayed by splenic plaque-forming cells), perhaps by biasing differentiation toward the erythroid series (98); cell differentiation is inextricably interwoven with stimulation by growth factors in vivo, and the result of stimulating stem cells into cycle cannot be predicted.

Further supporting evidence for a role for T cells in hematopoiesis comes from studies in vitro.

B. Effects of Lymphocytes on In-Vitro Hematopoiesis

With the knowledge that T cells elaborate numerous HPGFs, it would not be surprising if T cells were able to influence in-vitro colony assays, which are controllable and relatively easy to perform. However, several aspects need to be considered when using in vitro colonies to determine effects of T cells. First, non-T cells are able to elaborate HPGFs such as GM-CSF, G-CSF, and M-CSF. This may be stimulated by CSFs used to develop the assay (99). In certain situations, a complex self-stimulatory loop may be envisioned: GM-CSF may stimulate IL-1 production by neutrophils (100), which in turn stimulates both T cells and macrophages to produce more GM-CSF (101) and other factors. Second, evidence is accumulating that certain non-T marrow cells express growth factor receptors normally associated with T cells. Thus, early-stage myeloid cells have been reported to express the IL-2 α chain receptor (102) and may respond to IL-2 by suppressing growth (103). Mention has already been made of IL-4, which is both required as a growth factor and elaborated by a subset of T cells; IL-4 has significant synergistic effects on marrow cells where it augments responses to G-CSF and erythropoietin (Epo) (104). Experimental results may therefore depend partly on the content of Th2 cells in the marrow. Additionally, myeloid progenitor cells may share T-cell antigens, for example Thy1 (105), which may bias marrow samples treated to remove T cells.

Despite considerations such as these, T cells undoubtedly affect colony formation, particularly those from conditions in which an autoimmune origin of hemopoietic disease is suspected (106). Generally, T cells are suppressive, particularly to burst-forming units (BFU-E), but it is not clear whether this is due to direct interaction with progenitor cells or through elaboration of inhibitory factors such as IFN. Also, inhibition of colony growth can be effected by NK cells or large granular lymphocytes (LGLs). Of interest is the fact that thymomas may be accompanied by severe primary aplastic anemia, suggesting an involvement of the human thymus in hematopoiesis similar to that described in studies in rodents. Certainly, specific T suppressor cells with an Fc-γ-positive phenotype operative on CFU have been described (107) although the mechanism of cytotoxicity has not been elucidated.

The role of normal T cells in modulating normal marrow is similarly complex, but it does indicate some of the mechansism that may be functional or enhanced in autoimmune or aplastic conditions.

1. Effects of Normal T Cells on BFU-E

It is well known that the major source of burst-promoting activity (BPA) is the stimulated T cell (108). T cells can interact with BFU-E in both an MHC-restricted (109) and nonrestricted way (110), and may be suppressive or inhibitory. In the study of Estrov et al. (110), removal of marrow T cells, using CAMPATH depletion and T-cell rosetting, reduced by 50% marrow-derived BFU-E, which was restored by readdition of T cells. However, another similar study using T cells and hematopoietic progenitors derived from peripheral blood found no effect of T cells on BFU-E growth (111). The significance of this difference is unknown, but could reflect subtle differences between marrow and peripheral blood progenitors. For example, marrow and peripheral blood CFU-GM differ in their sensitivity to prostaglandins, marrow CFU-GM being inhibited by prostaglandin-E_1 at concentrations from 10^9 M to 10^{-5} M whereas peripheral CFU-GM are stimulated (112). In other studies, T cell stimulation of BFU-E requires the presence of monocytes/adherent cells (113,114), which may themselves increase BFU-E (115). These results may be attributed to the elaboration of BPA, which varies between different kinds of T cells, T-Ig Fc-μ-receptor-positive cells producing more than T-Ig Fc-γ phenotypes (116). A recent report (117) interestingly demonstrates that BPA can be stimulated by high concentrations of IL-2 on T and NK cells, suggesting that the β chain is competent alone to induce BPA expression.

The stimulatory effect of T cells in vitro appears to be paralleled by studies done in specific pathogen-free mice. These mice have significantly fewer marrow BFU-E, which is not affected by treatment with anti-thy antibody, and low marrow T cells; BFU-E are increased by addition of thymocytes, but the "helper" cell does not appear to be a thy-1 +ve cell (118). In contrast to these are various other studies showing that stimulated T lymphocytes are inhibitory to BFU-E; activation of T cells by either IL-2, after inducing IL-2r with anti-CD3 antibody, or by anti-T11 antibody that activates an alternative pathway of T stimulation (see Section III and Table 1), was profoundly suppressive to BFU-E (119,120), an effect mediated by IFN-γ.

2. Effects of Normal T Cells on CFU-GM

The effect of T cells on CFU-GM again appears to differ widely between reports, and depends on whether the T cells have been activated or are derived from peripheral blood or marrow. In some reports, marrow T cells appear to stimulate CFU-GM (121) in the presence of IL-2, although adherent cells also had a similar potentiating action. Such results are explicable on secretion of HPGF, such as GM-CSF, IL-3, M–CSF, or IL-1 by cooperative interaction of lymphokine cascades activated by cells within the assay. The potentiating action of human adherent cells is opposite to that of murine adherent cells, which inhibit murine CFU-c (122). Other studies (123) demonstrate that IL-2-stimulated T cells are inhibitory to CFU-GM. Also, peripheral blood T cells/NK cells/LGLs appear to inhibit normal CFU-GM (124,125). Overall, no definitive picture emerges from such studies on CFU-GM, perhaps because CFU-GM are so heterogeneous and respond to a variety of factors, most of which can be synthesized within the colony itself.

3. Possible Effector Mechanisms of T-Cell Modulation of Hematopoiesis In Vitro

First, despite the above observations, numerous if disparate observations demonstrate that mature T cells are not needed for hematopoiesis in vitro. Long-term marrow cultures can be maintained in the absence of T cells, and although depletion of thy-1 +ve cells has been reported to block initiation (126), long-term cultures can be initiated from nu/nu marrow (127); the effect of anti-thy-1 may therefore be more directly on progenitor cells themselves (105). Such studies do not rule out a role for immature or prethymocytes, however.

Several soluble modulators (besides HPGF) have been described that have a T-cell origin, or result from interaction of T cells with accessory cells. Numerous studies implicate IFN-γ as an important negative regulator of in-vitro colonies derived from all marrow lineages (128-131). Different lineages respond differently, human GEMM and BFU-E being equally susceptible to IFNs α, β, and γ, whereas CFU–GM are relatively insensitive to IFNs α and β (128). Interferon has been shown to decrease the self-renewal capacity of normal and leukemic CFU (131), which would be expected to reduce colony size. However, IFN does not completely block colony formation and very high doses have been used (1000 U/ml) to obtain high levels of colony inhibition. Other inhibitors, therefore, appear to be involved. Among them are prostaglandins (132) and leukotrienes (133); both may be stimulatory or inhibitory depending on concentration. T cells, PBLs, and T lines may produce a GM–CFU inhibitor lymphokine (134). In addition to these, there is an array of other lymphokines/growth factors, for example TNF-α, TGF-β, IL-6/IFN-β2, and neuropeptides (adrenocorticotropin, endorphins), all of which may be associated with T-cell modulation of hematopoietic responses by either acting directly on progenitors, if T-cell-derived, or stimulating/inhibiting T cells to elaborate known HPGF. For example, β-endorphin [a product of the proopio-melanocorticotrophin (POMC) gene] inhibits production of a T-cell-derived neutrophil chemotactic factor (135); at the same time, T lymphocytes themselves express the POMC gene when stimulated with mitogens and can be activated by opioid peptides (136,137). Furthermore, classic HPGF such as GM–CSF and IL-3 may directly upregulate IL-2 α-chain receptors on T cells (138), disproportionately increasing the potential for activating secondary mediators, particularly if IL-2 is used as an inducer or amplifier of T-cell function. In a recent study of the interaction of IL-2 with lymphocytes and marrow cells after allogeneic marrow transplant, T-cell-derived IFN-γ and TNF were significantly increased, and both the T cells and soluble products were markedly suppressive to CFU (139).

The principal difficulty in analyzing the role of T cells in hematopoiesis is the complexity of both the hematopoietic cell system, described in earlier chapters, and of the multiple array of HPGF that may be secondarily involved. This is well illustrated in the long-term marrow culture system, where although stromal cells may be able to produce classical HPGF, their role appears to be slight (140,141). Again, few studies have employed serum-free conditions; in the study by Sonoda et al. (142), null-cell, T-depleted, serum-free marrow was stimulated by recombinant HPGFs. Single HPGFs produced few colonies; mixtures of IL-3 with other CSFs (M-, G-, and GM-CSF) were the most effective, but no combination produced GEMM colonies, a result differing from many obtained in standard colony assays run in serum. In another study (143), human IL-3 was poorer at developing GM colonies but better at developing erythroid bursts in the absence of fetal bovine serum. Both studies emphasize the greater efficiency of colony induction when combinations of HPGF are used.

The exact role of intact T cells in progenitor cell growth and differentiation is thus unclear, and awaits further experimentation in which defined progenitor populations can be plated with pure factors in serum-free conditions, and soluble products from T cells or the HPGF genes expressed in them can be monitored. The effects of varying the amounts and nature of HPGF secreted by T cells clearly has important implications for the types of colony generated, and their size and capacity for self-renewal. The activation state of the T cells and their potential reactivity to various ligands thus becomes a deciding factor in the way they may be able to influence hematopoietic development. The experimental design, therefore, is difficult, and it is only recently that such experiments have become technically feasible.

C. T Cells in Bone Marrow Transplantation

Progenitor cells are compartmentalized between the marrow and peripheral blood. Progenitors in the periphery clearly have an opportunity to interact with T cells or their soluble products; is the same true for T cells in the marrow, which might be expected to have greater significance because the numbers of progenitor cells are vastly greater? First, the numbers of T cells with mature phenotype in human marrow is surprisingly high; 20-25% by cone biopsy, and up to 30% by suction aspiration (144,145). Ratios of Th to Ts are lower than in peripheral blood (0.6 in marrow, 1.6 in blood). Although most appear to have a mature phenotype, there is evidence that they are defective in providing T helper function, and by inference produce less lymphokines. However, IL-2 was originally discovered as a product of marrow T cells (146). A lowered ability to mediate in immune function is also reflected in peripheral blood after transplantation, when although T cell numbers and ratios reach normality fairly rapidly, they exhibit long-lasting functional defects (147,148). Whether this represents inability to fully process the antigen repertoire after the manipulations involved in transplantation, or is a defect at the T stem-cell level in marrow a long time after it has ceased embryonic/neonatal differentiation into various compartments of stem cells, is not known. However, as in the murine models above, T cells do appear to be involved in marrow engraftment. Experience with removing T cells from donor marrow to reduce GvH reactions is accompanied by significant reductions in graft "take" (149,150). Because the marrow environment after grafting is essentially host (151), and because donor stem cells appear to lodge within the marrow by a lectin-like recognition process (152), a direct contribution by T cells to lodgment seems unlikely. These observations suggest a "trophic" function for T cells, perhaps mediated by lymphokines generally and HPGF in particular.

D. Interrelationships Between T Cells and Accessory Cells

No discussion concerning T cells and their products is complete without reference to the mutual axes existing between accessory cells and T cells, and the role of other cytokines such as IL-1. This interaction is obligatory for T-dependent antigen proliferation, during which HPGFs are elaborated. The mechanism is not known, but it is possible that T cells signal macrophages to synthesize IL-1 (153). However, the effect of IL-1 on hematopoietic cells can be both indirect and direct. Interleukin-1 is a powerful radioprotectant in vivo (154) and induces stem-cell cycling (155). In vitro, IL-1 is a strong synergistic factor for high proliferative potential (HPP) stem cells, but again there is a division of HPP stem cells into those that respond to different combinations of HPGF (GM–CSF, M–CSF, and IL-3) (156) and those that respond to hematopoietin 1, which appears to be IL-6 (157,158). In addition, the effect of IL-1 in inducing other HPGFs from non-T cells must be taken into account (101). The combination of T cells, macrophages, and cross-reactive growth factors is thus a potentially very complicated network to unravel.

E. Evidence for Regulation of Progenitor Cells by T Cells In Vivo

If a role for T-cell-derived growth factors in hematopoiesis is postulated, it should be possible to identify conditions in which T cells make an observable contribution to differentiation or multiplication of precursors in vivo through secreting HPGF. Identification of IL-5 (see Chapter 8) was brought about through the use of T cells from animals displaying eosinophilia, in which the T cells appear to be specifically upregulated for IL-5. In the case of mast cells already discussed, evidence exists of the dual regulation of progenitor cells by both T cells and stromal factors. Thus, $S1/S1^d$ mice have a genetic defect making marrow stromal cells incompetent to sustain stem cells. They are also defective in skin mast cells. The defect in skin mast cells appears to be reversible by stromal elements from normal mice capable of migrating into skin (159). Evidence exists that a common progenitor cell can give rise to both connective-tissue type (e.g., skin) and gut-associated mast cells (160), of which different subpopulations respond to IL-3 in combination with IL-4 (connective-tissue type) or IgE (gut-associated type) (161,162). Although there are no data on the subsets of lymphocytes associated with mast-cell development or deficiencies and the spread of lymphokines they produce in vivo, gut T cells from normal mice infected with parasites produce increased amounts of IL-3 (163), which correlates with increased mucosal mast cells. The relationship between mast cells, marrow cells, T cells and W/W^v mouse thus appears to be that athymic mice lack T modulator/inducer cells that operate through the secretion of HPGF, W/W^v mice have a progenitor defect that cannot be corrected by HPGF (164). This cannot be the whole answer, however, because such mice can still generate mastlike IL-3-dependent lines from marrow in vitro (164; J. M. Garland, unpublished data).

VI. SUMMARY

It may be conclusively said that lymphocytes (T cells) produce an array of very powerful hematopoietic peptides, at least one of which (IL-3) appears to be exclusive. All are able to induce proliferation of stem cells of varying commitment, and in combination can reproduce virtually the entire sequence of hematopoietic cell production from the earliest stem cell, for example the HPP stem cells, which responds to T lymphokines in vitro or IL-3 in vivo (166,167), to a fully differentiated myeloid cell. Many T lymphokines can be demonstrated to exert highly significant changes in both hematopoiesis and lymphopoiesis in vivo. Thus, IL-3 suppresses NK development, thought to be partly responsible for hybrid resistance (168,169), as well as augmenting Ig responses [perhaps through an action on pre-B cells (170)] and stimulating hematopoiesis (171). The effect on hematopoiesis is probably direct, because infused IL-3 is readily detectable when bound to the surfaces of cells both in the marrow and spleen (172). It is therefore somewhat surprising that the regulatory role of T cells themselves is less well defined. This arises variously from the availability of purified or recombinant products that replace crude T cell supernates, the complexity of analyzing a multicell system in which very many growth factors cannot be controlled for experimentally, and a certain lag in understanding the basic biology of T cells. This last is perhaps due to the historical recognition of the T cell as an immunocyte, which had led it to be principally investigated in an immunological context. The advent of molecular genetics has resulted in a reappraisal of the interrelationships between many cells, and growth-factor technology has been led to the realization that immune reactivity is perhaps only one function of the T-cell system, and that the classic concept of "immune surveillance" is only a part of a perhaps much wider "surveillance" system in which peptide growth factors and T cells play an exceedingly important role. A pertinent observation concerns abrogation of renal allograft rejection in rats by preoperation injection of donor blood. Although the allografts survive well, the graft is infiltrated by allospecific and nonspecific cytotoxic T cells and expresses donor class I and II antigens (173) in a manner equal to that of a rejecting graft. What, therefore, is the role of lymphokines in this situation—Which of them are expressed, do they activate target cells, and what is the network that controls the macrophages and dendritic cells involved in allograft rejection?

It therefore may be expected that much closer attention will be given to the mechanisms by which T cells elaborate their hematopoietic effectors, and to under what circumstances (and how) they may modulate the known control of hematopoiesis. It is the progress in molecular and cellular biology of lymphocytes and CSFs that now makes the complexities open to analysis.

REFERENCES

1. Durand, D. B., M. R. Bush, J. G Morgan, A. Weiss, and G. R. Crabtree. A 275 base-pair region at the 5′ end of the IL-2 gene enhances expression from a heterologous promoter in response to signals from the T cell antigen receptor. *J. Exp. Med. 165*:395-407 (1987).
2. Meidel, M. L., J. D. Hulmes, D. R. Weber, P. Balon, and Y.-C. Pan. Structural analysis of the recombinant soluble human IL-2 receptor. Primary structure, assignment of disulfide bonds and core IL-2 binding structure. *Biochem. Biophys. Res. Commun. 154*:372-379 (1988).
3. Ishi, T., T. Takeshita, N. Numata, and K. Sugamura. Protein phosphorylation mediated by IL-2/IL-2 receptor β chain interaction. *J. Immunol. 141*: 177-179 (1988).
4. T. Tsuchida, and T. Sakane. Intracellular activation signals required for the induction of IL-2 responsiveness in resting T cell subsets in humans. *J. Immunol. 140*:3446-3449 (1988).

5. Nishi, M., Y. Ishida, and T. Honjo. Expression of functional Interleukin 2 receptors in human light-chain/tac transgenic mice. *Nature 331*:267-269 (1988).
6. Malek, T. R., T. D. Ashwell, R. N. Germain, E. M. Shevach, and J. Miller. The murine Interleukin 2 receptor: biochemical structure and regulation of expression. *Immunol. Rev. 92*:81-101 (1986).
7. Goldsmith, M. A., and A. Weiss. Early signal transduction by the antigen receptor without commitment to T cell activation. *Science 240*:1029-1031 (1988).
8. Greene, W. C., W. J. Leonard, Y. Wand, P. Svetlik, N. J. Peffer, J. G. Sodroski, C. A. Rosen, W. C. Goh, and W. A. Haseltine. The transactivator gene of HTLV II induces IL-2 receptor and IL-2 cellular gene expression. *Science 232*:877-880 (1986).
9. Farrar, W. J., D. K. Ferris, and A. Harel-Belan. Lymphokine molecular signal transduction (in press).
10. Möller, G. (Ed.) Activation antigens and signal transduction in lymphocyte activation. *Immunol Rev. 95* (1987).
11. Weiss, A., J. Imboden, K. Hardy, B. Manger, C. Terhorst, and J. Stobe. Role of T3/antigen receptor complex in T cell activation. *Ann. Rev. Immunol. 4*:593-619 (1986).
12. Berkhout, B., B. Alarcon, and C. Terhorst. Transfer of genes encoding the T cell receptor: association of the CD3 complex into cos cells results in assembly of the macromolecular structure. *J. Biol. Chem. 263*:8528-8536 (1988).
13. Samelson, I. F., M. D. Patel, A. M. Weissman, J. B. Harford, and R. D. Klausner. Antigen activation of murine T cells induces tyrosine phosphorylation of a polypeptide associated with the T cell antigen receptor. *Cell 46*: 1083-1090 (1986).
14. Scheurich, P., U. Ugur, M. Wrann, and F. Pfizenmaier. Early activation during primary activation of T cells: antigen receptor cross-linking and IL-1 initiate the proliferative response of human T cells. *Eur. J. Immunol. 15*: 1091-1095 (1985).
15. Alcover, A., D. Ramarli, N. E. Richardson, H. C. Chane, and G. C. Reinherz. Functional and molecular aspects of human T cell lymphocyte activation via T3/Ti and Th pathways. *Immunol. Rev. 95*:5-36 (1987).
16. Linch, D., D. C. Wallace, and K. O'Flynn. Signal transduction in human T cells. *Immunol. Rev. 95*:137-159 (1987).
17. Weiss, A., R. L. Wiskoril, and J. D. Stobo. Role of T3 surface molecules in the activation of human T cells. *J. Immunol. 133*:123-128 (1984).
18. Manger, B., A. Weiss, J. Imboden, T. Laine, and J. D. Stobo. Role of protein kinase c in transmembrane signaling by the T cell antigen receptor complex: effects of stimulation with soluble or immobilized CD3 antibody. *J. Immunol. 139*:2755-2760 (1987).
19. Tsuchido, T., and T. Sakane. Intracellular activation signals required for the induction of IL-2 responsiveness in resting T cells. *J. Immunol. 140*: 3446-3449 (1988).

20. Treves, S., F. DiVereiho, V. Cerudolo, P. Zanello, D. Colavo, and T. Pozzan. Calcium and inositol phosphates in the activation of T-cell mediated cytotoxicity. *J. Exp. Med. 166*:33-42 (1987).
21. Bismuth, G., F. Faure, I. Theodorou, P. Debre, and T. Hercend. Triggering of the phosphoinositide transduction pathway by a monoclonal antibody specific for the human γ/δ cell receptor. *J. Immunol. 138*:1135-1138 (1988).
22. Panteleo, G., D. Olive, A. Polgi, T. Pozzan, L. Moretta, and A. Moretta. Antibody-induced modulation of the CD3/Ti receptor complex causes T cell refractoriness by inhibiting the early metabolic steps involved in T cell activation. *J. Exp. Med. 166*:619-624 (1987).
23. Wakelam, J. J., S. A. Davies, M. D. Houslay, T. McKay, C. J. Marshal, and A. Hall. Normal p21-N-ras couples bombesin and other growth factor receptors to inositol phosphate production. *Nature 323*:173-176 (1986).
24. O'Shea, J. J., K. R. Urdal, H. T. Chong, T. M. Chused, L. F. Samelson, and R. D. Klausner. Aluminum fluoride induces phosphoinositide turnover, alteration in cytoplasmic free calcium and phosphorylation of the T cell antigen receptor in murine T cells. *J. Immunol. 139*:3463-3469 (1987).
25. Cockroft, S. Polyphosphoinositol phosphodiesterase: regulation by a novel guanine nucleotide binding protein Gp. *Trends Biochem. Sci. 12*:75-78 (1987).
26. Aussel, C., D. Mary, J. F. Peyron, C. Pelassy, B. Ferrua, and M. Fehlmann. Inhibition and activation of Interleukin 2 synthesis by direct modification of guanosine triphosphate-binding proteins. *J. Immunol. 140*:215-220 (1988).
27. Y. Nishizuka. The molecuar heterogeneity of protein kinase c and its implications for cellular regulation. *Nature 334*:661-665 (1988).
28. Shearman, M. S., N. Berry, T. Oda, K. Ase, U. Kikkawa, and Y. Nishizuka. Isolation of protein kinase c sub-species from preparations of human T lymphocytes. *FEBS Lett. 234*:387-391 (1988).
29. Chatila, T. A., and R. S. Geha. Phosphorylation of T cell membrane proteins by activators of protein kinase c. *J. Immunol. 140*:4308-4314 (1988).
30. Freidrich, B., and M. Gullberg. The role of protein kinase c in early activation versus growth of T lymphocytes. *Eur. J. Immunol. 18*:485-488 (1988).
31. Maddon, P. J., D. R. Littman, M. Godfrey, D. G. Maddon, L. Chess, and R. Axel. Isolation and nucleotide sequence of the cDNA encoding the T cell surface protein T4; a new member of the Ig gene family. *Cell 42*: 93-104 (1985).
32. Rudd, L. E., J. M. Trevillyan, J. D. Dasgupta, L. L. Wong, and S. F. Schlosman. The CD4 receptor is complexed in detergent lysates to a protein kinase (pp58) from human T cells. *Proc. Natl. Acad. Sci. USA 85*:5190-5194 (1988).
33. Blue, M. L., D. A. Hafler, K. A. Craig, H. Levine, and S. F. Schlossman. Phosphorylation of CD4 and CD8 molecules following T cell triggering. *J. Immunol. 139*:3949-3954 (1987).

34. Carrel, S., A. Moretta, G. Pantelo, G. Tambussi, P. Isler, B. Perussia, and J.-C. Cerotini. Stimulation and proliferation of CD4 positive peripheral blood T lymphocytes induced by an anti-CD4 monoclonal antibody. *Eur. J. Immunol. 18*:333-339 (1988).
35. Ledbetter, J. A., C. H. June, P. S. Rabinovitch, A. Grossman, T. T. Tsu, and J. B. Imboden. Signal transduction through the CD4 receptor: stimulatory versus inhibitory activity is regulated by CD4 proximity to the CD3/T cell receptor. *Eur. J. Immunol. 18*:525-532 (1988).
36. Plunkett, M. L., M. E. Sanders, P. Selvaraj, M. L. Dustin, and T. A. Springer. Rosetting of activated human T lymphocytes with autologous erythrocytes. Definition of the receptor and ligand molecules as CD2 and lymphocyte-function associated antigen 3. *J. Exp. Med. 165*:664-676 (1987).
37. Meier, S. C., R. E. Hussey, M. Fabri, D. Fox, O. Acuto, S. F. Schlossman, and E. L. Reinhertz. An alternative pathway of T cell activation: a functional role for the 50 KDa T11 sheep erythrocyte receptor protein. *Cell 36*:897-906 (1984).
38. Scott, C. F., J. M. Lambert, R. S. Kalish, C. T. Morimoto, and S. F. Schlossman. Human T cells can be directed to lyse tumour targets through the activation of the T11-E rosette receptor pathway. *J. Immunol. 140*:8–14 (1988).
39. Hu, J., C. Vaquero, S. Huet, A. Bernard, and G. Sterkers. Interleukin 2 upregulates its own production. *J. Immunol. 139*:4109-4115 (1987).
40. Alcover, A., H. C. Chang, P. H. Saire, R. E. Hussey, and E. Reinhertz. The T11 (CD2) cDNA encodes a transmembrane protein which encompasses T11-1, T11-2 and T11-3 epitopes but which does not independently mediate Ca^{++} influx. Analysis by gene transfer in a baculovirus system. *Eur. J. Immunol. 18*:363-367 (1988).
41. Carrera, A. C., M. Rincon, M. D. DeLandazuri, and M. Lopez-Botet. CD2 is involved in the regulation of cAMP levels in T cells. *Eur. J. Immunol. 18*:961-964 (1988).
42. Weiss, A., B. Manger, and J. Imboden. Synergy between the T3/antigen receptor and Tp44 in the activation of human T cells. *J. Immunol. 137*:819-825 (1986).
43. Moretta, A., Panteleo, G., M. Lopez-Botet, and L. Manger. Involvement of T44 molecules in an antigen-dependent pathway of T cell activation: analysis of the correlation to the T cell antigen receptor complex. *J. Exp. Med. 162*:823-838 (1985).
44. Pierres, A., M. Lopez, C. Cerdan, J. Nunes, D. Olive, and C. Mawas. Triggering of CD28 molecules synergises with the CD2 (T11-1 and T11-2)-mediated T cell activation. *Eur. J. Immunol. 18*:685-690 (1988).
45. Mills, G. B., D. J. Stewart, A. Mellors, and E. W. Gelfand. IL2 does not induce phosphoinositide hydrolysis in activated T cells. *J. Immunol. 136*:3019-3034 (1986).
46. Sich-Thuy, L., M. Dukovitch, N. J. Peffer, A. S. Fauci, J. H. Kehrl, and W. J. Greene. Direct activation of human resting T cells by IL2: the role

of an IL2 receptor distinct from the tac protein. *J. Immunol. 139*:1550-1556 (1987).

47. Depper, J. M., W. J. Leonard, C. Drogula, M. Kronke, T. A. Walmann, and W. C. Greene. Interleukin 2 augments transcription of the IL2 receptor gene. *Proc. Natl. Acad. Sci. USA 82*:4230-4334 (1985).
48. Kumagai, N., S. H. Benedict, G.-B. Mills, and E. W. Gelfand. Requirement for simultaneous presence of phorbol ester and calcium ionophore in the expression of human T lymphocyte proliferation and related genes. *J. Immunol. 139*:1393-1399 (1987).
49. Macdonald, H. R., and M. Nabholz. T cell activation. *Annu. Rev. Cell Biol. 2*:231-253 (1986).
50. Weiss, A., R. Shields, M. Newton, B. Manger, and J. Imboden. Ligand receptor interactions required for commitment to the activation of the IL2 gene. *J. Immunol. 138*:2169-2176 (1987).
51. Greene, W. C., and W. J. Leonard. The human IL2 receptor. *Annu. Rev. Immunol. 4*:69-95 (1986).
52. Farrar, W. L., J. L. Cleveland, S. Beckner, E. Bonvini, and S. W. Evans. Biochemical and molecular events associated with interleukin 2 regulation of lymphocyte proliferation. *Immunol. Rev. 92*:49-65 (1986).
53. Shannon, M. F., J. R. Gamble, and M. A. Vadas. Nuclear proteins interacting with the promoter regions of the human granulocyte-macrophage colony stimulating factor gene. *Proc. Natl. Acad. Sci. USA 85*:674-678 (1988).
54. Shannon, M. F. *VIth. International Lymphokine Workshop*, Evian (1988).
55. Cherwinski, H. M., J. H. Schumaker, K. D. Brown, and T. R. Mosmann. Two types of mouse helper T cell clones: III: further differences in lymphokine synthesis between Th1 and Th2 clones revealed by RNA hybridisation. *J. Exp. Med. 166*:1229-1244 (1987).
56. Kurt-Jones, E., S. Hamberg, J. O'Hara, W. Paul, and A. K. Abbas. Heterogeneity of helper/inducer T lymphocytes. I. Lymphokine production and lymphokine responsiveness. *J. Exp. Med. 166*:1774-1787 (1987).
57. Kelso, A. An assay for colony stimulating factor (CSF) produced by single T lymphocytes: establishment of the frequency of cells producing GM–CSF and multi-CSF within a T lymphocyte clone. *J. Immunol. 136*:2930-2937 (1986).
58. Prystowsky, M., J. M. Ely, D. Beller, G. Eisenberg, J. Goldman, F. Goldwasser, J. N. Ihle, J. Quintas, H. Remold, S. N. Vogfel, and F. Fitch. *J. Immunol. 129*:2337-2344 (1982).
59. Kelso, A., and A. L. Glasebrook. Secretion of interleukin 2, macrophage activating factor, interferon and colony-stimulating factor by alloreactive T lymphocyte clones. *J. Immunol. 132*:2924-2931 (1982).
60. Kelso, A., and N. Gough. Expression of hematopoietic growth factor genes in murine T lymphocytes. *Lymphokines 13*:209-238 (1987).
61. Powers, G., A. K. Abbas, and R. A. Miller. Frequency of IL2 and IL4 secreting T cells in naive and antigen-stimulated lymphoid populations. *J. Immunol. 140*:3352-3357 (1988).

62. Bertoglio, J. H., and E. Leroux. Differential effects of glucocorticoids on the proliferation of a murine helper and a cytolytic T cell clone in response to IL2 and IL4. *J. Immunol. 141*:1191-1196 (1988).
63. Maggi, G., G. DePrete, D. Macchia, P. Parronchi, A. Tiri, J. Chretien, M. Ricci, and S. Romagnani. Profiles of lymphokine activities and helper function for IgE in human T cell clones. *Eur. J. Immunol. 18*:1045-1050 (1988).
64. Nimer, S. D., E. P. Morita, M. J. Martis, W. Wachsman, and J. C. Gasson. Characterisation of the human GM–CSF promoter region by genetic analytical correlation with DNAse I footprinting. *Mol. Cell. Biol. 8*:1979-1984 (1988).
65. Goronzy, J., C. Weyand, J. Imboden, B. Manger, and C. G. Fathman. Heterogeneity of signals required in T cell activation within a panel of human proliferating T cell clones. *J. Immunol. 138*:3087-3093 (1987).
66. Granelli-Piperno, A., L. Andrus, and R. M. Steinman. Lymphokine and non-lymphokine mRNA levels in stimulated human T cells. Kinetics, mitogen requirement and effects of cyclosporin A. *J. Exp. Med. 43*:922-935 (1986).
67. Lu, L., E. F. Srour, D. J. Warren, D. L. Walker, L. D. Graham, E. B. Walker, J. Jansen, and H. Broxmeyer. Enhancement of release of G– and GM–CSF from PHA stimulated sorted subsets of human T lymphocytes by recombinant human TNFα; synergism with recombinant human IFNγ. *J. Immunol. 141*:201-207 (1988).
68. Kelso, A., D. Metcalf, and N. M. Gough. Independent regulation of GM–CSF and multilineage CSF production in T lymphocyte clones. *J. Immunol. 136*:1718-1725 (1986).
69. Herold, K. C., D. W. Lancki, D. E. Dunn, K. Arai, and F. Fitch. Activation of lymphokine genes during stimulation of T cell clones. *Eur. J. Immunol. 16*:1533-1538 (1986).
70. Lee, M. G., L. J. Norbury, N. K. Spurr, and P. Nurse. Regulated expression and phosphorylation of a possible mammalian cell cycle protein. *Nature 333*:676-679 (1988).
71. Baserga, R., D. E. Waechter, K. T. Soprano, and N. Galanti. Molecular biology of cell division. *Ann. N.Y. Acad. Sci. 397*:110-120 (1982).
72. Mukaida, N., H. Yagisawa, T. Kawai, and T. Kashara. Role of protein kinase c activation in signal transduction by IL2. *Biochem. Biophys. Res. Commun. 154*:187-193 (1988).
73. Borner, C., U. Eppenberger, R. Wyss, and D. Fabro. Continuous synthesis of protein kinase c related proteins after down-regulation by phorbol ester. *Proc. Natl. Acad. Sci. USA 85*:2110-2114 (1988).
74. Murray, A. M., A. Fournier, and S. J. Hardy. Proteolytic activation of protein kinase c: a physiological reaction. *Trends Biochem. Sci. 12*:53-54 (1987).
75. Kajikawa, N., A. Kishimoto, M. Shiota, and Y. Nishizuka. Ca^{++}-dependent neutral protease and proteolytic activation of Ca^{++}-activated, phospholipid dependent protein kinases. *Methods Enzymol. 102*:279-290 (1987).

76. Stanley, E., D. Metcalf, P. Sobieszcuk, N. M. Gough, and A. R. Dunn. Structure and expression of the murine gene encoding granulocyte-macrophage colony stimulating factor: evidence for utilisation of alternative promoters. *EMBO J. 4*:2569-2573 (1985).
77. Nabel, G. J., C. Gorka, and D. Baltimore. T cell specific expression of IL2; evidence for a negative regulatory site. *Proc. Natl. Acad. Sci. USA 85*:2934-2938 (1988).
78. Yokota, T., S. Miyatake, H. Hagiwara, T. Mosmann, J. Conaway, R. Conaway, A. Miyajima, Y. Takeba, N. Arai, F. Lee, and K. Arai. Isolation and characterisation of the mouse IL3 gene and its expression in activated T cells. *Lymphokines 15*:393-408 (1988).
79. Maruyama, M., H. Shibuya, H. Harada, M. Hatakeyama, J. Fujita, J. Inoue, M. Yoshida, and T. Taniguchi. Evidence for the aberrant activation of an IL2 autocrine loop by HTLV-I encoded p40x and Ts/Ti complex triggering. *Cell 48*:343-350 (1987).
80. Ruben, S., H. Poteat, T.-H. Tan, K. Kawagami, R. Roeder, W. Haseltine, and C. Rosen. Cellular transcription factors and regulation of the IL2 receptor gene expressed by HTLV-I tax gene product. *Science 241*:89-92 (1988).
81. Wiktor-Jedrzejczak, W., S. Sharkis, A. Ahmed, and K. W. Sell. Theta-sensitive cell and erythropoiesis: identification of a defect in W/W^v anaemic mice. *Science 196*:313-315 (1977).
82. Harrison, D. E., C. M. Astle, and J. A. DeLaittre. Processing by the thymus is not required for cells that cure and repopulate a W/W^v recipient. *Blood 54*:1152-1157 (1979).
83. Hara, H., K. Ohne, K. Noguchi, K. Nagai, K. Tsuyama, and Y. Kitamura. Presence of pluripotent hematopoietic presursors in vitro (CFU-mix) in hematopoietic tissues from mice of W/W^v phenotype. *Cell Tissue Kinet. 15*:25-30 (1983).
84. Sawada, U., and S. Adler. In vitro hematopoiesis of W/W^v and ++ marrow cells alone and with thymocytes. *Exp. Hematol. 8*:702-708 (1980).
85. Gregory, C. J., E. A. McCulloch, and J. E. Till. Transient erythroid spleen colonies: effects of erythropoietin in normal and genetically anaemic W/W^v mice. *J. Cell. Physiol. 86*:1-8 (1975)'
86. Dennert, G., and R. Hyman. Functional thy 1 +ve cells in cultures of spleen cells from nu/nu mice. *Eur. J. Immunol. 10*:583-588 (1980).
87. Gillis, S., and J. Watson. IL2 induction of antigen-specific cytolytic T cells in nude mice. *J. Immunol. 126*:1245-1248 (1981).
88. Sharkis, S. J., M. I. Collector, R. C. Meacher, and L. C. Sensenbrenner. Effect of neonatal thymectomy on proliferation of mouse bone marrow stem cells. *Exp. Hematol. 8*(suppl. 7):40-41 (1988).
89. Lipscombe, H. L., and J. G. Sharp. Alterations in bone marrow cellularity following thymectomy in rats. *Experientia 35*:1401-1406 (1979).
90. Lepault, F., M. P. Fache, and E. Frindel. Effect of thymectomy and splenectomy on pluripotent stem cells and progenitors of the myeloid and B lymphoid lineages. *Leuk. Res. 4*:663-670 (1980).

91. Petrov, R. Factors controlling stem cell differentiation. *Tsitologiia 5*: 602-609 (1979).
92. Resnitsky, P., D. Zipori, and N. Trainin. Effect of neonatal thymectomy on hematopoietic tissues in mice. *Blood 37*:634-646 (1971).
93. Goodman, J. W., N. L. Basford, and S. G. Shinpock. On the role of the thymus in hematopoietic differentiation. *Blood Cells 4*:53-64 (1978).
94. Kendall, M. D., and J. A. Frazier. Ultrastructural studies on erythropoiesis in the avian thymus. *Cell Tissue Res. 199*:37-61 (1979).
95. Zipori, D., and N. Trainin. Role of thymic humoral factors in the proliferation of bone marrow CFU-s from thymectomised mice. *Exp. Hematol. 3*:389-398 (1975).
96. Duprez, V., R. Mariarz, O. Weinberger, and S. J Burakoff. Thymectomised, irradiated and bone-marrow reconstituted chimeras have normal cytolytic T precursors but a defect in lymphokine production. *J. Immunol. 132*:2185-2189 (1984).
97. Koskov, V. A., I. A. Orlouskaya, and I. G. Tsyrcova. Effect of hematopoietic stem cell proliferation on the immune response in mice. *Cell Tissue Kinet. 20*:485-491 (1987).
98. Frindel, E., E. Leuchars, and A. J. S. Davies. Thymus dependence of bone marrow stem cell proliferation in response to certain antigens. *Exp. Hematol. 4*:275-281 (1976).
99. Horiguchi, J., M. K. Warren, and D. Kufe. Expression of the macrophage-specific colony stimulating factor in human marrows treated with granulocyte-macrophage colony stimulating factor. *Blood 69*:1259-1261 (1987).
100. Lindemann, A., D. Reidel, S. C. Oster, S. C. Meuer, R. M. Blohn, R. M. Mertelsmann, and F. Hermann. Granulocyte-macrophage colony stimulating factor induces IL1 production by human polymorphonuclear neutrophils. *J. Immunol. 140*:837-839 (1988).
101. Fibbbe, J. VanRamme, A. Biliau, P. J. Voogt, N. Kinkerken, P. M. Kluck, and J. H. Falkenburg. IL1 (22 Kd factor) induces the release of granulocyte-macrophage colony stimulating factor from human mononuclear phagocytes. *Blood 68*:1316-1321 (1986).
102. Armitage, R. J., A. P. Lai, P. J. Roberts, and J. C. Cawley. Certain myeloid cells express receptors for IL2. *Br. J. Haematol. 64*:799-807 (1986).
103. Naldini, A., W. R. Fleischmann, Z. K. Ballas, K. D. Klimpel, and G. R. Klimpel. IL2 inhibition of in vitro granulocyte-macrophage colony formation. *J. Immunol. 139*:1880-1884 (1987).
104. Peschel, C., W. E. Paul, J. O'Hara, and I. Green. Effects of B cell stimulating factor/IL4 on human hematopoietic progenitor cells. *Blood 70*: 254-263 (1987).
105. Miller, R. A., J. M. Lipton, D. L. Linch, S. J. Burakoff, and D. G. Nathan. Thy 1 is a differentiation antigen that characterises immature murine erythroid and myeloid hematopoietic progenitors. *J. Cell. Physiol. 123*: 25-32 (1985).

106. Mangan, K. F. Immune disregulation of hematopoiesis. *Annu. Rev. Med. 38*:61-70 (1987).
107. Bacigalupo, A., A. Podesta, M. C. Mingari, L. Moretta, G. Piaggio, M. J. VanLint, A. Durando, and A. M. Marmont. Generation of CFU-c suppressor cells in vitro: an experimental model for immune-mediated marrow failure. *Blood 57*:491-496 (1981).
108. Westbrook, C., J. C. Gasson, S. E. Gerber, M. E. Selsted, and D. W. Golde. Purification and characterisation of human T-lymphocyte derived erythroid potentiating activity. *J. Biol. Chem. 269*:9992-9996 (1984).
109. Torok-Storb, and J. A. Hansen. Modulation of in vitro BFU-e growth by normal Ia +ve T cells is restricted by HLA-DR. *Nature 298*:473-474 (1983).
110. Estrov, Z., C. Roifman, Y.-P. Wang, T. Grunberger, E. W. Gelfand, and M. H. Freedman. Regulatory role of IL2-responsive T lymphocytes on early and mature erythroid progenitor proliferation. *Blood 67*:1607-1610 (1986).
111. Nomdedeu, B., R. S. Gormus, and M. Banisadre. Human peripheral blood erythroid BFU: evidence against a T-lymphocyte requirement for proliferation in vitro. *Exp. Hematol. 8*:845-852 (1980).
112. Richman, C. M., and E. D. Johnson. Differences between peripheral and bone marrow CFU–GM: granulocyte and macrophage progenitor cells from peripheral blood and bone marrow differ in their responses to PG-E1. *Blood 70*:1792-1796 (1987).
113. Mangan, K. F., and J. F. Desforges. The role of T lymphocytes and monocytes in the regulation of human erythropoietic peripheral burst forming units. *Exp. Hematol. 8*:717-727 (1980).
114. Zuckerman, K. S. Human erythroid burst forming units: growth in vitro is dependent on macrophages but not on lymphocytes. *J. Clin. Invest. 67*: 702-709 (1981).
115. Murphy, M. J., and A. Urabe. In *In Vitro Aspects of Hematopoiesis*, M. J. Murphy (Ed.). Springer-Verlag, New York, p. 189 (1979).
116. Mangan, K. F., G. Chikkapa, and W. B. Scharfman. Modulation of erythropoietic peripheral BFU-e proliferation by T lymphocytes bearing Fc receptors for IgM or IgG. *Exp. Hematol. 8*(suppl. 7):45 (1980).
117. Skettino, S., J. Phillips, L. Camer, A. Nagler, and P. Greenberg. Selective generation of erythroid burst-promoting activity by recombinant human interleukin 2-stimulated human T lymphocytes. *Blood 71*:907-914 (1988).
118. Kanamura, A., S. Durban, M. Gallagher, S. Miller, and J. J. Trentin. Augmentation of erythroid burst formation by addition of thymocytes and other myelo-lymphoid cells. *J. Cell. Physiol. 104*:187-197 (1980).
119. Burdach, S., and L. J. Levitt. Receptor-specific inhibition of bone marrow erythropoiesis by recombinant DNA-derived interleukin 2. *Blood 69*:1368-1371 (1987).

120. Burdach, S., M. Shatsky, B. Wagenhort, and L. Levitt. The T cell CD2 determinant mediates inhibition of erythropoiesis by the lymphokine cascade. *Blood 72*:770-775 (1988).
121. Estrov, Z., C. Rofman, C. Mills,T. Grunberger, E. W. Gelfand, and M. H. Freedman. Regulatory role of IL2 responsive T lymphocytes in human marrow granulopoiesis. *Blood 69*:1161-1166 (1987).
122. Zipori, D., and J. Sasson. Adherent cells from mouse bone marrow inhibit formation of CSF-induced myeloid colonies. *Exp. Hematol. 8*:816-817 (1986).
123. Burdach, S., M. Shatsky, and L. J. Levitt. Receptor-specific modulation of myelopoiesys by the recombinant DNA-derived interleukin 2. *J. Immunol. 139*:452-457 (1987).
124. Frassoni, F., M. Podesta, M. T. VanLint, A. Marmont, E. Rossi, R. Vimercati, and A. Bacigalupo. GM–CSF growth in chronic granulocytic leukaemia is not affected by soluble inhibitors released by aplastic anaemia T cells or mitogen-primed normal T lymphocytes. *Br. J. Haematol. 50*: 647-653 (1982).
125. Kerndrup, G., and P. Hokland. NK cell mediated inhibition of bone marrow colony formation (CFU–GM) in refractory anaemia (pre-leukaemia): evidence for patient-specific cell populations. *Br. J. Haematol. 69*:457-462 (1988).
126. McMillan, M. A., and R. Simmons. A T-lymphocyte role in Dexter-type marrwo cultures. *Br. J. Haematol. 48*:171-172 (1981).
127. Garland, J. M., and T. M. Dexter. Expression of 20-α hydroxysteroid dehydrogenase in murine long-term bone marrow cultures. *Eur. J. Immunol. 12*:998-1001 (1982).
128. Broxmeyter, H. E., L. Lu, E. Platzer, L. Feit, L. Juliano, and B. Y. Rubin. Comparative analysis of the influences of human α, β and γ interferons on multipotential (CFU–GEMM), erythroid (BFU–E) and granulocyte-macrophage (CDF–GM) progenitor cells. *J. Immunol. 131*:1300-1305 (1983).
129. Rigby, W. F. C., E. D. Ball, P. M. Guyre, and M. W. Fanger. Effects or recombinant DNA-derived interferons on the growth of myeloid progenitor cells. *Blood 65*:858-863 (1985).
130. Van T'Hull, E., H. Shellekens, B. Lowenberg, and M. J. DeVries. Influence of interferon preparations on the proliferative capacity of human and mouse bone marrow cells in vitro. *Cancer Res. 38*:915-900 (1978).
131. Taetle, R., R. N. Buick, and E. A. McCulloch. Effect of interferon on colony formation in culture by blast cell progenitors in acute myeloid leukaemia. *Blood 56*:549-552 (1980).
132. Gomagne, J., M. Adolphe, M. Semichon, L. Zizine, and P. Lechat. Effects of in vivo treatment with indomethacin on mouse GM–CFC cells in culture (CFU-c). Possible role of prostaglandins. *Exp. Hematol. 8*:1157-1165 (1980).

133. Estrov, Z., D. S. Halperin, G. Coleani, and M. H. Freedman. Modulation of human hematopoiesis by leukotrienes in vitro. *Br. J. Haematol. 69*: 321-327 (1988).
134. Trull, O. N., G. Rovera, and D. Ferrerro. A novel lymphokine that inhibits hematopoietic progenitor cell proliferation. *Nature 309*:166-168 (1984).
135. Brown, S. L., and D. G. Van Epps. Suppression of a T lymphocyte derived chemototactic factor produced by the opioid peptides β-endorphin and met-enkephalin. *J. Immunol. 134*:3385-3390 (1985).
136. Oates, E. C., G. P. Allway, G. R. Armstrong, R. A. Boyajian, J. H. Kehrl, and B. S. Prabhakar. Human lymphocytes produce proopio-melanocortin related transcripts. *J. Biol. Chem. 263*:10041-10044 (1988).
137. Sibinga, N. E., and A. Goldstein. Opioid peptides and opioid receptors in cells of the immune system. *Annu. Rev. Immunol. 6*:219-249 (1988).
138. Santoli, D., S. C. Clark, B. L. Kreider, P. A. Maslin, and G. Rovera. Amplification of IL2-driven T-cell proliferation by recombinant human IL3 and GM–CSF. *J. Immunol. 141*:519-526 (1988).
139. Heslop, H. E., C. M. Price, H. G. Prentice, F. J. Cordingley, A. D. Webster, V. Hoffbrand, and M. K. Brenner. In vitro analysis of the interactions of recombinant IL2 with regenerating lymphoid and myeloid cells after allogeneic marrow transplant. *J. Immunol. 140*:3461-3466 (1988).
140. Alberico, T. A., J. N. Ihle, L-M. Liang, H. E. McGrath, and P. J. Quesenberry. Stromal growth factor production in irradiated lectin-exposed long-term bone marrow cultures. *Blood 69*:1120-1127 (1987).
141. Lipschite, D. A., K. B. Udupa, J. M. Taylor, R. K. Shadduck, and A. Waheed. Role of CSF in myelopoiesis in murine long-term bone marrow culture. *Blood 69*:1211-1217 (1987).
142. Sonada, Y., Y-C. Yang, G. C. Wong, S. C. Clark, and M. Ogawa. Analysis in serum-free culture of targets for recombinant human hematopoietic growth factors. IL3 and GM–CSF are specific for early developmental stages. *Proc. Natl. Acad. Sci. USA 85*:4360-4364 (1988).
143. Migliaccio, G., A. R. Migliaccio, and J. W. Adamson. In vitro differentiation of human granulocyte-macrophage and erythroid progenitors: comparative analysis of the influence of recombinant human erythropoietin, G–CSF, GM–CSF and IL3 in serum-supplemented and serum-deprived cultures. *Blood 72*:248-256 (1988).
144. Clark, P., D. E. Normansell, D. J. Innes, and C. E. Hess. Lymphocyte subsets in normal marrow. *Blood 67*:1600-1606 (1986).
145. DeGast, G. C., and T. A. Platts-Mills. Functional studies of T cells in adult human bone marrow. *Clin. Exp. Immunol. 38*:99-105 (1979).
146. Morgan, D. A., F. Ruscetti, and R. Gallo. Selective in vitro growth of T lymphocytes from normal human marrow. *Science 193*:1007-1009 (1976).
147. Lunn, L. G. The kinetics of immune reconstitution after human marrow transplantation. *Blood 69*:369-380 (1987).

148. Rozand, M. K., B. R. Smith, S. T. Burakoff, and R. A. Miller. Long-lasting deficit of functional T precursors in human bone marrow transplant recipients revealed by limiting dilution analysis. *J. Immunol 136*:4040-4048 (1986).
149. Hale, G., S. Cobbold, and H. Waldmann. T cell depletion with CAMPATH I in allogeneic bone marrow transplantation. *Transplantation 45*:753-759 (1988).
150. Ferrara, J. M., J. MIchaelson, S. J. Burakoff, and P. Mauch. Engraftment following T cell depleted bone marrow transplantation. *Transplantation 45*:948-952 (1988).
151. Laver, J., S. C. Dhanwar, R. J. O'Reilly, and H. Castro-Malaspina. Host origin of human hematopoietic microenvironment following allogeneic bone marrow transplantation. *Blood 70*:1966-1968 (1987).
152. Aizawa, S., and M. Tavassoli. Molecular basis of the recognition of intravenous transplanted hematopoietic cells by bone marrow. *Proc. Natl. Acad. Sci. USA 85*:3180-3183 (1988).
153. Weaver, C. T., and E. Unanue. T cell activation of membrane Interleukin 1 on macrophages. *J. Immunol. 137*:3868-3873 (1986).
154. Neta, R., S. Douches, and J. J. Oppenheim. Interleukin 1 is a radioprotectant. *J. Immunol. 136*:2483-2487 (1986).
155. Neta, R., M. B. Szetein, J. J. Oppenheim, S. Gillis, and S. D. Douches. The in vivo effects of Interleukin 1. I Bone marow cells are induced to cycle after administration of interleukin 1. *J. Immunol. 139*:1861-1865 (1987).
156. Zsebo, K. M., J. Wypych, V. N. Yuschenkoff, H. Lu, P. Hunt, P. P. Dukes, and K. E. Langley. Activities of hematopoietin 1 in IL1-induced activity on early hematopoietic cells of the bone marrow. *Blood 71*:962-968 (1988).
157. McNiece, I. K., B. E. Robinson, and P. J. Quesenberry. Stimulation of murine colony forming cells with high proliferative potential by the combination of GM–CSF and CSF-1. *Blood 72*:191-195 (1988).
158. Van Snick, J. V., S. Cayphas, J.-P. Szikora, J-C. Renauld, E. Van Roost, T. Boon, and R. K. J. Simpson. cDNA cloning of murine interleukin hematopoietin 1: homology with human IL6. *J. Exp. Med. 165*:891-907 (1987).
159. Matsuda, H., and Y. Kitamura. Migration of stromal cells supporting mast cell differentiation into open wounds produced in the skin of mice. *Exp. Hematol. 9*:38-43 (1981).
160. T. Kobayashi, T. Nakando, T. Nakahata, H. Asai, Y. Yagi, K. Tsuji, A. Komiyama, T. Akabane, S. Kojima, and Y. Kitamura. Formation of mast cell colonies in methyl-cellulose by mouse peritoneal cells and differentiation of these cloned cells in both the skin and gut mucosa of W/W^v mice. Evidence that a common precursor can give rise to both "connective tissue type" and "mucosal" mast cells. *J. Immunol. 136*: 1378-1384 (1986).

161. Kowanshi, H. Role of IgE as a mast cell development co-factor in the differentiation of murine gut-associated mast cells in vitro. *Eur. J. Immunol. 16*:689-692 (1986).
162. Hamaguchi, Y., Y. Kanakura, J. Fujita, S.-I. Takeda, J. Nakano, S. Tarui, T. Homjo, and Y. Kitamura. IL4 is an essential factor of in vitro clonal growth of connective-tissue type mast cells. *J. Exp. Med. 165*:268-274 (1987).
163. Guy-Grand, D., M. Dy, G. Luffau, and P. Vassalli. Gut mucosal mast cells: origin, traffic and differentiation. *J. Exp. Med. 160*:12-28 (1984).
164. Crowle, P. K., and N. D. Reed. Bone marrow origin of mucosal mast cells. *Int. Arch. Allergy Appl. Immunol. 73*:242-247 (1984).
165. Schrader, J. W., S. J. Lewis, I. Clark-Lewis, and J. G. Culvenor. The persisting (P) cell: histamine content, regulation by a T cell derived factor, origin from bone marrow precursors and relationship to mast cells. *Proc. Natl. Acad. Sci. USA 78*:323-327 (1981).
166. Nakahata, T., and M. Ogawa. Identification in culture of a class of hematopoietic cell progenitors with extensive capacity to self-renew and generate multi-potential hematopoietic colonies. *Proc. Natl. Acad. Sci. USA 79*: 3843-3847 (1982).
167. Williams, D. E., G. Hangoc, S. Cooper, H. S. Boswell, R. K. Shadduck, S. Gillis, A. Waheed, D. Urdal, and H. E. Broxmeyer. Effects of purified recombinant murine IL3 and/or purified natural murine CSF-1 in vivo on the proliferation of murine high- and low-proliferative potential colony forming cells: demonstration of synergy. *Blood 70*:410-413 (1987).
168. Kalland, T. Physiology of NK cells in vivo: regulation of progenitors by interleukin 3. *J. Immunol. 139*:367-375 (1987).
169. Melchner, H. V., and P. F. Bartlett. Mechanism of early allogeneic marrow graft rejection. *Immunol. Rev. 71*:31-56 (1983).
170. Palacios, R., H. Karasuyama, and A. Rolink. Ly-1 +ve PRO-B lymphocyte clones. Phenotype, growth requirements and differentiation in vitro and in vivo. *EMBO J. 6*:3687-3693 (1987).
171. Kimoto, M., V. Kindler, M. Higaki, L. Ody, S. Izui, and P. Vassali. Recombinant murine IL3 fails to stimulate T or B lymphopoiesis in vivo but enhances immune responses to T cell dependent antigena. *J. Immunol. 140*:1889-1894 (1988).
172. Metcalf, D., and N. A. Nicola. Tissue localisation and fate in mice of injection of multipotent CSF. *Proc. Natl. Acad. Sci. USA 85*:3160-3164 (1988).
173. Armstrong, H. E., E. M. Bolton, I. McMillan, S. C. Spencer, and J. A. Bradley. Prolonged survival of actively-enhanced rat renal allografts despite accelerated cell infiltrations and rapid induction of both class I and class II antigens. *Eur. J. Immunol. 18*:193-197 (1988).
174. Maizel, A. L., R. Shashikant, S. Mehta, D. Franzini, L. R. Lachman, and R. J. Ford. Human T lymphocyte-monocyte interaction in response to lectin: kinetics of entry into S phase. *J. Immunol. 127*:1058-1076 (1981).

175. Simmons, D., and B. Seed. The Fc-γ receptor is a phospholipid-linked membrane protein. *Nature 333*:568-570 (1988).
176. Noessal, C. V., F. Miedema, M. Brower, M. A. DeRie, C. A. Arden, and R. A. W. VanLier. Regulatory properties of LFA-1 α and β chains in human T lymphocyte activation. *Nature 333*:850-852 (1988).
177. McAteer, M. J., A. C. Legarde, H. M. Georgiou, and D. Bellgrau. A requirement for the CD5 antigen in T cell proliferation. *Eur. J. Immunol. 18*:1111-1117 (1988).
178. Hambor, J. E., M. L. Tykocinski, and D. R. Kaplan. Functional consequences of an anti-sense RNA mediated inhibition of CD8 surface expression in a human T cell clone. *J. Exp. Med. 168*:1237-1248 (1988).
179. Gunter, K., T. R. Malek, and E. M. Shevach. T cell activation properties of an anti-Thy-1 monoclonal antibody. Possible analogy to OKT/Leu 4. *J. Exp. Med. 159*:716-721 (1984).
180. Garman, R. D., K. A. Jacobs, S. C. Clark, and D. H. Raulet. B cell stimulating factor 2 (β-2 IFN) functions as a second signal for interleukin 2 production by rat and murine T cells. *Proc. Natl. Acad. Sci. USA 84*: 7629-7633 (1987).
181. Ranges, G. E., I. S. Figari, T. Espivik, and M. A. Palladino. Inhibition of cytolytic T cell development by transforming growth factor β and reversal by recombinant TNF-α. *J. Exp. Med. 166*:991-998 (1987).

12
Large-Scale Production of Hematopoietic Growth Factors

DAVID R. THATCHER
ICI Pharmaceuticals, Macclesfield, Cheshire, England

I. INTRODUCTION

The administration of preparations of colony-stimulating factor as medicines in the fight against disease has now become a reality with the approval of erythropoietin for use in the treatment of anemia associated with kidney failure (1). Several companies are also engaged in the clinical evaluation of related factors including granulocyte (G)–CSF (Amgen, Chugai, Kyowa Hakko Kogyo), granulocyte-macrophage (GM)–CSF (Genetics Institute/Sandoz, Immunex/Behringwerke, Glaxo), interleukin (IL)-3 (Schering-Plough, Genetics Institute) and M–CSF (Genetics Institute, Cetus). The initial results of these trials in both primates and humans are encouraging and the emergence of more of these factors as effective pharmaceutical medicines is certain. This chapter analyzes some of the major issues that are posed by the need to develop proteins of potential therapeutic value such as CSFs into safe, stable pharmaceutical preparations.

Therapeutic protein preparations are regarded by the pharmaceutical industry and governmental regulatory agencies as biologics, in contrast to chemically synthesized drugs. The latter are traditionally produced entirely or in part by organic synthesis and have structures that can be completely validated at atomic resolution by chemical and physical analysis. Biologics, on the other hand, are isolated from living organisms and in the past the active principles of such pharmaceutical preparations have been minor components in a milieu of contaminating biological products, too complex to be completely defined by existing analytical methodology. Recent advances in gene cloning and protein separation technology have now permitted the production of recombinant protein biologics at extremely high levels of purity. However, their formidible structural complexity remains an outstanding barrier to the rigorous definition of therapeutic proteins as chemical entities. To assure safety and reproducible efficacy, the actual methods of production and purification of protein therapeutics become part of the product definition. In consequence, major modifications and improvements in production strategy or separation chemistry must be further validated both by the analytical techniques available (intermediate and end-product analysis) and by further investigation of the toxicity, pharmacokinetics, and efficacy of the new preparation. Obviously, in a highly competitive commercial environment and where patent rulings are often unclear, the introduction of major changes in production methods can have serious consequences on a development program. It is judicious, therefore, to move into a clinical program only when significant scale-up and process development work has been completed. This process development research can be expensive, time consuming, and require specialized equipment and facilities. Such expenditure may be unavoidable because, except in cases in which the structure of an animal protein is close to that of the human equivalent (e.g., porcine and human insulin), immunological intolerance can preclude the preclinical evaluation of human proteins in many animal efficacy models. Many human protein factors have restricted species specificity and can only be really tested in vivo, using human volunteers. In the absence of adequate animal models, therefore, a major issue facing those companies developing hematopoietic growth factors as therapeutic agents is to decide what level of investment is required to ensure that such potential products can be rapidly evaluated in the clinic and at the same time proceed to commercial manufacture without a major revamp of production methods.

The various alternative production methods that are being used in the development of hematopoietic growth factors as therapeutics will be explored and the various logistical issues that may decide their eventual success will be discussed.

II. CHOICE OF A PRODUCTION STRATEGY

Sources of human or for that matter animal hematopoietic growth factors are extremely poor. These factors are all potent, being active at picomolar concentrations, and their expression levels may vary depending on the physiological state of the body. In fact, only two hematopoietic growth factors have been isolated in any significant quantity from human tissue: M–CSF (2) from normal human urine and erythropoietin (3), from patients with aplastic anemia. In both cases, levels are too low to be seriously considered as sources of factor for clinical development. For example, Xu Tao et al. (2) reported the isolation of only 8.5 mg of M–CSF from 10,000 L of urine.

Hematopoietic growth factors have also been isolated from the culture medium of human cells grown in vitro. Granulocyte-colony-stimulating factor, for example, has been purified from medium conditioned by the growth of a squamous-cell carcinoma cell line CHU2 (4) and GM–CSF has been purified from peripheral blood lymphocytes grown in culture (5). For several logistical reasons, production from sources such as these is too expensive to be commercially feasible. Also, even when cell lines have been immortalized these transformed cell lines often express several growth factors at low levels, creating formidible separation problems in downstream processing and exacerbating the constant problem of poor recovery.

Complete, de novo, chemical synthesis of IL-3 has been achieved by solid-phase peptide synthesis technology (6). However, problems of incomplete coupling and difficulties in intermediate purification exclude peptide synthesis as a serious route for the production of pharmaceutical-grade proteins of high molecular weight.

These sources of hematopoietic growth factors have now been superceded by our ability to isolate and manipulate the genetic material encoding their structure. A number of alternative recombinant microorganisms and cell lines can now be engineered for the high-level synthesis of these human growth factors.

This choice in recombinant production system is limited mainly by a tradeoff between structural complexity, the inherent stability of the biologically active protein itself, and process economics. Large (>30 kDa) proteins and proteins that require posttranslational modification for biological activity are most effectively produced by mammalian cell lines that are genetically engineered to secrete them. Smaller growth factors can be produced in a variety of cultured cells including those of mammalian, insect, mycetal fungal, yeast, and bacterial origin. Each approach has both advantages and disadvantages and these must be balanced against the required specification of the final dosage unit and economic constraints imposed by the commercial production of the drug. Although profit margins in the pharmaceutical industry are higher than in many other industries, research and development costs are huge and must be recouped and excessive production costs undoubtedly impede the development of an effective drug.

The least expensive and richest source of recombinant protein is through the genetic manipulation of *Escherichia coli* strains. Fermentation of these bacteria is well understood and they can grow rapidly on cheap, simple organic media. Using high copy number plasmid technology and high-level transcriptional promoters, *E. coli* can be induced to synthesize more than 50% of its total cell protein as a recombinant product. Intracellular synthesis of these recombinant proteins often results in their deposit in an insoluble state, forming morphologically distinct granules or inclusion bodies. The accumulation of recombinant proteins in these granules is probably due to the reduced solubility of these foreign proteins in the *E. coli* cytosol. Growth factors, for instance, have evolved as extracellular chemical messengers and do not accumulate at any significant level in any cell type. These proteins often also contain disulphide bridges and are naturally glycosylated—both are posttranslational modifications that *E. coli* cannot accomplish.

Another drawback to using *E. coli* as a source of recombinant protein is that in many cases this mode of production results in an *N*-terminally heterogenous protein product. Hematopoietic growth factors, being extracellular, are naturally secreted from the producer cell type under the direction of a mammalian leader peptide sequence that is precisely cleaved off during transport through the endoplasmic reticulum. Because the *E. coli* secretory apparatus does not recognize mammalian leader sequences, processing is performed by the molecular biologist who clones the final mature coding sequence into an expression vector and initiates translation with an initiator methionine codon immediately preceding this coding sequence. During what is now intracellular synthesis of a foreign protein sequence in *E. coli, N*-terminal heterogeniety may develop because the enzyme that normally removes initiator methionine from *E. coli* proteins, methionine aminopeptidase, may not necessarily recognize the new heterologous protein sequence. Recombinant proteins produced in *E. coli* therefore often have variable amounts of the *N*-terminal initiator methionyl protein contaminating the natural sequence.

Precise *N*-terminal processing and disulphide bridge formation have been achieved in *E. coli* by an alternative to the intracellular expression strategy. The natural mammalian leader sequence (e.g., GM–CSF) (7) is deleted from the gene and is repalced by an oligonucleotide coding for the expression of a bacterial leader sequence. On fermentation and induction, synthesis of the recombinant protein is directed to the periplasmic space, during which precise cleavage of the bacterial leader is achieved and disulphide bridge formation occurs.

If glycoslylation of the recombinant product is considered a necessary property, the eukaryotic expression systems must be used. Granulocyte-macrophage colony-stimulating factor (8), IL-3, and M–CSF (17) are naturally *N*-glycosylated whereas GM–CSF (9) and G–CSF are also *O*-glycosylated. The requirement for glycosylation is therefore a crucial issue in choosing a production system and the final answer may not be apparent until extensive pharmacokinetic data have accumulated.

Eukaryotic microorganisms such as the yeast *Saccharomyces cerevisiae* are able to process, glycosylate, and secrete recombinant mammalina gene products. A disadvantage of this system, however, is that when glycosylation occurs a large mannose-rich, branched-carbohydrate core is added at each *N*-glycosylation site. Because sialic acid residues are not found terminating these mannose side chains, clearance of yeast recombinant proteins by the liver can potentially lead to less favorable pharmacokinetic properties than those associated with either the non-glycosylated *E. coli* product or the product isolated from recombinant mammalian cell lines. It is possible that this problem can be overcome by enzymatic

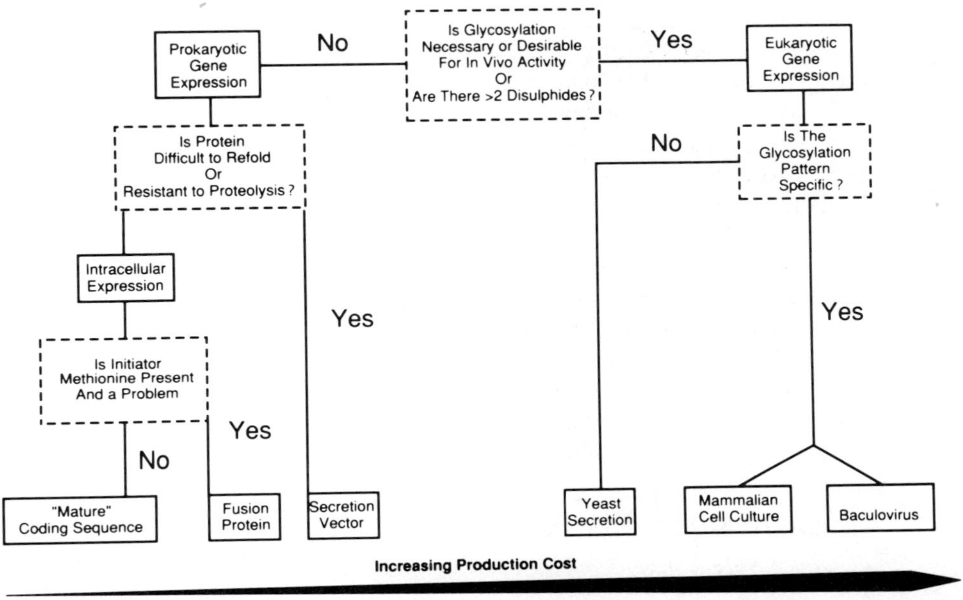

Figure 1 Decision tree for the selection of a recombinant protein production strategy.

removal of part of the mannose side chain residues of purified yeast products followed by artificially capping the remaining carbohydrate with a sialic acid terminal transferase (10).

Glycosylation patterns that are thought to be more akin to those occurring during natural synthesis may be achieved by expression in recombinant mammalian cell lines. Most large-scale production of hematopoietic growth factors has been performed using CHO cells originally derived from Chinese hamster ovaries, although recombinant material has been successfully isolated from recombinant monkey kidney and murine cell lines on a commercial scale.

The complexity of the posttranscriptional processing of some hematopoietic growth factors is best illustrated by the events that occur during the synthesis of M–CSF (Chapter 8). Although M–CSF is encoded by a single genetic locus located on chromosome 5, alternative splicing of the primary RNA transcript produces two M–RNA species (11). These two mRNAs ultimately yield two forms of M–CSF precursor on translation: a 61-kDa and a 28-kDa molecule. The larger precursor contains a 298 amino acid residue insert between residues 181 and 182 of the smaller sequence. The first 32 residues of both precursor molecules are identical and are the signal sequences that direct export of the proteins to the endoplasmic reticulum, before being removed by proteolysis. The longer precursor also undergoes *C*-terminal processing to give a 223 residue glycoprotein that is homodimeric in solution. The smaller protein is processed at the *C*-terminus, is also glycosylated and dimeric. It is remarkable that all of these processing events–splicing, secretion, disulphide bridge formation, glycosylation, and proteolytic truncation–can occur in genetically engineered host cells derived from species that are widely divergent in evolutionary terms.

At the present time mammalian cell culture provides a system that produces complex recombinant proteins closest in structure to their naturally occurring counterparts. Mammalian cell culture, however, is expensive and has several other disadvantages (Section III) that must be compared with the advantages of microbial production systems.

Figure 1 summarizes the issues pertaining to the choice of an expression system in the form of a decision tree.

III. LARGE-SCALE CELL CULTURE

A. Microbial Systems

1. Fermentation

The emerging biotechnology industry has been able to draw on several generations of fermentation process engineering experience. Bioreactors have been designed at between 2.0×10^5 and 1.5×10^6 l level for the fermentation of non-pharmaceutical products and antibiotic intermediates. The large body of theory and the empirical experience acquired during the development of these fermentation processes has been exploited during (and is in fact a major advantage of) scaling-up recombinant *E. coli* and yeast production systems.

The most commonly used fermenters are cylindrical, jacketed pressure vessels with an aspect ratio of approximately 3:1. Temperature control is provided by heat transfer coils and all ports are sealed to maintain asepsis. *Escherichia coli* is usually grown under aerobic conditions and much effort has been put into the design of aeration and agitation systems. The mass transfer of oxygen can often limit growth and it has sometimes been found necessary to feed the culture with pure oxygen or to continuously feed the carbon source to the fermentor (fed-batch processes) to prevent anaerobiosis. By the control of pH, aeration, and feed rate, extremely high rates of growth and productivity have been achieved. Fermenter productivity may be improved by strategies that increase the final levels of biomass attained per batch or by increasing the level of product accumulating per cell or by a combination of both approaches. Usually, a time course for a laboratory-scale fermentation is defined and the growth rate, pH, and medium composition are optimized. Scale-up correlations are then used to predict the aeration (mass transfer), heating and cooling (heat transfer) regimen for the full-scale fermentation.

Fermentation scale-up is emerging as one of the few areas of biotechnology in which adequate mathematical descriptions and simulations exist, and spectacular scale-up ratios have routinely been achieved. However, the physiology of host vector interactions are often poorly understood, and firm correlations between expression level and growth conditions must be established to assure successful scale-up.

A major advantage of microbial fermentation systems is the low cost of the growth media, *E. coli* usually being grown on glycerol (12) or glucose (13). Most hematopoietic growth factors expressied in *E. coli* have been produced under either constitutive control (tryptophan promoter) (13) or using the heat-inducible promoter (pL) (14-17). Expression vectors are usually derivatives of pBr322 or related high-copy-number plasmids.

The disadvantage of using *E. coli* production systems is that often the codon preference of the original mammalian gene prevents high levels of expression. For example, Devlin et al. (17) raised the AT/GC ratio of the G–CSF gene from 36 to 57% and increased recombinant *E. coli* expression from 0 to 17% of total cell protein. This effect is thought to be due more to the abolition of stable and unproductive secondary mRNA structures than to any effect on pools of translation factors. Other less well documented phenomena may make high-level synthesis less favorable for growth and cell division. Plasmid instability may result and becomes a problem that must be circumvented in the construction of a commercial production strain. This can be accomplished, in some instances by incorporating into the vector a gene coding for antibiotic resistance or a gene complementing the absence of a chromosomal function necessary for survival. Under such selection, retention of the expression vector is necessary for viability of the strain and any deleterious effects of high-level heterologous product synthesis will not result in plasmid-free cells dominating the culture. However, β-lactam antibiotics can give rise to highly immunogenic adducts and these antibiotics cannot be used to stabilize production vectors. Another disadvantage of *E. coli* as a production system, as discussed in the previous section, is that prokaroyotes are unable to glycosylate heterologous mammalian gene products.

The presence or absence of a suitable microbial leader sequence preceding the mature protein sequence determines whether the protein accumulates intracellularly or is secreted into the fermentation broth. These are the two major microbial production strategies and they dictate the type of downstream process that can be adopted.

2. Intracellular Synthesis

As discussed in the previous section, intracellular synthesis of heterologous proteins such as erythropoietin (18), GM–CS (19), and G–CSF (14) often results in their deposition as morphologically distinct granules or inclusion bodies. Although this compartmentalization has an advantage of sequestering the product from proteolytic degradation, inclusion body formation requires specialized approches to primary recovery (Section IV.A). The balance between rates of expression and rates of proteolytic degradation determine the overall level of product accumulation and in the case of GM–CSF (16), this level has been significantly increased by the use of the protease-deficient *E. coli* host mutation (Lon$^-$) coupled with a mutation in the heat-shock response (HpTR$^-$).

Specific proteolytic processing resulting in the removal of initator methionine residues from the *N*-terminal sequences of recombinant proteins is not always attainable. The specificity of the *E. coli* host methionine aminopeptidase is dependent on the size of the side chains of the two amino acid residues immediately following the initiator methionine. Some workers have truncated the poorly processed natural amino acid sequence [three residues in M–CSF (15), one residue in G–CSF (17)] to provide a sequence on which the enzyme will act, and in both cases the unnatural *N*-terminal methionine residue is removed. In the case of erythropoietin (18), synthesis in *E. coli* results in removal of both initiator methionine and the partial removal of the first alanine residue of the mature amino acid sequence. The intracellular synthetic route for heterologous products in *E. coli* can therefore lead to heterogeneity in covalent structure.

The intracellular milieu within the *E. coli* cytosol is thought to be reducing, and most heterologous proteins that normally have disulphide bridges in their structure are synthesized in the reduced form. These disulphides must then be formed by chemical oxidation during downstream processing, an operation that invariably leads to substantial losses due to interchain and mixed disulphide formation (Section IV.B.2).

3. Extracellular Secretion

Secretion systems have the advantage of allowing posttranslational processing and also enrichment of the specific activity of the product stream; *E. coli* and yeast naturally secreting few, potentially contaminating, extracellular proteins.

The two major drawbacks of intracellular *E. coli* expression, namely lack of disulphide bridge formation and incomplete *N*-terminal processing, can be overcome by replacing the natural mammalian leader sequence with a prokaryotic leader sequence. Libby et al. (7) cloned the OMP–A *E. coli* membrane protein gene and isolated the gene sequence encoding the signal peptide that directs synthesis to the inner membrane. This gene sequence was then placed upstream of the human GM–CSF mature coding sequence and control of the fusion protein placed under the *lpp* promoter. Induction by IPTG led to the synthesis of recombinant GM–CSF and the accumulation of the correctly processed, oxidized protein in the periplasmic space. The large-scale recovery of proteins from the periplasm, however, is complicated by variable levels of cell lysis, and systems that allow transport through both inner and outer membranes are preferable. This can be achieved by using leaky host strains expressing low levels of the *kil* gene (20) or by using hemolysin fusion proteins.

The expression of heterologous proteins in recombinant yeast strains results not only in disulphide bridge formation and clean *N*-terminal processing but also in *N*-linked and *O*-linked glycosylation (21).

Unfortunately, yeast has a different biochemical pathway for the synthesis of the carbohydrate moieties and these are often incompletely transferred to the protein glycosylation sites (21,22). Yeast carbohydrates contain a large complex mannose core that is not capped by sialic residues. These recombinant yeast glycosylation products are susceptible to rapid clearance by the liver and some workers have embarked on a strategy of eliminating glycosylation by site-specific mutagenesis of the protein recognition sites, the so-called "low cal" or "diet" second-generation recombinant yeast products.

The most successful yeast secretion systems fuse the gene of interest, e.g., GM–CSF (27,23) or G–CSF (24) to the *N*-terminal domain of the yeast alpha mating factor preprotein. This sequence, which can be truncated to as small as 19 residues (23), directs synthesis by a natural secretion mechanism that results in export of the fusion protein to the cell broth. The presequence is removed during synthesis and the residual alpha mating factor sequence is cleaved by the KEX2 protease. If the Arg-Lys KEX2 recognition site occurs within the heterologous sequence (e.g., G–CSF at Lys-22), yields can be improved and specific proteolysis can be reduced by their removal using site-specific mutagenesis (27). Incomplete processing of these fusion proteins can lead to substantial *N*-terminal heterogeneity (22) and partial degradation by exopeptidases (21). Selection of yeast production vectors is usually through complementation of amino acid auxotrophes. The GAL1 (25), G3P (12), and ADH2 (22) promoters have all been used to direct transcription of CSF genes in yeast.

In summary, yeast expression systems offer the advantages of microbial systems, high productivity and a wealth of fermentation experience, coupled with the advantages of posttranslational modification during secretion. The disadvantages are due to incomplete and inappropriate glycosylation coupled with potential problems of specific proteolytic degradation of some amino acid sequences.

B. Higher Eukaryotes

Precise glycosylation patterns have in some cases [e.g., erythropoietin (26) and GM–CSF (27)] been shown to be essential for high in vivo biological potency. At the present only expression in recombinant mammalian cell lines will achieve this near natural *O*-linked and *N*-linked glycosylation. From the process-development point of view there are five major disadvantages of using mammalian cell culture: (a) cell fragility, (b) low productivity, (c) media complexity, (d) the possibility of contamination by endogenous growth factors expressed by the host cell line, and (e) potential contamination of the final product with pathogenic viruses.

Compared with their microbial counterparts, mammalian cells are more susceptible to shear and when grown in suspension require specialized stirred-tank bioreactors or airlift fermenters. Cells that are anchorage dependent can be grown on microcarriers or in perfusion devices. In batch culture, mammalian cell systems grow slowly (doubling times 10-24 hr) and cell densities may be limited by the exhaustion of nutrients (such as glucose and glutamine), the accumulation of waste products (largely lactate and ammonia), or the depletion of trace amounts of growth factors from the medium. Large-scale fermenters for cell culture must therefore be of a more rigorous aseptic design than their microbial counterparts or contamination will result in the loss of a significant fraction of production runs. The reality of such slow growth of mammalian cells has resulted in the popularity of a continuous culture strategy, especially in combination with microcarrier and perfusion devices. Nutrients are continuously replenished, waste product concentrations are reduced, and continuous removal of the culture fluid minimizes possible proteolytic degradation. Once the optimal cell density has been attained, continuous production of such conditioned medium can be extended for several months.

A number of different cell lines have been used for the expression of hematopoietic growth factors. A major objective of fermentation research is to reduce the dependence of these cell lines on fetal calf serum supplements, the single most complicating factor in the downstream processing of mammalian cell products. Serum can be reduced by the use of cycles of serum-free nutrient (production phase) followed by cycles of serum-containing media (restoration phase) as in the production of erythropoietin (18,26) and G–CSF (28) or by inducing the host cells to grow on a serum-free cocktail of nutrients and growth factors. Whichever approach is adopted, the growth media used in the culture of mammalian cell lines is an order of magnitude more expensive than those used to grow *E. coli* or yeast.

Heterologous genes are usually introduced into mammalian cells via shuttle vectors that contain both an *E. coli* and a viral origin of replication. In some cells such as the green-monkey-derived COS lines, replication of these vectors, accompanied by transient low-level expression, occurs from 3 to 15 days followed by a loss in productivity. Stable expression, however, may be obtained by the use of a vector coding for dihydrofolate reductase (DHFR) complementing a host strain (e.g., CHO cell line) that lacks this enzyme (12,18,29). The expression level may then be amplified by gradually increasing media concentrations of the DHFR-inhibitor, methotrexate. When amplification has been completed the heterologous gene is found to be integrated into the genome set at multiple chromosomal sites. The best production strains are stable and retain this new karyology over many generations or passage numbers. Murine C127 (30), murine NIH/3T3 (26), and CV1 African green-monkey cells with plasmid vectors containing bovine papilloma virus promoters (28) and neomycin (15) resistance have also been used to generate stable production cell lines.

High levels of M–CSF (31) and erythropoietin (32) have been secreted by insect cell lines derived from *Spodoptera frugiperda* coinfected with polyhedrosis virus (baculovirus) DNA and a vector utilizing the polyhedron gene promoter. Homologous recombination results in a virus that synthesizes the recombinant gene product instead of the viral head protein. Although these insect lines can be grown on serum-free media, expression is transient and being dependent on transfection or infection for each batch, these systems are unlikely to fulfill the regulatory requirements necessary for the production and validation of a pharmaceutical drug. Also, insect cells do not appear to have the capacity to convert a high mannose core to complex *N*-linked oligosaccharides during glycosylation. If insect cells do lack the necessary sialotransferases, then baculovirus expression systems have little advantage over yeast for production purposes. The products of both will have reduced pharmacokinetic stability.

CHO and baby hamster kidney cells (BHK) have been experimentally demonstrated to *N*-glycosylate recombinant erythropoietin in a similar way to the natural urinary protein (33,34). Human erythropoietin has three possible *N*-glycosylation sites (Asn-X-(Ser/Thr) in the amino acid sequence and all are occupied in both the recombinant and urinary protein by a fucosylated, sialic-acid-capped tetraanternary complex oligosaccharide that together comprises 30% of the mass of the whole molecule. Both CHO and BHK-derived recombinant erythropoietin appear to be more uniform in their glycosylation pattern than the natural product. Granulocyte-colony-stimulating factor (4) and GM–CSF are both *N*- and *O*-glycosylated (26,27).

Thus, where glycosylation has been shown to be critical to the activity and in vivo performance of the product, mammalian cell expression is the only viable production alternative.

IV. PROTEIN ISOLATION

Unlike the pharmaceuticals derived by synthetic organic chemistry, therapeutic proteins are produced in comparatively dilute aqueous solution and are contaminated by (a) thousands of polymers of very similar structure (host and vector proteins) and (b) by other macromolecules of high molecular weight (nucleic acids, carbohydrates). Another disadvantage for the process chemist is that the range of extraction and purification methods that can be employed to recover therapeutic proteins is limited by the fragility of protein conformational structure, with the free energy difference between the stable native bioactive conformation of GM–CSF (35), for example, and the unfolded inactive state being on the order of only −21 kJ/mol. These specialized protein purification methods are still evolving, their theory often being poorly understood at a fundamental level, their scale-up remaining a matter of enlightened empiricism, and their automation virtually unexplored.

Many of the crucial separation techniques employed such as chromatography are fouled by nonproteinaceous contaminants such as nucleic acids and lipid, and protein production processes are usually divided into primary recovery, purification, and final polishing and formulation steps. Stable intermediates are conveniently held between these steps rather than continuous processing from broth to final formulated product (Fig. 2).

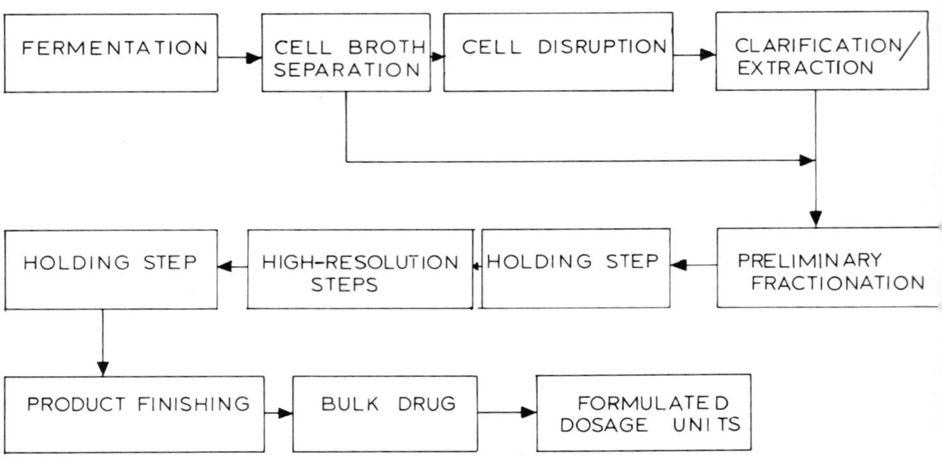

Figure 2 Downstream processing of protein therapeutics.

A. Primary Recovery

The objective of primary recovery operations is largely the removal of the bulk of the water and nonproteinaceous cellular constitutents from the process stream.

Initially, cells in the culture medium, whether the product is secreted or intracellular, must be concentrated by centrifugation or crossflow filtration. Centrifugation is most commonly used, and several continuous-flow designs adapted from the food processing and chemical industry are available. These centrifuges often subject the process stream to shear levels that can lyse fragile cells, and a new generation of less disruptive centrifuges is being designed specifically for harvesting higher eukaryotic cells. Because the microbial organisms and cell lines used in the production of growth factors are recombinant in origin, most environmental and worker protection legislation requires that these initial dewatering operations be performed in process equipment that is fully contained to minimize accidental release. For secreted products the clarified supernatants from initial centrifugation operations are concentrated further by crossflow filtration (12,30,36), then passed through a depth filter to remove residual particulates, and finally membrane-filter-sterilized before leaving containment. The fermenter, centrifuge, and associated vessels and ultrafilters are then sterilized.

Intracellular products, on the other hand, either may be derived from cells killed in the fermenter after which containment can be broken or alternatively, live cells may be processed until an intermediate stage can be chemically sterilized or membrane-filter-sterilized. Such processing by definition involves cell breakage or homogenization to release the intracellular product from the cytoplasm. Two cell-disruption techniques have found favor because of attractive economic performance and scalability: bead mills containing glass beads agitated at high speed and high-pressure homogenizers from the food processing industry. The performance of cell homogenizers is proportional to both the pressure drop across a constriction valve and the physical properties of the cell. Microbial cells, particularly when grown into stationary phase, may be more resistant to shear than are cells harvested in early log phase.

When the product is in the form of *E. coli* inclusion bodies, the purity of the recovered material will be cricially dependent on the level of cell disruption. Pretreatment of the starting material with a cell-wall-degrading enzyme such as lysozyme may be necessary to reduce the total number of passes through the homogenizer required to give 100% lysis. The product is then recovered by centrifugation of the rapidly sedimenting dense inclusion bodies under conditions in which the lighter cell debris remains in the supernatant along with most of the nucleic acid and other soluble components of the cell.

When the intracellular product is a soluble cytosolic protein the debris must be flocculated with an agent such as polyethyleneimine, which also preci-

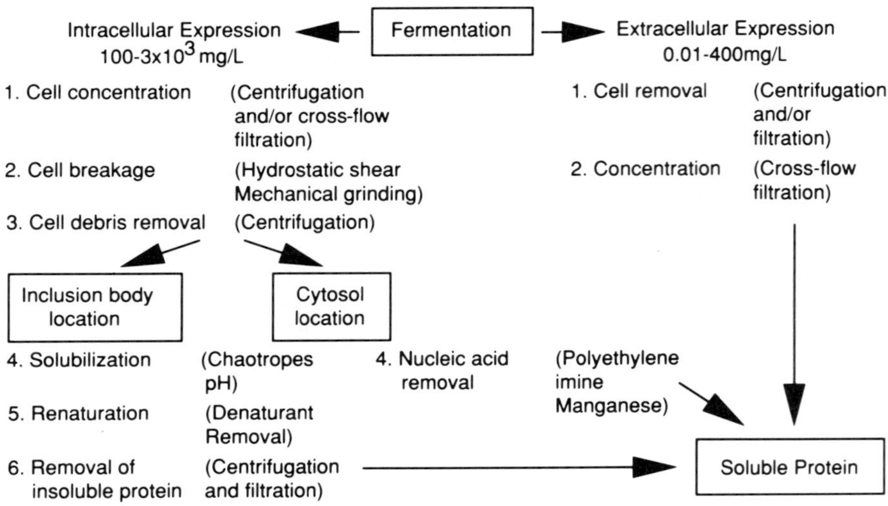

Figure 3 Primary fractionation and production systems.

pitates nuclei acids and lipopolysaccharides before being clarified by centrifugation and the product stream filter sterilized.

The steps involved in the primary recovery of proteins from intracellular and extracellular production systems are summarized in Fig. 3.

B. Purification Strategies

The overriding principle guiding the development of protein purification processes, whatever the primary recovery process and whatever the production system, is the selection of those separation technologies that most effectively exploit differences in physical properties between the product and the contaminating host proteins. Despite the claim by Shimamura et al. (37) that *E. coli*-derived G–CSF can be purified by the same method as recombinant prochymosin, there are really no generic methods for the purification of recombinant hematopoietic growth factors, and each procedure must be tailored to the properties of the product. Consequently, the more analytical data available on the physicochemical properties of the product, the more rational the basis for process design. During the early phases of a process development project much of this crucial information is not available and one can expect laboratory-scale procedures to be inefficient, laborious, and unscalable. As the behavior and stability of the purified product become better understood, the number of steps necessary to attain acceptable levels of purity should decrease and the yield of each unit operation should increase. The level of investment required to design and implement a dedicated commercial-scale process for the purification of a therapeutic protein is enormous and most companies do not undertake such an enterprise until there are at least strong indications of drug efficacy. The most commonly accepted solution to this problem in the biotechnology industry is to develop laboratory-scale procedures to the point at which 100 g to 10 kg of clinical-grade product can be manufactured per annum in a flexible, multipurpose pilot plant.

The details of these processes is necessarily rather confidential information. However, analysis of the patent literature shows that several major trends in the purification of recombinant growth factors are emerging.

The production system adopted for synthesis and the primary recovery steps selected for early downstream processing are the most important elements dictating the design of an optimal purification strategy. These factors decide which steps should be incorporated in the process and in which order they should be performed.

1. Holding Steps (Intermediates)

An industrial-scale process for the production of a therapeutic protein must have a number of holding steps which are equivalent to process intermediates. The logistics of running either a continuous purification process or batch process from biomass to final bulk drug substance as an uninterrupted sequence of consecutive unit operations are extremely complex and only an extremely well-engineered process would succeed reproducibly. Intermediates provide a mechanism for reducing the effect of biological variation on process performance. In particular, reproducible performance of high-resolution chromatographic separations requires the maintenance of similar feed characteristics (volume, concentration, and composition).

After primary recovery it is therefore convenient to establish a holding step, the product being held in a state (dried powder, frozen precipitate, or frozen liquid) that is stable for several months. In the case of the intracellular *E. coli* inclusion body route, the cell debris/inclusion body centrifuge pellet may be stored frozen at –70°C without degradation. Soluble products are more difficult to stabilize. Ammonium sulphate is a widely used precipitant in protein chemistry and is known to stabilize proteins in solution. Frozen ammonium sulphate precipitations are therefore a useful means of storing impure G–CSF for further processing (24,38,29).

2. Solubilization of Inclusion Bodies and Refolding

The exact chemical nature of proteins sequestered in these morphologically distinct granules formed by *E. coli* is not known. These inclusions may be solubilized by the use of acids, alkaline pH, detergents, or chaotropic solvents or mixtures of these solvents. Invariably solubilization is accompanied by complete denaturation of the protein product. For example, G–CSF (40), erythropoietin (18), and GM–CSF (16,41) inclusion bodies have been solubilized in 6 *M* GuHCl whereas M–CSF inclusions can be treated with 8 *M* urea or 7 *M* GuHCl (15). Granulocyte colony-stimulating factor inclusion bodies have also been solubilized by treatment with sodium dodecyl sulfate (17) or sarkosyl (14). The solubilization procedures employed for the extraction of the recombinant product from inclusion body preparations will also release product from whole cells, and the relative advantage of purifying an inclusion body intermediate or directly extracting unhomogenized cells with solvent must be evaluated. The decision must be balanced by the fact that large-scale homogenization and centrifugation unit operations are extremely expensive in energy requirement and capital costs, and when expression levels are >30% the maximum purification that can be achieved by inclusion body isolation is only 1.5-2-fold.

After solubilization the choice is either to partially purify the product in the denatured unfolded state or to refold the protein immediately by removal of the denaturing, solubilizing agent. The advantage of purifying the denatured protein before refolding is that cellular components that may interfere with refolding can be removed, with the added advantage that subsequent refolding can be monitored on-line by spectroscopic techniques. The disadvantage is that the purification method must be buffer independent, the most popular of which, gel filtration, is difficult to scale up (Section VI.B.3) and is consequently unsuitable as an initial fractionation step. The disadvantage of immediate refolding is that the renaturation process itself cannot be followed directly by any analytical method and must be optimized empirically. Renaturation of purified recombinant products rarely exceeds 50% in recovery. Much of the protein solubilized is either chemically modified in some way or forms insoluble aggregates during refolding, thereby diverting potentially useful product into a useless kinetically trapped state. Because aggregation is by definition a higher-order reaction, flux through this pathway can be reduced by dilution of both the product in the solubilizing solvent and during the refolding process. In fact, many refolding processes involve dilution of the solubilized extract into a suitable renaturing aqueous solvent (15,16). Many workers have found that yields may be improved by exchanging the initial solubilizing agent for a denaturing solvent of lower strength, which allows the native structure to form but keeps the poorly soluble, folding intermediates (which precipitate in aqueous solution) soluble, e.g., GM–CSF (41) and G–CSF (40). When refolding is complete the process stream can be returned to a conventional aqueous buffer system. The rationale for this strategy was first presented by Orisini and Goldberg (42) in 1978 using chymotrypsinogen and this work has remained a guiding principle for the design of many recombinant refolding processes.

Buffer exchange operations currently are performed at scale using various ultrafiltration membranes in various configurations such as hollow fibers and spiral cartridges. All of these devices require tangential flow across the membrane to reduce fouling. Nevertheless, because aggregation is a major byproduct of the refolding process, fouling is still a major challenge in scale-up design.

Hematopoietic growth factors such as G–CSF and GM–CSF have disulphide bridges as integral covalent bonds in the native structure. During intracellular synthesis in *E. coli* these bonds are not formed because of the net reducing environment of the *E. coli* cytosol. Depending on the influence of these bonds on the final native structure, the oxidation step necessary for their formation can be performed at various stages during the refolding unit operation. In particular the presence of naturally occurring free thiol residues, free to catalyze mixed disulphide conformers, and disulphide bridges in the same molecule (e.g., G–CSF) can exacerbate the refolding problem. Some companies have reduced this difficulty by mutating the free thiol group to methyl (alanine) or methoxy (serine) groups by genetic engineering techniques (24,37,39). Such analogue forms of G–CSF are fully active. Proteins such as GM–CSF require disulphide bridges to maintain a compact globular conformation, with reduction resulting in an unfolded intermediate that retains secondary structure (35). This species is soluble in 3 *M* urea, moreover, the fully folded disulphide bridged molecule is also stable under these conditions. Reduced protein obtained by recovery from *E. coli* could therefore be oxidized in 3 *M* urea by air (16), air in the presence of Cu^{2+} ions (14), or by using a redox buffer system (35). When fully oxidized, the second solubilization agent can be removed by buffer exchange and any insoluble byproducts such as residual reduced protein and other folding intermediates can be separated by centrifugation. Other proteins refold to a near-native soluble conformation even when the disulphide bridges are reduced. Oxidation steps may then be performed after final buffer exchange.

3. Separation Steps

Recombinant protein products usually share several biophysical properties with most of the major contaminating host proteins. Simple single-step operations such as precipitation and extraction are therefore not selective enough, and high-resolution chromatography is the only effective means of obtaining homogeneous proteins for therapeutic use. These chromatographic systems exploit subtle differences in ionic interactions, hydrophobic interactions, molecular weight differences, and specific affinity interactions.

Ion exchange chromatography is the most popular initial chromatography step. Many recombinant proteins expressed in *E. coli* that have isolectric points greater than 6.0, e.g., GM–CSF (7,41), and G–CSF (14) can be purified to a near-homogeneous state by combinations of ion-exchange chromatographic steps. This can occur because the majority of soluble proteins in this host organism are acidic in nature and have quite different charge interactions to the recombinant protein. Also, anion exchange chromatography usefully removes residual nucleic acid and lipopolysaccharide not eliminated during primary recovery. Ion exchange chromatography is also useful in partially separating secreted recombinant products from media components and the bulk of the serum protein additives (30,36,38,40).

Hydrophobic chromatography is the second most important technique available for protein purification. Because many small-scale purification methods incorporate reversed-phase high-performance systems, there has been a tendency to try to scale this operation in production. At present large scale HPLC is useful only for very high value added products as the low scale up ratios encountered translate into low throughput and poor process economics. Column packing technology is in its infancy and the high cost of modern matrices coupled with the scale and efficiency of industrial-scale high-performance liquid chromatography (HPLC) is limited. Erythropoietin (36,42), GM–CSF (7), and G–CSF (24), have been purified on silica-based matrices coated with C-8 and C-4 phases. Acetonitrile (7), which must be shown to be free of acrylonitrile, ethanol (36), and propanol (38) are the most usual organic modifiers in the mobile phase, with trifluoroacetic acid being the counterion of choice. Hydrophobic interaction chromatography, in which the hydrophobic ligand density is lower and the matrix is hydrophilic (e.g., Pharmacia's Phenyl Sepharose and Toyosoda's Butyl Toyopearl), can be used at nondenaturing pH, and is especially useful in situations in which the product is known to be contaminated with host proteins of a more hydrophobic nature, for example, after solubilization and extraction of cell debris in an inclusion body process. Hydrophobic chromatography also has application in the separation of the recombinant product from degraded species of related sequence generated by proteolytic nicking during biosynthesis or processing. The introduction of a break in the polypeptide chain on an exterior loop of the native structure can have (a) little effect on molecular weight of the species because the fragments are held together by noncovalent forces and (b) little effect on overall charge of the species because both a positively and a negatively charged group are introduced simultaneously. Therefore, both gel filtration and ion-exchange methods fail to separate these types of product-derived contaminant. However, the introduction of two charged residues into the structure often has a marked effect on overall hydrophobicity, the degraded species being significantly more hydrophilic and consequently having lowered affinity for the hydrophobic ligand. In these cases proteolytically nicked fragments are effectively separated by hydrophobic interaction chromatography.

Gel filtration, like high-performance hydrophobic chromatography, is a unit operation that has a severely limited scale-up ratio. There is a temptation to use this as an initial purification operation after primary recovery, but the consequences of adopting this strategy are that the ultimate batch size will be limited to $<$20 L of initial feed. Gel filtration, however, is extremely useful as a final desalting step where bulk product can be exchanged into a buffer suitable for storage or formulation, and if a wide-pore-size gel is used, high-molecular-weight aggregates can be removed (e.g., GM–CSF) (41).

To the unintiated, immunoaffinity chromatography is a universal solution to the challenge of protein purification. Using a purified, immobilized monoclonal antibody, enormous selectivity in separating a product from the background of contaminating proteins can be achieved. Under optimal conditions the product, e.g., erythropoietin (33) is so pure that current analytical technology is hard pressed to detect contamination. However, there are a number of less well publicized disadvantages of immunoaffinity chromatography. Because the immobilized antibody recognizes a small section of the target protein surface area or epitope, quite gross charges in conformation or covalent structure in other parts of the molecule will not affect binding. In other words, immunoaffinity chromatography has little selectivity between the product and its degraded forms. Also, the binding energy often may be so high that elution can only be achieved by disruption of the product structure with denaturing eluents such as low pH. Under such conditions, the purification step itself is liable to generate misfolded conformers. Generic methods of protein purification have been developed by a combination of affinity chromatography and genetic engineering. For example, a highly immunogenic affinity tag peptide was fused to the G–CSF sequence at the genetic level (24). The fusion protein was isolated by immunoaffinity chromatography and the affinity tag cleaved off by treatment with mucosal enterokinase. In a more complex system Smith and Johnson genetically fused glutathione S-transferase to GM–CSF (43). The fusion protein was isolated by affinity chromatography on glutathione agarose and the GM–CSF was recovered after treatment of the purified fusion protein with coagulation factor Xa.

Products of eukaryotic cell culture are often glycosylated and it is possible to use this as an affinity chromatographic handle by employing lentil lectin or *Concanavalin A* affinity chromatography [e.g., GM–CSF (19), erythropoietin (3)]. The affinity constant for these interactions, however, is much lower than the corresponding immunoaffinity-protein interaction. Consequently, lectin chromatography usually requires a concentrated, highly enriched product stream.

4. Integration of Separation Steps

An integrated purification process, in addition to allowing the isolation of the product in a minimum number of separation steps, reduces the number of buffer exchange operations and keeps the process stream as concentrated as possible. For example, ion exchange steps are often best optimized by elution through pH changes so that multiple ion-exchange operations may be performed successively. Ammonium sulphate precipitation as discussed above is a useful concentration and holding step. After solubilization, residual salt can be removed by hydrophobic interaction chromatography in which high ionic strength promotes binding and low ionic strength brings about elution. Such integration of process steps reduces the need for buffer exchange operations, which at scale must be performed by crossflow filtration or desalting gel filtration and invariably result in reduced overall yield.

C. Formulation

Hematopoietic growth factor preparations are administered as sterile injectable solutions. The final dosage form can be sterile solution or a sterile lyophilized powder, reconstituted in sterile water or saline immediately before use. Proteins are relatively fragile macromolecules and the formulation of therapeutic proteins for clinical use has added a challenging new dimension to pharmaceutical development. Also, many proteins of interest are extremely potent and doses may be $<200\ \mu g$. At these concentrations of the active substance, carrier proteins such as serum albumin must be added to prevent adsorption of the active principle to the walls of the vial, syringe, infusion pump tubing, etc. Also, lyophilized dosage forms usually require a bulking agent such as mannitol or sucrose and an amphoteric stabilizer such as glycine to delocalize possible charge interactions, minimize aggregation, and produce a cosmetically acceptable appearance.

The formulation of the protein is the critical stage of production and each excipient must be of assured clinical quality. The process itself and subsequent filling of the dosage units must be executed under enviromental conditions that preclude contamination by any particulate matter. The sterile filling air handling systems must be aseptic and validated before use and a substantial fraction of the total number of dosage units must be retained and tested for sterility and particle contamination.

It is essential that the stability of the final dosage form is known accurately and in the early stages of a development program, potency must be continually verified while a particular batch of product is in use.

V. QUALITY ASSURANCE AND CHARACTERIZATION

All pharmaceutical products must be produced under conditions that assure that the predicted potency of each dosage unit is close to specification and that the possibility of cross-contamination or mislabeling during manufacture is minimized. These conditions are documented and systematized in accepted industry norms of good manufacturing practice. Ensuring adherence to these principles is the role of quality assurance groups within each pharmaceutical company, and the maintenance of a uniform standard is the role of national regulatory authorities (e.g., U.S. Food and Drug Administration). Each batch of protein manufactured must be produced according to the same detailed set of processing instructions, the batch record, and each routine unit operation within that record must be defined by a validated standard operating procedure. Each piece of equipment must be qualified on installation and calibrated in use and all cleaning procedures must be validated to ensure that cross-contamination of subsequent batches cannot occur. All the chemicals, including most importantly the process water employed during manufacture, must be analyzed before use and their manufacturing specifications must be determined.

The quality of pharmaceutical drugs cannot be ascertained solely by rigorous end-product analysis or final quality control. This is especially true of biologics such as hematopoietic growth factors, for which it is impossible to define an exact structure. The product is therefore partially defined by the method of preparation and significant process changes or changes in formulation could potentially produce novel entities. Before commercial manufacture can proceed it is important to validate the production process itself. There are a number of factors that could potentially lead to variability in the potency of a therapeutic protein. These are summarized in Fig. 4. It is hoped that by reproducible process performance and careful process validation this variation will be eliminated. Validation is usually performed by production of six or more batches to a given batch record and by analyzing the process itself in tems of reproducibility and yield and the quality of each intermediate and the final dosage form. Only then can such a process be used to prepare material for human use.

The quality control of recombinant therapeutics is evolving into a specialist discipline in itself. The two major objectives of this work are to establish the identity of the final product and its purity. Colinearity of the protein amino acid sequence with the recombinant gene sequence must be determined. Each fermentation must be initiated from a vial of a cell bank that has been validated by complete DNA sequencing of the plasmid vector (in the case of microbes) or complete DNA sequencing of the coding region and integration sites (for mammalian cell lines). Figure 5 summarizes the steps necessary to validate a cell blank, which can be then used to supply inocula for production batches. After fermentation the stability of the production system at the genetic level must be validated by sequence analysis. There are many factors during biosynthesis and manufacture that could result in a protein structure different from that predicted from the genetic code of the recombinant insert, and it is vital to determine the exact structure of the final bulk drug substance. As yet there is no one

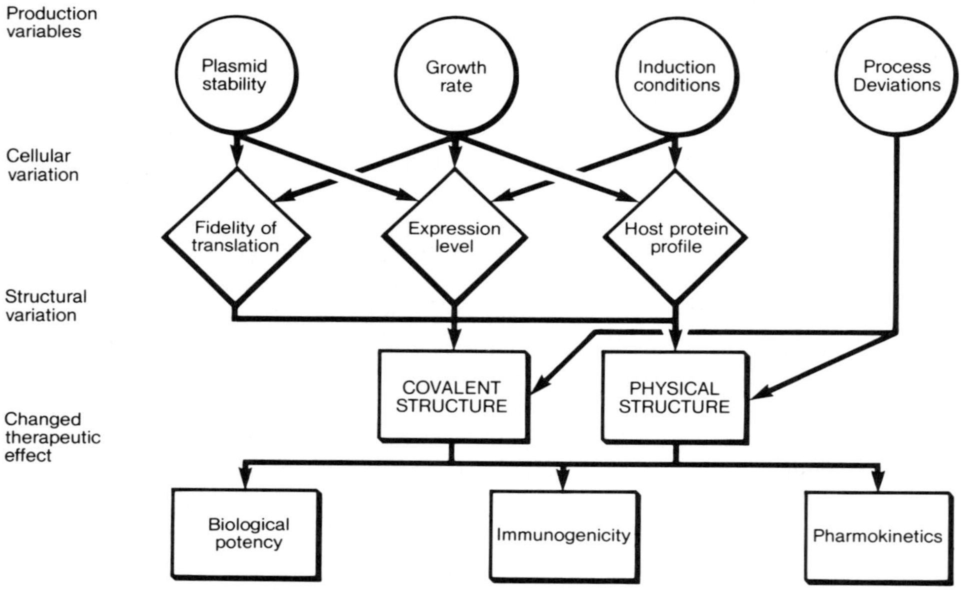

Figure 4 Potential sources of variability in the production of recombinant proteins.

technique that will give routine detailed information on the structure of a protein and a battery of less precise techniques (Table 1) must be used to establish an overall picture of product quality and therefore process reproducibility. Of these, biological activity is perhaps the most informative and as tight a specification as the bioassay will allow must be set to guarantee reproducible specific activity and accurate dispensing of the final dosage form. Colinearity of the protein covalent structure is determined by Edman gas phase amino acid sequencing in conjunction with amino acid analysis and peptide mapping after proteolytic digestion. *N*-terminal analysis and *C*-terminal analysis must be performed on several process batches to validate that consistent levels of initiator methionine are removed during biosynthesis and that no proteolytic truncation has occurred during processing. These methods have also been used to characterize and establish the disulphide bridging pattern of recombinant G–CSF (40,44) and GM–CSF (41).

Investigations of conformational structure include measures of ultraviolet absorption, fluorescence, and circular dichroism. The latter technique has been used to estimate levels of secondary structure in G–CSF (4) and GM–CSF (35) and has been shown to correlate with the pattern predicted from sequence data

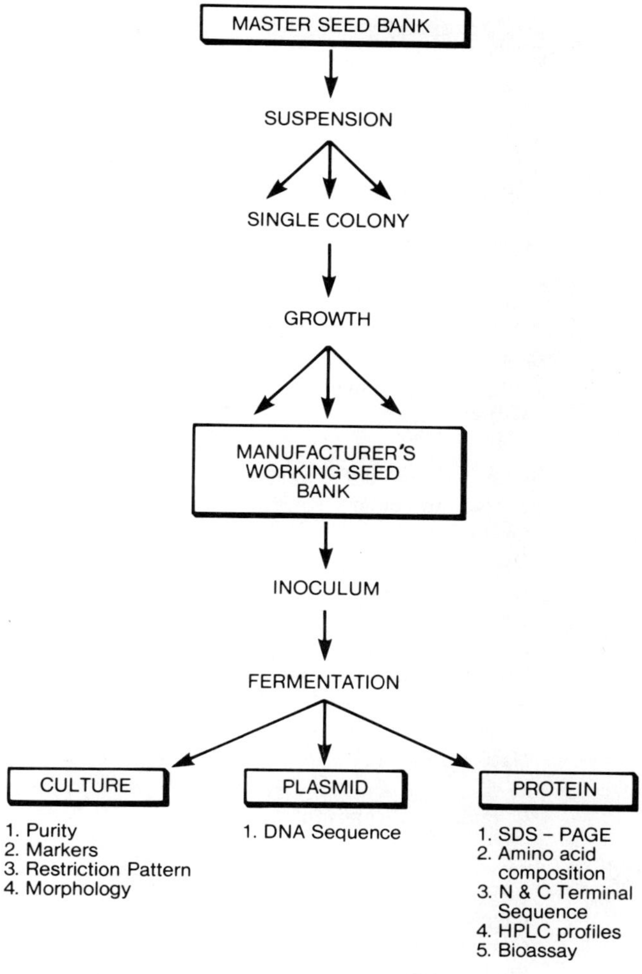

Figure 5 Validation of a production strain.

Table 1 Analytical Methods Used in the Validation of Processes for the Production of Therapeutic Proteins and Their Routine Analysis

Method	Property	Process validation	Routine analysis
Identity tests			
Potency			
activity assay	Biological activity	+	+
Chemical structure			
DNA sequencing	Sequence of gene in seed stock and after fermentation	+	–
Amino acid analysis	Chemical composition	+	+
Edman sequencing	*N*-terminal amino acid sequence	+	+
mass spectrometry	Chemical composition and *C*-terminal analysis	+	+
peptide analysis	Internal amino acid sequence	+	+
Physical structure			
calibrated SDS–PAGE	Covalent molecular weight	+	+
gel filtration	Native molecular weight	+	+
native PAGE	Overall charge	+	+
isoelectric focusing	Overall charge	+	+
reversed-phase HPLC	Overall hydrophobicity	+	+
circular dichroism	Secondary and tertiary structure	+	–
fluorescence	Tyrosine and tryptophan environment of tertiary structure	+	–
analytical ultra-centrifugation	Native molecular structure	+	–
proton NMR	Overall conformation	+	–
Purity tests			
SDS–PAGE	Contaminants differing in covalent molecular weight	+	+
reversed-phase HPLC	Contaminants differing in hydrophobicity	+	+
gel filtration	Contaminants differing in quarternary structure	+	+
ELISA	Contaminants differing in antigenic structure	+	+
DNA hybridization	DNA contamination	+	+

SDS–PAGE, sodium dodecyl sulfate-polyacrylamide gel electrophoresis; HPLC, high-performance liquid chromatography; NMR, nuclear magnetic resonance; ELISA, enzyme-linked immunosorbent assay.

alone. Mass spectrometry and proton nuclear magnetic resonance are both techniques that as yet have not been widely used in pharmaceutical biotechnology. However, they are the only techniques that have the capability of allowing rapid and precise estimation of covalent and conformational structure at the atomic level. Mass spectrometry has been revolutionized by volatilization of protein sample ions without prior derivitization using fast atom bombardment (FAB–MS) (45) and their analysis by high field magnetic and array detectors. Unlike gas phase sequence analysis, FAB–MS methods can also be used to determine most types of posttranslational modification and in the case of the glycosyl moieties of glycoproteins, determine the exact nature of the carbohydrate chains and linkage to the polypeptide chain (33).

Nuclear magnetic resonance (NMR) is also a technique that is being rapidly developed for the determination of the solution structure of proteins. Five-hundred-megahertz machines often give a complex proton NMR spectrum that can be used as a conformational fingerprint for comparing one batch of protein with a reference sample. For proteins of <12,000 molecular weight, interactions between peptide backbone protons and side chains can be identified by two-dimensional NMR methods and the overall conformation of the protein can be solved. Proton NMR has also been used to determine the structure of the glycosyl moieties of CHO cell derived recombinant G–CSF (46) and erythropoietin (34).

In addition to establishing identity of the product each production batch must be analyzed for purity. Purity is usually quantitated by gel electrophoresis, HPLC, and enzyme-linked immunosorbent assay for contaminating host cell proteins. Recombinant DNA methods and modern separation technology has allowed biotechnology products to be manufactured at >99% purity. High purities again increase the assurance of quality because the major contaminating species from biological production systems rarely can be characterized. Of particular concern is the transmission of potentially harmful genetic information through DNA contamination of final dosage units. A guideline from the regulatory authorities, therefore, is that the purification process must produce dosage forms containing <10 p*M* DNA per dose (equivalent to less than one complete host genome). Additionally, the processes used to prepare injectables from mammalian cell culture must be shown to inactivate or remove live virus. A degree of assurance that retroviruses and other potentially harmful pathogens will be eliminated during production can thus be gained indirectly.

VI. SUMMARY

Genetic engineering has given us the means to produce large quantities of hematopoietic growth factors for clinical evaluation as therapeutic agents. The production technology for exploiting advances in genetic engineering is itself a new and rapidly evolving field of endeavor that has added another layer of choice for production. Commercial manufacturers of growth factors are now faced with an ever-expanding variety of alternatives in primary production systems, from *E. coli* to transgenic cows and a variety of choices in separation technologies for the isolation of these factors in a pharmaceutically acceptable form. Overlaying this plethora of options is the possibility of using site-specific mutagenesis to change the natural structure of growth factors to produce new chemical entities with novel pharmaceutical properties. For example, soluble forms of M–CSF (47) have been created by specific deletion of hydrophobic transmembrane regions of the molecule and improved forms of G–CSF have been claimed to have been synthesized by directed amino acid substitution (47).

The utility of these approaches will ultimately be decided by the relative efficacy of each product in the clinic and to a lesser extent process economics in the commercial marketplace.

ACKNOWLEDGMENTS

I thank my many colleagues at the Biogen Research Corporation, Glaxo Institute of Molecular Biology, and ICI Pharmaceuticals for many hours of discussion on the subject of this chapter.

REFERENCES

1. Amgen received approval for the sale of erythropoietin in Switzerland during July 1988.
2. Tao, X., Gao, G., Zhang, H. Z., Boesma, A., Lamblin, G. and K. K. Han. Isolation and characterization of human urinary colony stimulating factor. *Bid. Chem. Hoppe Seyler 368*:187-194 (1987).
3. Yanagi, H., A. Soga, and S. Ogino. Purification of human urinary erythropoietin. *J. Chromatogr. 417*:178-182 (1987).
4. Nomura, H., I. Imazeki, M. Oheda, N. Kubota, M. Tamura, M. Ono, Y. Ueyama, and S. Asano. Purification and characterization of human granulocyte colony stimulating factor. *EMBO J. 5*:871-876 (1986).
5. Price, G. B., J. S. Senn, E. A. McCulloch, and J. E. Till. The isolation and properties of granulocytic colony stimulating activities from medium conditioned by human peripheral leucocytes. *Biochem. J. 148*:209-217 (1975).
6. Clark-Lewis, I., R. Aerbersold, M. Ziltener, J. W. Schrader, L. E. Hood, and S. B. Kent. Structure function studies of interleukin 3 using an automated peptide synthesis approach. *UCLA Symp. Mol. Cell. Biol. New Ser. 41*:323-334 (1987).
7. Libby, R. T., G. Braedt, S. R. Kronheim, C. J. March, D. L. Urdal, T. A. Chiaverotti, R. J. Tushinski, D. Y. Mochizuki, T. P. Hopp, and D. Cosman.

Expression and purification of native human granulocyte-macrophage colony stimulating factor from *E. coli* secretion vector. *DNA 6*:221-229 (1987).
8. Clark, S., and R. Kamen. Haematopoietic growth factors. *Science 236*: 1229-1237 (1987).
9. Stanley, E. R. The macrophage colony stimulating factor, CSF-1. *Methods Enzymol. 116*:564-587 (1985).
10. Rasmussen, J. (Personal communication).
11. Ladner, M. B., G. A. Martin, J. A. Noble, D. M. Nikoloff, R. Tal, E. S. Kawasaki, and T. J. White. Human CSF-1: gene structure and alternative splicing of mRNA precursors. *EMBO J. 6*:2693-2698 (1987).
12. International Patent Application Number PCT/EP85/00326 assigned to Sandoz AG, Basel.
13. Komatzu, Y., T. Matsumoto, T. Kuga, T. Nishi, S. Sekine, A. Saito, M. Okabe, M. Morimoto, and S. Itoh. Cloning of granulocyte colony stimulating factor cDNA from human macrophages and its expression in *E. coli. Jpn. J. Cancer Res. 78*:1179-1181 (1987).
14. International Patent Application Number PCT/US86/01708 assigned to Kirin/Amgen, Thousand Oaks, CA.
15. European Patent Application Number 87309409.8 assigned to Cetus Corp., Emeryville, CA.
16. Burgess, A. W., G. Begley, G. R. Johnson, A. F. Lopez, D. J. Williamson, J.-J. Mermod, R. J. Simpson, A. J. Schmitz, and J. F. Delmarter. Purification and properties of bacterially synthesized human granulocyte macrophage colony stimulating factor. *Blood 69*:43-51 (1987).
17. Devlin, P. E., R. J. Drummond, P. Toy, D. F. Mark, K. W. K. Watt, and J. J. Devlin. Alteration in amino terminal codons of human granulocyte colony stimulating factor increases expression levels and allows efficient processing by methionine aminopeptidase in *E. coli. Gene 65*:13-22 (1988).
18. International Patent Application Number PCT/US84/02021 assigned to Kirin/Amgen, Thousand Oaks, CA.
19. Donahue, R E., R. J. Wang, R. J. Kauffman, L. Fontch, A. C. Leary, J. S. Witek-Giannetti, M. Metzger, R. Hewick, D. R. Steinbrink, G. Shaw, R. Kamen, and S. C. Clark. Effects of N-linked carbohydrate on the in vivo properties of human GM–CSF. *Cold Spring Harbor Symp. Quant. Biol. 51*:685-692 (1986).
20. Kato, C., T. Kobayashi, T. Kudo, T. Furusato, Y. Murakami, T. Tanaka, H. Baba, T. Oishi, E. Ohtsuka, M. Ikehara, T. Yanagida, H. Kato, S. Moriyama, and K. Horikoshi. Construction of an excretion vector. *Gene 54*: 197-202 (1987).
21. Ernst, J. F., J.-J. Mermod, J. F. Delamarter, R. J. Mattaliano, and P. Moonen. O-glycosylation and novel processing events during secretion of α-factor/GM–CSF fusions by *Saccharomyces cerevisiae. Bio/Technology 5*:831-834 (1987).

22. Price, V., D. Mochizuki, C. J. March, D. Cosman, M. C. Deeley, R. Klinke, W. Clevenger, S. Gillis, P. Baker, and D. Urdal. Expression, purification and characterization of recombinant murine granulocyte macrophage colony stimulating factor. *Gene 55*:287-293 (1987).
23. Ernst, J. Efficient secretion and processing of heterologous proteins in *Saccharomyces cerevisiae* is mediated solely by the presegment of α-factor precursor. *DNA 7*:355-360 (1988).
24. European Patent Application Number 87303509.1 assigned to Immunex Corp., Seattle, WA.
25. Shaw, K. J., B. R. Frommer, J. A. Anagnost, S. Navula, and P. J. Leibowitz. Regulted secretion of GM–CSF in *Saccharomyces cerevisiae* via GAL1: MFα1 precursor. *DNA 7*:117-126 (1988).
26. Goto, M., K. Akai, A. Murakami, C. Hashimoto, E. Tsuda, M. Ueda, G. Kawanishi, N. Takahashi, A. Ishimoto, M. Chiba, and R. Sasaki. Production of recombinant human erythropoietic in mammalina cells: host cell dependency of the cloned glycoprotein. *Bio/Technology 6*:67-71 (1988).
27. Moonen, P., J.-J. Mermod, J. F. Ernst, M. Hirschi, and J. F. De Lamarter. Increased biological activity of deglycosylated recombinant human granulocyte macrophage colony stimulating factor produced by yeast or animal cells. *Proc. Natl. Acad. Sci. USA 84*:4428-4431 (1987).
28. Tsuchiya, M., H. Nomura, S. Asano, Y. Kaziro, and S. Nagato. Characterization of recombinant human granulocyte colony stimulating factor produced in mouse cells. *EMBO J. 6*:611-616 (1987).
29. European Patent Application Publication 169566 assigned to Chugai, Tokyo.
30. International Patent Application Number PCT/US85/02405 assigned to Genetics Institute.
31. Inlow, D., D. Marano, and B. Maiorella, Large scale insect cell culture for recombinant protein production. Americam Chemical Society 194th National Meeting on Strategies in Cell Culture Scale Up, Abstract 70 (1987).
32. Wojchowski, D. M., S. M. Orkin, and A. J Sytkowski. Active human erythropoietin expressed in insect cells using a baculovirus vector: a role for N-linked oligosaccharide. *Biochem. Biophys. Acta 910*:224-232 (1987).
33. Sasaki, H., B. Bothner, A. Dell, and M. Fukuda. Carbohydrate structure of erythropoietin expressed in CHO cells by a human erythropoietin cDNA. *J. Biol. Chem. 262*:12059-12076 (1987).
34. Tsuda, E., M. Goto, A. Murakami, K. Akai, M. Ueda, G. Kawanishi, N. Takahashi, R. Sasaki, H. Chiba, H. Ishihara, M. Mori, S. Tejima, S. Endo, and Y. Arato. Comparative structural study of N-linked oligosaccharides of urinary and recombinant erythropoietins. *Biochemistry 27*:5646-5654 (1988).
35. Wingfield, P., P. Graber, P. Moonen, S. Craig, and R. H. Pain. The conformation and stability of recombinant-derived granulocyte macrophage colony stimulating factors.
36. United States Patent Number 4,667,016 assigned to Kirin/Amgen, Thousand Oaks, CA.

37. European Patent Application 119157.3 assigned to Kyowa Hakko Kogyo Co. Ltd., Tokyo.
38. United States Patent Number 4,658,018 assigned to Immunex Corp., Seattle, WA.
39. European Patent Application 119157.3 assigned to Kyowa Hakko Kogyo Co. Ltd., Tokyo.
40. Wingfield, P., R. Benedict, G. Turcatti, B. Allet, J.-J. Mermod, J. J. De Lamarter, M. G. Simona, and K. Rose. Characterization of recombinant derived granulocyte colony stimulating factor (G–CSF). *Biochem. J. 256*: 213-218 (1988).
41. Schrimsher, J. L., K. Rose, M. G. Simona, and P. Wingfield. Characterization of human and mouse granulocyte macrophage colony stimulating factors derived from *E. coli. Biochem. J. 247*:195-199 (1987).
42. Orsini, G., and M. E. Goldberg. Refolding of reduced chymotrypsinogen. *J. Biol. Chem. 253*:3453-3458 (1978).
43. Smith, D. B., and K. S. Johnson. Single step purification of polypeptides expressed in *E. coli* as fusions with glutathione S transferase. *Gene 67*: 31-40 (1988).
44. Lu, M. S., T. C. Boone, L. M. Souza, and P-M. Lai. Disulphide and secondary structures of recombinat human granulocyte colony stimulating factor. *Arch. Biochem. Biophys. 268*:81–92 (1989).
45. Morris, H., and F. M. Greer. Mass spectrometry of natural and recombinant proteins and glycoproteins. *Trends Biotechnology 6*:140-147 (1988).
46. Oheda, M., S. Mase, M. Ono, and T. Ikenaka. Structures of the sugar chains of recombinant human granulocyte colony stimulating factor produced by Chinese hamster ovary cells. *J. Biochem. 103*:544–546 (1988).
47. Kuga, T., Komatsu, Y., Yamasaki, M., Sekine, S., Miyaji, H., Nishi, T., Sato, M., Yokoo, Y., Asano, M., Okabi, M., Morimoto, M., and S. Itoh. Mutagenesis of human granulocyte colony stimulating factor. *Biochem. Biophys. Res. Commun. 159*:103-111 (1989).

13
Clinical Application of Hematopoietic Growth Factors

WILLIAM P. STEWARD, J. HOWARD SCARFFE, T. MICHAEL DEXTER, and NYDIA G. TESTA
Paterson Institute for Cancer Research, Christie Hospital and Holt Radium Institute, Manchester, England

I. INTRODUCTION

The different types of mature cells that appear in the blood stream are continually produced during the process of hemopoiesis. The complexity of the control mechanisms required for this system is illustrated by the number of cells involved (there are 100 times more cells in the bone marrow of a single adult human than there are people in the world) and their rate of turnover (e.g., every hour 1.1×10^{10} erythrocytes come to the end of their lifespan) (1). Despite these formidable numbers, the hemopoietic system is able to maintain the different populations of cells at remarkably constant levels in health, and to respond rapidly under conditions of hemopoietic stress (e.g., increasing erythrocyte production after hemorrhage and lymphocyte numbers after viral infection).

Growth factors play a major role in the control of hemopoiesis, and recombinant DNA technology has recently allowed the production of sufficient amounts of five of these—erythropoietin, granulocyte-macrophage colony-stimulating factor (GM–CSF), granulocyte colony-stimulating factor (G–CSF), macrophage colony-stimulating factor (M–CSF), and interleukin-3 (IL-3)—for use in preclinical testing. As will be reviewed below, erythropoietin, GM–CSF, and G–CSF have already been used in clinical studies in humans. There is great interest in hemopoietic growth factors among a wide variety of clinicians because these agents have considerable potential value in a variety of diseases (Table 1).

The rational use of myeloid growth factors in vivo can be inferred from their observed in vitro effects (Chapter 1). When administered to humans, they might be expected to increase circulating granulocyte and/or macrophage num-

Table 1 Potential Clinical Applications of Myeloid Colony-Stimulating Factors

Treatment of bone marrow failure idiopathic neoplastic iatrogenic
Augment rate of recovery after bone marrow transplantation
Reduce duration or degree of leukopenia after chemotherapy
Increase granulocyte number and function (e.g., in patients with acquired immunodeficiency syndrome)
Treatment of established bacterial and fungal infections
Improve host defense against potential infection after major trauma (e.g., burns)
Treatment of leukemia—alter the rates of self-reproduction and differentiation
Treatment of myelodysplasia—increase normal differentiation and reduce blast population

bers. The new cells should have enhanced properties to counter infection and the latter, possibly, neoplasia. Such effects are of particular interest to oncologists. Many cytotoxic agents show a dose-response relationship in terms of their anticancer effects (2) but their major dose limiting toxicity is myelosuppression. Even with the myelosuppression caused by conventional chemotherapy regimens, many patients require hospitalization for prolonged antibiotic therapy after developing septicemias, and some die. It is likely that the myeloid CSFs could reduce the incidence and severity of such events and furthermore, could allow cytotoxic drugs to be given at a dose and/or frequency that is greater than is currently conventional, with the expectation of improving the response rate and survival of patients with cancer.

This chapter will summarize the information that has been gained about hemopoietic growth factors from preclinical testing in nonhuman primates and other experimental animals and will review the data that is available from clinical studies in humans. Several studies using myeloid growth factors in humans have been completed and many more are ongoing. The majority of trials were only established within the last year and, therefore, several of the results detailed below are from preliminary communications. The potential for the use of these agents in future studies, and the possible toxicity that could be encountered, will also be discussed.

II. EXPERIMENTAL IN VIVO STUDIES

A large amount of information about the effects of hemopoietic growth factors is available from their use in experimental animals and is summarized in Table 2.

Table 2 In Vivo Effects of Hemopoietic Growth Factors on Normal Experimental Animals

Factor	Species	Main effects	Refs.
Murine IL-3	Mouse	Increased spleen colony-forming cells and progenitor cells (mainly in spleen, less in bone marrow). Increased cycling of spleen colony-forming cells. Some increase in peripheral blood cells. Influx of mature cells to the site of infection.	41,42,43
Human GM–CSF Murine GM–CSF	Mouse, monkey	Marked leukocytosis (increase of neutrophils, monocytes, eosinophils). No increase in platelets. Lymphocytosis (monkey). Influx of mature cells to the site of infection.	44,45
Human G–CSF	Mouse, hamster, monkey	Marked neutrophilia. Increase of progenitor cells, mainly in spleen (mouse).	46,47,48
IL-3 + GM–CSF or + M–CSF	Mouse, monkey	Synergistic	40,49,50

IL, interleukin; G, granulocyte; M, macrophage; CSF, colony-stimulating factor.

The stimulation of red blood cell production by erythropoietin has been reviewed extensively in the literature (Chapter 9). Preclinical experience with G–CSF and GM–CSF in normal mammals demonstrated dose-dependent marked increases of peripheral leukocyte counts (up to 10-fold) but although GM–CSF resulted in increments of neutrophils, eosinophils, and macrophages, only an increase of neutrophils was seen with G–CSF. There were also increases in the number of progenitor cells (1), a response more noticeable in murine spleen than in bone marrow. After IL-3, marked increases in the numbers and cycling rate of spleen colony-forming cells of mice (a measure of the stem-cell population) as well as increases in progenitor cells were not translated into increases in peripheral blood cells of the magnitude that might be expected from the changes in the primitive cell populations. In all cases there was a quick return to normal values after the cessation of treatment.

There was an acceleration of leukocyte recovery induced after chemotherapy had been given: Severe neutropenia associated with administration of cyclophosphamide to mice was prevented by daily injections of G–CSF and, importantly, death in these animals caused by inoculation of bacteria (after cyclophosphamide) was also prevented by administration of growth factor. Enhanced and faster hemopoietic recovery of mice treated with 5-fluorouracil after administration of G–CSF has also been observed, shown by higher numbers of circulating cells and of progenitor cells in spleen and marrow (3,4). Administration of IL-1 in combination with G–CSF further enhanced recovery (4). The radioprotective effect that IL-1 exerts on lethally irradiated mice is also enhanced by administration of GM– or G–CSF. The use of these factors allows a lower dose of IL-1 to confer its optimal radioprotection (5). These agents should, therefore, have a useful role in reducing the toxicity associated with chemotherapy or radiotherapy. Recovery after autologous bone marrow transplantation in monkeys was accelerated by administration of GM–CSF (6,7).

III. CLINICAL TRIALS

A. Erythropoietin

Patients with end-stage renal failure become profoundly anemic mainly as a result of the reduced production of erythropoietin. They may require intermittent red blood cell transfusions to maintain hemoglobin at levels that allow them to carry out their usual daily activities. Unfortunately, repeated transfusions carry several risks including the transmission of infections, iron overload, and the development of cytotoxic antibodies that can interfere with subsequent renal transplantation. The hope that the administration of erythropoietin would abrogate the need for transfusions in these patients made them an ideal group to use for the initial clinical trials of this growth factor. The first study was opened in December 1985.

Reports of the initial two phase I/II studies of the use of erythropoietin in a total of 46 patients with end-stage renal failure demonstrated dose-dependent rises in hemoglobin and hematocrit levels (8,9). These findings were confirmed by a subsequent larger ongoing study for which a preliminary report of the first 250 patients is available (10). All but one of the patients treated to date has become transfusion independent and all have reported increased exercise tolerance and sense of well-being as the anemia was corrected. Some patients have received erythropoietin in excess of 2 years and they have remained sensitive to the hormone throughout that time without developing any detectable antibodies.

Acute side effects attributable to erythropoietin have been minimal. Transient mild dyspnea, arthralgia or myalgia, flushing sensation, and bone pain have occasionally been reported but always disappear after the first two or three doses have been given. Approximately 50% of patients experience increases in diastolic blood pressure and, early in the trials, some experienced seizures. With the realization that early institution of effective control of hypertension is required, this complication is becoming infrequent.

Studies have been established to examine the administration of erythropoietin to patients with anemia associated with chronic diseases (such as rheumatoid arthritis). Preliminary communications suggest encouraging results in these conditions. Neoplasia is often associated with anemia (either resulting from the underlying malignancy or from its treatment) and studies are planned to assess the role of erythropoietin in patients with a variety of neoplasms. A second indication for the use of erythropoietin in this group of patients may arise from the observation that at high doses it stimulates megakaryopoiesis in vitro and in vivo (11). The potential of this agent to alleviate the thrombocytopenia induced by chemotherapy and to increase the rate of recovery from bone marrow transplantation is obviously of great interest. To date, erythropoietin has been given by intermittent intravenous bolus injections (e.g., at a dose of 150 U/kg/week in three divided doses for patients with renal failure) and further investigations into alternative routes and schedules of administration (e.g., subcutaneous injections or continuous intravenous infusion) may reveal more effective dose-reponse relationships.

B. GM–CSF

Three formal phase I studies to determine the toxicity and biological effects of GM–CSF in patients with normal bone marrow reserves have been reported. Herrmann et al. (12), using glycosylated GM–CSF, compared the results of intravenous bolus and continuous infusion administration at six different dose levels (30-1000 $\mu g/m^2$/day). At 1000 $\mu g/m^2$/day there was only a modest (twofold) increment of the total white blood cell count (WBC) after bolus administration, but a 17-fold increase was seen during continuous infusion at the same dose. Side effects were reported as mild and included low grade fever, bone discomfort, myalgia, dyspnea, nausea, and headache. Similar disparities in response to the two methods of administration were seen in patients with primary or secondary bone marrow failure given GM–CSF as daily bolus injections or continuous intravenous infusions. No increase in leukocyte counts occurred with bolus injections but up to 8.5-fold rises in granulocyte counts resulted from the continuous infusion (13). Recent work from our own group has confirmed the superiority of continuous intravenous infusions of GM–CSF as compared with intermittent bolus injections, both in terms of hematological response (up to 10-fold increment of WBC at doses of 3 and 10 $\mu g/kg$) (Fig. 1) and lack of any toxicity.

A phase I study with 20 patients given nonglycosylated GM–CSF as daily half-hour intravenous bolus injections at doses of 0.3-60 $\mu g/kg$/day has recently been completed (14). Increases (up to fourfold rises) of total WBC, eosinophil, and neutrophil polymorphs were seen at dose levels of 10 $\mu g/kg$ and higher. Patients developed a neutropenia immediately after the start of administration

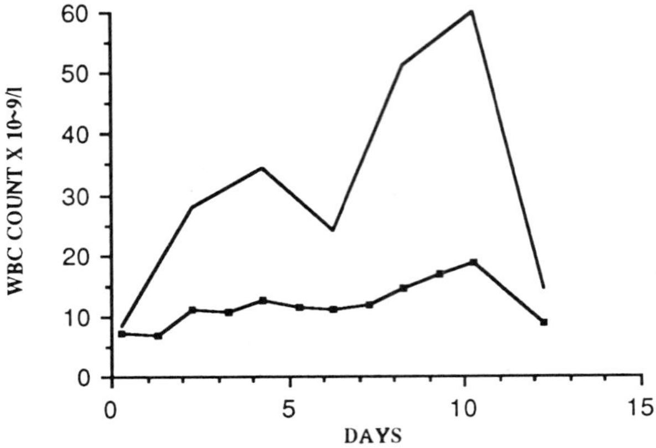

Figure 1 Profile of total leukocyte count in patient receiving granulocyte-macrophage colony-stimulating factor (GM–CSF) given by daily intravenous half-hour bolus injections (–•–) at a dose of 10 μg/kg/day and, 4 months later, as a continuous infusion (——) at a dose of 3 μg/kg/day. The triphasic increase of peripheral leukocyte count seen after the administration of GM–CSF is illustrated with an initial early rise due to dermargination of cells, a subsequent plateau phase, and a final phase of rapid rise due to the appearance of leukocytes produced as a result of proliferation of bone marrow progenitor cells. The two curves show the superiority of continuous infusions over bolus injections, with the former route producing a significantly higher rise of white blood cell count even though the dose of GM–CSF was lower.

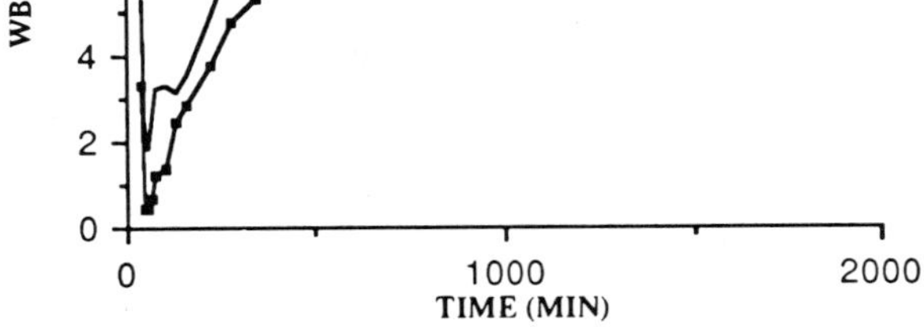

Figure 2 Profile of change in total leukocyte count (——) and neutrophils (–•–) for 18 hr after commencement of half-hour intravenous infusion of granulocyte-macrophage colony-stimulating factor. The counts fell rapidly within 5 min and subsequently returned to pretreatment levels 4 hr later. Higher values were found subsequently. (From Ref. 14, used with permission.)

of GM–CSF and the count returned to baseline levels 4 hr later (Fig. 2). The subsequent rise of the count during the 10 days of daily injections occurred in a triphasic fashion. There was an early increase, probably due to demargination of cells, a subsequent plateau phase, and a final rapid rise after the 8th day resulting from the appearance of leukocytes produced by cell proliferation induced in the bone marrow by GM–CSF. Toxic side effects included transient pyrexias, bone pain, and pruritus. Severe toxicity occurred in four of the five patients who received a dose of 60 μg/kg and included pericarditis and a "capillary leak" syndrome. One patient died of a pulmonary embolus although this may have been tumor related. Of great interest was the observation of significant tumor regression in one patient with a pretreated metastatic soft-tissue sarcoma—an observation that may have important clinical and scientific implications.

Responses to subcutaneous injections of nonglycosylated GM–CSF were examined in the third phase I study (15). This route of administration appears to be more effective than intermittent bolus intravenous injections resulting in fivefold rises of total leukocyte counts at a dose of 15 μg/kg given daily for up to 10 days. Administration by this route was generally well tolerated apart from occasional skin rashes at injection sites.

These phase I studies confirm in vitro observations that continuous exposure to CSFs is required to stimulate cell proliferation and functional activity. The half-life of intravenously injected GM–CSF is only 1-3 hr (16), whereas subcutaneous injections, even at doses of 3 μg/kg, produce serum levels >1 ng/ml (the concentration rquired to give >90% of maximal cell proliferation in vitro) for 10 hr or more (15). Optimal administration for future clinical trials will, therefore, have to be via the subcutaneous route (perhaps twice daily) or by continuous intravenous infusion.

Phase I/II studies in humans have been established to determine the effects of GM–CSF on patients with bone marrow suppression resulting from a variety of causes (chemotherapy, myelodysplasia, aplastic anemia) and encouraging results have been obtained. Multilineage responses to continuous infusions of this growth factor (at doses of 15-500 $\mu g/m^2$/day) were seen in 6 of 14 patients with cytopenias resulting from malignancy (either due to marrow infiltration or after therapy) (17). Up to 70-fold increases of the total leukocyte counts were recorded with minimal toxicity. The potential for reduction of chemotherapy-induced myelotoxicity by GM–CSF has been examined in 16 patients with soft-tissue sarcomas in a phase I/II study (18). Patients received 3-7 days of GM–CSF by continuous intravenous infusion during the phase I part of the trial and were then given a combination of ifosfamide, doxorubicin hydrochloride (Adriamycin), and dimethyltriazenoimidazole carboxamide (DTIC) with subsequent reintroduction of GM–CSF. A second cycle of chemotherapy followed, but without growth factor. There were significantly higher nadir total leukocyte and platelet counts after the first cycle (accompanied by GM–CSF) as compared with the second cycle, suggesting a role for this agent in protecting against cytotoxic-induced myelosuppression.

An increasingly employed method of treating patients with refractory or advanced malignancies is to give high doses of chemotherapy followed by autologous marrow rescue. Unfortunately, there is often considerable morbidity and mortality associated with this procedure and, at the least, the patient is often in hospital for a lengthy (and costly) period until the peripheral blood count returns to normal. A potential role for hemopoietic growth factors could, therefore, be their use to accelerate hemopoietic reconstitution after high-dose chemotherapy and/or radiotherapy with bone marrow transplantation. Brandt et al. (19) reported a study of 19 patients with refractory breast cancer or melanoma who were given high-dose combination chemotherapy (cyclophosphamide, carmustine, and cisplatin) followed by autologous bone marrow transplantation. Granulocyte-macrophage colony-stimulating factor was administered as a continuous intravenous infusion for 14 days starting 3 hr after bone marrow infusion. Differing doses of GM–CSF were used in sequential patient groups. The results were compared with those of four consecutively treated historical controls who received the same chemotherapy/transplant regimen. There was accelerated recovery of leukocytes and granulocytes in patients given GM–CSF (although this failed to reach levels of significance–probably because of the small numbers of patients involved). Those patients receiving doses of >8 μg/kg/day of GM–CSF had higher leukocyte counts 2 weeks after transplantation and more rapid recovery than those at lower doses but the higher doses were associated with more toxicity. Problems with edema, weight gain, and myalgia occurred at 32 μg/kg and in two patients a "capillary-leak" syndrome (with acute renal failure in one) occurred. Despite these side effects, the patients receiving GM–CSF generally tolerated the chemotherapy with much less subjective and objective toxicity than controls did, with fewer episodes of septicemia and less hepatotoxicity and nephrotoxicity. Three other studies have employed high-dose chemotherapy and/or total body irradiation followed by autologous bone marrow transplantation for a total of 36 patients with lymphoid malignancies (20-22). Granulocyte-macrophage colony-stimulating factor was given after transplant by daily short infusions (20,21) or continuous infusion (22) and at doses >60 $\mu g/m^2$/day neutrophil and platelet recovery was significantly accelerated (as compared with historical and concurrent controls). There was also a marked reduction of the days on which patients had fevers and of the duration of hospitalization.

Patients with acquired immunodeficiency syndrome (AIDS) often have a neutropenia in associated with reduced neutrophil function and lymphopenia, a combination that is responsible for the high incidence of life-threatening infections. An important study has investigated the use of GM–CSF in 16 such patients who were leukopenic (23). Varying doses were used intravenously as an initial 1-hr injection followed by a continuous infusion for 14 days. Dose-dependent increases of circulating neutrophils, eosinophils, and monocytes were induced. Neutrophil function tests have recently been reported and show that GM–CSF increased responses in the functional assays used for the majority of patients (24). Eight patients also exhibited increased lymphocyte counts. More prolonged treatment will be necessary in future studies to determine the effects of GM–CSF on morbidity and mortality.

Impressive responses to GM–CSF were seen in eight patients with the myelodysplastic syndrome (25) who were given doses of 50-500 $\mu g/m^2$/day by continuous intravenous infusion for 2 weeks every month. Dose-dependent increases of absolute neutrophil counts were observed in all patients (5-300-fold increments), as well as increases in eosinophils, monocytes, and lymphocytes. Interestingly, rises in platelet and reticulocyte counts were also noted and two patients who had previously required red blood cell and platelet transfusions became transfusion independent. Bone marrow examination after GM–CSF administration showed reduced blast cell counts and decreases in dysplastic changes in all patients. Recent work from this group, using cytogenetic analysis, has demonstrated that mature granulocytes appearing in the peripheral blood after administration of GM–CSF, were derived not only from normal progenitors but also from the myelodysplastic clone (26). Two other studies (27,28) have included a total of 18 patients with myelodysplasia and 12 patients with aplastic anemia. Granulocyte-macrophage colony-stimulating factor was given as short daily intravenous infusions for 7-day periods only (in both studies, the treatment period was lengthened to 14 days in a small number of later patients). Dose-dependent rises of leukocyte counts (predominantly granulocytes and monocytes) were observed in the majority of patients but only one had an increase of platelets. There was a more pronounced effect in patients with myelodysplasia than in those with aplastic anemia. Unfortunately, the bone marrow blast count inceased in four patients, requiring discontinuation of GM–CSF.

C. Granulocyte Colony-Stimulating Factor

Granulocyte colony-stimulating factor has not been administered to patients with normal bone marrow reserves in a formal phase I study, but rather has been used in phase I/II trials in combination with chemotherapy (29-31). The phase I components of these trials demonstrate the activity of this agent in producing rapid increases of the leukocyte counts at doses of 1 $\mu g/kg$/day and higher. No serious toxicity attributed to G–CSF has been seen. The first phase I/II study of the use of G–CSF in humans included 12 patients with small-cell lung cancer

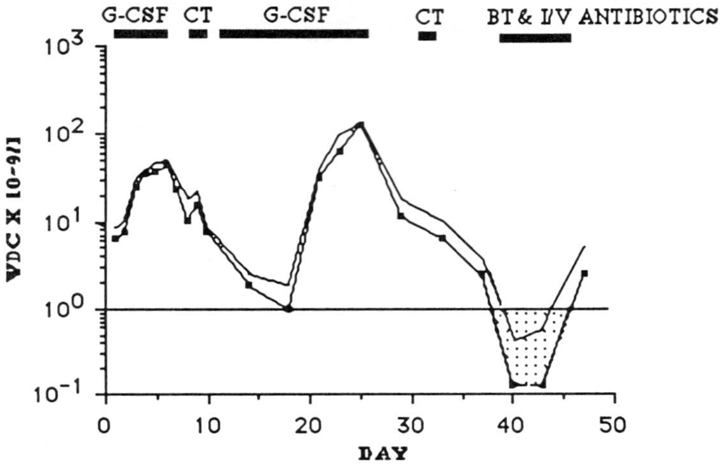

Figure 3 Profile of hematological parameters in one patient treated with recombinant granulocyte colony-stimulating factor (G–CSF) (as a continuous intravenous infusion) and combination chemotherapy. The time course from the start of therapy (in days) is shown on the lower x-axis and the periods during which the patient received the growth factor (G–CSF) at a dose of 10 μg/kg/day, chemotherapy (CT), intravenous antibiotics (I.V. antibiotics), and blood transfusion (BT) is indicated along the top of the diagram. The increase of total white blood cell count (—) produced by G–CSF was due to a selective increase of peripheral neutrophils (–■–). When chemotherapy was followed by G–CSF, the period of absolute neutropenia (count $<1 \times 10^9$/L)–indicated by the shaded area–was almost completely abolished. In contrast, when the patient received the same chemotherapy without G–CSF, she became severely neutropenic and required hospital admission for intravenous antibiotics and a blood transfusion. (Figure courtesy of Dr. M. Bronchud from the study in Ref. 29.)

(29). Granulocyte colony-stimulating factor was given by continuous intravenous infusion, initially for 5 days with increasing doses (1-30 μg/kg) being administered to sequential patients groups. This was followed by combination chemotherapy using etoposide, ifosfamide and doxorubicin hydrochloride. Patients were then assigned sequentially to receive G–CSF on odd or even chemotherapy cycles (commencing 24 hr after the end of chemotherapy and continuing for 14 days). Significant rises in total leukocyte counts (predominantly neutrophils) were seen at all dose levels of G–CSF and the count rose within 48 hr of commencing the infusion. The eosinophilia seen after GM–CSF administration does not occur. The neutrophils were shown to be normal from tests of their phagocytic activity and mobility (32). Although the depth of the neutrophil nadir after chemotherapy was not always reduced, the period of absolute neutropenia was significantly shortened (by a median of 80%) in those cycles followed by G–CSF with a return to normal (or above normal) counts within 14 days of starting cytotoxic administration (Fig. 3). Importantly, the incidence of severe infections was markedly lower when G–CSF was given. Bone marrow trephines showed a 20% increase in cellularity during G–CSF administration.

A phase I/II study using G–CSF given as twice-daily short intravenous infusions at doses of 1-60 μg/kg/day was reported by Morstyn et al. (30). Granulocyte colony-stimulating factor was given for 5 days before and 10 days after melphalan (25 mg/m^2) to 15 patients with a variety of advanced malignancies. There was a dose-dependent rise (up to 10-fold increase) of neutrophils during the first phase of G–CSF. The duration of neutropenia after melphalan decreased with increasing doses of G–CSF, lasting only 1 day at a dose of 30 μg/kg. Pharmacokinetics performed during this study revealed a biphasic clearance of the growth factor with a $t_{1/2}$-alpha of 8 min and a $t_{1/2}$-beta of 110 min.

A third study has shown a protective effect for G–CSF on chemotherapy-induced myelosuppression (31). Twenty-seven patients receiving doxorubicin, cisplatin, vinblastine, and methotrexate for transitional-cell carcinoma of the bladder were given G–CSF (as once-daily half-hour intravenous infusions) after the first or second courses of chemotherapy. Administration of growth factor significantly reduced the duration of neutropenia and the number of days on which antibiotics had to be given after cytotoxic therapy. The addition of G–CSF increased the percentage of patients able to receive their subsequent planned course of chemotherapy without delay from 29 to 100% and significantly reduced the incidence of chemotherapy-related mucositis.

IV. FUTURE CLINICAL APPLICATIONS

Although only a limited number of clinical studies have been reported to date, it is already apparent that erythropoietin, G–CSF, and GM–CSF have important therapeutic indications. Erythroid, leukocyte and, possibly, platelet counts can be increased and these effects have been used to the benefit of patients. The debilitating effects of anemia associated with a variety of diseases can be corrected by erythropoietin. The duration of leukopenia and toxicity of chemotherapy can be decreased by both G– and GM–CSF. The optimal scheduling and route of administration of erythropoietin have not yet been determined but it appears that G–CSF and GM–CSF should be given by continuous intravenous infusion or subcutaneous injection at doses of between 3 and 10 μg/kg/day to obtain the maximal biological effect.

Further studies are now necessary to determine whether the dosages of chemotherapeutic agents can be increased and/or their scheduling can be intensified by safely reducing myelosuppression, the previous major dose-liming toxicity. Both maneuvers may, theoretically, be expected to improve response rates and survival for patients with a variety of neoplasms. Another area of potential clinical application of hemopoietic growth factors is to administer them after high-dose chemotheapy and/or radiotherapy with allogeneic bone marrow transplantation. Such allogeneic transplants are increasingly used for patients with hematological malignancies (particularly leukemias and lymphomas) and the addition of growth factors should reduce the time to full engraftment. Unfortunately, other toxicities of chemotherapy—e.g., mucositis, cardiotoxicity, renal impairment—will remain and may become dose limiting. The observation of tumor regression in a phase I study of GM–CSF (14) increases the potential value of this agent in future studies when it is combined with cytotoxic drugs.

An area of considerable interest for the future use of myeloid growth factors is in patients with acute and chronic myelogenous leukemias. There is in vitro evidence that use of these agents could be contraindicated in such patients: Leukemic populations are dependent on the CSFs for proliferation, the quantitative responsiveness being similar to that of normal cells (33). In addition, two of the CSF genes have been demonstrated to act as proto-oncogenes capable of transforming murine preleukemic cells to fully leukemic cells (34). On the other hand, GM–CSF and G–CSF may induce differentiation commitment in responding neoplastic populations. For example, exposure of WEHI-3B leukemia cell lines to G–CSF leads to differentiated postmitotic granulocytes and monocytes (35). More importantly, the self-generation of myeloid leukemic stem cells can be suppressed by G–CSF in vitro or in animals transplanted with leukemic cells and eventually the leukemic population is completely extinguished by exposure to this growth factor (36). Similar results are obtained using GM–CSF or G–CSF in human promyelocytic leukemia cell lines (37). This field has been reviewed in Chapters 3 and 4. Therefore, although there remains the concern that administration of CSFs to patients with leukemia could accelerate the progress of their disease, there is the intriguing possibility that these agents may have the opposite effect and suppress leukemic populations by favoring differentiation over self-renewal. Anecdotal evidence exists to suggest that it may be reasonable to be optimistic about the effect that myeloid growth factors could have. Buchner et al. (38) administered GM–CSF to six patients with acute leukemia who were at high risk of death and who were aplastic after chemotherapy and noted a more rapid recovery of neutrophil count than in controls. There was no evidence for stimulation of the leukemic population in any of the patients. An alternative approach to the use of myeloid growth factors in leukemia could be to combine them with chemotherapy because there is in vitro evidence that GM–CSF enhances the sensitivity of leukemic clonogenic cells to long-term low-dose cytosine arabinoside with sparing of normal clonogenic cells (39). Such an approach may be safer, at least in initial studies, than giving CSFs alone.

The majority of studies with hemopoietic growth factors have involved patients with malignant diseases. Studies in nonneoplastic conditions (other than renal failure) are still at an early stage but it must be a priority to develop the initial promising work with GM–CSF in patients with AIDS. It will be important to explore the potential for these agents to reduce the morbidity and mortality of patients whose immune function is suppressed as a result of major trauma (e.g., to prevent or reduce the severity of infections that frequently occur in burns victims or in patients who have undergone major surgery). Granulocyte colony-stimulating factor and GM–CSF could also be used for patients with aplastic anemia or familial neutropenias during episodes of infection; this might be a more appropriate use of these agents than giving them for a prolonged period, at least at the present time.

The imminent availability for clinical testing of IL-3 and M–CSF raises the possibility of potentiating the effects of those factors currently available by administering them in combination. Interleukin-3 increases the eosinophil, platelet, and reticulocyte counts in primates and a synergistic effect is noted if GM–CSF is infused after the IL-3 (40). High-dose erythropoietin could be combined with IL-3 and/or GM–CSF to promote increases of leukocytes, platelets, and erythroid cells in patients with pancytopenias.

The normal in vivo response to infection is the elevation of the levels of more than one type of CSF. It would therefore seem logical to assume that optimal results in terms of the prevention of sepsis in humans could be achieved by administering two or more CSFs and a combination of G–CSF and GM–CSF would be one sensible choice. In view of the influx of mature cells to the site of injection (Table 2), local application of growth factors to localized infected areas (for example, peritoneum) may be considered. Although only limited data are at present available on the in vivo effects of combinations of hemopoietic growth factors (49,50), or the combined effects of these plus other bioregulatory molecules (such as IL-1, tumor necrosis factor, transforming growth factor and others), in vitro data (reviewed in Chapter 1) leads to the expectation that combinations of factors may be tailored to specific therapeutic needs. The need to rapidly expand the numbers of mature functional cells (e. g, after chemotherapy or radiotherapy, or to combat infection) would indicate the use of GM– or G–CSF, whereas the need to expand the primitive cell populations (for example, to correct aplastic anemia) may require the use of IL-3 or other factors that will preferentially stimulate the stem cells. These choices will be dictated by the windows of action of the growth factors and their specific effects on cell proliferation and differentiation.

V. POSSIBLE RISKS OF HEMOPOIETIC GROWTH FACTORS

The design of future studies will need to bear in mind some *theoretical* risks of the administration of CSFs. Each factor could divert hemopoietic stem cells into a specific lineage of differentiation and reduce the production of other cells (for example G–CSF, by stimulating myelopoiesis, could lead to reduced erythropoiesis). The CSFs can stimulate mature leukocytes to produce potentially toxic products (e.g., prostaglandins, tumor necrosis factor, fibroblast growth factor, interferons) (51) and can increase the expression of cell surface antigens (53), which may enhance local inflammatory responses. A final concern is that the administration of chemotherapy to patients given CSFs (whose hemopoietic progenitor cells may directly or indirectly be recruited into cell cycle) could lead to marrow aplasia if the timing and extent of this treatment is not carefully considered. Experimental work designed to investigate these possible problems is at present being carried out.

Transgenic mice that constitutively produce 40-fold elevations of GM–CSF provide a model of potential toxicicity of long-term administration of myeloid growth factors (53). The mice become blind, accumulate activated macrophages in the peritoneal and pleural cavities, develop muscle wasting, and ultimately die prematurely. Such effects may be related to macrophage stimulation during critical remodeling of tissues and to continuous high levels of locally produced factor and may, therefore, not be relevant for a patient receiving injected CSF. Nevertheless, it will be important to carefully monitor patients who receive long-term administrations of myeloid growth factors. It would seem sensible to choose dosages of CSFs that do not elevate leukocyte counts to unnecessarily high levels to minimize possible toxicity.

VI. SUMMARY

Although many questions remain about the clinical use of hemopoietic growth factors, it is remarkable how much information has been gained in such a short time. There is enormous interest in these agents among physicians from a variety of specialties—all providing new ideas for potential studies. As the supplies increase, many new trials can be opened. It is important to design these studies with due regard to the results of in vitro and in vivo preclinical experiments—preferably after discussion with experimental hematologists—or vital information will be missed. There can have been few more exciting developments in clinical medicine in the recent past than the availability of the hemopoietic growth factors and, hopefully, it will not be long before the preliminary trials are completed and these agents are accepted for routine use to the benefit of a large number of patients.

REFERENCES

1. Metcalf, D. The granulocyte-macrophage colony-stimulating factors. *Science 229*:16-22 (1985).
2. DeVita, V. T. Dose-response is alive and well. *J. Clin. Oncol. 4*:1157-1159 (1986).
3. Shimamura, M., Y. Kobayashi, A. Yuo, A. Urabe, T. Okabe, Y. Komatsu, S. Itoh, and F. Takaku. Effect of human recombinant granulocyte colony-stimulating factor on hematopoietic injury in mice induced by 5-fluorouracil. *Blood 69*:353-355 (1987).
4. Moore, M. A. S., and D. J. Warren. Synergy of interleukin-1 and granulocyte colony-stimulating factor: in vivo stimulation of stem cell recovery and hematopoietic regeneration following 5-fluorouracil treatment in mice. *Proc. Natl. Acad. Sci. USA 84*:7134-7139 (1987).
5. Neta, R., J. J. Oppenheim, and S. D. Douche. Interdependence of the radioprotective effects of human interleukin alpha, granulocyte colony-stimulating factor and murine recombinant granulocyte-macrophage colony-stimulating factor. *J. Immunol. 140*:108-111 (1988).
6. Monroy, R. L., R. R. Skelly, T. J. MacVittie, T. A. Davis, J. J. Sauber, S. C. Clark, and R. E. Donahue. The effect of recombinant GM–CSF on the recovery of monkeys transplanted with autologous bone marrow. *Blood 70*: 1696-1699 (1987).
7. Neinhuis, A. W., R. E. Donahue, S. Karlsson, S. C. Clark, B. Agricola, N. Antinoff, S. E. Pierce, P. Turner, W. F. Anderson, and D. G. Nathan. Recombinant human granulocyte-macrophage colony-stimulating factor (GM–CSF) shortens the period of neutropenia after autologous bone marrow transplantation in a primate model. *J. Clin. Invest. 80*:573-577 (1988).
8. Winearls, C. G., M. J. Pippard, M. Downing, D. O. Oliver, C. Reid, and P. M. Cotes. Effect of human erythropoietin derived from recombinant DNA on the anemia of patients maintained by chronic hemodialysis. *Lancet 2*:1175-1178 (1986).
9. Esbach, J. W., J. C. Egrie, M. R. Downing, J. K. Browne, and J. W. Adamson. Correction of the anemia of end-stage renal disease with recombinant human erythropoietin. Results of a phase I and II clinical trial. *N. Engl. J. Med. 316*:73-78 (1987).

10. Adamson, J. W., J. C. Egrie, J. K. Browne, M. R. Downing, and W. Esbach. The use of recombinant human erythropoietin (EPO) to correct the anemia of end-stage renal disease: a progress report. *Behring Inst. Mitt. 83*:188-192 (1988).
11. McDonald, T. P., M. B. Cottrell, R. E. Clift, W. C. Cullen, and F. K. Lin. High doses of recombinant erythropoietin stimulate platelet production in mice. *Exp. Hematol. 15*:719-721 (1987).
12. Herrmann, F., G. Schulz, A. Lindemann, W. Meyenburg, W. Oster, D. Krumwieh, and R. Mertelsmann. Yeast expressed granulocyte-macrophage colony-stimulating factor in cancer patients: a phase Ib clinical study. *Behring Inst. Mitt. 83*:107-118 (1988).
13. Rifkin, R. M., E. M. Hersh, and S. E. Salmon. A phase I study of therapy with recombinant granulocyte-macrophage colony-stimulating factor administered by IV bolus or continuous infusion. *Behring Inst. Mitt. 83*: 125-133 (1988).
14. Steward, W. P., J. H. Scarffe, R. Austen, D. Crowther, and P. Loynds. Phase I study of recombinant DNA granulocyte-macrophage colony-stimulating factor (rGM–CSF). *Proc. Am. Soc. Clin. Oncol. 7*:614 (1988).
15. Morstyn, G., G. Lieschke, J. Cebon, D. Maher, J. L. Villeval, U. Duhrsen, K. McGrath, A. Boyd, M. O'Connor, N. A. Nicola, M. Green, W. Sheridan, D. Metcalf, R. Fox, M. Rallings, R. Spiegel, and E. Bonnem. Phase I study of bacterially-synthesized granulocyte-macrophage colony-stimulating factor. Recent developments with hematopoietic growth factors and intron A. *XII Congress International Society of Hematology Milan*, Abstract 6 (1988).
16. Metcalf, D. *The Hemopoietic Colony Stimulating Factors*. Amsterdam, Elsevier (1984).
17. Vadhan-Raj, S., S. Buescher, A. LeMaistre, M. Keating, R. Walters, C. Ventura, W. Hittelman, H. E. Broxmeyer, and J. U. Gutterman. Stimulation of hematopoiesis in patients with bone marrow failure and in patients with malignancy by recombinant human granulocyte-macrophage colony-stimulating factor. *Blood 72*:134-141 (1988).
18. Antman, K., J. Griffin, and A. Elias. Effect of rGM–CSF on chemotherapy induced myelosuppression in sarcoma patients. *Proc. Am. Soc. Clin. Oncol. 7*:160 (1988).
19. Brandt, S. J., W. P. Peters, S. K. Atwater, J. Kurtzberg, M. J. Borowitz, R. B. Jones, E. J. Shpall, R. C. Bast, J. G. Colleen, and D. H. Oette. Effect of recombinant human granulocyte-macrophage colony-stimulating factor on hematopoietic reconstitution after high-dose chemotherapy and autologous bone marrow transplantation. *N. Engl. J. Med. 318*:869-876 (1988).
20. Nemunaitis, J., J. W. Singer, C. D. Buckner, R. Hill, R. Storb, E. D. Thomas, and F. R. Appelbaum. Use of recombinant human granulocyte-macrophage colony-stimulating factor in autologous marrow transplantation for lymphoid malignancies. *Blood 72*:834-836 (1988).
21. Blazar, B. R., M. B. Widmer, J. H. Kersey, N. K. C. Ramsay, P. B. McGlave, D. L. Urdal, S. Gillis, C. Henney, and D. A. Vallera. Recombinant human

granulocyte-macrophage colony-stimulating factor in human and murine bone marrow transplantation. *Behring Inst. Mitt. 83*:170-180 (1988)

22. Link, H., M. Freund, H. Kirchner, M. Stoll, H. Schmid, P. Bucksy, J. Seidel, G. Schulz, R. E. Schmidt, H. Riehm, H. Poliwoda, and K. Welte. Recombinant human granulocyte-macrophage colony-stimulating factor (rhGM–CSF) after bone marrow transplantation. *Behring Inst. Mitt. 83*:313-319 (1988).
23. Groopman, J. E., R. T. Mitsuyasu, M. J. DeLeo, D. H. Oette, and D. W. Golde. Effect of human granulocyte-macrophage colony-stimulating factor on myelopoiesis in the acquired immunodeficiency syndrome. *N. Engl. J. Med. 317*:593-598 (1987).
24. Baldwin, G. C., J. C. Gasson, S. G. Quan, J. Fleischmann, R. Weisbart, D. Oette, R. T. Mitsuyasu, and D. W. Golde. Granulocyte-macrophage colony-stimulating factor enhances neutrophil function in acquired immunodeficiency syndrome patients. *Proc. Natl. Acad. Sci. USA 85*:2763-2766 (1988).
25. Vadhan-Raj, S., M. Keating, A. LeMaistre, W. N. Hittelman, K. McCredie, J. M. Trujillo, H. E. Broxmeyer, C. Henney, and J. U. Gutterman. Effects of human granulocyte-macrophage colony-stimulating factor in patients with myelodysplastic syndromes. *N. Engl. J. Med. 317*:1545-1552 (1987).
26. Hittelman, W. N., and S. Vadhan-Raj. Cytogenetic evidence for in vivo maturation of the abnormal clone after treatment of patients with recombinant human granulocyte macrophage colony-stimulating factor (GM–CSF). *3rd Conference on Differentiation Therapy, Sardinia*, Abstract 97 (1988).
27. Hoelzer, D., A. Ganser, J. Greher, B. Volkers, and F. Walther. GM–CSF in patients with myelodysplastic syndromes—a Phase I/II study. *Third Conference on Differentiation Therapy, Sardinia*, Abstract 57 (1988).
28. Antin, J. H., B. R. Smith, W. Holmes, and D. S. Rosenthal. Phase I/II study of recombinant human granulocyte-macrophage colony-stimulating factor (GM–CSF) in aplastic anemia and myelodysplastic syndrome. *Blood 72*:705-713 (1988).
29. Bronchud, M. H., J. H. Scarffe, N. Thatcher, G. Morgenstern, D. Crowther, L. M. Souza, N. K. Alton, N. G. Testa, and T. M. Dexter. Phase I/II study of recombinant human granulocyte colony-stimulating factor in patients receiving intensive chemotherapy for small cell lung cancer. *Br. J. Cancer 56*:809-813 (1987).
30. Morstyn, G., L. Campbell, L. M. Souza, N. K. Alton, J. Keech, M. Green, W. Sheridan, D. Metcalf, and R. Fox. Effect of granulocyte colony-stimulating factor on neutropenia induced by cytotoxic chemotherapy. *Lancet 1*:667-672 (1988).
31. Gabrilove, J. L., A. Jakubowski, H. Scher, C. Sternberg, G. Wong, J. Grous, A. Yagoda, K. Fain, M. A. S. Moore, B. Clarkson, H. F. Oettgen, K. Alton, K. Welte, and L. Souza. Effect of granulocyte colony-stimulating factor on neutropenia and associated morbidity due to transitional cell carcinoma of the urothelium. *N. Engl. J. Med. 318*:1414-1422 (1988).

32. Bronchud, M. H., M. Potter, G. Morgenstern, M. J. Blasco, J. H. Scarffe, N. Thatcher, D. Crowther, L. M. Souza, N. K. Alton, N. G. Testa, and T. M. Dexter. In vitro and in vivo analysis of the effects of recombinant granulocyte colony-stimulating factor in patients. *Br. J. Cancer 58*:64-69 (1988).
33. Metcalf, D. The molecular biology and functions of the granulocyte-macrophage colony-stimulating factors. *Blood 67*:257-267 (1986).
34. Lang, R. A., D. Metcalf, N. M. Gough, A. R. Dunn, and T. J. Gonda. Expression of a hematopoietic growth factor cDNA in a factor-dependent cell line results in autonomous growth and tumorigenicity. *Cell 45*:531-542 (1985).
35. Metcalf, D., and N. A. Nicola. Autoinduction of differentiation in WEHI-3B leukemia cells. *Int. J. Cancer 30*:773-777 (1982).
36. Metcalf, D. Regulator-induced suppression of myelomonocytic leukemia cells: clonal analysis of early cellular events. *Int. J. Cancer 30*:203-207 (1982).
37. Begley, C. G., D. Metcalf, and N. A. Nicola. Purified colony-stimulating factors (G–CSF and GM–CSF) induce differentiation in human HL-60 leukemic cells with suppression of clonogenicity. *Int. J. Cancer 39*:99-106 (1987).
38. Buchner, T. H., W. Hiddemann, M. Zuhlsdorf, M. Koenigsmann, A. Bockmann, J. van de Loo, and G. Schulz. Human recombinant granulocyte-macrophage colony-stimulating factor (GM–CSF) treatment of patients with acute leukemias in aplasia and at high risk of early death. *Behring Inst. Mitt. 83*.308-312 (1988).
39. deWitte, T., P. Muus, C. Haanen, N. van der Lely, E. Koekman, A. van der Locht, G. Blankenborg, and J. Wessels. GM–CSF enhances sensitivity of leukemic clonogenic cells to long-term low dose cytosine arabinoside with sparing of the normal clonogenic cells. *Behring Inst. Mitt. 83*:301-307 (1988).
40. Donahue, R. E., J. Seehra, C. Norton, K. Turner, M. Metzger, B. Rock, F. Carbone, R. Sieghal, Y. C. Young, and S. Clark. Stimulation of hematopoiesis in primates with human interleukin-3 and granulocyte-macrophage colony-stimulating factor. [Abstract]. *Blood 70*:133a (1987).
41. Kindler, V., B. Thoreus, S. DeKossodo, B. Allet, J. F. Eliason, D. Thatcher, N. Farber, and P. Vassali. Stimulation of hemopoiesis in vivo by recombinant bacterial murine interluekin-3. *Proc. Natl. Acad. Sci. USA 83*:1001-1005 (1986).
42. Lord, B. I., G. Molineux, N. G. Testa, M. Kelly, E. Spooncer, and T. M. Dexter. The kinetic response of hemopoietic precursor cells, in vivo, to highly purified recombinant interleukin-3. *Lymphokine Res. 5*:97-104 (1986).
43. Metcalf, D., C. G. Begley, G. R. Johnson, N. A. Nicola, A. F. Lopez, and D. J. Williamson. Effects of purified bacterially synthesised murine multi-CSF (IL-3) on hematopoietisis in normal adult mice. *Blood 68*:46-57 (1986).

44. Metcalf, D., C. G. Begley, D. J. Williamson, E. C. Nice, J. DeLemarter, J. J. Mermod, D. Thatcher, and A. Schmidt. Hemopoietic responses in mice injected with purified recombinant murine GM–CSF. *Exp. Hematol. 15*:1-9 (1987).
45. Donahue, R. E., E. A. Wang, D. K. Stone, R. Kamen, G. G. Wong, P. K. Sehgal, D. G. Nathan, and S. C. Clark. Stimulation of haematopoiesis in primates by continuous infusion of recombinant human GM–CSF. *Nature 321*:872-875 (1986).
46. Welte, K., M. A. Bonilla, A. P. Gillio, T. C. Boone, G. K. Potter, J. L. Gabrilove, M. A. S. Moore, R. J. O'Reilly, and L. M. Souza. Recombinant human granulocyte colony-stimulating factor: effects on hematopoiesis in normal and cyclophosphamide-treated primates. *J. Exp. Med. 165*:941-948 (1987).
47. Cohen, A. M. K. M. Zsebo, H. Inoue, D. Ming, T. C. Boone, V. R. Chazin, L. Tsai, T. Ritch, and L. M. Souza. In vitro stimulation of granulopoiesis by recombinant human granulocyte colony-stimulating factor. *Proc. Natl. Acad. Sci. USA 84*:2484-2488 (1987).
48. Tamura, M., K. Hattori, H. Nomura, M. Oheda, N. Kubota, I. Imazebi, M. Ono, Y. Oeyama, S. Nagata, N. Shinafuji, and S. Asano. Induction of neutrophilic granulocytosis in mice by administration of purified human native granulocyte colony-stimulating factor (G–CSF). *Biochem. Biophys. Res. Commun. 142*:454-460 (1987).
49. Broxmeyer, H. E., D. E. Williams, G. Hangoc, S. Cooper, R. K. Gillis, R. K. Shadduck, and D. C. Bicknell. Synergistic myelopoietic actions in vivo after administration to mice of combinations of purified natural murine colony-stimulating factor-1, recombinant murine interleukin-3, and recombinant murine granulocyte-macrophage colony-stimulating factor. *Proc. Natl. Acad. Sci. USA 84*:3871-3875 (1987).
50. Broxmeyer, H. E., D. E. Williams, S. Cooper, R. K. Shadduck, S. Gillis, A. Waheed, D. L. Urdal, and D. C. Bicknell. Comparative effects in vivo of recombinant interleukin-3, natural murine colony-stimulating factor-1, and recombinant murine granulocyte-macrophage colony-stimulating factor on myelopoiesis in mice. *J. Clin. Invest. 79*:721-730 (1987).
51. Kurland, J. I., L. M. Pelus, P. Ralph, R. S. Bockman, and M. A. S. Moore. Induction of prostaglandin E synthesis in normal and neoplastic macrophages: role for colony-stimulating factor(s) distinct from effects on myeloid progenitor cell proliferation. *Proc. Natl. Acad. Sci. USA 76*:2326-2330 (1979).
52. Lopez, A. F., D. J. Williamson, J. R. Gamble, C. G. Begley, J. M. Harlan, S. J. Klebanoff, A. Waltersdorph, G. Wong, S. C. Clark, and M. A. Vadas. Recombinant human granulocyte-macrophage colony-stimulating factor stimulates in vitro mature human neutrophil and eosinophil function, surface receptor expression, and survival. *J. Clin. Invest. 78*:1220-1228 (1986).
53. Lang, R. A., D. Metcalf, R. A. Cuthbertson, I. Lyons, E. Stanley, A. Kelso, G. Kannourakis, D. J. Williamson, G. K. Klintworth, T. J. Gonda, and A. R.

Dunn. Transgenic mice expressing a hemopoietic growth factor gene (GM–CSF) develop accumulations of macrophages, blindness, and a fatal syndrome of tissue damage. *Cell 51*:675-686 (1987).

14

Hematopoietic Growth-Factor Signal Transduction and Regulation of Gene Expression

WILLIAM L. FARRAR, ANNA T. BRINI, and ANNICK HAREL-BELLAN*
National Cancer Institute–Frederick Cancer Research Facility, Frederick, Maryland

MARIE KORNER and DOUGLAS K. FERRIS
Program Resources, Inc., National Cancer Institute–Frederick Cancer Research Facility, Frederick, Maryland

I. THE HEMATOPOIETIC GROWTH FACTORS AND THEIR RECEPTORS

The differentiation and maturation of cells of the myeloid lineage is under the control of a family of soluble, polypeptidic factors, collectively referred to as colony-stimulating factors (CSFs) (1). In the mouse, four of these factors have been isolated, cloned, and their molecular and biological characteristics have been determined. The human counterparts of these murine factors have also been cloned and their biological characteristics have been studied. This section will focus on the molecular characteristics of both human and murine CSFs, namely interleukin-3 (IL-3), also called multipotential colony-stimulating factor (multi-CSF), granulocyte-macrophage colony-stimulating factor (GM–CSF), granulocyte colony-stimulating factor (G–CSF), and macrophage colony-stimulating factor (M–CSF), also called CSF-1.

Two of the four CSFs, G–CSF and M–CSF, proved to be relatively lineage-specific. Granulocyte colony-stimulating factor preferentially stimulates colony formation of neutrophilic granulocytes whereas M–CSF acts on macrophage precursors. In contrast, the colony-stimulating action of GM–CSF and multi-CSF (IL-3) leads to the formation of colonies of macrophages and neutrophils as well as eosinophils, erythroid cells, megakaryocytes, and mast cells in the case of IL-3 (1).

Each of these factors possesses very high biological activity (picomolar range), and each exists only in trace amounts in physiological fluids. The very low concentrations of these factors in tissues and body fluids rendered their purification

*Present affiliation: Bâtiment de Recherches, Villejuif, France.

extremely difficult. However, most of them are purified from either cell-conditioned media (2) or from natural body fluids.

Although the murine IL-3 has been purified from WEHI-3 cell-line-conditioned medium, human IL-3 is still not purified. However, Yang et al. (3), using a gibbon IL-3 cDNA as hybridization probe, isolated the human IL-3 gene. The amino acid sequence of the human IL-3 deduced from the DNA sequence presented only 29% structural homology with its murine counterpart, but differed only by 11 amino acids with the gibbon recombinant IL-3, which was shown to present most of the biological activities attributed to murine IL-3 (3).

In contrast, murine IL-3 has been purified to homogeneity (4), its cDNA cloned (5,6), and its chromosomal gene was isolated and characterized (7). Both the human and murine IL-3 genes consist of five relatively small exons separated by one large and three small introns. The regions of highest sequence identity were found in the putative promoter region (59%). The apparent molecular weight of purified murine IL-3 was 28,000 (4). It consisted of a single glycosylated polypeptide chain. The analysis of the human IL-3 nucleotide sequence also indicates two potential *N*-glycosylation sites (3). The expected mass for human IL-3 is 14-15 kDa. The biological importance of IL-3 glycosylations is not known. In a recent review Clark and Kamen (8) discussed the biological importance of glycosylation of hemopoietic CSFs. They pointed out that IL-3 and GM–CSF possess many common structural and biological features. Granulocyte-macrophage CSF, like IL-3, is produced by T cells (but also by endotherlial cells and fibroblasts), and as mentioned above they both stimulate colony formation of granulocytes and macrophages.

Murine GM–CSF was originally purified from mouse-lung conditioned medium (2) where the specific activity appeared to be very high (60,000 colonies/mg of protein) as compared with other sources (for example, 890 colonies/mg of protein in endotoxin-treated mice sera). Like IL-3, murine GM–CSF is a monomeric glycoprotein. It has an apparent molecular weight of 23,000 (2). The primary structure of the protein deduced from the nucleotide sequence has a molecular weight of 13,000, which is in good agreement with the apparent molecular weight of 16,800 found after extensive deglycosylation with endoglycosidase F (9). In the case of GM–CSF, the oligosaccharide moieties seem to reduce growth-factor activity (8). Furthermore, murine GM–CSF was found to be encoded by a unique gene, located on chromosome 11 (9). Therefore, the physical and biological differences found between GM–CSF from different organs can be probably attributed to posttranscriptional modifications (10).

The natural human GM–CSF, also called CSF-2, was initially purified from MO-T-lymphoblast cell-line-conditioned medium (11). Its molecular weight ranged from 18,000 to 25,000 (12), reflecting the heterogeneous glycosylation of this hemopoietin. The natural human GM–CSF was partially sequenced (12).

The cDNA for human GM–CSF was cloned (13) and its sequence predicted a 144 amino acid polypeptide. Comparison of murine and human GM–CSF primary sequences deduced from their respective cDNA showed a 60% sequence homology (13). The complete human gene was 2.5 kilobases (kb) in length, contained three introns, and was present as a single copy in the human genome (14). Interestingly, the location of the human GM–CSF was found to be on the long arm of chromosome 5 (15), which was also the location of the IL-3 gene (8), the M–CSF gene (16), and its receptor, the c-*fms* gene (15). The concentration of these genes on chromosome 5 and their spacial proximity is certainly intriguing and raises provocative questions about the regulation of expression of these genes.

Although IL-3 and GM–CSFs present overlapping biological activities and possess identical *N*-terminal dipeptide (Ala- Pro), there is no significant amino acid sequence identity between these hemopoietins.

As mentioned above, G–CSF and M–CSF almost exclusively stimulate colony formation of, respectively, neutrophilic granulocytes and macrophages from committed precursor cells in semi-solid agar culture (17-19). Murine G–CSF was initially purified from endotoxin-stimulated mice-lung cell-conditioned medium and possessed a molecular weight of 24,000-25,000 (20). Human G–CSF was purified from the conditioned medium of the human squamous carcinoma cell line CHU2, which produces G–CSF constitutively (21). The molecular weight of G–CSF from CHV2 cells was 18,000-22,000. Both human and murine G–CSF are monomeric glycoproteins. Human G–CSF was partially sequenced, which allowed chemical synthesis of an oligonucleotide probe (22) and the final identification of the G–CSF gene (23). The human G–CSF gene consists of five exons separated by four introns. An alternative use of 5′ splice donor sequence in intron 2 might be a reason for the production of two polypeptides, which were found to differ by a deletion/insertion of three amino acids at the position 36-38 (23). Whether there are differences in the activity and production of these two forms of G–CSF remains unclear.

The murine G–CSF cDNA was isolated and characterized using the human G–CSF cDNA as probe (24). The mature G–CSF sequence consisted of 178 amino acids corresponding to a molecular weight of 19,061. Recombinant human and murine G–CSF were 72.6% identical in amino acid sequence. This important structural homology is in agreement with the full two-way biological cross-reactivity of murine and human G–CSFs (25-27). No *N*-glycosylation sites were found in either human or murine G–CSF sequences but because some oligosaccharide moieties were found in the purified human G–CSF (21) and the natural murine G–CSF molecular weight is 24,000-25,000, it is likely that these hemopoietins are *O*-glycosylated (23,24). Furthermore, unlike glycosylation of GM–CSF, which reduces its activity, the glycosylation of G–CSF does not interfere with its biological activities, because recombinant G–CSF produced in *Escherichia coli* proved to be highly active (28).

Macrophage colony-stimulating factor (alternatively named CSF-1) is the only one of the four major hemopoietins to be homodimeric. The two subunits are linked by a disulfide bond and the molecule loses its biological activity when it is reduced into two separated monomers (29,18). Both murine and human M–CSFs are heavily glycosylated (~40% of the total mass is due to *N*-glycosylations) (29) but the sugar moieties do not interfere with receptor binding and biological activities (29). Murine M–CSF was purified from L-cell-conditioned medium and was characterized as a heterogenous sized molecule (18,30). The size varied between 65,000 and 70,000 when examined in nonreducing conditions. Human CSF-1 was purified from human urine (31) and presented substantial size heterogeneity: 40,000-70,000 (32) and 85,000 when analyzed by gel filtration (33). The diversity of M–CSF molecular weight was attributed to heterogeneity in *N*-glycosylation. Indeed, treatment with endo β-*N*-acetylglucosaminidase D produced a much smaller active species (29). A rough estimate of the molecular weight of the polypeptide portions of both murine and human M–CSF subunits was 14,500 (29). Recently, however, the study of the biosynthesis of M–CSF raised the possibility of the coexistence of at least two structurally related proteins (34,35). Indeed, two different mRNAs both encoding M–CSF precursor peptides were isolated. The "small" 1.5-2.0 kb mRNA encodes for a 26-kDa "small" precursor, whereas the "large" (~4-kb mRNA) encodes for a "large" 61-kDa precursor. The two mRNAs are transcripts of the same gene, but with differential splicing. The coding region of the "large" sequence differs from the coding region of the "small" sequence by an insertion of 894 base pairs between nucleotides 721 and 722 of the latter (35). This insert preserves the reading frame in such manner that the resulting precursors possess the same amino acids sequence except for an insert of 298 residues between the residue 181 and 182 of the "small" precursor. Both precursors are then further processed by proteolytic cleavages to yield mature polypeptides of 26 kDa (223 residues) and probably a 16-kDa polypeptide (145 residues) (8,35). Analysis of mRNA from different cellular sources of M–CSF revealed that the most abundant form was approximately 4 kb in size (35). The M–CSF precursors both contain a *C*-terminal hydrophobic domain reminiscent of a membrane "anchor." It was therefore suggested that secreted M–CSF might be a proteolytic product of a membrane-bound precursor (8,34). However, mutated cDNAs for M–CSF, lacking the transmembrane domain, have been found to express a secreted form of M–CSF when transfected into COS1 cells (8). Therefore, the mechanism of M–CSF secretion remains to be determined.

Colony-stimulating factors exert their biological activities by binding to specific plasma membrane receptors (Table 1). The receptors were found both on

Table 1 Colony-Stimulating Factors (CSFs) and Their Receptors

Name[a]	CSF M_r[b] (10^{-3}) Murine	Human	Subunits	Receptor M_r (10^{-3})	K_d (M)
IL-3	28	14-15	1	60-70	1.1 (10^{-10})
GM–CSF	23	18-25	1	84-130	10^{-8}-10^{-9} 3 (10^{-11})
G–CSF	24-25	18-22	1	150	6-9 (10^{-11})
M–CSF	65-70	40 70	2 2	165	4 (10^{-10})

[a]IL-3, interleukin-3 (also called multi-CSF or CSF-2α); G, granulocyte; M, macrophage; GM–CSF also called CSF-2, M–CSF also called CSF-1, G–CSF also called CSF-β.
[b]Sizes of natural CSF except for human IL 3.

tissue cells and in vitro cultured cell lines. The density of CSF receptors on cells is variable but generally very low [50-500 sites for G–CSF per normal mature murine granulocyte (36,37) and 50,000-150,000 sites for M–CSF per murine bone-marrow-derived macrophages (38)]. The receptor-CSF interaction is generally followed by internalization of the complex and its subsequent degradation in the cell (27). The affinities of CSF receptors for their respective ligands were determined by binding studies using ^{125}I-labeled natural or recombinant CSFs. The equilibrium dissociation constants (K_d) of CSF receptors-CSFs interactions are in picomolar to nanomolar range at 37°C. The binding sites appear generally as homogeneous populations, except in the case of GM–CSF receptors. Indeed, using natural purified GM–CSF, Walker and Burgess (39) saw evidence of two populations of GM–CSF binding sites on mouse neutrophils. The major populaton possessed a "low" apparent affinity ($K_d = 10^{-9}$ *M*) whereas a minor population (3-16% of the total binding sites) appeared as a "higher" affinity site ($K_d = 2.10^{-11}$ *M*). However, Park et al. (40), using recombinant GM–CSF, identified a single population of sites on seven different cell lines with an apparent affinity close to the "low"-affinity sites ($K_d = 10^{-8}$-10^{-9} *M*). Whether the two populations of binding sites evidenced with natural GM–CSF by Walker and Burgess are due to a heterogeneity of GM–CSF glycosylation or correspond to molecular receptor entities is not clear. Recently, however, Di Persio and coworkers (41) examined the binding of human recombinant GM–CSF to a variety of human cells and cell lines. Like Park et al. (40) they found a single population of binding sites but with much higher affinity ($K_d = 4.10^{-11}$ *M*).

Very little is known about the structure of CSF receptors and even less is known about the mechanism by which CSF signals are transduced.

The CSF receptors are glycoproteins with molecular weights ranging from 60,000-70,000 for the IL-3 receptor (42,43) to 165,000 for the M–CSF receptor (44,45). The receptor for GM–CSF is a 89,000 (41) to 130,000 kDa molecule (40), which is rapidly degraded into a much smaller 70,000 molecule in the absence of protease inhibitors. The G–CSF receptor is 150,000 kDa (46) (Table 1).

By contrast with what is known about the IL-3, GM–CSF, and G–CSF receptors, knowledge of the M–CSF receptor is much more advanced. The M–CSF receptor was recently identified as product of the c-*fms* protooncogene (47). Like receptors for platelet-derived growth factor (PDGF) and epidermal growth factor (EGF), the M–CSF receptor is a monomeric transmembrane glycoprotein possessing an extracellular ligand binding site and an intracellular *C*-terminal enzymatic domain that specifies tyrosine-specific protein kinase activity (47). The *C*-terminus of the molecule possess an essential tyrosine at position 969 that when phosphorylated reduces the protein kinase activity. In contrast to c-*fms* protooncogene, the v-*fms* gene 3′ coding sequence is truncated and therefore the protein does not contain the 969 tyrosine (48). The mutation of tyrosine 969 to phenylalanine and the mutation of the M–CSF binding site are responsible for tumorigeneity of the cells expressing these receptors. Interestingly, the transformation of murine IL-3-dependent cell lines by the v-*fms* gene leads to factor independence and tumorigeneity without affecting the affinity or density of IL-3 receptors on these cells (49). Therefore, v-*fms* can transform IL-3-dependent cells by a nonautocrine mechanism. Even though IL-3 receptor tyrosine kinase activity has not been established, IL-3 induces tyrosine phosphorylation of a 70-kDa protein, which could be the IL-3 receptor itself, in IL-3-dependent murine cell line FDC-PI cells (50). These observations point out the importance of tyrosine phosphorylation as a regulatory event of proliferation and differentiation of cells from the hemopoietic lineage.

II. GROWTH-FACTOR-INDUCED PROTEIN PHOSPHORYLATION

The biological responses of lymphoid and myeloid cells to lymphokines and CSFs have been extensively described in the literature. In contrast to the vast amount of work detailing those responses, however, very little effort has been made toward an understanding of the biochemical mechanisms involved in signal transduction initiated by the respective receptors. What little work has been accomplished in this area has focused mainly on the IL-2 receptor of activated T lymphocytes. Thus, any description of the probable biochemical mechanisms of actions of CSFs must rely in part on analogies with IL-2 and even in fairly large measure, by comparison with what is known of signal mechanisms mediated by other growth factors such as EGF, PDGF, and insulin. It must be born in mind, however, that analysis of the latter factors and their mechanisms of action are not complicated by the issues of survival and differentiation. Fibroblasts, for instance, do not die when EGF is removed; they merely cease growing and when restimulated with EGF they simply renew proliferation. The CSFs, on the other hand, promote both the survival and proliferation as well as the differentiation of the vast majority of their normal target cells (19,51). For this reason it is at least possible that their signaling mechanisms are at some level different than those for other growth factors.

Growth-factor receptors can, at present, be divided into two classes: those that have intrinsic tyrosine kinase activity, such as the receptors for CSF-1 (52), EGF (53), PDGF (54,55), and insulin (56), and those for which no intrinsic kinase activity of any type has been identified, such as receptors for GM–CSF, IL-3, and IL-2. The receptors that have intrinsic tyrosine kinase activity are generally relatively large proteins, with molecular masses from 135 kDa for the bovine-brain-derived growth-factor receptor (57) to 165 kDa for CSF-1r (52), and 170 kDa for EGFr (53). Structurally these receptors consist of, at minimum, an external ligand binding site, a membrane-spanning region, and a cytoplasmic kinase domain that is activated on ligand binding.

Although no intrinsic kinase activity has been found for a number of growth-factor receptors, it is not yet certain that none will be found. In the case of the the IL-2 receptor it was previously shown that antibodies to the Tac antigen could effectively block IL-2 binding and function, and it was assumed that Tac represented the entirety of the IL-2 receptor. Several groups demonstrated that Tac could be phosphorylated both in vitro (58) and in situ (59) and for a time it seemed possible that Tac itself might be a protein kinase. The molecular cloning and sequencing of Tac, however, demonstrated conclusively that the intra-cytoplasmic portion of Tac wa not large enough to contain protein kinase activity (6). More recently it has been shown that the high-affinity IL-2 receptor consists of both Tac, now called the IL-2r α chain, and a 70-kDa protein termed the IL-2r β chain (61-64). Therefore, it remains possible that the high-affinity IL-2 receptor, which is responsible for most of the biological effects of IL-2, may have a kinase domain, because the β chain has not been thoroughly characterized. Similar revelations could be forthcoming for other receptors and it would not be surprising to find that every growth-factor receptor either has intrinisic kinase activity or is very closely associated with a protein that does.

The binding of a growth factor to its receptor initiates a signal transduction pathway that traverses the cell membrane, passes through the cytoplasm into the nucleus, culminating in altered transcription of certain genes and eventual cell division, or in the case of CSF stimulation, cell division with concomitant differentiation. Although the entire process may take many hours, the initial signal mechanisms must be quite rapid because changes in transcription of some genes, such as c-*fos*, can be detected within minutes of growth-factor binding (65,66). In fact, several rapid intracellular events take place almost universally when cells are stimulated by growth factors, regardless of cell type or to which growth factor they are responding. These rapid consensus events include changes in membrane potentials, cytoplasmic alkalinization, ion fluxes, and alterations in protein phosphorylation. Other rapid events, such as hydrolysis of membrane phospholipids or alterations in cyclic nucleotide levels, are very common but not apparently universal (67-69). Any, and potentially all, of these rapid events might be involved in the signaling mechanism of a growth factor that elicits them; however, some (or all) could conceivably be irrelevant in the absence of data proving their necessity.

Protein phosphorylation has been one of the major focal points in efforts to understand growth-factor-induced signal transduction. Indeed, it has been recognized for many years that reverisble phosphorylation of proteins is probably the single most important mechanism regulating cell metabolism and is involved at some level in virtually every cellular process. The discoveries during the last 8 years that an entire class of protooncogene products and growth-factor receptors possess tyrosine kinase activity have served to support the fundamental involvement of protein phosphorylation in signal transduction and growth regulation (70-72). In contradiction to this idea, studies using antibodies directed against the EGF receptor have suggested that some of the effects of EGF such as calcium flux and alkalinization were independent of the kinase function of the receptor (73). More recently, however, functional analyses of EGF receptors mutated at a site required for phosphate transfer have indicated tht the biological effects of EGF require the tyrosine kinase activity of its receptor (74).

A. Phosphorylation Induced by IL-2

Interleukin-2, the first cytokine to be molecularly studied, has preceded many of the CSFs in studies regarding its mechansm of action. We will summarize here some of the known biochemical and molecular effects of IL-2 on lymphocyte cell lines in comparison with the rapid advances being made with CSFs.

The signal process initiated by binding of IL-2 to its high-affinity receptor has been investigated by several groups. Among the rapid events that have been reported are: increased GTP binding and hydrolysis (75), activation and translocation of protein kinase C (PK-C) (76), activation of S6 kinase (77), and phosphorylation of a variety of cellular proteins on serine or threonine residues. We reported that both IL-2 and a pharmacological activator of PK-C, synthetic diacylglycerol (1-olyeol-2-acetyl-glyceral (OAG)) stimulated the phosphorylation of proteins of 30 and 68 kDa (78). Phosphorylation of a 67/68 kDa protein was also reported by Isshi et al. (79). In addition to these proteins, IL-2 stimulated the phosphorylation of two additional proteins, p60 and p80, which were not phosphorylated in response to OAG (78). This indicated that multiple kinase systems were probably activated by the growth factor. This probability was confirmed when we found that p30 and the S6 protein of the 40S ribosomal subunit were identical and that phosphorylation in situ in response to IL-2 or OAG produced a different tryptic phosphopeptide pattern than was produced from S6 phosphorylated in vitro by purified PK-C. In fact, we were able to isolate an S6 kinase activity from cell extracts that was free of PK-C activity, and this purified kinase-phosphorylated S6 in vitro to generate a phosphopeptide map identical to that generated when S6 was phosphorylated in situ in response to either IL-2 or OAG (77).

B. CSF-Induced Phosphorylation

In an earlier study we described that IL-3 stimulated activation and translocation of PK-C that was analogous to results seen with IL-2 (80). We have also examined phosphorylation induced by myeloid growth factors in a variety of cell lines. We found that IL-3 and G–CSF induced the phosphorylation of a 68 kDa protein in myeloid cell lines that proved to be identical to the p68 phosphorylated in response to IL-2 in lymphoid cell lines (81,82). The phosphorylation of p68 in both lymphoid and myeloid cell lines responding to appropriate growth factors suggested that it represented a point of convergence for several signal-transduction pathways. Figure 1 shows two-dimensional NEPHGE gel analysis of the total cellular extracts of ^{32}P-labeled FDC-P1 cells stimulated by IL-3 or PK-C activator OAG. Interleukin-3 stimulates the phosphorylation of p68, p80, and p20. Dephosphorylation of p110 protein was also seen with IL-3. Synthetic diacylglyceral (OAG) stimulated phosphorylation of p68, p20, and p80 but did not affect dephosphorylation of p110. The ability of synthetic diacylglyceral to affect the phosphorylation of some, but not all, of the proteins regulated by IL-3 suggested that an enzyme system identical or analogous in substrate specificity to PK-C may play a part of the spectrum of kinases activated by the growth ligand. In addition to our work, Garland (83) has reported that a 33-kDa protein was rapidly phosphorylated in two different murine myeloid cell lines, AC2 and 123, in response to IL-3. In addition, it was found that in a spontaneous IL-3-independent clone of AC2, p33 was constitutively phosphorylated. Further study revealed that p33 was located primarily in the endoplasmic reticulum and was phosphorylated entirely on serine residues (J. Garland, personal communication).

We and others have begun to investigate the involvement of tyrosine phosphorylation in CSF-initiated signal transduction. Koyasu et al. (84) reported on the IL-3-stimulated tyrosine and serine phosphorylation of a membrane glycoprotein of 150 kDa. Using an antiphosphotyrosine antibody coupled to sepharose to enrich for tyrosine-phosphorylated proteins and high-resolution two-dimensional NEPHGE, we have identified two proteins of 50 and 70 kDa that are rapidly phosphorylated in response to IL-3 (Fig. 2). Phospho-amino-acid

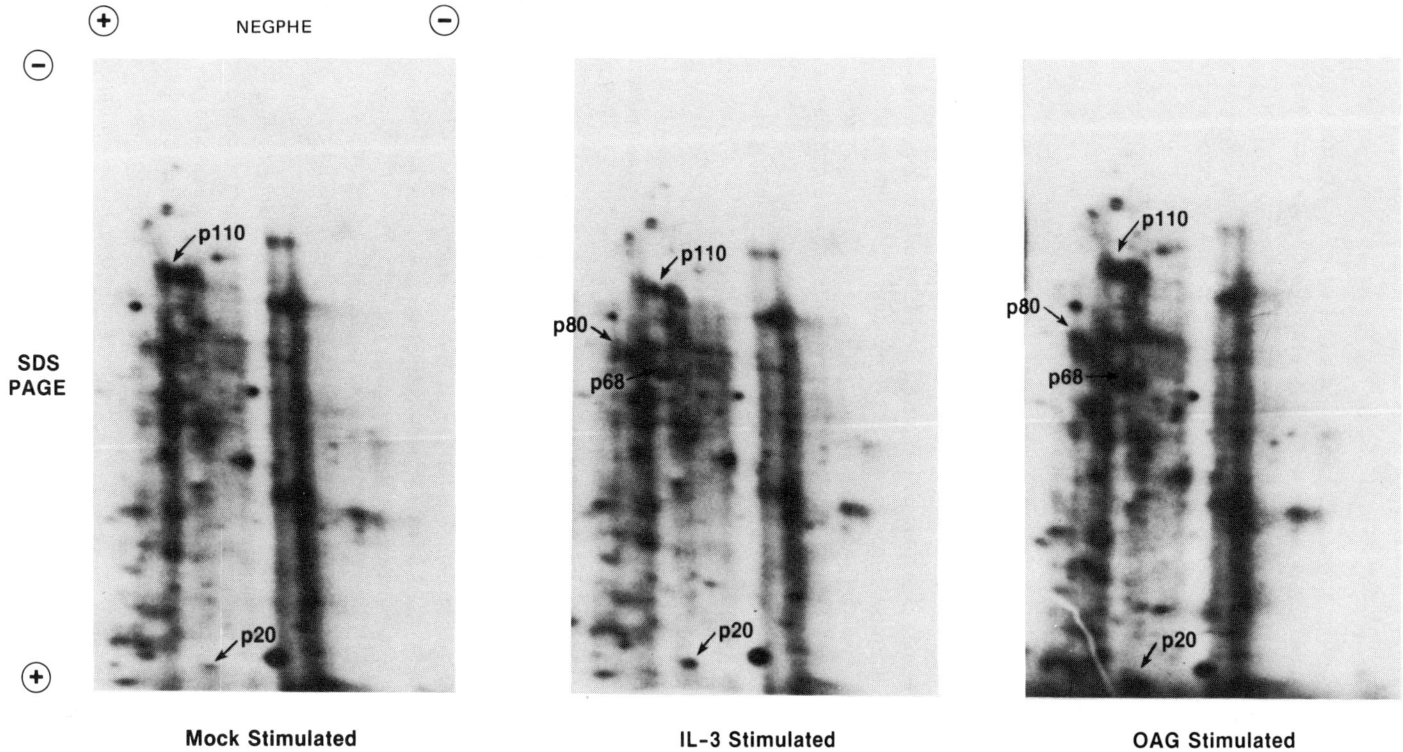

Figure 1 Interleukin-3 (IL-3) and diacylglyceral (OAG) stimulate phosphorylation of p68. Two-dimensional NEPHGE gel analysis of FCD-P1 cells stimulated with indicated treatment.

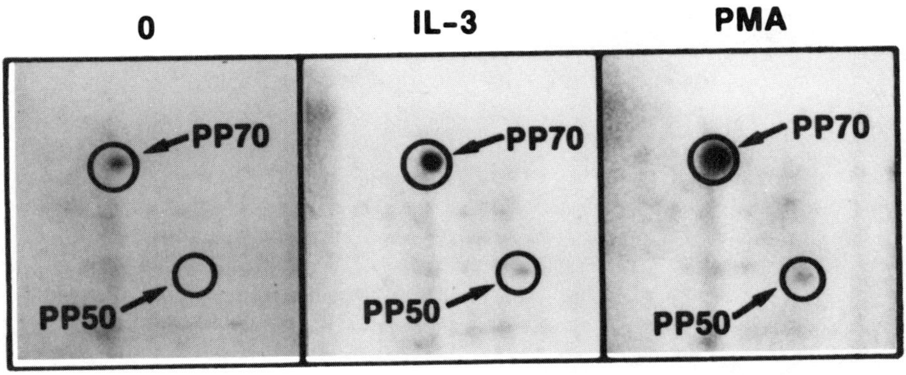

Figure 2 Two-dimensional NEPHGE analysis of antiphosphotyrosine affinity purified phosphoproteins from FDC-P1 cells. The phosphate-labeled, growth-arrested cells are stimulated with interleukin (IL-3) or phorobol myristic acid (PMA) for 5 min. The cells were lysed and the proteins were purified using antiphosphotyrosine coupled to sepharose beads. The purified proteins were analyzed by two-dimensional NEPHGE and autoradiography. The locations of two prominent phosphoproteins modulated by IL-3 or PMA are indicated.

analysis confirmed that both p50 and p70 contained phosphotyrosine as well as phosphoserine (Fig. 3). The size of p70 and kinetics of p70 phosphorylation are particularly interesting in the light of crosslinking studies that have suggested that the IL-3 receptor is a protein of 65-75 kDa (40,42). These results together with our previous work suggest that IL-3, and probably other myeloid growth factors, coactivate both multiple serine/threonine kinase systems and one or more tyrosine kinases. Recently, Morla et al. (85) have shown that stimulation of IC-2.9 cells, which can proliferate in response to IL-3, IL-4, or GM–CSF, with IL-3 (2 n*M*) caused increased tyrosine phosphorylation of proteins with M_r of 160, 95, 90, 70, and 55. Under the same conditions, stimulation with IL-4 induced increased tyrosine phosphorylation of proteins with M_r of 170 and 110 and GM–CSF stimulated phosphorylation increases of proteins with M_r of 150, 92, and 72. The receptors for IL-3 and GM–CSF are 68-75 kDa and 51 kDa, respectively, which is probably too small for intrinsic kinase domains. The binding of the ligands nevertheless triggers the activation of several classes of kinase activities, possibly through direct receptor-kinase interaction or coupling via GTP-binding proteins.

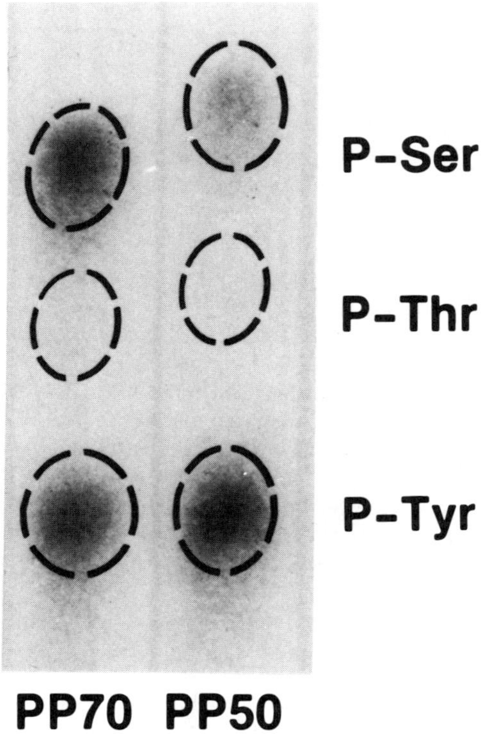

Figure 3 Phosphoamino acid analysis. Portions of the gel containing pp50 and pp70 were excised and the proteins were extracted, hydrolyzed, and analyzed by single-dimension, thin-layer, high-voltage electrophoresis and autoradiography. The positions of phosphoamino acid standards located with triketohydrindene hydrate are indicated by the dashed circles.

III. GENE REGULATION BY CSFs

As part of the growth factor family, CSFs induce a variety of changes in the genetic activity of their target cells. Like fibroblastic or lymphoid growth polypeptides, the addition of CSF to growth-arrested CSF-dependent cell lines starts a "consensus" program of gene expression, which is common to most of the studied growth-factor/target-cell interaction. However, it must be kept in mind that unlike most of the other growth factors, CSFs exert, beside their proliferation inducer function, a survival role. Deprived of CSF, the CSF-dependent cell lines cannot survive more than a few hours. Therefore, a growth-arrested CSF-dependent cell population is not a bona fide synchronized population. For this technical reason, caution is needed in interpreting the data in those systems. As part of the "consensus" program of gene induction by growth factors, the short half-lived messages for the nuclear protooncogenes c-*fos* and c-*myc* and ornithine decarboxylase have clearly been shown to be modulated by CSFs (86,87). In the murine myeloid cell line NSF-60.8, for example, which requires either GM–CSF, G–CSF, or multi-CSF, all three CSFs induce the expression of c-*fos* and c-*myc*, with identical kinetics (87). The induction of c-*fos* is, as in most systems, very transient, with a peak at 30 min of induction, and returning to background levels after ~1 hr. The induction of c-*myc* is more stable, and steady-state levels of c-*myc* messages remain high throughout the cell cycle (86,87). Study of nuclear protooncogene regulation by CSF-1 in the more differentiated monocyte system is complicated by the fact that a simple adherence step can activate the cells, at least for c-*fos* expression (88).

Besides the "consensus" gene expression program, CSFs also control a tissue-specific gene expression program. Granulocyte-macrophage colony-stimulating factor, for example, induces the expression of tumor necrosis factor (TNF) in a human myeloid cell line (89). Colony-stimulating factor 1 stimulates the expression of interferon (IFN), TNF, and other CSFs in normal monocytes (90). However, the study of normal monocytes is again complicated by the fact that a simple adherence step activates these genes at the level of message expression, although a second signal is necessary for the actual secretion of the product (88).

The signal transduction pathways involved in gene regulation by CSFs are far from being understood. Phorbol myristic acid, the pharamcological activator of the Ca^{2+}/phospholipid-dependent PK-C, induces c-*fos* and c-*myc* in growth-arrested CSF-dependent myeloid cell lines (91), suggesting a possible involvement of the enzyme in the signal transduction pathway leading to nuclear protooncogene induction in myeloid cell lines. However, the activity of another kinase, cyclic adenosine monophosphate (cAMP)-dependent protein kinase A (PK-A), has the opposite effects on c-*fos* and c-*myc* induction. Activation of PK-A by cAMP analogue induces c-*fos* message in murine myeloid cells (91) and in macrophages (92), and superinduces c-*fos*, in the presence of GM–CSF. In contrast, c-*myc* message is slightly inhibited by the activation of PK-A (91). These results suggest some differences in the pathways resulting in the induction of c-*fos* and c-*myc* in these cell lines (91).

The actual mechanism of regulation of steady-state mRNA levels in myeloid cell lines is, as in most systems, a combination between transcriptional and post-transcriptional regulation (92). In fibroblasts, the induction of c-*fos* mRNAs seems to be transcriptional, as assessed by nuclear run-off experiments (93). The existence of upstream regulatory elements in the c-*fos* gene, which mediates the serum induction, confirms that transcription initiation plays a role in the induction (94). However, the stability of the message is certainly important for the transient expression of c-*fos* in all systems (95). The situation for the c-*myc* gene is more controversial, with some authors reporting that transcriptional regulation does not play any role in c-*myc* induction in fibroblasts (96). In most cases however, c-*myc* has been reported to be regulated at the level of the transcription by growth factors. But transcription initiation is not the only step at which c-*myc* expression can be regulated. In fact, c-*myc* genes seem to be regulated at all steps of the processing, transcription initiation (93). Elongation of transcription (97), stability of the message (98), and translation of the message (99). This tight regulation suggests a very important role for c-*myc* protein in the control of cell proliferation. A transcriptional regulation of c-*myc* gene by GM–CSF has been demonstrated in the murine myeloid cell line NSF-60.8 (100). However, in myeloid cells, posttranscription regulation also seems to play a very important role. For example, IFN-γ, which is a potent proliferative inhibitor of various myeloid cells, inhibits c-*myc* gene expression, in most cases by posttranscriptional mechanisms. In the human myeloid tumor cell line Daudi, IFN destabilizes c-*myc* message (101). In the murine CSF-dependent myeloid cell line NSF-60.8, IFN-γ inhibits c-*myc* message induction by GM–CSF mostly by impairing the splicing of an intermediate precursor (100).

IV. HEAT-SHOCK PROTEINS IN MYELOID CELLS

The cellular transition from the resting state to active growth requires expression of several genes whose products seem to regulate events during the cell cycle. Examples of proteins whose synthesis is specifically induced in all cell types in response to growth-factor stimulation are protooncogene products and heat-shock proteins (HSPs). The last-mentioned class is a unique set of highly conserved proteins. They are induced by heat and other stresses including ethanol, glucose-deprivation, anoxia, sulphydryl reagents, heavy metals, and amino acid analogues (102,103).

The expression of HSPs in response to stress and even their amino acid sequences are highly conserved throughout evolution, implying that they may have some universally important functions. It is thought that they function in protecting cells during stress but it is also now widely believed that they serve vital roles during normal cellular processes. Heat-shock proteins have been classified in three main size classes: 20-30, 70, and 83-90 kDa. Recently much attention has focused on the major heat-shock protein HSP70 and its role in cell growth. Recent evidence suggests that HSP70 is involved in regulating mammalian cell growth. In fact, HSP70 has been found to be associated with two oncogene products as well as the immortalizing protein of a DNA tumor virus. Both c-*myc* protein (104) and the adenovirus E1A 13S gene product seem to be capable of specifically stimulating expression of the HSP70 gene (105,106). Furthermore, it has been shown that a large proportion of p53 oncogene protein was complexed with nuclear HSP70 in certain transformed cells (107) and that the ability of mutant p53 molecules to cause transformation was correlated with their ability to bind to HSP70 (108). Other evidence for a role for HSP70 in cell growth comes from studies showing serum-induced HSP70 mRNA and protein synthesis in quiescent HeLa cells (109,110). Additionally, after serum stimulation a large proportion of HSP70 protein becomes associated with the nucleus during the S phase of the cell cycle, whereas at other phase it is more diffusely distributed in both the nucleus and the cytoplasm (110,111).

Additionally, data suggest that HSP90 may be involved in important processes related to growth and differentiation as well. HSP90 is a major stress portein in most cell types; however, as with HSP70, synthesis of HSP90 not only occurs in response to stress, but also during normal growth (112,113). However, other data have recently shown that HSP90 mRNA levels do not change during the cell cycle in HeLa cells (114) where the HSP90 protein represent at least 1% of the total cellular proteins. Expression of HSP90 is developmentally regulated in the mouse embryo (112) and it has been found in molecular complexes with pp-60-*src* and other tyrosine kinase oncogene products (115,116). A variety of studies have also shown that HSP90 is closely associated with steroid receptors and forms biomolecular complexes with the 90-kDa untransformed progesterone receptor, apparently preventing association of the receptor with DNA in the absence of the hormone (117,118). These findings have led to suggestions that HSP90 may have a role in protein transport. A possibility that could explain how HSP90 might be regulated is that it is regulated by reversible covalent modification, because it is known to be heavily phosphorylated protein. In fact, a double stranded DNA-dependent protein kinase activity has been described that phosphorylates HSP90 in vitro, but it is not clear what effect, if any, phosphorylation or dephosphorylation have on HSP90 (119).

In one of our previous studies (120) we have examined the roles of HSP70 and HSP90 in human T lymphocytes during the cell cycle. We found that during cellular activation, which is essentially a process of differentiation, HSP70 mRNA was not modulated nor was there a specific induction of HSP70 protein synthesis. The synthesis of HSP90 protein, however, rapidly and specifically increased during the first hour of activation and then dropped while synthesis of all other proteins continued to rise. In contrast, during cell-cycle progression

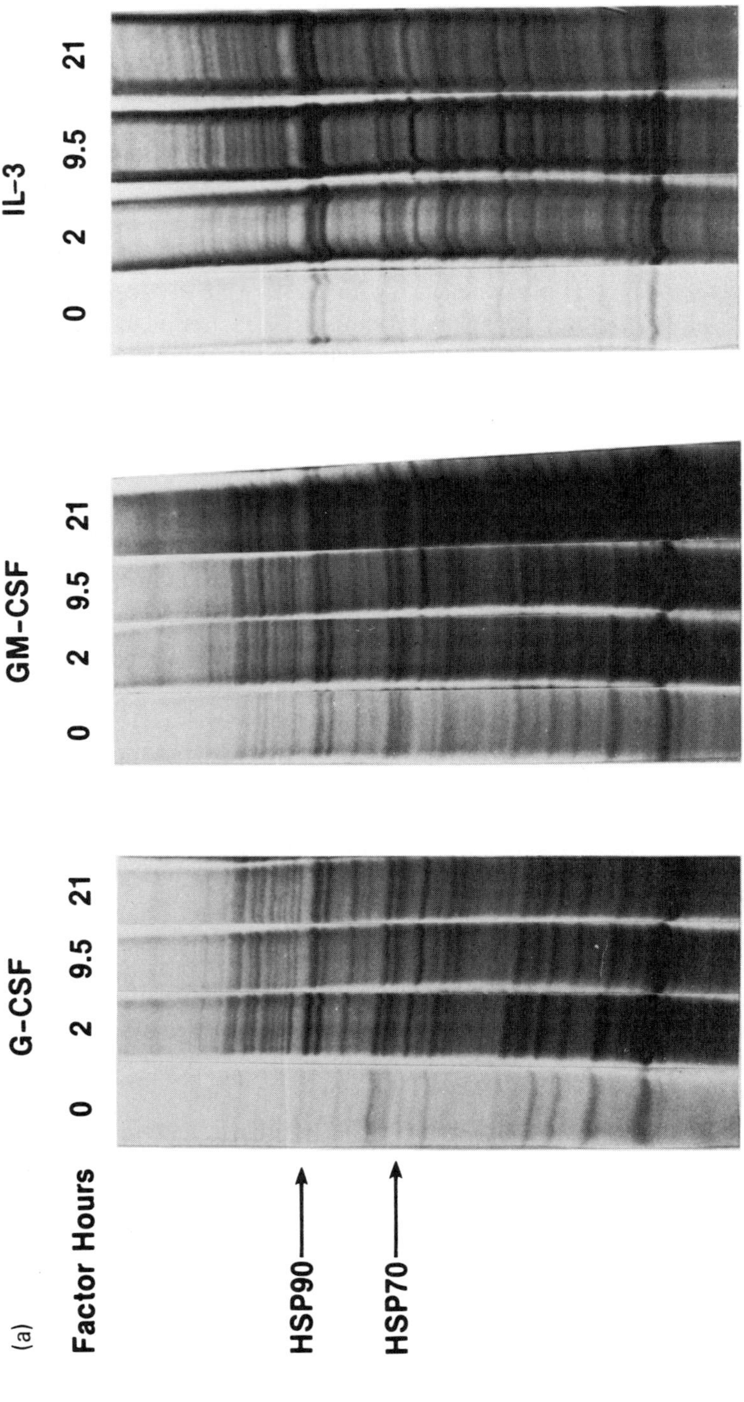
(a)
Factor Hours
G-CSF
0 2 9.5 21
GM-CSF
0 2 9.5 21
IL-3
0 2 9.5 21
HSP90
HSP70

(b)

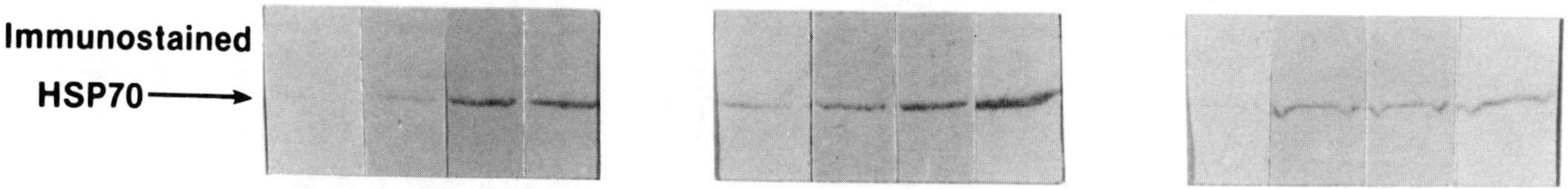

Figure 4 Multiple growth factors stimulate HSP synthesis in NSF-60.8. Cells that had been grown for 1 week in the presence of the indicated colony-stimulating factor (CSF) were washed free of growth factor and incubated for 4 hr before restimulation with the same CSF in which they had been grown. At the indicated times, actively synthesized proteins were labeled with ^{35}S-methionine and were analyzed by sodium dodecyl sulfate gel electrophoresis, Western blotting, and autoradiography. Equal amounts of cell lysates were loaded on each lane. (a) Autoradiography. (b) Immunostained HSP70.

induced by IL-2, synthesis of HSP70 mRNA and protein was increased while HSP90 did not appear to be specifically modulated. These results suggested that in T lymphocytes HSP70 has a role in the progression through the cell cycle, while HSP90 seems to be more important in process of differentiation. To further study the importance of HSPs during cell growth, we decided to examine the synthesis of HSPs in myeloid cells in response to different CSFs.

We examined (121) the relationship of CSF stimulation on the synthesis of HSP70 and HSP90 the growth-factor-dependent murine myeloid cell line NSF-60.8. This cell line represents an early progenitor in the myeloid lineage that is blocked in its ability to differentiate further but it is totally dependent for viability and proliferation on any one of several CSFs, IL-3, G–CSF, or GM–CSF. We have found that CSF stimulation induced rapid and striking increases in HSP70 protein synthesis (Fig. 4). In contrast, CSF treatment did not specifically elevate synthesis of HSP90 protein, which was high in unstimulated cells. These data suggested that HSP70 synthesis may be a common factor in the biochemical responses of hemopoietic cells to specific growth factors.

V. CSF SIGNAL TRANSDUCTION AND ONCOGENE TRANSFORMATION

So far we have reviewed two critical areas of biochemical and molecular events controlled by the hemopoietic growth factors. Colony-stimulating factors clearly regulate protein phosphorylation by the activation of both serine-threonine and tyrosine kinases. The CSFs also modulate common nuclear protooncogenes such as c-*myc* and c-*myb*. Oncogenic proteins representing membrane-bound tyrosine kinases, such as v-*abl*, v-*erb*B, and v-*src* (122) and the nuclear oncogene v-*myc* (123), when introduced by either transfection or retrovirus vectors, abrogate dependence for myeloid or lymphoid growth factors. Although these studies support earlier findings in fibroblastic cell lines, the conclusions nevertheless point to the interpretation that oncogenic proteins subserve or functionally intersect molecular pathways stimulated by the growth ligand. This interpretation may now be tested. Because we know that IL-3 stimulates tyrosine kinase activity in dependent-cell lines, a comparison of the tyrosine kinase substrates that are phosphorylated in response to the physiological ligand and those phosphorylated by transforming oncogene kinases can be made. Figure 5 compares the phosphotyrosine substrates stimulated in 32D cells with the same cells transformed with v-*src* or v-*erb*-B. Interleukin-3 stimulated the tyrosine phosphorylation of a number of proteins (lane 2) when compared with unstimulated 32D cells (lane 1). Interleukin-3-independent v-*src* 32D (lane 3) and v-*erb*-B (lane 4) transformed 32D cells showed high tyrosine phosphorylation of a number of similar substrates to that seen with IL-3 stimulation of the

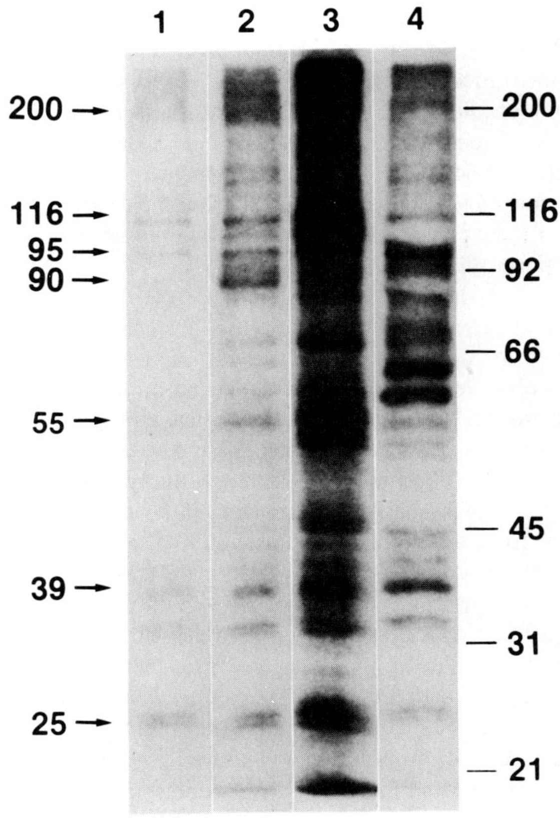

Figure 5 Interleukin-3 (IL-3) stimulates tyrosine phosphorylation of proteins also phosphorylated in *src* and *erb*-B transformed cells. Single-dimension analysis of phosphoproteins purified using antiphosphotyrosine antibodies coupled to sepharose beads. Lane 1, unstimulated 32D; lane 2, IL-3 stimulated 32D; lane 3, *src* transformed 32D; lane 4, *erb*-B transformed 32D. Molecular weight standards are indicated at the right in kilodaltons. Positions and approximate molecular weight proteins phosphorylated in response to IL-3 and in the transformed cell lines are as indicated by arrows.

parental cell line. We observed IL-3-stimulated phosphorylation of proteins of 220, 200, 140, 116, 100, 92, 70, 55, 38, and 33 kDa that are also phosphorylated in the *src* and *erb*-B transformed 32 D cells.

The demonstration that the physiological ligand stimulates the tyrosine phosphorylation of substrates, which are also phosphorylated by members of the tyrosine kinase oncogene family, provides direct evidence of biochemical convergence of the activity of transforming proteins and normal growth pathways. The identification and function of these substrates will provide many insights fundamental to the mechanism(s) controlling growth.

VI. SUMMARY

Hemopoietic cells have an absolute requirement for survival and proliferation for specific growth factors. The growth factors maintain the critical vitality of the cells by stimulating adenosine triphosphate (ATP) synthesis and hexose transport (124). Intracellular alkalinization also occurs rapidly through the stimulation of the Na^+/H^+ antiporter (125). These immediate metabolic events, not initiated by serum components, appear to be necessary for the integrity of cellular viability (Fig. 6). Interleukin-3 has been shown to induce the activation of PK-C through a mechanism(s) not requiring the hydrolysis of phosphoinositol 4,5 bisphosphate (126). A role for Ca^{2+} influx or intracellular release in the action of CSFs or interleukins has not been shown. Although downregulation of cAMP has been reported in response to IL-2 (127), the signal tranduction process of CSFs and IL-2 appears not to be mediated by upregulation of cyclic nucleotide metabolism or "classical" phospholipid degradative pathways.

Protein phosphorylation is clearly modulated by the hemopoietic cytokines, yet only the CSF-1 receptor has any known intrinsic kinase activity. Instead, the IL-3, GM–CSF receptors, and perhaps G–CSF appear to be coupling to kinases of both tyrosine and serine specificities. This may be a direct allosteric interaction with membrane-associated kinases or transduced through an intermediate protein such as those using GTP. Such is the case for many hormone receptors that couple to amplifying "second messenger" enzyme systems (i.e., adenylate cyclase, phospholipase C) or members of the insulin growth factor family that couple to tyrosine kinases in proximity to the receptors (IGF-II). One of the kinase systems that IL-2, IL-3, and other CSFs stimulate appears to have some characteristics similar to PK-C. Direct activators of PK-C stimulate some similar serine-threonine phosphorylation and perhaps even tyrosine phosphorylation. The hemopoietic growth factors, however, stimulate tyrosine phosphorylation of some proteins that are not phosphorylated in response to PK-C activators, suggesting that these kinase systems are independently regulated. Although phorbol esters stimulate many of the same metabolic activities (ATP synthesis in myeloid and lymphoid cell lines), growth-factor abrogation is clearly

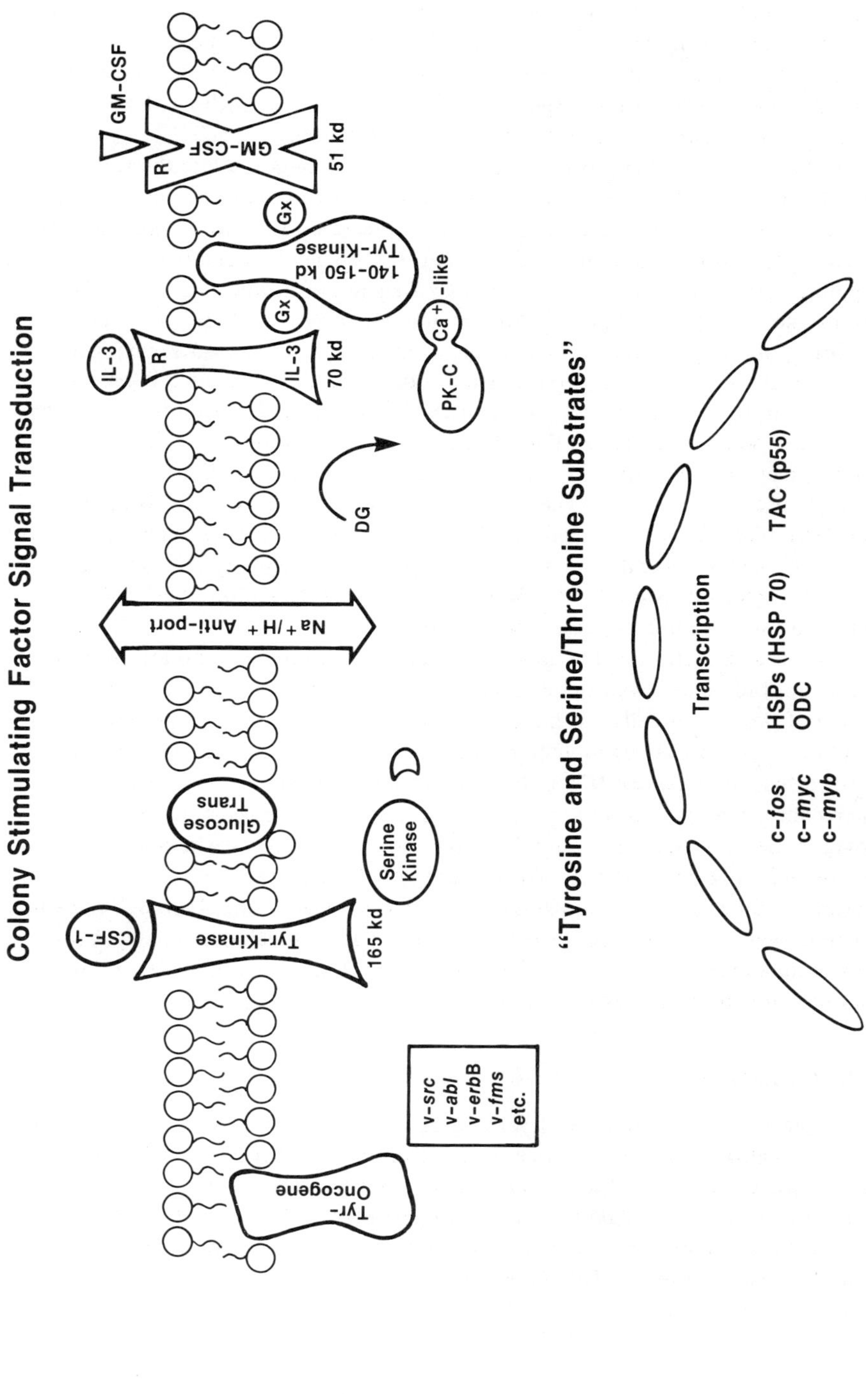

Figure 6 Proposed models of colony-stimulating factor (CSF) signal transduction and the regulation of gene transcription.

associated with the action of tyrosine kinase oncogenes or the nuclear oncogene effectors such as v-*myc* (122,123). It is likely, therefore, that tyrosine kinases are playing a critical role in the control of proliferation although the dominant amount of cellular protein phosphorylations are on serine. Both classes of kinases are apparently required for growth-factor action.

All the hemopoietic growth factors examined thus far stimulate the steady-state accumulation of the nuclear protooncogenes. Considerable attention has been given to c-*myc* expression because v-*myc* inserted retroviruses can also factor-abrogate lymphoid and myeloid IL-2 and IL-3-dependent cell lines (123). Oligonucleotide "antisense" to c-*myc* mRNA also inhibits mitogen and IL-2-driven T-cell proliferation (128). Antiproliferative agents such as cyclic AMP and IFN-γ also appear to target c-*myc* gene expression. Cyclic AMP analogues have been shown to inhibit c-*myc* transcription in IL-2-dependent cell lines (129) and in the human monocytic cell line HL-60 (130). Interferon-γ, a potent inhibitor of some myeloid cell lines, has also been shown to inhibit the maturation of precursor c-*myc* in GM–CSF-dependent myeloid cells. Other genes regulated are also sensitive to the effects of anitproliferative signals. While cAMP stimulates c-*fos*, c-*myb* and IL-2rα gene expression in IL-2-dependent cytotoxic T-cell clones, it also inhibits ornithine decarboxylase (ODC) and c-*myc* mRNA accumulation (131). Both IFN-γ and cAMP inhibit ODC mRNA accumulation in CSF-dependent myeloid cell (132).

The characterization of the molecular loci regulated by hemopoietic growth factors not only aids in our understanding of the mechanism(s) controlling growth of progenitor cells and mature lineages, but also serves as a potential target of antisignals of the biological response. Specific examples can be seen by the effects of cAMP or IFN-γ on proliferation genes such as c-*myc* and ODC. The recent advances in receptor chemistry and molecular biology have now provided the tools to examine fundamental molecular processes involved in growth and antigrowth signals. The integrity of these signals are particularly relevant to abberant regulation and oncogenic transformation. Although the study of hemopoietic growth-factor signal transduction is in its relative infancy, progress to date encourages the use of biochemical and molecular approaches to examine fundamental questions common to all mammalian cells.

ACKNOWLEDGMENT

This project has been funded at least in part with Federal funds from the Department of Health and Human Services under contract number N01-CO-74102 with Program Resources, Inc. The content of this publication does not necessarily reflect the views or policies of the Department of Health and Human Services, nor does mention of trade names, commercial products, or organizations imply endorsement by the U.S. Government.

REFERENCES

1. Metcalf, D. The hemopoietic colony stimulating factors. Elsevier Science Publishers, Amsterdam (1984).
2. Burgess, A. W., J. Camakais, and D. Metcalf. Purification and properties of colony-stimulating factor from mouse lung conditined medium. *J. Biol. Chem. 252*:1998-2003 (1977).
3. Yang, Y.-Ch., A. B. Garletta, P. A. Temple, M. P. Chung, S. Kovacie, J. A. Witek-Gianotti, A. C. Leary, R. Kuiz, R. E. Donahue, G. G. Wong, and S. C. Clark. Human IL 3 (multi-CSF): identification by expression cloning of a novel hematopoietic growth factor related to murine IL 3. *Cell 47*:3-10 (1986).
4. Ihle, J. N., J. Keller, L. Henderson, H. Klein, and E. Palaozynski. Procedures for the purification of interleukin 3 to homogeneity. *J. Immunol. 129*: 2431-2436 (1982).
5. Fung, M. C., A. J. Hapel, S. Ymer, D. R. Cohen, R. M. Johnson, H. D. Campbell, and I. E. Young. Molecular cloning of cDNA for murine interleukin 3. *Nature 307*:233-237 (1984).
6. Yakota, T., F. Lee, D. Rennick, C. Hall, N. Avai, T. Mosmann, G. Nabel, H. Cantor, and K. I. Arai. Isolation and characterization of mouse cDNA clone that express mast-cell growth-factor activity in monkey cells. *Proc. Natl. Acad. Sci. USA 81*:1070-1074 (1984).
7. Mitayake, S., T. Yokota, F. Lee, and K. I. Arai. Structure of the chromosomal gene for murine interleukin 3. *Proc. Natl. Acad. Sci. USA 82*:316-320 (1985).
8. Clark, S. C., and R. Kamen. The human hematopoietic colony-stimulating factors. *Science 236*:1229-1237 (1987).
9. Gough, N. M., J. Gough, D. Metcalf, A. Kelso, D. Grail, N. C. Nicola, A. W. Burgess, and A. R. Dunn. Molecular cloning of cDNA encoding a murine haematopoietic growth regulator, granulocyte-macrophage colony stimulating factor. *Nature 309*:763-767 (1984).
10. Nicola, N. A., A. W. Burgess, and D. Metcalf. Similar molecular properties of granulocyte-macrophage colony stimulating factors produced by different mouse organs in vitro and in vivo. *J. Biol. Chem. 254*:5290-5299 (1979).
11. Gasson, J. C., R. H. Weisbart, S. E. Kaufman, S. C. Clark, R. M. Hewick, G. G. Wong, and D. W. Golde. Purified human granulocyte macrophage colony stimulating factor: direct action on neutrophils. *Science 226*: 1229-1342 (1984).
12. Wong, G. G., J. S. Witek, P. A. Temple, K. M. Wilkens, A. C. Leary, D. P. Luxenberg, S. S. Jones, E. L. Brown, R. M. Kay, E. C. Orr, C. Shoemaker, D. W. Golde, R. J. Kaufman, R. M. Hewick, S. C. Clark, and E. A. Wang. Molecular cloning of human and gibbon T-cell-derived GM–CSF cDNAs and purification of the natural and recombinant human proteins. In *Cancer Cells in Growth Factors and Transformation*, Vol. 3. Cold Spring Harbor Laboratory, Cold Spring Harbor, New York, pp. 235-242 (1985).

13. Wong, G. G., J. S. Witek, P. A. Temple, K. M. Wilkens, A. C. Leary, D. P. Luxenberg, S. S. Jones, E. L. Brown, R. M. Kay, E. C. Orr, C. Shoemaker, D. W. Golde, R. J. Kaufman, R. M. Hewick, E. A. Wang, and S. C. Clark. Human GM–CSF: molecular cloning of the complementary DNA and purification of the natural and recombinant proteins. *Science 228*:810-815 (1958).
14. Kaushansky, K., P. J. O'Hara, K. Berkner, G. M. Segal, F. S. Hagen, and J. W. Adamson. Genomic cloning, characterization, and multilineage growth-promoting activity of human granulocyte-macrophage colony-stimulating factor. *Proc. Natl. Acad. Sci. USA 83*:3101-3105 (1986).
15. Le Beau, M. M., C. A. Westbrook, M. O. Diaz, R. A. Larson, J. D. Rowley, J. C. Gasson, D. W. Golde, and C. J. Sherr. Evidence for the involvement of GM–CSF and FMS in the deletion (5q) in myeloid disorders. *Science 231*:984-987 (1986).
16. Pettenati, M. J., M. M. Le Beau, R. S. Leomone, E. A. Shima, E. S. Kawasaki, R. A. Larson, C. J. Sherr, M. O. Diaz, and J. D. Rowley. Assignment of CSF-1 to 5g33.1: evidence for clustering of genes regulating hematopoiesis and for their involvement in the deletion of the long arm of chromosome 5 in myeloid disorders. *Proc. Natl. Acad. Sci. USA 84*:2970-2974 (1987).
17. Metcalf, D., and J. Nicola. Proliferative effects of purified granulocyte colony-stimulating factor (G–CSF) on normal mouse hemopoietic cells. *J. Cell Physiol. 116*:198-206 (1983).
18. Stanley, E. R., and P. M. Heard. Factors regulating macrophage production and growth. *J. Biol. Chem. 252*:4305-4312 (1977).
19. Metcalf, D. The granulocyte-macrophage colony-stimulating factors. *Science 229*:16-22 (1985).
20. Micola, N. A., D. Metcalf, M. Matsumoto, and G. R. Johnson. Purification of a factor inducing differentiation in murine meylomonocytic leukemia cells. *J. Biol. Chem. 258*:9017-9023 (1983).
21. Nomura, H., I. Imazaki, M. Oheda, N. Kubota, and S. Asano. Purification and characterization of human granulocyte colony-stimulating factor (G–CSF). *EMBO J. 5*:871-876 (1986).
22. Nagata, S., M. Tsuchiya, S. Asano, Y. Kaziro, T. Yamazaki, O. Yamamoto, Y. Hirata, N. Kubota, M. Oheda, H. Nomura, and M. Ono. Molecular cloning and expression of cDNA for human granulocyte colony-stimulating factor. *Nature 319*:415-418 (1986).
23. Nagata, S., M. Tsuchiya, S. Asano, O. Yamamoto, Y. Hirata, Y. Kubote, N. Oheda, H. Nomura, and T. Yamazaki. The chromosomal gene structure and two mRNAs for human granulocyte colony-stimulating factor. *EMBO J. 5*:575-581 (1986).
24. Tsuchiya, M., S. Asano, Y. Kaziro, and S. Nagata. Isolation and characterization of the cDNA for murine granulocyte colony-stimulating factor. *Proc. Natl. Acad. Sci. USA 83*:7633–7637 (1986).
25. Nicola, N. A., C. G. Bergley, and D. Metcalf. Identification of the human

analogue of a regulator that induces differentation in murine leukemic cells. *Nature 314*:625-628 (1985).

26. Bergley, C. G., D. Metcalf, and N. A. Nicola. Primary human myeloid leukemia cells: comparative responsiveness to proliferative stimulation by GM–CSF on G–CSF and membrane expression of CSF receptors. *Leukemia 1*:1-8 (1986).
27. Nicola, N. A. Hemopoietic growth factors and their interactions with specific receptors. *J. Cell. Physiol. 5*:9-14 (1987).
28. Welte, K., M. A. Bonilla, A. P. Gillis, T. C. Boone, G. K. Potter, J. L. Gabrilove, M. A. S. Moore, R. J. O'Reilly, and L. M. Souza. Recombinant human granulocyte colony-stimulating factor. *J. Exp. Med. 165*:941-948 (1987).
29. Das, S. K., and E. R. Stanley. Structure-function studies of a colony stimulating factor (CSF-1). *J. Biol. Chem. 257*:13679-13684 (1982).
30. Burgess, A. W., D. Metcalf, I. J. Kozka, R. J. Simpson, G. Vairo, J. A. Hamilton, and E. C. Nice. Purification of two forms of colony-stimulating factor from mouse L-cell-conditioned medium. *J. Biol. Chem. 260*:16004-16011 (1985).
31. Das, S. K., E. R. Stanley, L. J. Guilbert, and L. W. Foreman. Human colony-stimulating factor (CSF-1) radioimmunoassay: resultion of three subclasses of human colony-stimulating factors. *Blood 58*:630-641 (1981).
32. Stanley, E. R., G. Hansen, J. Woodcock, and D. Metcalf. Colony stimulating factor and the regulation of granulopoiesis and macrophage production. *Fed. Proc. 34*:2272-2278 (1975).
33. Mokoyoshi, K., T. Sude, K. Kuzumoto, F. Takaku, and Y. Minna. Granulocyte-macrophage colony-stimulating and binding activities of purified human urinary colony-stimulating factor to murine and human bond marrow cells. *Blood 60*:1378-1386 (1982).
34. Kawasaki, E. S., M. B. Ladner, A. M. Wang, J. VanArsdell, M. K. Warren, M. Y. Coyne, V. L. Schweickart, M. T. Lee, K. J. Wilson, A. Boosman, E. R. Stanely, P. Ralph, and D. F. Mark. Molecular cloning of a complementary DNA encoding human macrophage-specific colony-stimulating factor (CSF-1). *Science 230*:291-296 (1985).
35. Wong, G. G. P. A. Temple, A. C. Leary, J. A. S. Witek-Giannotti, Y.-Ch. Yang, A. B. Ciavletta, M. Chung, P. Murtha, R. Kriz, R. J. Kaufman, C. R. Feruz, B. S. Sibley, K. J. Turner, R. M. Hewick, S. C. Clark, N. Yanai, H. Yokote, M. Yamada, M. Saito, K. Motoyoshi, and F. Takaker. Human CSF-1: molecular cloning and expression of 4-kb cDNA encoding the human urinary protein. *Science 235*:1504-1508 (1987).
36. Nicola, N. A., and D. Metcalf. Binding of the differentiation-induce granulocyte-colony-stimulating factor, to responsive but not unresponsive leukemic cell lines. *Proc. Natl. Acad. Sci. USA 81*:3765-3769 (1984).
37. Nicola, N. A., and D. Metcalf. Binding of ^{125}I-labeled granulocyte colony-stimulating factor to normal murine hematopoietic cells. *J. Cell. Physiol. 124*:313-321 (1985).
38. Guilbert, L. J., and E. R. Stanley. The interaction of ^{125}I-colony-stimulat-

ing factor-1 with bone marrow-derived macrophages. *J. Biol. Chem. 261*: 4024-4032 (1986).
39. Walker, F., and A. W. Burgess. Specific binding of radioiodinated granulocyte-macrophage colony-stimulating factor to hemopoietic cells. *EMBO J. 4*:933-939 (1985).
40. Park, L. S., D. Friend, S. Gillis, and D. L. Urdal. Characterization of the cell surface receptor for granulocyte-macrophage colony-stimulating factor. *J. Biol. Chem. 261*:4177-4183 (1986).
41. Di Persio, J., P. Billing, S. Kaufman, P. Eghtesady, R. E. Williams, and J. C. Gasson. Characterization of the human granulocyte-macrophage colony-stimulating factor receptor. *J. Biol. Chem. 263*:1834-1841 (1988).
42. May, W. S., and J. N. Ihle. Affinity isolation of interleukin-3 surface receptor. *Biochem. Biophys. Res. Commun. 135*:870-879 (1986).
43. Park, L. S., D. Friend, S. Gillis, and D. L. Urdal. Characterization of the cell surface receptor for a multi-lineage colony-stimulating factor (CSF-2α). *J. Biol. Chem. 261*:205-210 (1986).
44. Morgan, C. J., and E. R. Stanley. Chemical crosslinking of the mononuclear phagocyte specific growth factor CSF-1 to its receptor at the cell surface. *Biochem. Biophys. Res. Commun. 119*:35-41 (1984).
45. Sherr, C. J., C. W. Rettenmier, R. Sacca, M. F. Roussel, A. T. Look, and E. R. Stanley. The c-fms proto-oncogene product is related to the receptor for the mononuclear phagocyte growth factor, CSF-1. *Cell 41*:665-676 (1985).
46. Nicola, N. A., and L. Peterson. Identification of distinct receptors for two hemopoietic growth factors (granulocyte colony-stimulating factor and multi-potential colony-stimulating factor) by chemical cross-linking. *J. Biol. Chem. 261*:12384-12389 (1986).
47. Sherr, C. J. Fibroblast and hematopoietic cell transformation by the fms oncogene (CSF-1 receptor). *J. Cell. Physiol. 5*:83-87 (1987).
48. Coussens, L., C. Van Beverena, D. Smith, E. Chen, R. L. Mitchell, C. M. Isacke, I. M. Verma, and A. Ullrich. Structural alteration of viral homologue of receptor proto-oncogene fms at carboxyl terminus. *Nature 320*: 277-281 (1986).
49. Wheeler, E. F., D. Askew, S. May, J. N. Ihle, and C. J. Sherr. The v-fms oncogene induces factor-independent growth and transformation of the interleukin-3-dependent myeloid cell line FDC-P. *Mol. Cell. Biol. 7*:1673-1680 (1987).
50. Ferris, D. K., J. Willette-Brown, T. Martensen, and W. L. Farrar. Interleukin 3 stimulation of tyrosine kinase activity in FDC-P1 cells. *Biochem. Biophys. Res. Commun. 154*:991-996 (1988).
51. Nicola, N. A., and M. Vadas. Hemopoietic colony-stimulating factors. *Immunol. Today 5*:76-80 (1984).
52. Yeung, Y. G., P. T. Jubinsky, A. Sengupta, D. C. Y. Yeung, and E. R. Stanley. Purification of the colony-stimulating factor 1 receptor and demonstration of its tyrosine kinase activity. *Proc. Natl. Acad. Sci. USA 84*:1268-1271 (1987).

53. Gill, G. N., P. J Bertics, and J. R. Santon. Epidermal growth factor and its receptor. *Mol. Cell. Endocrinol. 51*:169-186 (1987).
54. Ronnstrand, L., M. P. Beckmann, B. Fanlders, A. Ostman, B. Ek, and C. H. Heldin. Purification of the receptor for platelet-derived growth factor from porcine uterus. *J. Biochem. 7*:2929-2932 (1987).
55. Beshayee, S., A. H. Ross, R. Womer, and C. D. Scher. Purified human platelet-derived growth factor receptor has ligand-stimulated tyrosine kinase activity. *Proc. Natl. Acad. Sci. USA 83*:6756-6760 (1986).
56. Rosen, O. M. After insulin binds. *Science 237*:1452-1458 (1987).
57. Huang, S. S., and J. S. Huang. Association of bovine brain-derived growth factor receptor with protein tyrosine kinase activity. *J. Biol. Chem. 261*: 9568-9571 (1986).
58. Taguchi, M., T. P. Thomas, W. B. Anderson, and W. L. Farrar. Direct phosphorylation of the IL 2 receptor TAC antigen epitope by protein kinase C. *Biochem. Biophys. Res. Commun. 135*:239-242 (1986).
59. Shakelford, D. A., and I. S. Trowbridge. Induction of expression and phosphorylation of the human interleukin 2 receptor by a phorbol ester. *J. Biol. Chem. 259*:11706-11709 (1984).
60. Leonard. W. J., J. M. Depper, G. R. Crabtree, S. Rudikoff, J. Pumphrey, R. J. Robb, M. Kronke, P. B. Svetlik, N. J. Peffer, T. A. Waldman, and W. C. Greene. Molecular cloning and expression of cDNA for the human interleukin 2 receptor. *Nature 311*:612-619 (1984).
61. Sharon, M., R. D. Klausner, B. R. Cullen, R. Chizzonite, and W. J. Leonard. Novel interleukin 2 subunit detected by cross-linking under high affinity conditions. *Science 234*:859-864 (1986).
62. Robb, R. J., C. M. Rusk, J. Yodoi, and W. C. Green. Interleukin 2 binding molecule distinct from the TAC protein: analysis of ot's role in the formation of high affinity receptors. *Proc. Natl. Acad. Sci. USA 84*:2002-2007 (1987).
63. Teshigawara, K., H. M. Wang, K. Kato, and K. A. Smith. Interleukin 2 high affinity receptors requires two distinct binding proteins. *J. Exp. Med. 165*: 223-228 (1987).
64. Dukovich, M., Y. Wano, L-T. B. Thuy, P. Katz, B. R. Cullen, J. H. Kehrl, and W. C. Greene. A second human interleukin 2 binding protein that may be a component of high affinity IL 2 receptors. *Nature 327*:518-521 (1987).
65. Kruijer, W., J. A. Cooper, T. Hunter, and I. M. Verma. Platelet-derived growth factor induces rapid but transient expression of the c-*fos* gene and protein. *Nature 312*:711-717 (1984).
66. Greenberg, M. E., and E. B. Ziff. Stimulation of 3T3 cells induces transcription of c-*fos* proto-oncogene. *Nature 311*:433-438 (1984).
67. Besterman, H. M., S. P. Watson, and P. Cuatrecasas. Lack of association of epidermal growth factor; insulin; and serum-induced mitogenesis with stimulation of phosphoinositide degradation in BALB/c 3T3 fibroblasts. *J. Biol. Chem. 261*:723-728 (1986).
68. Moenner, M., I. Magnaldo, G. L'Allemain, D. Barritault, and J. Pouyssegur. Early and late mitogenic events induced by FGF on bovine epithelial lens

cells are not triggered by hydrolysis of polyphosphoinositides. *Biochem. Biophys. Res. Commun. 146*:32-44 (1987).
69. Justement, L., Z. Z. Chem, L. K. Harris, J. T. Ransom, V. S. Sandoval, C. Smith, D. Rennick, N. Roehm, and J. Cambier. BSF1 induces membrane protein phosphorylation but not phosphoinositide metabolism. Ca^{++} mobilization, protein kinase C translocation, or membrane depolarization in resting murine B lymphocytes. *J. Immunol. 137*:3664-3667 (1986).
70. Kolata, G. Is tyrosine the key to growth control? *Science 219*:377-340 (1983).
71. Hunter, T. The proteins of oncogenes. *Sci. Am. 251*:70-79 (1984).
72. Hunter, T., and J. A. Cooper. Protein-tyrosine kinases. *Annu. Rev. Biochem. 54*:897-902 (1985).
73. Defize, L. H., W. H. Moolenaar, P. T. Van der Saag, and S. W. de Laat. Dissociation of cellular responses to epidermal growth factor using antireceptor monoclonal antibodies. *EMBO J. 5*:1187-1192 (1986).
74. Chen, W. S., C. S. Lazar, M. Poenie, R. Y. Tsien, G. N. Gill, and M. G. Rosefeld. Requirement for intrinsic protein tyrosine kinase in the immediate and late actions of the EFG receptor. *Nature 2*:328 (1987).
75. Evans, S. W., S. K. Beckner, and W. L. Farrar. Interleukin 2 stimulates specific GTP binding and hydrolysis activities in lymphocyte membrane. *Nature 325*:166-169 (1987).
76. Farrar, W. L., and W. B. Andersen. Interleukin 2 stimulates association of protein kinase C with plasma membrane. *Nature 315*:233-238 (1985).
77. Evans, S. W., and W. L. Farrar. Interleukin 2 and diacylglycerol stimulate phosphorylation of 40S ribosomal S6 protein. *J. Biol. Chem. 262*:4624-4627 (1987).
78. Evans, S. W., and W. L. Farrar. Identity of common phosphoprotein substrates stimulated by interleukin 2 and diacylglycerol suggests the role of protein kinase C for IL2 signal transduction. *J. Cell. Biochem. 34*:47-53 (1987).
79. Ishii, T., M. Kohno, M. Nakamura, Y. Hinuma, and K. Sugamura. Characterization of interleukin 2-stimulated phosphorylation of 67 and 63 kDa proteins in human T-cells. *Biochem. J. 242*:211-214 (1987).
80. Farrar, W. L., P. T. Thomas, and W. B. Anderson. Altered cytosol/membrane enzyme redistribution on interleukin 3 activation of protein kinase C. *Nature 315*:235-238 (1985).
81. Evans, S. W., D. Rennick, and W. L. Farrar. Identification of a signal-transduction pathway shared by haemotopoietic growth factors with diverse biological specificity. *Biochem. J. 244*:683-687 (1987).
82. Evans, S. W., D. Rennick, and W. L. Farrar. Multilineage hematopoietic growth factor interleukin 3 and direct activators of protein kinase C stimulate phosphorylation of common substances. *Blood 68*:906-909 (1986).
83. Garland, J. M. Rapid phosphorylation of a specific 33-kDa protein (p33) associated with growth stimulation by murine and rat IL3 in different IL 3-dependent cell lines, and its constitutive expression in a malignant independent clone. *Leukemia 2*:94-102 (1988).

84. Koyasu, S., A. Toji, A. Miyajima, T. Akiyama, M. Kasuga, A. Urabe, J. Schrews, K. Arai, F. Takaku, and I. Yahara. IL 3-specific tyrosine phosphorylation of a membrane glycoprotein of Mr 150,000 in multi-factor dependent myeloid cell lines. *EMBO J. 6*:3979-3984 (1987).
85. Morla, A. O., J. Schreurs, J. Miyajima, and J. Y. J. Wang. Hematopoietic growth factors activate the tyrosine phosphorylation of distinct sets of proteins in interleukin-3-dependent murine cell lines. *Mol. Cell. Biol. 8*: 2214-2216 (1988).
86. Conscience, J., F. B. Verrier, and G. Martin. IL 3 dependent expression of the c-fos and c-myc proto-oncogene in hemopoietic cell lines. *EMBO J. 5*:317-319 (1986).
87. Harel-Bellan, A., and W. L. Farrar. Modulatin of proto-oncogene expression by colony stimulating factors. *Biochem. Biol. Res. Commun. 148*: 1001-1007 (1987).
88. Haskill, S., C. Johnson, D. Eierman, S. Becker, and K. Warren. Adherence induces selective mRNA expression of monocyte mediators and oncogenes. *J. Immunol. 140*:1690-1695 (1988).
89. Cannistra, S. A., A. Rambaldi, D. Spriggs, F. Herrman, D. Kufe, and J. D. Griffin. Human GM–CSF induces expression of the tumor necrosis factor gene by the U 937 cell lines and by normal human monocytes. *J. Clin. Invest. 79*:1720-1727 (1987).
90. Warren, M. K., and P. Ralph. Macrophage growth factor CSF 1 stimulates human monocyte production of interferon, tumor necrosis factor and colony stimulating factor activity. *J. Immunol. 137*:2281-2286 (1986).
91. Harel-Bellan, A., and W. L. Farrar. Regulation of proliferation in a murine CSF dependent myeloid cell line: superinduction of c-*fos* by the growth inhibitor 8-Br-cAMP. *J. Cell. Biochem. 38*:145-153 (1988).
92. Bravo, R., M. Neuberg, J. Burckhardt, J. Almendral, R. Wallich, and R. Muller. Involvement of common and cell type-specific pathways in c-fos gene control. Stable induction by cAMP in macrophages. *Cell 48*:251-253 (1987).
93. Greenberg, M. E., and E. B. Ziff. Stimulation of 3T3 cells induces transcription of the c-*fos* proto-oncogene. *Nature 311*:433-437 (1984).
94. Treisman, R. Identification of a protein binding site that mediates transcriptional response of c-*fos* gene to serum factors. *Cell 46*:567-571 (1986).
95. Van Beveren, C., F. Van Straaten, T. Curran, R. Muller, and I. M. Verma. Analysis of FBJ-MuSV provirus and c-*fos* (mouse) gene reveals that viral and cellula fos gene products have different carboxy termini. *Cell 32*: 1241-1247 (1983).
96. Blachard, J. M., M. Piechazick, C. Dani, J. C. Chambard, A. Franchi, J. Pouyssegur, and P. Jeanteur. C-myc gene is transcribed at high rate in GO arrested fibroblasts and is post-transcriptionally regulated in response to growth factors. *Nature 317*:443-448 (1985).
97. Bentley, D. L., and M. Groudine. A block to elongation is largely responsible for decreased transcription of c-*myc* gene in differentiated HL 60 cells. *Nature 321*:702-705 (1986).

98. Rabbits, P. H., A. Forster, M. A. Stinson, and T. H. Rabbits. Truncation of exon 1 from the c-*myc* gene results inprolonged c-*myc* stability. *EMBO J. 4*:3727-3731 (1985).
99. Darveau, A., J. Pelletier, and N. Sonenberg. Differential efficiencies of in vitro translation of mouse c-myc transcripts differing in the 5′ untranslated region. *Proc. Natl. Acad. Sci. USA 82*:2315-2318 (1985).
100. Harel-Bellan, A., A. T. Brini, and W. L. Farrar. Gamma interferon inhibits c-myc gene expression by impairing the splicing process in a CSF dependent murine myeloid cell line. *J. Immunol. 141*:1012-1017 (1988).
101. Knight, E., Jr., E. D. Anton, D. Fahey, B. Friedland, and G. J. Jonak. Interferon regulates c-*myc* gene expression in Daudi cells at the post-transcriptional level. *Proc. Natl. Acad. Sci. USA 82*:1151-1157 (1984).
102. Lindquist, S. The heat-shock response. *Annu. Rev. Biochem. 55*:1151-1153 (1986).
103. Lanks, K. W. Review article: modulators of the eukaryotic heat shock response. *Exp. Cell. Res. 165*:1-30 (1986).
104. Kingston, R. E., A. S. Baldwin, Jr., and P. A. Sharp. Regulation of heat shock protein 70 gene expression by c-myc. *Nature 312*:280-285 (1984).
105. Wu, B. J., H. C. Hurst, N. C. Jones, and R. I. Morimoto. The E1A 135 product of adenovirus 5 activates transcription of the cellular human HSP70 gene. *Mol. Cell. Biol. 6*:2994-2999 (1986).
106. Nevins, J. R. Induction of the synthesis of a 70,000 dalton mammalian heat shock protein by the adenovirus E1A gene product. *Cell 29*:913-917 (1982).
107. Pinshasi-Kimhi, O., D. Michalovitz, A. Ben-Zeev, and M. Oren. Specific interaction between the p53 cellular tumour antigen and major heat shock proteins. *Nature 320*:182-184 (1986).
108. Hinds, P. W., C. A. Finlay, F. B. Frey, and A. J. Levine. Immunological evidence for the association of p53 with a heat shock protein, hsc70, in p53 + ras transformed cell lines. *Mol. Cell. Biol. 7*:2863-2869 (1987).
109. Wu, B. J., and R. I. Morimoto. Transcription of the human hsp70 gene is induced by serum stimulation. *Proc. Natl. Acad. Sci. USA 82*:6070-6072 (1985).
110. Milarski, K. L., and R. I. Morimoto. Expression of human HSP70 during the synthetic phase of the cell cycle. *Proc. Natl. Acad. Sci. USA 83*: 9517-9521 (1986).
111. Welch, W. J., and J. R. Feramisco. Nuclear and nucleolar localization of the 72,000-dalton heat shock protein in heat shocked mammalian cells. *J. Biol. Chem. 259*:4501-4507 (1984).
112. Barnier, J. V., O. Bensaude, M. Morange, and C. Babinet. Mouse 89 kDa heat shock proteins: two polypeptides with distinct developmental regulation. *Exp. Cell. Res. 170*:186-189 (1987).
113. Kurtz, S., J. Rossi, L. Petko, and S. Lindquist. An ancient developmental induction: heat shock proteins induced in sporulation and oogenesis. *Science 231*:1154-1158 (1986).

114. Snow, C. M., K. Kitchever, H. T. Kas, E. Hickey, L. Weber, V. H. Heintz, and J. R. Nevins. Selective induction of human heat shock gene transcription by adenovirus E1A gene products, including the 125 EIA product. *Mol. Cell. Biol.* *7*:2884-2890 (1987).
115. Oppermann, H., W. Levinson, and J. M. Bishop. A cellular protein that associates with the transforming protein of Rous sarcoma virus is also a heat shock protein. *Proc. Natl. Acad. Sci. USA* *78*:1067-1070 (1981).
116. Ziemiecki, A., M. Catelli, I. Joab, and B. Moncharmont. Association of the heat shock protein HSP90 with steroid hormone receptors and tyrosine kinase oncogene products. *Biochem. Biophys. Res. Commun.* *138*: 1298-1301 (1986).
117. Schuh, S., W. Yonemoto, J. Brugge, V. J. Bauer, R. M. Riehl, W. P. Sullivan, and D. O. Toft. A 90,000-dalton binding protein common to both steroid receptors and the Rous sarcoma virus transforming protein, pp60v-src. *J. Biol. Chem.* *260*:14292-14299 (1985).
118. Catelli, M. G., N. Binart, J. R. Feramisco, and D. M. Helman. Cloning of the chick hsp90 cDNA in expression vector. *Nucleic Acids Res.* *13*: 6035-6037 (1985).
119. Walker, A. I., T. Hunt, R. J. Jackson, and C. W. Anderson. Double-stranded DNA induces the phosphorylation of several proteins including the 90,000 mol. wt. heat-shock protein in animal cell extracts. *EMBO J.* *4*;139-143 (1985).
120. Ferris, D. K., A. Harel-Bellan, R. Morimoto, W. J. Welch, and W. L. Farrar. Mitogen and lymphokine stimulation of heat shock proteins in T lymphocytes. *PNAS* *85*:3850-3854 (1988).
121. Ferris, D. K., A. T. Brini, A. Harel-Bellan, L. Webber, and W. L. Farrar. Stimulation of heat shock protein synthesis by a recombinant colony stimulating factor. *JOURNAL* (submitted).
122. Pierce, J. H., P. P. Di Fiore, S. A. Aaronson, M. Potter, J. Pumphrey, A. Scott, and J. N. Ihle. Neoplastic transformation of mast cells by Abelson-MuLV: abrogation of IL3 dependence by a nonautocrine mechanism. *Cell* *41*:685-687 (1985).
123. Cleveland, J. J., U. R. Rapp, and W. L. Farrar. Role of c-myc and other genes in IL 2 regulated CT6 T lymphocytes and their malignant variants. *J. Immunol.* *138*:3495-3504 (1987).
124. Whetton, A. D., C. M. Heyworth, and T. M. Dexter. Phorbol esters activate protein kinase C and glucose transport and can replace the requirement for growth factor in interleukin-3-dependent multipotent stem cells. *J. Cell. Sci.* *84*:93-97 (1986).
125. Mills, G. B., E. J. Crague, E. G., and S. Grinstein. IL 2 induces a rapid increase in intracellular pH through activation of a Na+/H+ anti-port. *J. Biol. Chem.* *260*:12500-12506 (1984).
126. Whetton, A. D., P. N. Monk, S. D. Consalvey, S. J. Huang, T. M. Dexter, and C. P. Downes. Interleukin 3 stimulates proliferation via protein kinase C. Activation without increasing inositol lipid turnover. *Proc. Natl. Acad. Sci. USA* *85*:3284–3289 (1988).

127. Beckner, S. K., and W. L. Farrar. Inhibition of adenylate cyclase by IL 2 in human T lymphocytes is mediated by protein kinase C. *Biochem. Biophys. Res. Commun. 145*:176-182 (1987).
128. Harel-Bellan, A., D. K. Ferris, M. Vinocour, J. T. Holt, and W. L. Farrar. Specific inhibition of c-*myc* protein biosynthesis using an antisense synthetic deoxy-oligonucleotide in human T lymphocytes. *J. Immunol. 140*: 2431-2435 (1988).
129. Farrar, W. L., S. W. Evans, U. R. Rapp, and J. L. Cleveland. Effects of antiproliferative cyclic adenosine 3′5′ monophosphate on interleukin 2 stimulated gene expression. *J. Immunol. 139*:2075-2080 (1987).
130. Blomhoff, H. K., E. Smeland, K. Beiske, et al. Cyclic AMP-mediated suppression of normal and neoplastic B cell proliferation is associated with regulation of c-myc and ha-ras proto-oncogenes. *J. Cell Physiol. 131*:426-431 (1987).
131. Farrar, W. L., M. Vinocour, J. L. Cleveland, and A. Harel-Bellan. Regulation of ornithine decarboxylase activity by IL 2 and cyclic AMP. *J. Immunol. 141*:967-971 (1988).
132. Farrar, W. L., and A. Harel-Bellan. Regulation of ornithine decarboxylase activity by colony stimulating factors and gamma interferon. *Blood 73*:1468-1475 (1989).

15

Molecular Mechanisms Involved in the Regulation of Gene Expression During Cell Differentiation and Development

PAUL R. HARRISON
Beatson Institute for Cancer Research, Glasgow, Scotland

I. INTRODUCTION

This chapter discusses some of the principles and evidence to emerge from recent research into the molecular mechanisms involved in gene regulation during cell differentiation and development. The topics chosen reflect the interests of the author and are in no way meant to cover comprehensively what is becoming an enormous subject. Nevertheless, it is hoped that the points discussed will represent a critical and coherent consideration of the main ideas and principles of the field overall as it stands at present.

II. EARLY EVENTS IN GENE ACTIVATION

A. Determination Versus Differentiation

To understand development, we need to understand how decisions that can create clonal precursors, stem cells, and stably maintained sublineages are controlled. Historically, a distinction has been made between the stable commitment or determination of cells to certain restricted lineages or fates as opposed to the overt differentiation of the cells when characteristic cell phenotypes are actually expressed. One classic example of this is the set of imaginal disks of *Drosophila* larvae that are "undifferentiated" groups of cells programmed or determined to produce different specific adult structures after development is completed (1): Cells from imaginal disks can proliferate extensively without differentiating when transplanted into the abdominal cavity of the adult; yet when cells from an individual disk are transferred back into the embryo they develop into the original adult structure, except in rare cases in which a different adult structure is produced similar to that normally produced by a different imaginal disk ("transdetermination") [see (2) for a recent outline of this evidence]. Another example is early mouse development, in which by the 16-cell stage the individual cells in the embryo are no longer totipotent but restricted in their developmental fates (3). A further well-studied situation is the way in which the multipotential hematopoietic stem cell gives rise to a hierarchy of progenitor cells committed to different blood cell lineages that proliferate and differentiate into recognizable blood cells under the influence of specific growth factors and microenvironmental influences [reviewed recently in (4,5)].

It has sometimes been supposed that the molecular mechanisms involved in the establishment of the determined/commited state will prove to be fundamentally different from those involved in the regulation of gene expression during overt differentiation, perhaps using a special class of regulatory gene. However, it is arguable that this classical distinction is more apparent than real because determined cells undoubtedly have a functionally changed phenotype and pattern of gene expression, but it is not so "obvious" biochemically by the relatively crude methods of analysis presently available. It is therefore, perhaps, more realistic to envisage development and differentiation as being regulated by a complex but unitary interlocking network of genes that interact progressively by a variety of molecular mechanisms, involving both intra- and intercellular factors, so as to produce a range of stable differentiated phenotypes via a multitude of metastable intermediate epigenetic states. In some cases, for example *Drosophila* and mammals, the primary developmental "decision" is the production of the three germ layers and subsequent development involves complex interactions between these germ layers, whereas in the nematode, for example, the formation of the germ layers is cell-autonomous and can occur relatively late in development. Nevertheless, in both situations it is becoming clear that specific genes are involved in the regulation of early development, for example certain homoeotic genes of *Drosophila* (6) and the nematode (7). Successful mouse development also requires the complementary activities of both paternal and maternal genes (8-11), but whether these are regulatory genes or encode essential cellular proteins or enzymes, for example, is unclear.

B. Genomic Imprinting and DNA Methylation

One of the most interesting questions is how such intermediate epigenetic states of development are set up in a stable way, a phenomenon known as "genomic imprinting." To account for such genomic imprinting, it is supposed that some mechanism of stable self-propagation of an altered DNA or chromosome structure is needed. In yeast, for example, it has recently been clearly demonstrated that mating-type interconversion uses asymmetry of the DNA strands to generate a regular lineage of cellular differentiation (12). How the two DNA strands are "imprinted" is not known but DNA methylation has the potential for such a purpose because it is known that a specific pattern of methylated DNA sequences is duplicated after DNA replication; cellular methylases recognize and faithfully duplicate hemimethylated DNA sequences (13). Indeed, studies of genomic imprinting exploiting transgenic mice have demonstrated that exogenous DNA sequences ("transgenes") inserted into certain chromosomal loci have different methylation patterns according to the parental origin of the transgene and that the methylation pattern is faithfully reversed on each germline transmission to the opposite sex (14-16). In at least one instance, this switching in methylation pattern according to the parental origin of the transgene correlates with its pattern of expression in the offspring (16).

However, what is not yet clear is whether specific patterns of methylation of DNA are the cause of epigenetic imprinting per se or the means whereby the epigenetic changes are "locked into" the chromosome subsequently. DNA methylation is known to be associated with the inactivation of genes such as HPRT on certain X-chromosomes [reviewed recently by Monk (17)], but in this case the 5′ region of the HPRT allele on the inactive X chromosome in female rodent embryos does not become fully methylated until some time after inactivation is believed to occur (18,19).

There is much other evidence that methylation of specific DNA sequences is involved in the regulation of gene activity: Bird and coworkers (20-22) have discovered that mammalian genomes contain ~30,000 localized "islands" of DNA 500-2000 base pairs (bp) in size that contain the bulk of the genomic CpG residues in unmethylated form; significantly, these CpG-rich islands are mostly associated with genes expressed constitutively and not with genes expressed in a tissue-specific manner (21). The main functions of DNA methylation changes may therefore be to keep "housekeeping" genes permanently active and certain X-chromosomes and other loci permanently repressed throughout development (21). However, the expression of tissue-specific genes does in many instances correlate with hypomethylation of certain regions of the genes, usually in the 5′ flanking regions (22); and treatment of cells with azacytidine (which is thought to result in hypomethylation of DNA) can lead to activation of genes normally silent in particular cell types [reviewed by Jones (23)]; but whether such changes are causally involved in activation of the genes is open to question: the experiments designed to test the effects of methylation on the expression of tissue-specific genes have for technical reasons involved transfecting the genes into cells in which the genes are not normally expressed, so that the experiments may be measuring the effects of methylation on the "artificial" regulation of those genes by factors that normally keep "housekeeping" genes constitutively active (22).

C. Chromosome Changes: DNAse I Hypersensitive Sites

Some of the earliest events associated with gene activation are changes in the chromosome structure that result in unfolding of the chromosome at specific sites resulting in increased sensitivity to DNAse I and other agents (24-26). In general, large regions (or "domains") of chromosomal DNA flanking genes become more sensitive to DNAse I in cell types in which the genes are active: in addition, particular sites within the active domain or at its boundaries (27) become even more DNAse-sensitive, the so-called DNAse I hypersensitive sites (DHSS). Perhaps the most convincing evidence that DHSS are important in gene control is that DHSS upstream of the *Drosophila* glue protein gene, *Sgs-4*, are found in regions known by mutational analysis to be involved in the tissue-specific regulation of the gene (28,29). Much other evidence shows that some tissue- or developmental-specific DHSS are found to be located at or close to functionally important *cis*-control elements as defined by gene transfer experiments [reviewed by Eissenberg and co-workers (30)]. Other DHSS seem to mark domain boundaries surrounding genes, for example the human β- globin gene locus (27).

One point of relevance to the question of epigenetic imprinting to emerge from such studies is that certain tissue-specific DHSS are created early in cell differentiation before transcription of the relevant gene and that they may persist during cell proliferation after the cessation of transcription. Perhaps the clearest example of this emerges from studies of globin gene activation in cells infected with temperature-sensitive (Ts) retroviruses. When chicken embryo fibroblasts are infected with Ts-Rous sarcoma virus at the permissive temperature, the cells become transformed and begin to transcribe globin mRNA (but at a much lower level than in erythroblasts) and this transcription is preceded by the formation of DHSS in both the α- and β-globin gene domains at the same sites as those found in normal red blood cells (31): When the v-*src* gene is inactivated subsequently by a temperature shift, the infected cells can be grown for 20 doublings without loss of the red-blood-cell-specific DHSS although the globin genes are no longer actively transcribed. In a similar manner, in certain erythroblast cell lines derived from avian bone marrow cells infected with avian erythroblastosis virus, the normal red-blood-cell-specific DHSS occur upstream of the globin genes without concomitant globin gene transcription (32).

Another instance in which cells are known to retain an epigenetic memory relates to steroid hormone induction of genes. On first exposure to steroid hormone, cells undergo a lag period before the transcription of the responsive genes is initiated, whereas after withdrawal of hormone followed by a second treatment a considerable time later, transcription of the responsive genes commences more or less immediately. Activation of such genes is known to be associated with the formation of certain DHSS, for example, in the case of induction of the lysozyme and vitellogenin genes by estradiol (33,34), and it was thought that the persistence of a specific DHSS might be responsible for the memory of prior exposure to hormone. However, more recent work showing that the induced DHSS is not as stable as the memory phenomenon itself has cast some doubt on this interpetation (35).

Thus, to conclude this section, it is clear that DNA methylation and enhanced DNAse I sensitivity/DHSS are both associated in some way with stable programming of either housekeeping or tissue-specific genes to be active or inactive, as the case may be. Indeed, the two changes in DNA/chromosome structure are probably intimately interconnected, because DNA methylation affects the formation of active chromosome structures as determined by DNAse I sensitivity (36). However, the crucial question as to how these chromosome modifications are established at highly specific sites involved in the regulation of genes remains to be answered. The binding of factors to specific DNA regions might be involved: for example, binding of the transcription factor TFIIIA to the internal control region of the ribosomal 5S gene is known to prevent the condensation and inactivation of the surrounding chromosome by histone H1 (37). However, the problem is how such factors gain access to the DNA in the chromosome in the first place. Perhaps crucial regulatory genes are maintained in an active state in the egg and then these, in response to fertilization, set up a cascade of events that initiate the progressive cycling of the overall regulatory gene network from one epigenetic state to the next that results in development. Although it has not yet proved possible to study or in some cases even identify such regulatory genes in mammals, work with some of the homoeotic genes of *Drosophila*, to be discussed later, suggests that they may be programmed and regulated using the same variety of transcriptional and posttranscriptional control mechanisms found to be involved in the regulation of other genes expressed during later stages of development and cell differentiation. Thus, the interest of early development and the early "commitment/determination" decisions may lie in elucidating the logic of the regulatory network and hierarchy of regulatory genes rather than in discovering novel molecular mechanisms per se.

The rest of this chapter will therefore review the types of molecular mechanisms involved in the tissue-specific expression of genes during development and cell differentiation: A major point to become apparent will be the flexibility in which different molecular control mechanisms are linked together to achieve a multitude of modes of regulation to suit different requirements.

III. *CIS* REGULATION OF GENES

A. Experimental Approaches

Cis-control regions involved in the regulation of cellular genes can be identified by mutating cloned genes and then studying the effect on their expression after reintroduction into cells. In some experiments the expression of the introduced genes is studied before their integration (transient expression): In this case there are usually large numbers of copies of the introduced DNA interacting with whatever relevant factors are present in the nucleus. Obviously, where the amounts of such factors are limiting, then only a small fraction of the exogenous DNA will be able to bind the factor and respond. In the second type of approach, the introduced genes are allowed to become integrated into the genomic DNA: usually, this involves integration of a large block of DNA [~1000 kilobases (kb)] as a concatamer, except for retrovirus-mediated transfer where usually only single copies become integrated. Clearly, in such experiments it would be desirable ideally to be able to introduce such mutated genes into their normal locus in the germ-line, in place of the normal wild-type gene. Although such homologous recombination is possible in yeast, it has only recently proved possible in higher eukaryotes. Although it is now possible to introduce genes into the germ-line of transgenic mice by injection into eggs, it has not been possible to introduce the genes into their normal locus. However, under exceptional circumstances it has proved possible in tissue-culture cells. For example, Smithies and co-workers (38) detected insertion of β-globin gene sequences into the human chromosomal β-globin locus in tissue-culture cells by homologous recombination, and Thomas and co-workers (39,40) have demonstrated a relatively high frequency of targeting of sequences containing the hypoxanthine phosphoribosyl transferase gene or the gene encoding resistance to noemycin to homologous but mutated sequences previously inserted into the genome of tissue-culture cells.

Thus, as a general rule, all gene-transfer experiments in mammalian cells are subject to influence by the site of integration, which cannot be controlled. In transgenic mice, the introduced genes are often regulated with the correct cell-type or developmental specificity but rarely as quantitatively as the normal endogenous genes. Moreover, there is some evidence that genes transfected into tissue culture cells may be preferentially integrated into active gene regions because most genes are in fact expressed at a low level, whatever the cell type transfected. One example to illustrate this concerns introduction of globin genes into erythroleukemia cell lines, which can be induced to differentiate and express erythroid markers by various means. When single *chromosomes* containing either the α- or β-globin loci from either human erythroid or non-erythroid cells are introduced into the mouse erythroleukemia cells, transcription of the exogenous globin genes, like the endogenous mouse counterparts, is increased greatly as the cells are induced to differentiate (41-45). In contrast when plasmids or cosmids containing *fragments of isolated genomic DNA* containing globin genes are transfected into erythroleukemia cells, the β-globin gene expression increases as the cells differentiate, whereas the α-globin gene is expressed at a relatively high level before differentiation and only increases slightly during erythroid differentiation (46). This is true even when α-globin gene-containing cosmids containing up to 48 kb of flanking sequence are transfected. These experiments clearly demonstrate important differences in the way the α- and β-globin genes are regulated when introduced into erythroleukemia cells as chromosomes, as opposed to isolated DNA. This might be because the *cis*-control elements regulating the α-globin gene are situated at very large distances and therefore missing from the 50-kb genomic DNA fragments transfected, or that control sequences in the vectors used for cloning the globin gene fragments induce the transcription of the α- but not β-globin gene in undifferentiated erythroleukemia cells. Alternatively, the globin genes may need to be in their normal chromosome location to be regulated correctly. Presumably this might not be necessary if the exogenous DNA contained within it the information to establish independently the correct chromosomal domain structure. Some recent evidence is very pertinent to this interesting possibility: Grosveld and colleagues (47) have measured the level of erythroid-specific expression of the human β-globin gene in transgenic mice when it is linked to fragments of the genomic flanking DNA from the 5′ to 3′ extremities of the entire β-globin domain as defined by its sensitivity to DNAse I, at approximately 50 kb 5′ and 20 kb 3′ to the gene, respectively. Their results show that such fragments containing the globin-gene-domain boundaries are regulated quantitatively to the same levels as the endogenous mouse β-globin gene in erythroid cells, whereas the level of expression of short globin gene fragments is very much lower. It will be very interesting to discover whether such globin-gene-domain boundary regions confer erythroid specificity to the enclosed domain or whether they are control sequences used generally to mark the boundaries of all active gene domains. In the latter case, the domain boundary regions would presumably be themselves under the control of control elements internal to the domain.

B. General *Cis*-Control Elements

It is clear then that there remain formidable problems in obtaining normal regulation of genes in gene transfer experiments. Nevertheless, introduction of cloned genes into transgenic mice or into tissue culture cells has afforded important, even if partial, insights into some aspects of gene regulation during cell differentiation and development. Deletion and single-base mutation analysis of many genes have identified a series of common sequence motifs that are necessary for proper transcription after introduction of certain genes into cells: The TATA consensus sequence is thought to specify the correct initiation site for transcription in genes that contain it; other sequence motifs are involved in regulating the rate of transcription, for example, the CAAT motif [see (48) for a recent compilation of such sequences], the CACCC motif upstream of the adult α- and β-globin genes (49) and the consensus motif, G/TGGGCGGPuPuPy, recognized by the transcription factor Sp1 (50). In general, such sequence motifs are thought to act as constitutive regulatory elements. However, many so-called "housekeeping" genes do not have TATA and/or CAAT motifs, for example, dihydrofolate reductase (51), adenosine deaminase (52), hypoxanthine phosphoribosyltransferase (53), HMG CoA reductase (54,55), β-tubulin (56), and glutathione peroxidase [which we have recently cloned (57)]. Generally, where such genes lack a TATA motif they are characterized by multiple sites of transcription initiation. As noted earlier (Section II.B), "housekeeping" genes are also generally associated with unmethylated CpG-rich islands, which may be responsible for keeping this class of genes constitutively active (22).

C. Cell-Type Specific *Cis*-Control Regions

In addition to these constitutive elements, a series of sequence motifs are continually being discovered that confer cell type- or developmental-specific *cis*-regulation of genes or response to specific environmental stimuli. A well-studied example is the "octamer" sequence present in the promoters [and also enhancers (see Section III.D)] of immunoglobulin genes that is involved in regulating their transcription specifically in lymphoid cells (58-62). Thus, the octamer oligonucleotide alone coupled to a TATA motif is sufficient for lymphoid-specific promoter activity (63). Nevertheless, the octamer sequence cannot be the only element regulating the immunoglobulin genes because it is also found in other genes not expressed in lymphoid cells, e.g., the U1 and U2 small nuclear RNA genes and the histone H2B gene. However, in these genes the octamer sequence is found in conjuction with other canonical control elements that are conserved among different U1/U2 or histone genes, respectively (64-67): these presumably cause it to become active in the absence of lymphoid cell factors.

Other *cis*-control elements that have been well defined include: a sequence motif responsible for the transcriptional response of the metallothionein genes to heavy metals (68,69) or interferon (70): and an element that mediates the response of genes to steroid hormones (68,71-75). Another interesting case is that of the family of heat-shock genes, which are heat-inducible by virtue of possessing (usually multiple copies of) an upstream consensus sequence element, CTnGAAnnTTCnAG (76). In addition, different members of the gene family are expressed in different ways during development, e.g., the *Drosophila* hsp26 and 28 heat-shock genes, which are expressed in ovarian nurse cells (77), and this is due to other *cis*-control elements. The 5′ flanking sequences conferring increased cell-type specific transcription of a variety of cellular genes have also been defined, for example: human cardiac actin (78), α- and δ-crystallin (79, 80), albumin (81,82), α_1-antitrypsin gene (83,84), and α-fetoprotein (85,86).

An important general point to emerge from such studies is that different genes are regulated by different combinations of such mocular elements, which can often operate independently (87). This is illustrated in the case of the metallothionein gene in Fig. 1.

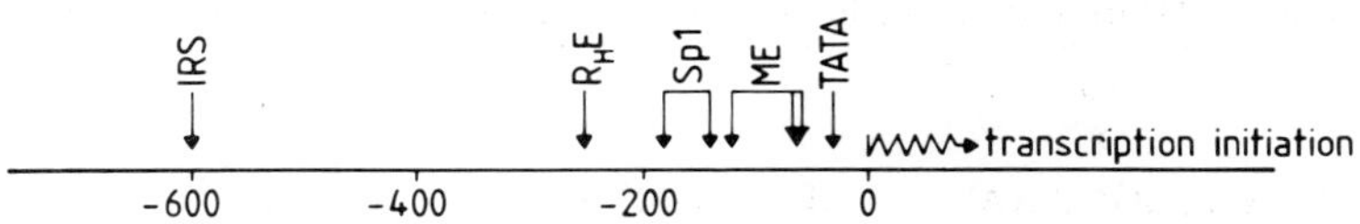

Figure 1 *Cis*-control regions involved in the regulation of metallothionein genes. The position of various control elements is shown relative to the site of transcription initiation. IRS, Interferon-responsive element; R_HE, steroid-hormone-receptor binding element; Sp1, two sites for binding of the transcription factor Sp1; ME, three copies of a control element mediating induction by heavy metals; TATA, sequence responsible for correct position of the site of transcription initiation.

D. Enhancers

In some instances, the *cis*-control regions defined by experiments such as those described above have the properties of so-called "enhancers": i.e., the *cis*-control elements are effective in increasing transcription of the gene when placed at considerable distances either 5′ or 3′ to the promoter, and in either orientation. Identified first in the 5′ control regions of eukaryotic viruses, examples of 5′ cell-type-specific enhancers have been found associated with the genes encoding lysozyme, α-fetoprotein, insulin, chymotrypsin, β-interferon, antithrombin, placental lactogen, α-crystallin, interleukin-2, elastase, and complement protein factor B [see (88) for a short review; also (86,89-92)]. A 5′ enhancer is also responsible for the light-induced regulation of certain plant genes (93). In other cases, the enhancer lies 3′ to the promoter, for example in the heavy and light-chain immunoglobulin genes (94-96) and, more recently, in the chicken and human adult β-globin genes (97-100), the human γ-A globin gene (101), in the erythroid-specific chicken histone H5 gene (102), and in the mouse α_2 (type I) collagen gene (103). Some cellular enhancers show homology to viral enhancers: the β-interferon enhancer, for example, shows homologies with both adenovirus/polyoma and SV40 type enhancer core motifs (104) (Table 1).

In many instances, such enhancers or other control elements can operate as independent units irrespective of the promoter or gene to which they are attached. However, this is not invariably so: for example, in the case of the immunoglobulin heavy-chain gene (105) and the β-globin gene (100) there seems to be a synergism between the enhancers and their natural promoters; this might be due, for example, to cooperative interactions between two molecules of the same protein bound to both of the regulatory elements. Mutational analysis of the β-globin and immunoglobulin gene enhancers has revealed that in fact they are complex regulatory units, like the SV40 enhancer (106-110), containing sub-elements that bind different proteins, only perhaps one of which is cell-type-

Table 1 Sequence Homologies to the β-Interferon (IFN) Gene Control Element

−76 −59		
AGAAGTGAAAGTGGGAAA	Human β-IFN	
GGAAGTGAA	Adenovirus 5 E1a enhancer	9/10
GGAAGTACT	Polyoma enhancer	7/10
AGAAGGGAAA	Human HSP70 promoter	9/10
GGACCTGAAA	Friend virus LTR	7/10
GTGGAAAG	SV40 "core" sequence	6/8
TTGGAAAG	CMV enhancer	5/8
GTGGCAAG	IgH gene enhancer	6/8
GTGGAAAC	c-*fos* gene regulatory element	6/8
GTGGTAAA	Adenovirus 2 E1a enhancer	7/8
−91 −67		
GGAAAACTGAA−GGGAGAAGTGAAA	Human β-IFN	
GGAAAACTGAATAAGAGGAAGTGAA	Adenovirus 5 E1a−200 region	

The human β-IFN constitutive transcription element (nucleotides −77 to −55) was compared with functionally important sequence elements from a number of other enhancer and promoter regions. The extent of homology is indicated at the right. CMV, cytomegalovirus; HSP, heat-shock protein; IgH, immunoglobulin heavy chain; LTR, long-terminal repeat. The sequences at the bottom indicate that the homology of the β-IFN 5′ flanking region with the adenovirus E1a enhancer can be extended.
Source; Data from Table 2 of Ref. 104.

specific. Thus, within the minimal region of the adult chicken β-globin gene enhancer required for full expression of enhancer activity, there are at least found binding sites for proteins that have different tissue distributions (99): Two are found only in cells actually transcribing the adult β-globin gene (adult erythroblasts), whereas one is also found in embryonic erythroblasts in which the adult β-globin gene is inactive. The protein binding to the fourth site is ubiquitous and probably corresponds to a cellular factor that also stimulates adenovirus replication (nuclear factor I).

A similar conclusion emerges from extensive in-vitro footprinting studies of the immunoglobulin gene enhancer (61,111-121). B-Lymphocyte-specific interactions of the immunoglobulin enhancer with cellular factor(s) has also been confirmed by in vivo footprinting methods (122). Moreover, there is evidence that one of the proteins that binds to the kappa immunoglobulin gene enhancer (NF-kappaB) after induction of lymphoid cells by lipopolysaccharide is in fact not lymphoid-cell-specific: This factor seems to be posttranslationally activated transiently during B-cell maturation (123) via a mechanism involving protein kinase C that also can be activated in nonlymphoid cells, for example, by phorbol esters (124). This implies that the interaction of this factor with one of the subelements of the immunoglobulin gene enhancer is a posttranslational regulatory mechanism that is not dedicated to immunoglobulin gene transcription per se but is used more widely in transcriptional regulatory circuits.

E. Negative Control of Gene Regulation

For some time, it was assumed that the effects of such control regions in increasing transcription meant that they were activated by some positively acting factor(s) only present in a particular cell type or developmental situation. However, recent work suggests that this is an oversimplification and that in many cases genes are regulated by tissue- or developmental-specific repressors binding to specific sequences that in turn control the activity of a constitutive enhancer or other positively acting element. The evidence supporting this conclusion has come from two approaches: first, finer dissection of control regions by deletion and single base-pair mutation; and second, by competition-type experiments (125). In the latter type of experiment, increasing amounts of the control region to be tested are cotransfected with the complete gene and the transcription of the gene monitored in different cell types. If the control region tested acts positively, then it will decrease the transcription of the test gene when both are transfected into a high-expressing cell type; however, if it acts as a negative control element, then the transcription of the test gene will be increased in the nonexpressing cell type due to removal of negatively-acting factors.

Examples of such negative regulation include the gene encoding cytochrome P450 (126), β-interferon (104,127,128), α-fetoprotein (86,129), c-*myc* (130, 131), rat insulin-1 (132), retinol-binding protein (133), immunoglobulin heavy chain (134), lysozyme (135), β-globin (99), rat growth hormone (136), low-density lipoprotein receptor (137), testis histone H2B-1 (138), ovalbumin (139), c-*fos* (140), HMG CoA reductase by cholesterol (55), and an adipocyte lipid-binding protein (141). A diagrammatic summary of some examples of such genes is presented in Fig. 2. In some cases, the repressor-binding and enhancer elements are separate, and in others adjacent (e.g., β-interferon) or overlapping (e.g., the testis histone H2B-1 gene).

In most of the examples cited, the positive and negative elements have been defined approximately by the use of deletion mutants in transfection experiments. However, the existence of *trans*-acting factors that recognize such elements has been demonstrated directly in some cases by DNAse-footprinting and

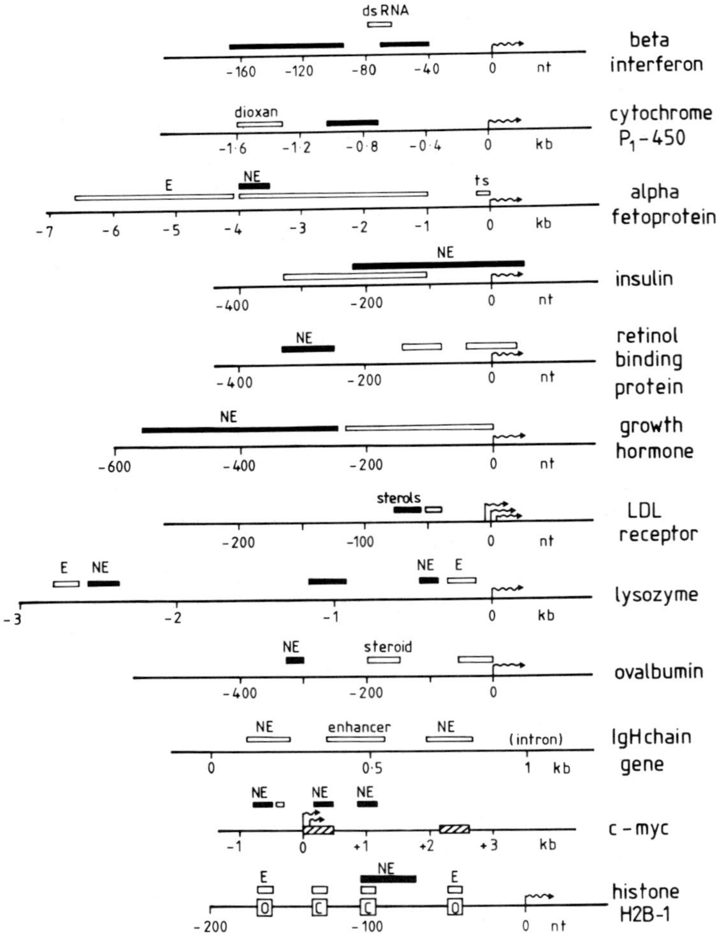

Figure 2 Examples of negatively acting *cis*-control regions and their interactions with positive *cis*-control regions. The diagram shows the locations of positive (open boxes) and negative (closed boxes) regulator regions for a variety of genes. In many cases, the locations of the control elements have been defined only approximately within the regions shown. E and NE denote that the positive or negative element functions only in cells expressing, or not expressing, the gene, respectively. The arrows denote the site of transcription initiation. In the case of the c-*myc* gene, exons 1 and 2 are shown as hatched boxes. C and O on the histone H2B-1 gene map refer to the CAAT and octamer sequences, respectively. In certain cases, certain elements function in response to known agents, e.g., dsRNA, dioxan, sterols and steroid hormones in the case of β-interferon, cytochrome P_1-450, LDL receptor, and ovalbumin, respectively. Negatively acting regions that are active in all cell types can presumably be overcome by cell-type-specific activation of positive control regions. Data from references cited in the text.

gel-retardation assays and some of the factors have been partially purified by affinity purification or other means (see Section IV). The general model to emerge from these studies is that specific genes are repressed in most tissues and become active in certain tissues where the repressor is no longer synthesized or is inactivated by binding to some developmental-specific signal. A puzzling parody of this model is the way in which adenovirus type 2 E1A gene products repress the immunoglobulin heavy-chain enhancer in lymphoid cells (142), whereas in fibroblasts they stimulate it (143): In this example, the E1A products seem somehow to regulate the immunoglobulin gene enhancer in precisely the opposite way from the normal situation.

These recent experiments help to explain at the molecular level the results of older cell-fusion experiments showing "extinction" of specific differentiated functions when a particular differentiated cell is fused to another cell type [see, for example, (144-148) for various aspects of this older work]. More recently, it has been demonstrated that such extinction of specific functions operates at the level of the changes in chromatin structure associated with the preactivation of the structural genes involved, as monitored by the formation of cell-type-specific DHSS (149,150).

This cell-fusion approach has now been extended using segregating interspecies hybrids to show that individual chromosomes can repress specific structural genes in a dominant, cell-autonomous manner; thus, genes on two different chromosomes from fibroblasts specifically repress the production of either albumin or tyrosine aminotransferase respectively when introduced as single chromosomes into hepatoma cells, and this repression is maintained as a stable characteristic during subsequent cell proliferation (151-153). Presumably, these genes on these individual chromosomes encode the *trans*-acting repressor molecules identified by the molecular approaches described above.

F. Mechanism of Action of Enhancers

How enhancers and other *cis*-control elements function to increase transcription from the promoter is not clear. Evidence discussed below shows that *cis*-control regions interact with nuclear proteins. One model of enhancer function envisages them as entry sites for transcription factors that are then able to interact with RNA polymerase and other factors to set up a functional transcription complex; this might be by scanning the DNA until a promoter is found or by causing the DNA to loop out so that the factors bound to the enhancer and promoter can interact. Earlier work suggesting that enhancers had greatest effect on the most proximal promoter was considered to support the view that enhancers act as bidirectional entry sites for transcription factors, which then scan the DNA for nearby promoters (154). However, more recent evidence shows that tandem immunoglobulin promoters are equally active in the presence of the immunoglobulin enhancer, suggesting that the enhancer exerts its influence uniformly over large distances and independently of the presence of intervening promoters (155). Another general model for enhancer function considers that they are somehow involved in organizing the surrounding chromatin into an active conformation. This model is supported by the extensive data considered previously that cell-type-specific enhancers are almost invariably associated with DHSS, which are thought to reflect regions of nucleosome-free DNA in the chromosome. The SV40 enhancer, for example, is associated with a stretch of nucleosome-free DNA several hundred base pairs in length even if transposed to an abnormal position in the viral genome (156,157). Moreover, experiments in which the DNA separating the SV40 enhancer from a linked β-globin gene is modified by an intercalating agent, psoralen, show that the structure of the DNA flanking the enhancer and gene is important for the enhancer-mediated transcription of the gene to occur (158).

One particular mechanism that has been advanced whereby enhancers might be able to interact with distant promoters in such active chromatin regions is by looping out of the intervening DNA. There is considerable evidence in prokaryotes and yeast to support the idea that transcription factors can interact at some distance with the promoter by looping out of the intervening DNA (159). One of the main facts supporting this model is that cooperative binding between, for example, lambda repressors only occurs if the repressors can interact on the same face of the DNA helix (159). Similarly, a single protein binds to both a promoter and enhancer region of the long-terminal repeat (LTR) of the murine sarcoma virus: Interestingly, mutations in the promoter that affect binding of the protein also affect its binding to the enhancer, implying cooperativity (160). Another finding consistent with the looping-out model is that the DNA-binding and activator functions of certain prokaryotic or yeast transcription factors can be separated and that one activator component can activate a different promoter if linked to the DNA-binding component recognizing the control region of the second gene [reviewed recently by Robertson (161)]. However, it is presently unclear whether interactions between regulatory enhancers and promoters could operate by DNA looping in instances where they are known to be separated by several kilobases.

A further model of enhancer function that has been proposed is that the enhancers affect distant genes by altering the twist of the DNA in the region, thereby increasing the negative supercoiling of an entire DNA domain and so increasing the transcription of supercoiling-dependent promoters within this domain. Indeed, the association of topoisomerase II cleavage sites with some enhancers might support this model (162-164). However, a recent experiment shows that a β-globin gene can be stimulated by an SV40 enhancer to which it is physically linked but topologically uncoupled by virtue of being situated in the hairpin loop of a tailed-circle construct (165). This seemingly definitive experiment seems to exclude the sequence-specific gyrase model for enhancer action, at least as far as the SV40 enhancer is concerned.

A final model for enhancer function that should perhaps be considered is that enhancers could act by targeting the template to a particular nuclear locale where transcription could then occur (the so-called "nuclear address" model). Various lines of evidence suggest that transcription of genes occurs at the nuclear matrix (also known as the nucleoskeleton or scaffold), the structure to which the chromosomal DNA domains or loops of chromatin are attached (166). Recent work has attempted to map the attachment sites of DNA to the nuclear matrix with respect to specific genes, for example, the *Drosophila* histone, hsp70 heat-shock, alcohol dehydrogenase and glue-protein (*sgs*-4) genes (163, 167) and the mouse immunoglobulin kappa gene (164): The results presented seem to show that the attachment sites are generally associated with the enhancers that regulate the genes, but this does not seem to be the case for the heat-shock hsp70 gene. However, such experiments are subject to serious difficulties in terms of eliminating artefactual associations on the one hand, and, on the other, rearrangement of DNA-protein complexes during the extraction process. It will therefore be very interesting to see whether these potentially exciting experiments prove to be corroborated by further work.

There is one further important fact that has implications concerning how enhancers function. During differentiation of pre-B cells into mature lymphocytes, the immunoglobulin genes are rearranged to give actively transcribed functional units and this process involves deletion of the tissue-specific enhancer, although the octamer sequence in the promoter is, of course, retained. (Indeed, this may be precisely why the octamer sequence is duplicated in the promoter of these genes.) This strongly implies that the immunoglobulin enhancer is involved somehow in establishing the active immunoglobulin gene domain and is not needed for subsequent active transcription of the gene. Because it is well established that chromosome changes involved in the early activation of genes can be stably maintained irrespective of whether transcription is actually occurring, the loss of the immunoglobulin enhancer during differentiation seems to be most consistent with the chromatin model of enhancer function described above. If the enhancer were involved directly in maintaining the supercoiling or scaffold attachment of the gene, its loss during differentiation would presumably prevent continued transcription.

IV. *TRANS*-ACTING FACTORS

A. In Yeast

The most extensively studied *trans*-acting factors that regulate genes are from prokaryotes or yeast, but some regulatory proteins from higher eukaryotes are beginning to be purified and their mechanisms of action understood. It should be emphasized that not all regulatory proteins involved in activating specific genes will necessarily be DNA-binding proteins. Nevertheless, the general approach taken in this area of research has naturally focused on identifying proteins that interact specifically with proven gene-control regions.

Perhaps the best understood examples of DNA-protein interactions involved in gene regulation are the lambda repressor and the *Escherichia coli trp* gene repressor where the mechanisms of action have been worked out in detail [see (161) for a recent short review]. In these cases, the repressor binds to the major groove of the DNA by means of a helix-turn-helix motif. On the basis of amino-acid sequence considerations, the yeast mating type repressor, α-2, and the protein domain encoded by the homeobox region of certain *Drosophila* homoeotic genes and their mammalian analogues may also bind to their DNA target sequences by a similar helix-turn-helix motif (168,169).

However, transcriptional activators must be inherently more complex than repressors because the activators not only have to bind to specific DNA sequences but also establish or interact with a functional transcriptional complex. It has been demonstrated in the case of two yeast gene regulatory proteins, GCN4 (which regulates the promoters of amino acid biosynthetic genes), and GAL4 (which mediates the galactose regulation of the GAL1, 7, and 10 genes), that the specific DNA-binding and activation domains are separate and can operate independently [(170-172); reviewed by Struhl (173)]. In these cases, the DNA-binding sequence is conserved between the set of regulated genes. However, this is not invariably so: for example, the yeast HAP1 activator binds to entirely different DNA sequences upstream of the CYC1 and CYC7 gene which it regulates (174).

In general, the activation domains of yeast activators are short acidic regions of perhaps 50 amino acids, but there is no discernible homology between the equivalent regions from both GCN4 and GAL4. This suggests that the yeast activation domains are short and not defined by a specific primary sequence but a more general topological feature.

B. *Trans*-Acting Factors Employing "Metal-Finger"-Type DNA Binding Domains

1. Ribosomal 5S Genes

One of the best studied gene-specific transcription factors from higher eukaryotes is the factor TFIIIA that controls *Xenopus* ribosomal 5S gene transcription by binding to an internal control region (ICR). In fact, *Xenopus* has two types of 5S genes, one expressed in the oocyte and the other during embryonic development. It is now known that the differential expression of the oocyte and somatic ribosomal 5S genes is due to TFIIIA binding with higher affinity to the ICR within the somatic gene relative to that in the oocyte gene: Early in development, the concentration of TFIIIA is sufficient to activate both 5S gene types, but as it decreases as development proceeds, only the somatic type remains active (175,176). This has been confirmed recently by showing that oocyte 5S genes can be transiently activated in *Xenopus* embryos by raising the level of TFIIIA (177). Other experiments have shown that TFIIIA keeps the somatic type genes active by preventing their repression by histone H1 (37).

The detailed mechanism of binding of TFIIIA to the 5S gene ICR is now beginning to be understood. The ICR is in fact composed of three distinct sequence elements organized as two functional domains, one of which interacts with the general polymerase III transcription factor TFIIIC whereas the other binds TFIIIA. Recent evidence shows that interactions between the two transcription factors is important, because altering the spacing between the two binding domains by –3 to +10 nucleotides increases their transcriptional efficiency and binding of TFIIIC seems to be dependent on the prior binding of TFIIIA to is domain (178,179).

The region of TFIIIA responsible for activation of transcription is localized to the *C*-terminus of the polypeptide, but TFIIIA contains an approximately 30 amino-acid residue motif repeated nine times, probably corresponding to nine zinc-binding domains or "fingers" formed by interaction of two invariant pairs of cysteines and histidines with an atom of zinc (180). In fact, the internal control region of the 5S gene also contains a DNA motif repeated nine times: This repeat is not immediately obvious from the DNA sequence per se, but it has been revealed by Fourier analysis of the cleavage pattern of the transcription complex by DNAse I (181) and its crystal structure (182).

2. Other Examples

Since then, striking similarities with the TFIIIA zinc finger motif have been noted in several yeast and *Drosophila* regulatory proteins, such as the yeast regulatory protein, ADR1 (required for transcriptional activation of the alcohol dehydrogenase gene), and the polypeptides encoded by the *Drosophila* genes Kruppel (whose function is required for the formation of the thoracic and abdominal segments in the embryo) and Serendipity (which is expressed during oogenesis and early development) [reviewed by Vincent (183)]. However, in most cases it has not been demonstrated unequivocally that the metal involved is actually zinc. Recently, evidence has been presented for a mammalian multigene family encoding several polypeptides containing "metal-finger"-like structures and these are differentially regulated during differentiation of teratocarcinoma cells (184).

3. Steroid-Hormone Receptors

The recent cloning of the glucocorticoid estrogen, progesterone, and thyroid hormone (c-*erb*-A) receptors has shown that this class of *trans*-acting factors also probably binds to sequences upstream of their target genes by means of metal-finger domains [reviewed by Green and Chambon (185)] ; however, in these cases the putative binding domains involve two pairs of cysteines rather than pairs of cysteines and histidines (186,187). A region of the receptor toward the carboxy terminus is responsible for binding steroid hormone; evidence is conflicting as to whether the transcription activation function of the receptor resides in the *N*-terminal region.

C. Other Eukaryotic Transcription Factors

Interactions of specific proteins with defined *cis*-control regions can be identified both in vivo, by the formation of DHSS or by in vivo DNAse I "footprinting" techniques, and in vitro by "footprinting" and gel retardation techniques. Factors shown to interact with some of the specific gene control regions considered in a previous section include: Sp1, the protein that binds to the constitutive control element of consensus sequence G/TGGGCGGPuPuPy (188); a TATA box-binding factor (189) nuclear factors I and III, which have binding sites in mammalian genomes but also stimulate adenovirus replication (190,191); a family of proteins that interact with various CAAT motifs (138,192-194); and a factor interacting with the heat-shock gene consensus sequence, both in vitro (195,196) and in vivo (197). In addition, factors such as AP1 and AP2 have been found that interact with control regions involved in the response of certain genes to phorbol esters or cyclic adenosine monophosphate-elevating agents, for example the metallothionein, growth hormone, c-*myc*, H-2K^b genes, and the SV40 and bovine papilloma virus enhancers (198-202). The factors that interact with the immunoglobulin gene promoter octamer motif and the enhancer have been discussed previously. However, only in a few instances have the regulatory proteins been purified and characterized.

Recent work has shown that the way in which such factors interact with the appropriate gene control regions can be quite complex. As noted earlier in connection with at least one of the factors binding to the immunoglobulin gene octamer motif, cell-type-specific regulation of a gene by its transcription factor is not necessarily due to the factor being cell-type-specific but rather that the factor is only converted to an active form in a particular cell type. A similar situation has been found in the case of inducible binding of a factor to the c-*fos* enhancer (203). Furthermore, what were originally thought to be constitutive control elements binding a single factor may prove to be functionally flexible due to multiplicity in the factors that interact with them. This is true for example for the "CCAAT" box: At least four different factors have been detected in different cell types that interact with the CCAAT motif associated with different genes, for example, the α-globin gene, thymidine kinase gene, retroviral long-terminal-repeat, sperm-specific histone H2B-1 gene, and histocompatibility complex class II genes (138, 192-194,204,205). There is also evidence that a CCAAT-like sequence upstream of various actin genes may play a role in tissue-specific transcription rather than simply as part of a basal promoter sequence (78,206). A similar conclusion arises from the finding that a natural single-base mutation in the CCAAT box of a fetal globin gene affects its developmental as well as basal expression (207). Work on the sperm histone H2B-1 gene suggests that inactivation of "CCAAT" box-containing promoters in certain cell types is due to binding of a repressor, the so-called CCAAT displacement protein, to sites overlapping the CCAAT motif and thus preventing binding of the appropriate CCAAT activator protein (138). A similar phenomenon also seems to operate in the regulation of the yeast CYC1 gene, where a repressor whose synthesis is controlled by heme binds competitively to the same site in the upstream control region of the gene as the transcriptional activator, HAP1 (325).

One final general point to be made concerning the mechanisms of action of *trans*-acting factors in regulating *cis*-control regions in a cell-type-specific manner is that some factors may act via protein-protein interactions with other sequence-specific DNA binding proteins. This has now been demonstrated for the regulation of cell-type-specific genes by the α-1 product of the yeast mating-type locus (208) and the transcriptional regulation of the yeast galactose regulon by an interplay involving protein-protein interactions between the positively-acting regulatory protein, GAL4 (that binds to the control region of the GAL1 gene), and the negative regulatory protein, GAL80 (209). Other similar examples from higher eukaryotes are the way in which SV40 T antigen regulates the binding of the AP2 transcription factor to the SV40 and metallothionein promoters by protein-protein interactions independently of its binding to DNA (200), and the situation with eukaryotic ribosomal RNA genes, in which a transcription initiation factor directs the polymerase to the correct position on the promoter solely by protein-protein contacts (210). Another example is the interaction between a DNA-binding transcription factor and a non-DNA-binding factor necessary for transcription of the ovalbumin gene in vitro (211).

V. POSTTRANSCRIPTIONAL CONTROL MECHANISMS

A. Differential Processing of mRNA Precursors

In virtually all eukaryotic structural genes transcribed by RNA polymerase II, the primary transcript is modified in three different ways in the course of producing a mature mRNA. First, the 5′ terminus of the transcript is modified (or "capped") by addition of a 5′-methyl guanosine residue, the transcript is cleaved at a site downstream of a specific signal [reviewed by Birnstiel and co-workers (212)] and (usually) the 3′ terminus is then polyadenylated; and finally, introns are spliced out to form the mature mRNA. This process is somehow linked with the transport of the RNA from the nucleus to cytoplasm, possibly involving the nuclear matrix (213). Histone genes are of course an exception to this in that they do not contain introns and histone mRNAs are usually not polyadenylated.

There is increasing evidence accumulating that this processing of the primary transcript is an important mechanism of cell-type-specific control of gene expression, in addition, it should be emphasized, to transcriptional control of the gene in many cases. There are basically two ways in which posttranscriptional control can operate: first, by selecting that mRNAs are transported out of the nucleus; and second, by producing two or more different mRNAs from the same primary transcript, either by differential splicing of exons or use of different polyadenylation sites [see (2,214) for general reviews].

With regard to the first possibility, some studies are consistent with mRNA regulation involving control at the level of transport of their precursors from the nucleus, for example certain liver mRNAs regulated by steroids (215). Perhaps the best example of this kind of posttranscriptional control is in sea urchin development, where there is evidence that the precursors of some mRNAs expressed only in the egg and in very early development persist in the nuclei late into development, even as late as adult somatic tissues (216).

Examples of cell-type-specific splicing of exons to yield different mRNAs during cell differentiation or development are increasing rapidly. The amylase genes were one of the first to be elucidated in detail. In fact, there are two amylase genes in the mouse: *Amy-1*, is expressed at a 100-fold level in the parotid gland compared with most other tissues (e.g., liver) but the protein is identical in all cell types; the other, *Amy-2*, exists as multiple, very similar copies in the mouse genome that are expressed exclusively in the pancreas at a much higher level than *Amy-1* is, even in the parotid. Although the *Amy-2* gene is apparently controlled entirely at the level of transcription (217), the *Amy-1* locus is controlled in a more complex way. In fact, the parotid- and non-cell-type-specific *Amy-1* mRNAs are produced by a combination of transcriptional and posttranscriptional control mechanisms: Two different promoters are used in the different cell types and then the two alternative promoter exons are spliced onto a common set of coding exons (218-220). Other examples of cell-type-specific posttranscriptional regulation of the production of related mRNAs from a single gene by differential splicing of alternative promoter or internal exons include: different human α_1-antitrypsin mRNAs in macrophages and hepatocytes (221,222); different rat or chicken myosin light chains (223-225); three different rat or human fibronectin mRNAs (226-228); the two fibrinogen mRNAs encoding the γA- and γB-chains (229); different chicken or hamster vimentin mRNAs (230-233); different *Drosophila* alcohol dehydrogenase mRNAs during development (234); tissue-specific generation of two preprotachykinin mRNAs (235); and different isoforms of the Ly-5 surface determinant distinguishing hemopoietic cell lineages (236,237).

A further way in which different mRNAs can be produced posttranscriptionally is by differential splicing generated by use of alternative polyadenylation sites that remove certain exon/intron splice sites. Examples of this type include: the developmental switch from membrane-bound to secreted forms of the immunoglobulin heavy chains (240); different calcitonin gene-related peptides in the thyroid gland and brain (241-243); mouse dihydrofolate reductase mRNAs (244); and platelet-derived growth factor A-chains in normal and tumor cells (245,246).

How different combinations of exon/intron splice sites are recognized for splicing in different cell types is not at all clear. However, it must involve cell-specific *trans*-acting processing factors: for example, when troponin genes are transfected into either muscle or non-muscle cells, then correct cell-type-specific processing of the exogenous gene occurs (247). Similarly, the appropriate cell-type-specific processing occurs when either the α_1-antitrypsin gene or a metallothionein promoter-calcitonin fusion gene is introduced into transgenic mice (222,243).

Finally, developmental regulation of the proteins encoded by certain *Drosophila* regulatory genes seems also to involve posttrancriptional as well as transcriptional control mechanisms. The homeotic genes *bithorax* and *Antennapedia* somehow govern the various structures associated with different segments, and they are transcriptionally activated soon after fertilization of the egg, from the time of blastoderm formation onward. Two other examples are the *Kruppel* gene, which is involved in the normal establishment of adjacent segments in embryos, and the *transformer* gene, which is one of the set of genes regulating somatic sexual differentiation. In all these instances, different mRNAs are produced in different tissues or segments by splicing of different exons together or use of different polyadenylation signals (238,239,329,330). Because such genes appear to have a regulatory function [the homeotic genes cited contain "homeoboxes" that encode DNA-binding domains (328) and the *Kruppel* gene encodes a "metal-finger"-type DNA binding domain (238,331)], it is clear that posttranscriptional control mechanisms are involved in the regulation of regulatory as well as the "simple" structural genes discussed earlier.

B. Changes in mRNA Stability

Another mechanism used by cells in regulating the expression of proteins in response to changes in the environment or during cell differentiation is to change the stability of their mRNAs [reviewed recently by Brawerman (248)]. One important class of mRNAs controlled in this way is those encoding proteins regulated by steroid hormones: Thus, stabilization of casein mRNA is induced by prolactin in breast tissue (249), as are the mRNAs encoding the egg-yolk protein precursor vitellogenin, and the egg-white proteins ovalbumin and conalbumin by estrogen in oviduct (250). Typically, the half-lives of such mRNAs change by a factor of 10-20 in these various circumstances. However, it should be stressed that this effect of steroids on the stability of these mRNAs is in addition to their effects on the transcription of the corresponding genes. Another example is the change in the stability of β-tubulin mRNAs during liver regeneration: In this case, it appears that the levels of unpolymerized β-tubulin subunits govern the stability of the mRNA via an interaction with sequences lying within the first 16 translated codons of the mRNA and thereby autoregulate their own synthesis (251,252). Differential stabilization of globin versus non-globin mRNAs has also been postulated to explain the massive increase in the relative concentration of globin mRNAs during the late stages of erythroid cell differentiation after transcription of the genes has ceased (253-255), although this evidence has been challenged by Volloch and Housman (256) on the grounds of cell, rather than mRNA, instability during erythroid differentiation under most culture conditions.

C. Translational Control

The final major step in the control of protein synthesis is binding of the mRNA to the ribosome and its translation. In some instances, cells use this stage to regulate how much protein is produced in different circumstances. Perhaps one of the clearest examples is control of the translation of the mRNA encoding one of the large ribosomal subunit polypeptides (L5) by one of the polypeptides present in the small ribosomal subunit (S8): this seems to be due to S8 recognizing similar stem-loop structures in the 16S ribosomal RNA and in the L5 mRNA in the region of the translation initiation codon (257). Thus, when S8 is in excess over the 16 S rRNA, it binds to the L5 mRNA and inhibits its translation. A somewhat similar mechanism may operation in the regulation of ferritin synthesis: Ferritin mRNA translation seems to be controlled in some way by iron (possibly by its recruitment to the polysomes), and it has been suggested that this is also mediated by a stem-loop structure in the 5′ untranslated region of the mRNA [reviewed recently by Munro and co-workers (258)].

Another fascinating example of translational control is the coordinate regulation of amino acid biosynthetic genes in yeast during amino acid starvation [reviewed by Hunt (259)]. As discussed in a previous section (IV.A), this set of genes is coordinately regulated by a positive transcription factor, GCN4, which is in turn regulated translationally by the product of the GCD1 gene (173).

This is achieved by the presence of four closely-spaced AUG codons upstream of the functional translation-initiation codon: Under conditions of excess amino acids, these upstream AUGs engage the ribosomes preferentially in nonproductive cycles of polypeptide initiation, whereas under conditions of amino-acid starvation, this preemptive binding of ribosomes to the upstream AUG codons is abolished, provided the GCD4 gene is functional. A similar mechanism seems to operate in the case of translational repression by arginine of the yeast CPA1 gene, encoding the glutaminase subunit of the arginine pathway carbamoyl-phosphate synthetase: the CPA1 gene also contains a short upstream open reading frame that has been shown to be essential for its repression by arginine (260). In mammalian cells, similar translational control seems to occur during hormonal induction of ornithine decarboxylase (ODC), an enzyme involved in polyamine biosynthesis. Like the other examples cited, the ODC mRNA also contains four AUG codons upstream of the long open reading frame encoding the enzyme (261).

One of the classic examples of translational control is the storage of maternal mRNAs during oocyte maturation before their translational activation after fertilization [see, for example, (262,263)]. The mechanism of translational activation of maternal mRNAs is probably different for different classes of mRNA: in sea urchins, histone mRNAs (comprising ~5% of the total) are localized in the pronucleus before fertilization and require "cap" methylation to be translated (264). There is other evidence in sea urchins that some maternal mRNAs are associated with ribonuclear protein complexes in the egg and are then released (by protease cleavage?) at fertilization (265,266). However, this conclusion has been challenged more recently by Moon and co-workers (267), who found no difference in the translatability of sea urchin RNAs isolated from functional polysomes and from nonpolysomal egg fractions. From studies with *Xenopus*, it it now thought likely that most of the maternal mRNAs are not translated in the egg due to limitations in its translation capacity, because addition of exogenous mRNAs but not polysomes depresses the synthesis of endogenous egg proteins (268,269). In addition, there is another class of non-polysomal poly A-containing RNAs that are untranslated in sea urchin eggs and disappear quickly during development: these RNAs contain transcripts of repetitive as well as unique DNA sequences and form RNA duplexes both in vivo and in vitro, they seem to be inherently untranslatable and are unlikely ever to function as bona-fide mRNAs, but their function is unknown (270).

VI. MECHANISMS OF COORDINATE GENE EXPRESSION

A. Shared *Cis*-Control Elements

One of the major questions of developmental biology is how particular sets of genes come to be activated and coordinately expressed in certain cell types.

Presumably, some common features of such subsets of genes must be recognized by the environments of the cells in which they are coexpressed. On the basis of prokaryotic and yeast examples, one fairly simple model for the regulation of coordinate gene expression is to suppose that sets of coexpressed genes share common *cis*-control regions recognized by a single *trans*-acting factor. Such a mechanism has been shown to operate in yeast, in the regulation of cell-type-specific genes by the mating-type locus (271,272) [reviewed by Dranginis (273)] ; for the galactose regulation of the GAL1, 7 and 10 genes by the GAL4 regulatory gene protein (171); and for the transcriptional induction of many amino-acid biosynthesis genes during amino-acid starvation mediated by the GCN4 regulatory gene protein (274).

Similar comparisons of sets of coexpressed genes in higher eukaryotes has in some cases revealed some interesting shared sequence motifs: for example, the consensus sequence present upstream of all heat-shock genes; the octamer motif present in the kappa and heavy-chain immunoglobulin gene promoters and the heavy-chain gene enhancer; and the consensus sequence present upstream of all steroid-regulated genes, to mention examples already discussed in a different context. Other examples of such shared homolgies between coexpressed or related genes are: the 15 bp homology upstream of the α-, β-, and γ-fibrinogen genes (275); a common binding site for a factor shared between the α- and β-interferon genes in their 5′ flanking regions (276); the consensus sequence, GPuGPuTTPyCAPy, upstream of certain hematopoietic growth factor genes [GM–CSF, IL-3, IL-2) (277)] ; a consensus sequence, AG/CAAAG/CCAGGAAAACTG, in the 5′ flanking region of the IL-2 qnd IL-2 receptor genes, the γ-interferon gene and the human immune deficiency virus LTR (90); a 20-bp consensus sequence shared between the γ-actin gene and the serum-responsive *cis*-control element of the c-*fos* gene (278), and a 5′ flanking sequence of 28 bp shared by the metallothionein, histocompatibility (H2/HLA) and (possibly) oligoadenylate synthetase genes, all of which are regulated by interferon (70). However, it should be stressed that these interesting sequence homologies should be interpreted cautiously until their functional roles have been established. For example, an (AT)-rich sequence, TTATTTAT, found, often in multiple copies, situated in the 3′-flanking region of genes involved in inflammatory responses [tumor necrosis factor, lymphotoxin, the hematopoietic growth factors GM–CSF and IL-1, fibronectin, and most interferons (279)] may have a role in regulating the stability or translation of the corresponding mRNAs in different situations (280,281).

B. Liver-Specific Functions

One model for control of cell differentiation, based on experiments with cell hybrids, suggests that sets of genes expressed only in a certain cell type are coordinately repressed in nonexpressing cell types by the products of one or more regulatory genes, which are themselves repressed only in the expressing cell type in question. As discussed previously (Section III.E), single-chromosome-transfer experiments with liver cells are pertinent to this question: tyrosine aminotransferase (TAT), alcohol dehydrogenase, and albumin are each repressed in nonliver cells by specific factors encoded by genes located on separate chromosomes (called tissue-specific extinguishers, Tse's) (151-153). In each case, all other liver functions tested are unaffected: when TAT is repressed by Tse-1, alanine aminotransferase, alcohol dehydrogenase, aldehyde dehydrogenase, and aldolase B are not; and when albumin is repressed by Tse-2, fibronectin, transferrin, complement factor C3, fructose 1.6-bisphosphatase, and phospho*enol*pyruvate carboxykinase are unaffected. Thus, the function of individual Tse's seems to be to repress individual rather than sets of liver-specific genes in non-liver cells, at least as far as the markers that have been investigated are concerned. There is as yet, no evidence that Tse's are involved in coordinate regulation of gene expression.

Nevertheless, other work seems to imply that such regulatory genes do exist because *cis*-control regions regulating some liver-specific genes do contain related sequence motifs. For example, comparison of the proven upstream control regions of the α_1-antitrypsin, haptoglobin, and retinol-binding protein genes shows that the α_1-antitrypsin gene control region contains an SV40-type core enhancer motif that is flanked on either side with 10-bp motifs, each of which shared with elements in the *cis*-control regions of one of the other two genes (83). Unfortunately, it is not possible to relate the latter experiments to those defining the Tse's because the liver functions studied in each case are different. As yet there is little evidence of similar shared control regions in sets of genes coexpressed in other differentiating systems, such as muscle or lens, or among the genes encoding egg-white or egg-yolk proteins, for example.

C. Erythroid Cell-Specific Gene Expression

Apart from perhaps B-lymphocytes and liver, one of the most extensively studied examples of cell differentiation is the erythroid system. However, despite intensive research little is understood as to how the α- and β-globin genes come to be coexpressed exclusively in the erythroid lineage. Two points have to be distinguished: the coordinate activation of the α- and β-globin gene clusters overall; and changes in the transcription of particular members of each cluster during development. Little if anything is known about the first question as to how the two globin gene domains are established in erythroid cells although, as mentioned earlier, recent work suggests that sequences close to the domain boundary may be important (47). However, it is clear that the particular α- and β-globin genes expressed at one stage of development are not necessarily coordinately regulated throughout development: For example, in humans the "adult" α-globin gene is first expressed during fetal erythropoiesis whereas the adult β-globin gene is not transcribed until the adult bone marrow stage of erythropoiesis.

In fact, such as is known about the mechanisms of regulation of the two sets of globin genes only serves to illustrate their differences. For example, in contrast to β-globin genes, some α-like globin genes are often associated with unmethylated CpG-rich islands, usually considered a characteristic of constitutively expressed genes (22). Moreover, the adult α- and β-globin genes respond differently in transfection experiments to the effects of viral enhancers linked in cis, to active replication origins nearby, and to viral *trans*-acting factors such as the adenovirus E1A products (282-286) [reviewed previously by this author (287)]. Furthermore, as mentioned previously (Section III.A), the human α-globin gene is expressed constitutively when transfected into murine erythroleukemia cells, whereas transcription of the β-globin gene is increased when the erythroleukemia cells are induced to differentiate (46).

The molecular basis of the tissue- and developmental-specific regulation of various members of the β-globin cluster has been investigated in considerable detail. Thus, genomic DNA fragments containing the human β-, γ-, and ε-genes have been found to be expressed in the appropriate manner both when transfected into erythroid or nonerythroid cells (288-291) and when introduced into transgenic mice (292-295). When human β-, γ- or ε-globin genes are transfected into mouse erythroleukemia cells (which themselves express adult mouse globins), transcription of only the exogenous adult human β- but not the fetal γ-globin gene is increased during erythroleukemia cell differentiation (290,296, 297). (The evidence as to whether the embryonic human ε-globin gene is increased during mouse erythroleukemia differentiation is conflicting, but most of the evidence suggests that it is not.) Similarly, when human β-, or ε-globin genes are introduced into the human erythroleukemia cell line, K562 (which itself expresses fetal and embryonic γ- and ε-globins but not adult β-globin), the γ- and ε-globin genes are transcribed at high levels, whereas the adult β-globin gene is not (297-301). Similar transfection experiments have demonstrated that the sequences responsible for the reduced expression of human δ-globin compared with β-globin in adults reside in the second intron of the δ-globin gene (302).

Various lines of evidence have shown unequivocally that the correct regulation of the various β-globin genes is mediated by multiple control elements in both 5′ and 3′ flanking regions. As far as 5′ tissue-specific *cis*-control elements are concerned, these have generally been located within a few hundred nucleotides upstream of the cap site [see, for example, (291)], although, as noted earlier, regions far upstream may also be important in establishing the active β-globin locus in erythroid cells (47). The immediate upstream flanking regions of the β-globin genes contain sites for binding erythroid-specific proteins, both in vivo as demonstrated by the presence of DHSS and in-vivo "footprinting," an also by in-vitro "footprinting" studies (303-307). This has been investigated most thoroughly so far in the chicken: For example, the erythroid-specific DHSS upstream of both the β^{adult}- and $\beta^{hatching}$-globin genes contain two binding sites for nuclear proteins, one at a consensus sequence for what seems to be the chicken equivalent of nuclear factor I, and the other near the functionally important CACC motif identified originally by Dierks and co-workers (49). In addition, the chicken β^{A}-globin gene contains a string of G-residues from −177 to −195 nt that binds an erythroid-specific protein found in mature erythrocytes but not in earlier erythroblasts (306). However, the G-string sequence is present neither upstream of the chicken β^{H}-globin gene nor any of the mammalian β-globin genes; its function is presently unclear.

As far as the upstream region of mammalian β-globin genes is concerned, we have postulated that a region at approximately −150 nt might be one such control element because it contains a 13-nt sequence that is conserved among adult β-globin genes and it is a site of upstream transcription that increases in parallel with the cap-site transcription during erythroid cell differentiation (the "B box," Fig. 3) (308,309); this has now been confirmed by transfection experiments by Grosveld and co-workers (310). However, this may not be an erythroid-gene-specific effect but rather a response of genes containing particular types of control elements, because transcription from integrated retroviral LTRs is also increased as erythroleukemia cells differentiate, provided the LTRs contain a functional enhancer (311). Recently, we have identified another region from −182 to −346 nt upstream of the mouse β^{major}-gene that confers a 10-fold increase in transcription of a linked CAT gene in uninduced erythroleukemia cells compared with fibroblasts using a transient transfection assay. Moreover, this region contains two to three binding sites for nuclear proteins as determined by DNAse I footprinting, one of which is a GATAAG sequence motif (Fig. 3A—discussed more fully below) that binds an erythroid-cell-specific factor (M. Plumb, J. Frampton, K. McLeod, M. Walker, H. Wainwright, and P. R. Harrison, 332, and unpublished data).

However, 5′ control regions are not the only elements involved in the regulation of the β-globin locus. By linking parts of the β-globin gene to a histocompatibility gene or α-globin gene body, it has been inferred that the increase in transcription of the β-globin gene during erythroleukemia cell differentiation can be mediated separately by sequences from either the upstream flanking region or 3′ flanking sequences (46,312). This conclusion is also reinforced by transgenic mouse experiments in which an adult human β-globin gene fragment truncated to within 48 nt of the cap site was found to be correctly regulated (292). The 3′ control regions regulating the human β- and γ-globin genes and the chicken β^{A}-gene have now been identified as erythroid-specific enhancers (Fig. 3a and c) (97-101, 327), as discussed earlier (Section III.D). It is interesting

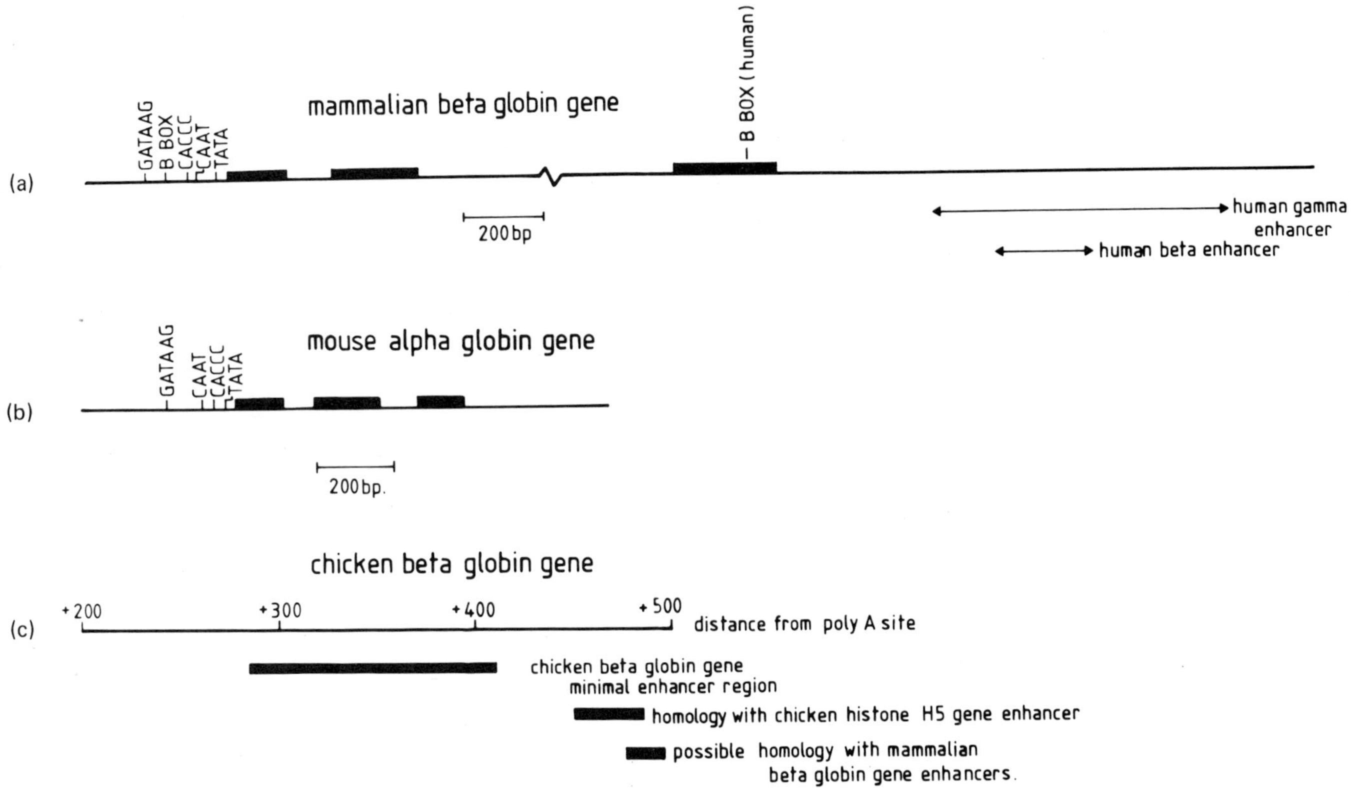

Figure 3 Consensus motifs and control regions associated with globin genes. (a) and (b). Maps of β- and α-globin genes respectively, with the exons marked as black boxes. The positions of various consensus motifs mentioned in the text are noted (TATA, CAAT, CACCC, B-BOX, GATAAG). The positions of the human gamma and beta 3′ enhancers are shown relative to the end of the third exon, to allow for variations in the lengths of the second intron in the various β-globin genes. (c) Map of the 3′ flanking region of chicken β-globin gene. The black boxes denote the position of the enhancer region and homologies with other β-globin genes or the chicken histone H5 3′ enhancer.

that the 13-nt conserved sequence in the 5′ control element 150 nt upstream of adult β-globin genes mentioned earlier is duplicated in the 3′ flanking region of the human β-globin gene (Fig. 3a), where it functions as an enhancer (310,327).

There has been little success reported in defining *cis*-control regions responsible for the tissue-specific expression of the α-globin genes. Recently, the 555-nt upstream region of the human ζ-globin gene has been shown by stable transfection experiments to be sufficient for its tissue-specific expression, but the sequences involved have not been defined any more precisely (291). As yet, no 3′ enhancer has been identified in respect of any α-globin gene. In our own studies, we recently have found that the 4-kb of DNA upstream of the mouse adult α-globin gene increases transcription of a linked CAT gene in a tissue-specific manner, by a factor of 50-fold in erythroblasts (Friend cells) compared with fibroblasts. Deletion analysis has shown that this erythroid-specific effect is mediated mainly by two regions at about −190 nt and −1.9 kb; in both cases a sequence motif, GATAAG, binding an erythroid-specific factor (EF1), has been implicated in their functional activity (332,333) (J. Frampton, M. Walker, M. Plumb, and P. R. Harrison, submitted for publication). As the distal element is not functional when placed upstream of the proximal element lacking the GATAAG element, it suggests that the two regulatory elements cooperate and that the proximal GATAAG motif is required for this cooperativity. Other proteins also bind to sequences close to the GATAAG motifs in both proximal and distal control regions, and the action of EF1 in controlling erythroid-specific transcription is probably mediated by interactions with some of these other transcription factors. The transcription factor, EF1, is present in erythroblasts (Friend cells) prior to onset of high-level globin gene transcription and therefore may play a role in activation of genes during commitment of stem cells to the erythroid pathway.

GATAAG sequence motifs are found also upstream of certain other globin genes; generally they are found bound to erythroid-specific proteins in erythroblasts in vivo, as evidenced by the formation of erythroid-specific DHSS at such GATAAG motifs. Thus, to date GATAAG sequences have been identified upstream of the embryonic chicken α^{D}, α^{π} and β^{ρ} globin genes (313), the human δ- and ζ-globin genes and, as noted above, the mouse $\beta^{major}/\beta^{minor}$-globin genes. Why only these particular globin genes possess upstream GATAAG sequences is unclear, but it may be significant that most of these genes are expressed in fetal or embryonic erythroid cells. Indeed, a very recent report indicates that a region −170 to −190 nt upstream of the human γ-globin gene is a binding site for an erythroid-specific protein: Significantly, this region contains on the noncoding strand the sequence GATAGAG close to an immunoglobulin gene octamer consensus sequence (326). There is also a GATAAG sequence 70 nt upstream of the erythroid-specific promoter for the human porphobilinogen deaminase gene (M. Goossens, personal communication), and we have shown that this sequence also binds the same erythroid-specific factor as binds to the mouse α- and β-globin genes (332). However, a GATAAG sequence is not found in the 300 nt upstream of the red-blood-cell-specific lipoxygenase gene, which we have cloned (314,334,335). Quite possibly, the erythroid-specific GATAAG-binding protein recognizes a broader consensus sequence than we can yet discern; until this is clarified, it is difficult to assess what the role of the GATAAG sequence might be in the regulation of erythroid-specific functions generally.

In terms of coordinate regulation of erythroid-specific genes, some tantalizing sequence similarities have emerged from comparisons of the 3′ enhancer regions regulating the globin and the erythroid-specific histone H5 genes: These share a related sequence extending over ~34 nt (99,102) that overlaps a 17-nt partially conserved region in the 3′ flanking regions of the chicken, human, mouse, and rabbit adult β-globin genes (100). However, the 3′ flanking sequence shared by the chicken β-globin and histone H5 genes does not lie directly within the minimal β-globin gene enhancer region possessing full enhancer activity but rather slightly 3′ to it (Fig. 3c). Insufficient information is available at present to assess whether other erythroid-specific genes, like those coding for the lipoxygenase or carbonic anhydrase I, are regulated by the 3′ enhancers containing a similar sequence, but it may be significant that glutathione peroxidase has erythroid-specific DHSS in the 3′ flanking DNA, which might mean that there is a 3′ regulatory element in that region [(149); I. Chambers and P. R. Harrison, unpublished data].

A particularly relevant example illustrating the flexibility employed by cells in regulating genes encoding a functionally interdependent set of polypeptides is the heme pathway enzymes. The first enzyme in the pathway, 5-aminolevulinic acid synthase, is repressed in liver by heme and stimulated by porphyrinogenic agents, whereas the erythroid enzyme is not: It appears that this is probably due to tissue-specific regulation of related members of a multigene family, rather than to posttranscriptional control of a single gene (315,316). In contrast, the third enzyme in the pathway, porphobilinogen deaminase, is greatly elevated in erythroid cells compared with other cell types (e.g., liver), and this has been shown to be due to posttranscriptional control involving the splicing of two different promoter exons into a common set of coding exons, as in the case of the amylase gene (317). Finally, the fifth heme pathway enzyme, uroporphyrinogen decarboxylase, is also elevated in erythroid cells, but in this case the mechanism appears to be entirely transcriptional control of a single promoter (318).

Thus, taking the work on B-lymphocyte, liver, and erythroid cell differentiation together, there are intriguing clues that coexpressed genes may share some related *cis*-control motifs, but what their precise roles are is far from clear. Certainly, the situation is not a straightforward as (say) the control of cell-type-specific genes by the mating-type locus in yeast. Indeed, it should be borne in mind that a shared *cis*-control region involved in regulating the coordinate expression of a set of genes need not necessarily operate by binding a tissue-specific factor itself: as the work with immunoglobulin and other genes suggests, it may be the precise *overall combination* of *cis*-control elements that gives the gene its unique pattern of expression, even though each individual control element lacks this specificity.

VII. SUMMARY

One general point to emerge from this review is that each step in the pathway for gene expression from DNA to final polypeptide via chromosome structure, primary transcript, mRNA processing, stability, and translation is exploited in some developmental context as a means of regulation. Moreover, there is little evidence that the regulation of different stages in this pathway is associated consistently with particular classes of genes or developmental processes: for example, homeotic and other regulatory genes, which are important in early development and in determining the overall structures of organisms, are regulated by both transcriptional and posttranscriptional mechanisms, as are many other genes encoding cell-type-specific proteins expressed late during terminal differentiation. There are also instances in which transcription factors regulating tissue-specific transcription are present constitutively but are activated only in certain cell types by posttranslational events, and in yeast, coordinate regulation of amino-acid biosynthesis genes involves the translational control of a transcription factor.

Second, as far as transcriptional control is concerned, it is now clear that in most instances cell-type specificity arises from the activities of multiple *cis*-control elements that can act either positively or negatively: The precise way in which the gene functions depends on what combination of *cis*-control elements it possesses.

A third somewhat negative point is that in general the mechanisms responsible for the coordinate expression of genes during cell differentiation and development are not yet understood. In view of the complexities of possible mechanisms noted above, it may be that closely related genes (such as those encoding the α- and β-globins, for example) in fact come to be coordinately expressed via a complex, indirect route. Indeed, it is quite possible that such mechanisms will not be revealed simply by a reductionist approach studying relatively short regions of DNA and that some form of overall "systems analysis" will be required.

Finally, what we know relates mainly to expression of genes within particular cell types. How groups of cells are organized into complex structures remains a mystery at the molecular level, although progress into the genetic principles involved in the regulation of such "compartments" is being made, mainly in insects (319-322), reviewed recently by Brower (323), although to some extent also in the mouse [discussed by Hogan and co-workers (324)].

ACKNOWLEDGMENTS

This work was supported by the Cancer Research Campaign and Medical Research Council. The author is grateful to various members of the Institute, in particular John Paul, John Wyke, John Pitts, Bob Brown, and Mary Harper, and members of his own group, Mark Plumb, Jon Frampton, Ian Anton, Dave Llewyllyn, John Chester, and Kay McLeod, for stimulating discussions about various aspects of the research covered by this review and for constructive comments on the draft manuscript. The author is also grateful to Rae Fergusson for secretarial assistance and to Dave Tallach for artwork.

REFERENCES

1. Hadorn, E. Problems of determination and transdetermination in genetic control of differentiation. *Brookhaven Symp. Biol. 18*:148-161 (1965).
2. Darnell, J., H. Lodish, and D. Baltimore. *Molecular Cell Biology*. Scientific American Books, New York (1986).
3. Kelly, S. J. Studies on the developmental potential of 4- and 8-cell stage mouse blastomeres. *J. Exp. Zool. 200*:365-376 (1979).
4. Whetton, A. D., and T. M. Dexter. Haemopoietic growth factors. *Trends Biochem. Sci. 11*:207-211 (1986).
5. Clark, S. C., and R. Kamen. The human hemopoietic colony-stimulating factors. *Science 236*:1229-1237 (1987).
6. Struhl, G. A gene product required for correct initiation of segmental determination in Drosophila. *Nature 293*:36-41 (1981).
7. Sternberg, P. W., and H. R. Horvitz. The genetic control of cell lineage during nematode development. *Annu. Rev. Genet. 18*:489-524 (1984).
8. Surani, M. A. H., S. C. Barton, and M. L. Norris. Influence of parental chromosomes on spatial specificity in androgenetic/parthogenetic chimaeras in the mouse. *Nature 326*:395-397 (1987).
9. Surani, M. A. H., S. C. Barton, and M. L. Norris. Development of reconstituted mouse eggs suggests imprinting of the genome during gametogenesis. *Nature 308*:548-550 (1984).
10. Surani, M. A. H., S. C. Barton, and M. L. Norris. Nuclear-transplantation in the mouse; heritable differences between parental genomes after activation of the embryonic genome. *Cell 45*:127-136 (1986).
11. Cattanach, B. M., and M. Kirk. Differential activity of maternally and paternally derived chromosome regions in mice. *Nature 315*:496-498 (1985).
12. Klar, A. J. S. Differentiated parental DNA strands confer developmental asymmetry on daughter cells in fission yeast. *Nature 326*:466-470 (1987).

13. Razin, A., and A. D. Riggs. DNA methylation and gene function. *Science 210*:604-610 (1980).
14. Reik, W., A. Collick, M. L. Norris, S. C. Barton, and M. A. Surani. Genomic imprinting determines methylation of parental alleles in transgenic mice. *Nature 328*:248-251 (1987).
15. Sapienza, C., A. C. Peterson, J. Rossant, and R. Balling. Degree of methylation of transgenes is dependent on gamete of origin. *Nature 328*:251-254 (1987).
16. Swain, J. L., T. A. Stewart, and P. Leder. Parental legacy determines methylation and expression of an autosomal transgene: a molecular mechanism for parental imprinting. *Cell 50*:719-727 (1987).
17. Monk, M. Methylation and the X chromosome. *BioEssays 4*:204-208 (1986).
18. Lock, L. F., N. Takagi, and G. R. Martin. Methylation of the HPRT gene on the inactive X occurs after chromosome inactivation. *Cell 48*:39-46 (1987).
19. Kaslow, D. C., and B. R. Migeon. DNA methylation stabilizes X chromosome inactivation in eutherians but not in marsupials: evidence for multistep maintenance of mammalian X chromosome dosage compensation. *Proc. Natl. Acad. Sci. USA 84*:6210-6214 (1987).
20. Bird, A., M. Taggart, M. Frommer, O. J. Miller, and D. MacLeod. A fraction of the mouse genome that is derived from islands of nonmethylated, CpG-rich DNA. *Cell 40*:91-99 (1985).
21. Bird, A. P., DNA methylation–how important in gene control? *Nature 307*:503-504 (1984).
22. Bird, A. P. CpG-rich islands and the function of methylation. *Nature 321*: 209-213 (1986).
23. Jones, P. A. Altering gene expression with 5-azacytidine. *Cell 40*:485-486 (1985).
24. Weinbraub, H., and M. Groudine. Chromosomal subunits in active genes have an altered conformation. *Science 193*:848-856 (1976).
25. Wu, C., and W. Gilbert. Tissue-specific exposure of chromatin structure at the 5′ terminus of the rat preproinsulin II gene. *Proc. Natl. Acad. Sci. USA 78*:1577-1580 (1981).
26. Stalder, J., A. Larsen, J. D. Engel, M. Dolan, M. Groudine, and H. Weintraub. Tissue-specific DNA cleavages in the globin chromatin domain introduced by DNAase I. *Cell 20*:451-460 (1980).
27. Tuan, D., and I. M. London. Mapping of DNAse I-hypersensitive sites in the upstream DNA of human embryonic epsilon-globin gene in K562 leukemia cells. *Proc. Natl. Acad. Sci. USA 81*:2718-2722 (1984).
28. Shermoen, A. W., and S. K. Beckendorf. A complex of interacting DNAse I-hypersensitive sites near the Drosophila glue protein gene, Sgs4. *Cell 79*: 601-607 (1982).
29. McGinnis, W., A. W. Shermoen, J. Heemskerk, and S. K. Beckendorf. DNA sequence changes in an upstream DNAse I-hypersensitive region are corre-

lated with reduced gene expression. *Proc. Natl. Acad. Sci. USA 80*:1063-1067 (1983).
30. Eissenberg, J. C., I. L. Cartwright, G. H. Thomas, and S. C. R. Elgin. Selected topics in chromatin structure. *Annu. Rev. Genet. 19*:485-536 (1985).
31. Groudine, M., and H. Weinbraub. Propagation of globin DNAase 1-hypersensitive sites in absence of factors required for induction: a possible mechanism for determination. *Cell 30*:131-139 (1982).
32. Weinbraub, H., H. Beug, M. Groudine, and T. Graf. Temperature-sensitive changes in the structure of globin chromatin in lines of red cell precursors transformed by ts-AEV. *Cell 28*:931-940 (1981).
33. Burch, J. B. E., and H. Weintraub. Temporal order of chromatin structural changes associated with activation of the major chicken vitellogenin gene. *Cell 33*:65-76 (1983).
34. Fritton, H., T. Igo-Kemenes, J. Norwock, U. Strech-Jurk, M. Thiesen, and A. E. Sippel. Alternative sets of DNAse I-hypersensitive sites characterized the various function states of the chicken lysozyme gene. *Nature 311*: 163-165 (1984).
35. Burch, J. B. E., and M. I. Evans. Chromatin structural transitions and the phenomenon of vitellogenin gene memory in chickens. *Mol. Cell. Biol. 6*: 1886-1892 (1986).
36. Keshet, I., J. Lieman-Hurwitz, and H. Cedar. DNA methylation affects the formation of active chromatin. *Cell 44*:535-543 (1986).
37. Schlissel, M. S., and D. D. Brown. The transcriptional regulation of Xenopus 5S RNA genes in chromatin: the roles of active stable transcription complexes and histone H1. *Cell 37*:903-913 (1984).
38. Smithies, O., R. G. Gregg, S. S. Boggs, M. A. Koralewski, and R. S. Kucherlapati. Insertion of DNA sequences into the human chromosomal β-globin locus by homologous recombination. *Nature 317*:230-234 (1985).
39. Thomas, K. R., K. R. Folger, and M. R. Capecchi. High frequency targeting of genes to specific sites in the mammalian genome. *Cell 44*:419–428 (1986).
40. Thomas. K. R., and M. R. Capecchi. Site-directed mutagenesis by gene targeting in mouse embryo-derived stem cells. *Cell 51*:503-512 (1987).
41. Deisseroth, A., U. Bode, J. Fontana, and D. Hendrick. Expression of human α-globin genes in hybrid mouse erythroleukaemia cells depends on differentiated state of human donor cell. *Nature 285*:36-38 (1980).
42. Deisseroth, A., and D. Hendrick. Activation of phenotypic expression of human globin genes from nonerythroid cells by chromosome-dependent transfer to tetraploid mouse erythroleukemia cells. *Proc. Natl. Acad. Sci. USA 76*:2185-2189 (1979).
43. Ley, T. J., Y. L. Chiang, D. Haidaris, N. P. Anagnou, U. L. Wilson, and W. French Anderson. DNA methylation and regulation of the human β-globin-like genes in mouse erythroleukemia cells containing human chromosome 11. *Proc. Natl. Acad. Sci. USA 81*:6618-6622 (1984).

44. Anagnou, N. P., T. Y. Lim, E. Yuan, J. Wieder, S. Helder, D. Glaister, B. Marks, A. Wang, D. Colbert, and A. Deisseroth. Regulatory factors specific for adult and embryonic globin genes may govern their expression in erythroleukemia cells. *Blood 65*:705-712 (1985).
45. Papayannopoulou, T., D. Lindsley, S. Kurachi, K. Lewison, T. Hemenway, M. Melis, N. P. Anagnou, and V. Najfield. Adult and fetal human globin genes are expressed following chromosomal transfer into MEL cells. *Proc. Natl. Acad. Sci. USA 82*:780-784 (1985).
46. Charnay, P., R. Treisman, P. Mellon, M. Chao, R. Axel, and T. Maniatis. Differences in human alpha- and beta-globin gene expression in mouse erythroleukemia cells: the role of intragenic sequences. *Cell 38*:251-263 (1984).
47. Grosveld, F., G. B. van Assendelft, D. R. Greaves, and G. Kollias. Position independent high level expression of the human beta globin gene in transgenic mice. *Cell 51*:975-985 (1987).
48. Bucher, P., and E. N. Trifonov. Compilation and analysis of eukaryotic POL II promoter sequence. *Nucleic Acids Res. 14*:10009-10027 (1986).
49. Dierks, P., A. Von Ooyen, M. D. Cochran, C. Dobkin, J. Reiser, and C. Weissmann. Three regions upstream from the cap site are required for efficient and accurate transcription of the rabbit beta-globin gene in mouse 3T6 cells. *Cell 32*:695-706 (1983).
50. Kadonaga, J. T., K. A. Jones, and R. Tjian. Promoter-specific activation of RNA polymerase II transcription by Sp1. *Trends Biochem. Sci. 11*: 20-23 (1986).
51. Masters, J. N., and G. Attardi. Discrete human dihydrofolate reductase gene transcripts present in polysomal RNA map with their 5′ ends several hundred nucleotides upstream of the main mRNA start site. *Mol. Cell. Biol. 5*:493-500 (1985).
52. Valerio, D., M. G. C. Duyvesteyn, B. M. M. Dekker, G. Weeda, T. H. M. Berkvens, L. Van der Voorn, H. Van Ormondt, and A. J. Van der Eb. Adenosine deaminase: characterization and expression of a gene with a remarkable promoter. *EMBO J. 4*:437-443 (1985).
53. Melton, D. W., D. S. Koneki, J. Brennand, and C. T. Caskey. Structure, expression and mutation of the hypoxanthine phosphoribosyltransferase gene. *Proc. Natl. Acad. Sci. USA 81*:2147-2151 (1984).
54. Reynolds, G. A., S. K. Basu, T. F. Osborne, D. J. Chin, G. Gil, M. S. Brown, J. L. Goldstein, and K. L. Luskey. HMG CoA reductase: a negatively regulated gene with unusual promoter and 5′ untranslated regions. *Cell 38*:275-285 (1984).
55. Osborne, T. F., J. L. Goldstein, and M. S. Brown. 5′ end of HMG CoA reductase gene contains sequences responsible for cholesterol-mediated inhibition of transcription. *Cell 42*:203-212 (1985).
56. Lee, M. G-S., S. A. Lewis, C. D. Wilde, and N. J. Cowan. Evolutionary history of a multigene family: an expressed human beta-tubulin gene and three processed pseudogenes. *Cell 33*:477-487 (1983).

57. Chambers, I., J. Frampton, P. Goldfarb, N. Affara, W. McBain, and P. R. Harrison. The structure of the mouse glutathione peroxidase gene: the selenocysteine in the active site is encoded by the "termination" codon, TGA. *EMBO J. 5*:1221-1227 (1986).
58. Faulkner, F. G., and H. G. Zachau. Correct transcription of an immunoglobulin kappa gene requires an upstream fragment containing conserved sequence elements. *Nature 310*:71-74 (1984).
59. Bergman, Y., D. Rice, R. Grosschedl, and D. Baltimore. Two regulatory elements for kappa immunoglobulin gene expression. *Proc. Natl. Acad. Sci. USA 81*:7041-7045 (1984).
60. Mason, J. O., G. T. Williams, and M. S. Neuberger. Transcription cell type specificity is conferred by an immunoglobulin VH gene promoter that includes a functional consensus sequence. *Cell 41*:479-487 (1985).
61. Grosschedl, R. and D. Baltimore. Cell-type specificity of immunoglobulin gene expression is regulated by at least three DNA sequence elements. *Cell 41*:885-897 (1985).
62. Foster, J., J. Stafford, and C. Queen. An immunoglobulin promoter displays cell type specificity independently of the enhancer. *Nature 315*:423-425 (1985).
63. Wirth, T., L. Staudt, and D. Baltimore. An octamer oligonucleotide upstream of a TATA motif is sufficient for lymphoid-specific promoter activity. *Nature 329*:174-178 (1987).
64. Murphy, J. T., R. R. Burgess, J. E. Dahlberg, and E. Lund. Transcription of a gene for human U1 small nuclear RNA. *Cell 29*:265-274 (1982).
65. Mattaj, I. W., S. Lienhard, J. Jiricny, and E. M. DeRobertis. An enhancer-like sequence within the Xenopus U2 gene promoter facilitates the formation of stable transcription complexes. *Nature 316*:163-167 (1985).
66. Ciliberto, G., R. Buckland, R. Cortese, and L. Philipson. Transcription signals in embryonic Xenopus Laevis U1 RNA genes. *EMBO J. 4*:1537-1543 (1985).
67. Sive, H. L., N. Heintz, and R. G. Roeder. Multiple sequence elements are required for maximal in vitro transcription of a human histone H2B gene. *Mol. Cell. Biol. 6*:3329-3340 (1986).
68. Karin, M., A. Haslinger, H. Holtgreve, R. I. Richards, P. Krauter, H. W. Westphal, and M. Beato. Characterization of DNA sequences through which cadmium and glucocorticoid hormones induce human metallothionein-IIA gene. *Nature 308*:513-519 (1984).
69. Stuart, G. W., P. F. Searle, H. Y. Chen, R. L. Brinster, and R. D. Palmiter. A 12-base-pair DNA motif that is repeated several times in metallothionein gene promoters confers metal regulation to a heterologous gene. *Proc. Natl. Acad. Sci. USA 81*:7318-7322 (1984).
70. Friedman, R. L., and G. R. Stark. Alpha-interferon-induced transcription of HeLa and metallothionein genes containing homologous upstream sequences. *Nature 314*:637-639 (1985).
71. Renkawitz, R., G. Schutz, D. Von der Ahe, and M. Beato. Sequences in

the promoter region of the chicken lysozyme gene required for steroid regulation and receptor binding. *Cell 37*:503-510 (1984).
72. Dean, D. C., B. J. Knoll, M. E. Riser, and B. W. O'Malley. A 5'-flanking sequence essential for progesterone regulation of an ovalbumin fusion gene. *Nature 305*:551-554 (1983).
73. Geise, S., C. Scheidereit, H. M. Westphal, N. E. Hynes, B. Groner, and M. Beato. Glucocorticoid receptors recognize DNA sequences in and around murine mammary tumour virus DNA. *EMBO J. 1*:1613-1619 (1982).
74. Payvar, F., D. DeFranco, G. L. Firestone, B. Edgar, O. Wrange, S. Okret, J-A. Gustafsson, and K. R. Yamamoto. Sequence-specific binding of glucocorticoid receptor to MTV DNA at sites within and upstream of the transcribed region. *Cell 35*:381-392 (1983).
75. Scheidereit, C., S. Geisse, H. M. Westphal, and M. Beato. The glucocorticoid receptor binds to defined nucleotide sequences near the promoter of mouse mammary tumour virus. *Nature 304*:749-752 (1983).
76. Pelham, H. R. B. A regulatory upstream promoter element in the Drosophila hsp 70 heat-shock gene. *Cell 30*:517-528 (1982).
77. Cohen, R. S., and M. Meselson. Separate regulatory elements for the heat-inducible and ovarian expression of the Drosophila hsp26 gene. *Cell 43*: 737-746 (1985).
78. Minty, A., and L. Kedes. Upstream regions of the human cardiac actin gene that modulate its transcription in muscle cells: presence of an evolutionary conserved motif. *Mol. Cell. Biol. 6*:2125-2136 (1986).
79. Okazaki, K., K. Yasuda, H. Kondoh, and T. S. Okada. DNA sequences responsible for tissue-specific expression of a chicken alpha-crystallin gene in mouse lens cells. *EMBO J. 4*:2589-2595 (1985).
80. Hayashi, S., H. Kondoh, K. Yasuda, G-I. Soma, Y. Ikawa, and T. S. Okada. Tissue-specific regulation of a chicken delta-crystallin gene in mouse cells: involvement of the 5' end region. *EMBO J. 4*:2201-2207 (1985).
81. Ott, M-O., L. Sperling, P. Herbomel, M. Yaniv, and M. Weiss. Tissue-specific expression is conferred by a sequence from the 5' end of the rat albumin gene. *EMBO J. 3*:2205-2210 (1984).
82. Heard, J-M., P. Herbomel, M.-O. Ott, A. Mottura-Rollier, M. Weiss, and M. Yaniv. Determinants of rat albumin promoter tissue specificity analyzed by an improved transient expression system. *Mol. Cell. Biol. 7*:2425-2434 (1987).
83. DeSimone, V., G. Ciliberto, E. Hardon, G. Paonessa, F. Palla, L. Lundberg, and R. Cortese. Cis- and trans-acting elements responsible for the cell-specific expression of the alpha1-antitrypsin gene. *EMBO J. 6*:2759-2766 (1987).
84. Shen, R-F., Y. Li, R. N. Sifers, H. Wang, C. Hardick, S. Y. Tsai, and S. L. C. Woo. Tissue-specific expression of the human alpha1-antitrypsin gene is controlled by multiple cis- regulatory regions. *Nucleic Acids Res. 15*:8399-8415 (1987).

85. Widen, S. G., and J. Papaconstantinou. Extinction of alpha-fetoprotein gene expression in somatic cell hybrids involves cis acting DNA elements. *Mol. Cell. Biol.* 7:2606-2609 (1987).
86. Hammer, R. A., R. Krumlauf, S. A. Camper, R. L. Brinster, and S. M. Tilghman. Diversity of alpha-fetoprotein gene expression in mice is generated by a combination of separate enhancer elements. *Science 235*: 53-58 (1987).
87. Serfling, E., M. Jasin, and W. Schaffner. Enhancers and eukaryotic gene transcription. *Trends Genet. 1*:224-230 (1985).
88. Rogers, B. L., and G. F. Saunders. Transcriptional enhancers play a major role in gene expression. *Bioassays 4*:62-65 (1986).
89. Hamer, R. E., G. H. Swift, D. M. Ornitz, C. J. Quaife, R. D. Palmiter, R. L. Brinster, and R. J. MacDonald. The rat elastase I regulatory element is an enhancer that directs correct cell specificity and developmental onset of expression in transgenic mice. *Mol. Cell. Biol.* 7:2956-2967 (1987).
90. Fujita, T., H. Shibuya, T. Ohashi, K. Yamanishi, and T. Taniguchi. Regulation of human interleukin-2 gene: functional DNA sequences in the 5′ flanking region for the gene expression in activated T lymphocytes. *Cell 46*: 401-407 (1986).
91. Theisen, M., A. Stief, and A. E. Sippel. The lysozyme enhancer: cell-specific activation of the chicken lysozyme gene by far-upstream DNA element. *EMBO J. 5*:719-724 (1986).
92. Wu, L.-C., B. J. Morley, and R. D. Campbell. Cell-specific expression of the human complement protein factor B gene: evidence for the role of two distinct 5′-flanking elements. *Cell 48*:331-342 (1987).
93. Fluhr, R., C. Kuhlemeier, F. Nagy, and N.-H. Chua. Organ-specific and light-induced expression of plant genes. *Science 232*:1106-1111 (1986).
94. Queen, C., and D. Baltimore. Immunoglobulin gene transcription is activated by downstream sequence elements. *Cell 33*:741–748 (1983).
95. Gillies, S. D., S. L. Morrison, V. T. Oi, and S. Tonegawa. A tissue-specific transcription enhancer element is located in the major intron of a rearranged immunoglobulin heavy chain gene. *Cell 33*:717-728 (1983).
96. Banerji, J., L. Olson, and W. Schaffner. A lymphocyte-specific cellular enhancer is located downstream of the joining region in immunoglobulin heavy chain genes. *Cell 33*:729-740 (1983).
97. Choi, O-R., and J. D. Engel. A 3′ enhancer is required for temporal and tissue-specific transcriptional activation of the chicken adult beta-globin gene. *Nature 323*:731-734 (1986).
98. Hesse, J. E., J. M. Nickol, M. R. Lieber, and G. Felsenfeld. Regulated gene expression in transfected primary chicken erythrocytes. *Proc. Natl. Acad. Sci. USA 83*:4312-4316 (1986).
99. Emerson, B. M., J. M. Nickol, P. D. Jackson, and G. Felsenfeld. Analysis of the tissue-specific enhancer at the 3′ end of the chicken adult beta-globin gene. *Proc. Natl. Acad. Sci. USA 84*:4786-4790 (1987).

100. Kollias, G., J. Hurst, E. DeBoer, and F. Grosveld. The human beta-globin gene contains a downstream developmental specific enhancer. *Nucleic Acids Res. 15*:5739-5747 (1987).
101. Bodine, D. M., and T. J. Ley. An enhancer element lies 3′ to the human gamma-A globin gene. *EMBO J. 6*:2997-3004 (1987).
102. Trainor, C. D., S. J. Stamler, and J. D. Engel. Erythroid-specific transcription of the chicken histone H5 gene is directed by a 3′ enhancer. *Nature 328*:827-830 (1987).
103. Rossi, P., and B. De Crombrugghe. Identification of a cell-specific transcriptional enhancer in the first intron of the mouse alpha-2 (type I) collagen gene. *Proc. Natl. Acad. Sci. USA 84*:5590-5594 (1987).
104. Goodburn, S., H. Burstein, and T. Maniatis. The human beta-interferon gene enhancer is under negative control. *Cell 45*:601-610 (1986).
105. Garcia, J. V., L. Bich-Thuy, J. Stafford, and C. Queen. Synergism between immunoglobulin enhancers and promoters. *Nature 322*:383-385 (1986).
106. Firak, T. A., and K. N. Subramanian. Minimal transcriptional enhancer of simian virus 40 is a 74-base-pair sequence that has interacting domains. *Mol. Cel.. Biol. 6*:3667-3676 (1986).
107. Herr, W., and J. Clarke. The SV40 enhancer is composed of multiple functional elements that can compensate for one another. *Cell 45*:461-470 (1986).
108. Ondek, B., A. Shepard, and W. Herr. Discrete elements within the SV40 enhancer region display different cell-specific enhancer activities. *EMBO J. 6*:1017-1025 (1987).
109. Xiao, J. H., I. Davidson, D. Ferrandon, R. Rosales, M. Vigneron, M. Macchi, F. Ruffenach, and P. Chambon. One cell-specific and three ubiquitous nuclear proteins bind in vitro to overlapping motifs in the domain B1 of the SV40 enhancer. *EMBO J. 6*:3005-3013 (1987).
110. Rosales, R., M. Vigneron, M. Macchi, I. Davidson, J. H. Xiao, and P. Chambon. In vitro binding of cell-specific and ubiquitous nuclear proteins to the octamer motif of the SV40 enhancer and related motifs present in other promoters and enhancers. *EMBO J. 6*:3015-3025 (1987).
111. Mercola, M., J. Goverman, C. Mirell, and K. Calame. Immunoglobulin heavy-chain enhancer requires one or more tissue-specific factors. *Science 227*:266-270 (1985).
112. Peterson, C. L., K. Orth, and K. L. Calame. Binding in vitro of multiple cellular proteins to immunoglobulin heavy-chain enhancer DNA. *Mol. Cell. Biol. 6*:4168-4178 (1986).
113 Singh, H., R. Sen, D. Baltimore, and P. A. Sharp. A nuclear factor that binds to a conserved sequence motif in transcriptional control elements of immunoglobulin genes. *Nature 319*:154-158 (1986).
114. Kadesh, T., P. Zervos, and D. Ruezinsky. Functional analysis of the murine IgH enhancer: evidence for negative control of cell-type specificity. *Nucleic Acids Res. 14*:8209-8221 (1986).
115. Sen, R., and D. Baltimore. Multiple nuclear factors interact with the immunoglobulin enhancer sequences. *Cell 46*:705-716 (1986).

116. Weinberger, J., D. Baltimore, and P. A. Sharp. Distinct factors bind to apparently homologous sequences in the immunoglobulin heavy-chain enhancer. *Nature 322*:846-848 (1986).
117. Schlokat, U., D. Bohmann, H. Scholer, and P. Gruss. Nuclear factors binding specific sequences within the immunoglobulin enhancer interact differentially with other enhancer elements. *EMBO J. 5*:3251-3258 (1986).
118. Augerau, P., and P. Chambon. The mouse immunoglobulin heavy-chain enhancer: effect on transcription in vitro and binding of proteins present in HeLa and lymphoid B cell extracts. *EMBO J. 5*:1791-1797 (1986).
119. Wasylyk, C., and B. Wasylyk. The immunoglobulin heavy-chain B-lymphocyte enhancer efficiently stimulates transcription in non-lymphoid cells. *EMBO J. 5*:553-560 (1986).
120. Gerster, T., P. Matthias, M. Yhali, J. Jiricny, and W. Schaffner. Cell-type specificity elements of the immunoglobulin heavy chain gene enhancer. *EMBO J. 6*:1323-1330 (1987).
121. Lenardo, M., J. W. Pierce, and D. Baltimore. Protein-binding sites in Ig gene enhancers determine the transcriptional activity and inducibility. *Science 236*:1573-1577 (1987).
122. Ephrussi, A., G. M. Church, S. Tonegawa, and W. Gilbert. B lineage-specific interactions of an immunoglobulin enhancer with cellular factors in vivo. *Science 227*:134-140 (1985).
123. Atchison, M. L., and R P. Perry. The role of the kappa enhancer and its binding factor NF-kappaB in the developmental regulation of kappa gene transcription. *Cell 48*:121-128 (1987).
124. Sen, R., and D. Baltimore. Inducibility of K immunoglobulin enhancer-binding protein NF-KB by a posttranslational mechanism. *Cell 47*:921-928 (1986).
125. Scholer, H. R., and P. Gruss. Specific interaction between enhancer-containing molecules and cellular components. *Cell 36*:403-411 (1984).
126. Jones, P. B. C., D. R. Galeazzi, J. M. Fisher, and J. P. Whitlock. Control of cytochrome P1-450 gene expression by dioxin. *Science 227*:1499-1502 (1985).
127. Zinn, K., and T. Maniatis. Detection of factors that interact with the human beta-interferon regulatory region in vivo by DNAase I footprinting. *Cell 45*:611-618 (1986).
128. Kuhl, D., J. De la Fuente, M. Chaturvedi, S. Parimoo, J. Ryals, F. Meyer, and C. Weissmann. Reversible silencing of enhancers by sequences derived from the human IFN-alpha promoter. *Cell 50*:1057-1069 (1987).
129. Muglia, L., and L. B. Rothman-Denes. Cell type-specific negative regulatory element in the control region of the rat alpha-fetoprotein gene. *Proc. Natl. Acad. Sci. USA 83*:7653-7657 (1986).
130. Chung, J., E. Sinn, R. R. Reed, and P. Leder. Trans-acting elements modulate expression of the human c-myc gene in Burkitt lymphoma cells. *Proc. Natl. Acad. Sci. USA 83*:7918-7922 (1986).

131. Remmers, E. F., J.-Q. Yang, and K. B. Marcu. A negative transcriptional control element located upstream of the murine c-myc gene. *EMBO J. 5*: 899-904 (1986).
132. Nir, U., M. D. Walker, and W. J. Rutter. Regulation of rat insulin 1 gene expression: evidence for negative regulation in nonpancreatic cells. *Proc. Natl. Acad. Sci. USA 83*:3180-3184 (1986).
133. Colantuoni, V., A. Pirozzi, C. Blance, and R. Cortese. Negative control of liver-specific gene expression: cloned human retinol-binding protein gene is repressed in HeLa cells. *EMBO J. 6*:631-636 (1987).
134. Imler, J.-L., C. Lemaire, C. Wasylyk, and B. Wasylyk. Negative regulation contributes to tissue specificity of the immunoglobulin heavy-chain enhancer. *Mol. Cell. Biol. 7*:2558-2567 (1987).
135. Steiner, C., M. Muller, A. Baniahmad, and R. Renkawitz. Lysozyme gene activity in chicken macrophages is controlled by positive and negative regulatory elements. *Nucleic Acids Res. 15*:4163-4178 (1987).
136. Larsen, P. R., J. W. Harney, and D. D. Moore. Repression mediates cell-type-specific expression of the rat growth hormone gene. *Proc. Natl. Acad. Sci. USA 83*:8283-8287 (1987).
137. Sudhof, T. C., D. W. Russell, M. S. Brown, and J. L. Goldstein. 42 bp element from LDL receptor gene confers end-product repression by sterols when inserted into viral TK promoter. *Cell 48*:1061-1069 (1987).
138. Barberis, A., G. Superti-Furga, and M. Busslinger. Mutually exclusive interaction of the CCAAT-binding factor and of a displacement protein with overlapping sequences of a histone gene promoter. *Cell 50*:347-359 (1987).
139. Gaub, M.-P., A. Dierich, D. Astinotti, I. Touitou, and P. Chambon. The chicken ovalbumin promoter is under negative control which is relieved by steroid hormones. *EMBO J. 6*:2313-2320 (1987).
140. Sassone-Corsi, P., and I. M. Verma. Modulation of c-fos gene transcription by negative and positive cellular factors. *Nature 326*:507-510 (1987).
141. Distel, R. J., H-S. Ro, B. S. Rosen, D. L. Groves, and B. M. Spiegelman. Nucleoprotein complexes that regulate gene expression in adipocyte differentiation: direct participation of c-fos. *Cell 49*:835-844 (1987).
142. Hen, R., E. Borelli, and P. Chambon. Repression of the immunoglobulin heavy chain enhancer by the adenovirus-2 E1A products. *Science 230*: 1391-1394 (1985).
143. Borelli, E., R. Hen, C. Wasylyk, B. Wasylyk, and P. Chambon. The immunoglobulin heavy chain enhancer is stimulated by the adenovirus type 2 E1A products in mouse fibroblasts. *Proc. Natl. Acad. Sci. USA 83*: 2846-2849 (1986).
144. Davidson, R. L. Gene expression in somatic cell hybrids. *Annu. Rev. Genet. 8*:195-218 (1974).
145. Ringertz, N., and R. Savage. *Cell Hybrids*. Academic Press, New York (1976).
146. Fougere, C., and M. C. Weiss. Phenotypic exclusion in mouse melanoma-

rat hepatoma hybrid cells: pigment and albumin production are not re-expressed simultaneously. *Cell 15*:843-854 (1978).
147. Allan, M., and P. Harrison. Co-expression of differentiation markers in hybrids between Friend cells and lymphoid cells and the influence of cell shape. *Cell 19*:437-447 (1980).
148. Goldfarb, P., W. Ostertag, D. Conkie, J. Eskdale, G. Coletta, D. Frisby, and P. Harrison. Erythroid differentiation and Friend virus expression in Friend cell hybrids. *In Vivo and In Vitro Erythropoiesis: The Friend System*, G. B. Rossi (Ed.). Elsevier/North-Holland, Amsterdam, pp. 383-390 (1980).
149. Affara, N., J. Fleming, P. S. Goldfarb, E. Black, B. Thiele, and P. R. Harrison. Analysis of chromatin changes associated with the expression of globin and non-globin genes in cell hybrids between erythroid and other cells. *Nucleic Acids Res. 13*:5629-5644 (1985).
150. Dyson, P. J., P. R. Cook, S. Searle, and J. A. Wyke. The chromatin structure of Rous sarcoma proviruses is changed by factors that act in trans in cell hybrids. *EMBO J. 4*:413-420 (1985).
151. Killary, A. M., and R. E. K. Fournier. A genetic analysis of extinction: trans-dominant loci regulate expression of liver-specific traits in hepatoma hybrids. *Cell 38*:523-534 (1984).
152. Peterson, T. C., A. M. Killary, and R.E. K. Fournier. Chromosomal assignment and trans regulation of the tyrosine aminotransferase structural gene in hepatoma cell hybrids. *Mol. Cell. Biol. 5*:2491-2494 (1985).
153. Petit, C., J. Levilliers, M.-O. Ott, and M. C. Weiss. Tissue-specific expression of the rat albumin gene: genetic control of its extinction in microcell hybrids. *Proc. Natl. Acad. Sci. USA 83*:2561-2565 (1986).
154. Wasylyk, B., C. Wasylyk, P. Augereau, and P. Chambon. The SV40 72 bp repeat preferentially potentiates transcription starting from proximal natural or substitute promoter elements. *Cell 32*:503-514 (1983).
155. Atchison, M. L., and R. P. Perry. Tandem kappa immunoglobulin promoters are equally active in the presence of the kappa enhancer: implications for models of enhancer function. *Cell 46*:253-262 (1986).
156. Sargosti, S., G. Moyne, and M. Yaniv. Absence of nucleosomes in a fraction of SV40 chromatin between the origin of replication and the region coding for the late leader RNA. *Cell 20*:65-73 (1980).
157. Jongstra, J., T. L. Reudelhuber, P. Ouswr, C. Benoist, C.-B. Chae, J.-M. Jelsch, D. J. Mathis, and P. Chambon. Induction of altered chromatin structures by simian virus 40 enhancer and promoter elements. *Nature 307*:708-714 (1984).
158. Courey, A. J., S. E. Plon, and J. C. Wang. The use of psoralen-modified DNA to probe the mechanism of enhancer action. *Cell 45*:567-574 (1986).
159. Ptashne, M. Gene regulation by proteins acting nearby and at a distance. *Nature 322*:697-701 (1986).
160. Johnson, P. F., W. H. Landschutz, B. J. Graves, and S. L. McKnight.

Identification of a rat liver nuclear protein that binds to the enhancer core element of three animal viruses. *Genes Devel. 1*:133-146 (1987).
161. Robertson, M. Specificity and flexibility. *Nature 327*:464-466 (1987).
162. Yang, L., T. C. Rowe, E. M. Nelson, and L. F. Liu. In vivo mapping of topisomerase II-specific cleavage sites on SV40 chromatin. *Cell 41*:127-132 (1985).
163. Gasser, S. M., and U. K. Laemmli. Cohabitation of scaffold binding regions with upstream/enhancer elements of three developmentally regulated genes of D. melanogaster. *Cell 46*:521-530 (1986).
164. Cockerill, P. N., and W. T. Garrard. Chromosomal loop anchorage of the kappa immunoglobulin gene occurs next to the enhancer in a region containing topoisomerase II sites. *Cell 44*:273-282 (1986).
165. Plon, S. E., and J. C. Wang. Transcription of the human beta-globin gene is stimulated by an SV40 enhancer to which it is physically linked but topologically uncoupled. *Cell 45*:575-580 (1986).
166. Jackson, D. A., and P. R. Cook. Transcription occurs at a nucleoskeleton. *EMBO J. 4*:919-925 (1985).
167. Mirkovitch, J., M.-E. Mirault, and U. K. Laemmli. Organization of the higher-order chromatin loop: specific DNA attachment sites on nuclear scaffold. *Cell 39*:223-232 (1984).
168. Shepherd, J. C. W., W. McGinnis, A. E. Carrasco, E. M. DeRobertis, and W. J. Gehring. Fly and frog homoeo domains show homologies with yeast mating type regulatory proteins. *Nature 310*:70-71 (1984).
169. Gehring, W. J. The homeobox: a key to the understanding of development. *Cell 40*:3-5 (1985).
170. Hope, I.A., and K. Struhl. Functional dissection of a eukaryotic transcriptional activator protein GCN4 of yeast. *Cell 46*:885-894 (1986).
171. Giniger, E., S. M. Varnum, and M. Ptashne. Specific DNA binding of GAL4, a positive regulatory protein of yeast. *Cell 40*:767-774 (1985).
172. Ma, J., and M. Ptashne. Deletion analysis of GAL4 defines two transcriptional activating segments. *Cell 48*:847-853 (1987).
173. Struhl, K. Promoters, activator proteins, and the mechansim of transcriptional initiation in yeast. *Cell 49*:295-297 (1987).
174. Pfeifer, K., T. Prezant, and L. Guarente. Yeast HAP1 activator binds to two upstream activation sites of different sequences. *Cell 49*:19-27 (1987).
175. Brown, D. D. The role of stable complexes that repress and activate eukaryotic genes. *Cell 37*:359-365 (1984).
176. Brown, D. D., and M. S. Schlissel. A positive transcription factor controls the differential expression of two 5S RNA genes. *Cell 42*:759-767 (1985).
177. Andrews, M. T., and D. D. Borwn. Transient activation of oocyte 5S RNA genes in Xenopus embryos by raising the level of the trans-acting factor TFIIIA. *Cell 51*:445-453 (1987).
178. Majowski, K., H. Mentzel, and T. Pieler. A split binding site for TFIIIC on the Xenopus 5S gene. *EMBO J. 6*:3057-3063 (1987).

179. Pieler, T., J. Hamm, and R. G. Roeder. The 5S gene internal control region is composed of three distinct sequence elements, organised as two functional domains with variable spacing. *Cell 48*:91-100 (1987).
180. Miller, J., A. McLachan, and A. Klug. Repetitive zinc-binding domains in the protein transcription factor IIIA from Xenopus oocytes. *EMBO J. 4*:1609-1614 (1985).
181. Rhodes, D., and A. Klug. An underlying repeat in some transcriptional control sequences corresponding to half a double helical turn of DNA. *Cell 46*:123-132 (1986).
182. McCall, M., T. Brown, W. N. Hunter, and O. Kennard. The crystal structure of d(GGATGGGAG) forms an essential part of the binding site for transcription factor IIIA. *Nature 322*:661-664 (1986).
183. Vincent, A. TFIIIA and homologous genes. The "finger" proteins. *Nucleic Acids Res. 14*:4385-4391 (1986).
184. Dhowdhury, K., U. Deutsch, and P. Gruss. A multigene family encoding several "finger" structures is present and differentially active in mammalian genomes. *Cell 48*:771-778 (1986).
185. Green, S., and P. Chambon. A superfamily of potentially oncogenic hormone receptors. *Nature 324*:615-617 (1986).
186. Weinberger, C., S. M. Hollenberg, M. G. Rosenfeld, and R. M. Evans. Domain structure of human glucocorticoid receptor and its relationship to the v-erb-A oncogene product. *Nature 318*:670-672 (1985).
187. Krust, A., S. Green, P. Argos, V. Kumar, P. Walter, J.-M. Bornert, and P. Chambon. The chicken oestrogen receptor sequence: homology with v-erbA and the human oestrogen and glucocorticoid receptors. *EMBO J. 5*.891-897 (1986).
188. Briggs, M. R., J. T. Kadonaga, S. P. Bell, and R. Tjian. Purification and biochemical characterization of the promoter-specific transcription factor, Sp1. *Science 234*:47-52 (1986).
189. Parker, C. S., and J. Topol. A Drosophila RNA polymerase II transcription factor contains a promoter-region-specific DNA-binding activity. *Cell 36*:357-369 (1984).
190. Leegwater, P. A. J., W. van Driel, and P. C. van der Vliet. Recognition site of nuclear factor I, a sequence-specific binding protein from HeLa cells that stimulates adenovirus DNA replication. *EMBO J. 4*:1515-1521 (1985).
191. Prujin, G. J. M., W. VanDriel, and P. C. Van der Vliet. Nuclear factor III, a novel sequence-specific DNA-binding protein from HeLa cells stimulating adenovirus DNA replication. *Nature 322*:656-659 (1986).
192 Morgan, W. D., G. T. Williams, R. I. Morimoto, J. Greene, R. E. Kingston, and R. Tjian. Two transcriptional activators, CCAAT-box-binding transcription factor and heat shock transcription factor, interact with a human hsp70 gene promoter. *Mol. Cell. Biol. 7*:1129-1138 (1987).
193. Dorn, A., J. Bollekens, A. Staub, C. Benoist, and D. Mathis. A multiplicity of CCAAT box-binding proteins. *Cell 50*:863-872 (1987).

194. Jones, K. A., J. T. Kadonaga, P. J. Rosenfeld, T. J. Kelly, and R. Tjian. A cellular DNA-binding protein that activates eukaroytic transcription and DNA replication. *Cell 48*:79-89 (1987).
195. Parker, C. S., and J. Topol. A Drosophila RNA polymerase II transcription factor binds to the regulatory site of an hsp 70 gene. *Cell 37*:273-283 (1984).
196. Topol., J., D. M. Ruden, and C. Parker. Sequences required for in vitro transcriptional activation of a Drosophila HSP 70 gene. *Cell 42*:527-537 (1985).
197. Wu, C. Two protein-binding sites in chromatin implicated in the activation of heat-shock genes. *Nature 309*:229-234 (1984).
198. Lee, W., P. Mitchell, and R. Tjian. Purified transcription factor AP-1 interacts with TPA-inducible enhancer elements. *Cell 49*:741-752 (1987).
199. Angel, P., M. Imagawa, R. Chiu, B. Stein, R. J. Imbra, H. J. Rahmsdorf, C. Jonat, M. Herrlich, and P. Karin. Phorbol ester-inducible genes contain a common Cis element recognized by a TPA-modulated trans-acting factor. *Cell 49*:729-739 (1987).
200. Mitchell, P., J. C. Wang, and R. Tjian. Positve and negative regulation of transcription in vitro: enhancer-binding protein AP-2 is inhibited by SV40 T antigen. *Cell 50*:847-861 (1987).
201. Chiu, R., M. Imagawa, R. J. Imbra, J. R. Bockoven, and M. Karin. Multiple cis- and trans-acting elements mediate the transcriptional response to phorbol esters. *Nature 329*:648-651 (1987).
202. Imagawa, M., R. Chiu, and M. Karin. Transcription factor AP-2 mediates induction by two different signal-transduction pathways: protein kinase C and cAMP. *Cell 51*:251-260 (1987).
203. Prywes, R., and R. G. Roeder. Inducible binding of a factor to the c-fos enhancer. *Cell 47*:777-784 (1986).
204. Cohen, R. B., M. Sheffery, and C. G. Kim. Partial purification of a nuclear protein that binds to the CCAAT box of the mouse alpha1-globin gene. *Mol. Cell. Biol. 6*:821-832 (1986).
205. Graves, B. J., P. F. Johnson, and S. L. McKnight. Homologous recognition of a promoter domain common to the MSV LTR and the HSV tk gene. *Cell 44*:565-576 (1986).
206. Minty, A., H. Blau, and L. Kedes. Two-level regulation of cardiac actin gene transcription: muscle-specific modulating factors can accumulate before gene activation. *Mol. Cell. Biol. 6*:2137-2148 (1986).
207. Gelinas, R., B. Endlich, C. Pfeiffer, M. Yagi, and G. Stamatoyannopoulos. G to A substitution in the distal CCAAT box of the A-gamma globin gene in Greek heriditary persistence of fetal haemoglobin. *Nature 313*:323-325 (1985).
208. Bender, A., and G. F. Sprague. MAT alpha1 protein, a yeast transcription activator, binds synergistically with a second protein to a set of cell-type-specific genes. *Cell 50*:681-691 (1987).

209. Johnson, S. A., J. M. Salmeron, and S. S. Dincher. Interaction of positive and negative regulatory proteins in the galactose regulon of yeast. *Cell 50*: 143-146 (1987).
210. Kownin, P., E. Bateman, and M. R. Paule. Eukaryotic RNA polymerase I promoter binding is directed by protein contacts with the transcription initiation factor and is DNA sequence-independent. *Cell 50*:693-699 (1987).
211. Tsai, S. Y., I. Sagami, H. Wang, M.-J. Tsai, and B. W. O'Malley. Interactions between a DNA-binding transcription factor (COUP) and a non-DNA binding factor (S300-II). *Cell 50*:701-709 (1987).
212. Birnstiel, M. L., M. Busslinger, and K. Strub. Transcription termination and 3′ processing: the end is in site. *Cell 41*:349-359 (1985).
213. Zeitlin, S., A. Parent, S. Silverstein, and A. Efstradiatis. Pre mRNA splicing and the nuclear matrix. *Mol. Cell. Biol. 7*:111-120 (1987).
214. Alberts, B., D. Bray, J. Lewis, M. Raff, K. Roberts, and J. D. Watson. *Molecular Biology of the Cell.* Garland, New York (1983).
215. Fulton, R., G. D. Birnie, and J. T. Knowler. Post-transcriptional regulation of rat liver gene expression by glucocorticoids. *Nucleic Acids Res. 13*:6467-6482 (1985).
216. Davidson, E. H., B. R. Hough-Evans, and R. J. Britten. Molecular biology of the embryo. *Science 217*:17-26 (1982).
217. Hagenbuchle, O., U. Schliber, S. Petrucco, G. C. VanTuyle, and P. K. Wellauer. Expression of mouse Amy-2a alpha-amylase genes is regulated by strong pancreas-specific promoters. *J. Mol. Biol. 185*:285-293 (1985).
218. Young, R. A., O. Hagenbuchle, and U. Schibler. A single mouse alpha-amylase gene specifies two different tissue-specific mRNAs. *Cell 23*: 451-458 (1981).
219. Schibler, U., A. Pittet, R. Young, O Hagenbuchle, M. Tosi, S. Gelman, and P. Wellauer. The mouse alpha-amylase multigene family: sequence organization of members expressed in the pancreas, salivary gland and liver. *J. Mol. Biol. 155*:247-266 (1982).
220. Schibler, U., O. Hagenbuchle, P. K. Wellauer, and A. Pittet. Two promoters of different strengths control the transcription of the mouse alpha-amylase gene Amy-1a in the parotid gland and the liver. *Cell 33*:501-508 (1983).
221. Perlino, E., R. Cortese, and G. Cilberto. The human alpha1-antitrypsin gene is transcribed from two different promoters in macrophages and hepatocytes. *EMBO J. 6*:2767-2771 (1987).
222. Rither, U., M. Tripodi, R. Cortese, and E. F. Wagner. The human alpha-1 antitrypsin gene is efficiently expressed from two tissue-specific promoters in transgenic mice. *Nucleic Acids Res. 15*:7519-7529 (1987).
223. Nabeshima, Y.-I., Y. Fujii-Kuriyama, M. Muramatsu, and K. Ogata. Alternative transcription and two modes of splicing result in two myosin light chains from one gene. *Nature 308*:333-338 (1984).
224. Rozek, C. E., and N. Davidson. Drosophila has one myosin heavy-chain gene with three developmentally regulated transcripts. *Cell 32*:23-34 (1983).

225. Robert, B., P. Daubas, M. A. Akimenko, A. Cohen, I. Garner, J. L. Guenet, and M. Buckingham. A single locus in the mouse encodes both myosin light chains 1 and 3. *Cell 39*:129-140 (1984).
226. Tamkun, J. W., J. E. Schwarzbauer, and R. O. Hynes. A single rat fibronectin gene generates three different mRNAs by alternative splicing of a complex exon. *Proc. Natl. Acad. Sci. USA 81*:5140-5144 (1984).
227. Schwarzbauer, J. E., J. W. Tamkun, I. R. Lemischka, and R. O. Hynes. Three different fibronectin mRNAs arise by alternative splicing within the coding region. *Cell 35*:421-431 (1983).
228. Vibe-Pedersen, K., A. R. Kornblihtt, and F. E. Baralle. Expression of a human alpha-globin/fibronectin gene hybrid generates two mRNAs by alternative splicing. *EMBO J. 3*:2511-2516 (1984).
229. Crabtree, G. R., and J. A. Kant. Organization of the rat gamma-fibrinogen gene: alternative mRNA splice patterns produce the gamma A and gamma B (gamma′) chains of fibrinogen. *Cell 31*:159-166 (1982).
230. Zehner, Z. E., and B. M. Paterson. Characterization of the chicken vimentin gene: single copy gene producing multiple mRNAs. *Proc. Natl. Acad. Sci. USA 80*:911-915 (1983).
231. Zehner, Z. E., and B. M. Paterson. Vimentin gene expression during myogenesis: two functional transcripts from a single copy gene. *Nucleic Acids Res. 11*:8317-8332 (1983).
232. Quax, W., W. V. Egberts, W. Hendriks, Y. Quax-Jeuken, and H. Bleomendal. The structure of the vimentin gene. *Cell 35*:215-223 (1983).
233. Capetanaki, Y. G., J. Ngai, C. N. Flytzanis, and E. Lazarides. Tissue-specific expression of two mRNA species transcribed from a single vimentin gene. *Cell 35*:411-420 (1983).
234. Benyajati, C., N. Spoerel, H. Haymerle, and M. Ashburner. The messenger RNA for alcohol dehydrogenase in Drosophila melanogaster differs in its 5′ end in different developmental stages. *Cell 33*:125-133 (1983).
235. Nawa, H., H. Kotani, and S. Nakanishi. Tissue-specific generation of two preprotachykinin mRNAs from one gene by alternative RNA splicing. *Nature 312*:729-734 (1984).
236. Saga, Y., J.-S. Tung, F.-W. Shen, and E. A. Boyse. Alternative use of 5′ exons in the specification of Ly-5 isoforms distinguishing hematopoietic lineages. *Proc. Natl. Acad. Sci. USA 84*:5364-5368 (1987).
237. Thomas, M. L., P. J. Reynolds, A. Chain, Y. Ben-Neriah, and I. S. Trowbridge. B-cell variant of mouse T200 (ly-5): evidence for alternative mRNA splicing. *Proc. Natl. Acad. Sci. USA 84*:5360-5363 (1987).
238. Gaul, U., E. Seifert, R. Schuh, and H. Jackle. Analysis of Kruppel protein distribution during early Drosophila development reveals posttranscriptional regulation. *Cell 50*:639-647 (1987).
239. Boggs, R. T., P. Gregor, S. Idriss, J. M. Belote, and M. McKeown. Regulation of sexual differentiation in D. melanogaster via alternative splicing of RNA from the transformer gene. *Cell 50*:739-747 (1987).
240. Early, P., J. Rogers, M. Davis, K. Calame, M. Bond, R. Wall, and L. Hood.

Two mRNAs can be produced from a single immunoglobulin mu gene by alternative processing pathways. *Cell 20*:313-319 (1980).
241. Amara, S. G., V. Jonas, M. G. Rosenfeld, E. S. Ong, and R. M. Evans. Alternative RNA processing in calcitonin gene expression generates mRNAs encoding different polypeptide products. *Nature 298*:204-244 (1982).
242. Amara, S. G., R. M. Evans, and M. G. Rosenfeld. Calcitonin/calcitonin gene-related peptide transcription unit:tissue-specific expression involves selective use of alternative polyadenylation sites. *Mol. Cell. Biol. 4*:2151-2160 (1984).
243. Crenshaw, E. B., A. F. Russo, L. W. Swanson, and M. G. Rosenfeld. Neuron-specific alternative RNA processing in transgenic mice expressing a metallothionein-calcitonin fusion gene. *Cell 49*:389-398 (1987).
244. Setzer, D. R., M. McGrogan, J. H. Nunberg, and R. T. Schimke. Size heterogeneity in the 3′ end of dihydrofolate reductase messenger RNA in mouse cells. *Cell 22*:361-370 (1980).
245. Collins, T., D. T. Bonthron, and S. H. Orkin. Alternative RNA splicing affects function of encoded platelet-derived growth factor A chain. *Nature 328*:621-624 (1987).
246. Tong, B. D., D. E. Auer, M. Jaye, J. M. Kaplow, G. Ricca, E. McConnathy, W. Drohan, and F. Duell. cDNA clones reveal differences between human glial and endothelial cell platelet-derived growth factor A-chains. *Nature 328*:629-631 (1987).
247. Breitbard, R. E., and B. Nadal-Ginard. Developmentally induced, muscle-specific trans factors control the differential splicing of alternative and constitutive troponin T exons. *Cell 49*:793-803 (1987).
248. Brawerman, G. Determinants of messenger RNA stability. *Cell 48*:5-6 (1987).
249. Guyette, W. A., R. J. Matusik, and J. M. Rosen. Prolactin mediated transcriptional and post-transcriptional control of casein gene expression. *Cell 17*:1013-1023 (1979).
250. Brock, M. L., and D. J. Shapiro. Estrogen stabilises vitellogenin mRNA against cytoplasmic degradation. *Cell 34*:207-214 (1983).
251. Gay, D. A., T. J. Yen, J. T. Y. Lau, and D. W. Cleveland. Sequences that confer beta-tubulin autoregulation through modulated mRNA stability reside within exon 1 of a beta-tubulin mRNA. *Cell 50*:671-676 (1987).
252. Pachter, J. S., T. J. Yen, and D. W. Cleveland. Autoregulation of tubulin expression is achieved through specific degradation of polysomal tubulin mRNAs. *Cell 51*:283-292 (1987).
253. Aviv, H., Z. Volloch, R. Bastos, and S. Levy. Biosynthesis and stability of globin mRNA in cultured erythroleukemic Friend cells. *Cell 8*:495-503 (1976).
254. Lowenhaupt, K., C. Trent, and J. J. B. Lingrel. Mechanisms for accumulation of globin mRNA during dimethyl sulphoxide induction of murine erythroleukaemia cells: synthesis of precursors and mature mRNA. *Dev. Biol. 63*:441-454 (1978).

255. Lowenhaupt, K., and J. B. Lingrel. Synthesis and turnover of globin mRNA in murine erythroleukemia cells induced with haemin. *Proc. Natl. Acad. Sci. USA 76*:5173-5177 (1979).
256. Volloch, V., and D. Housman. Stability of globin mRNA in terminally differentiating murine erythroleukemia cells. *Cell 23*:509-514 (1981).
257. Olins, P. O., and M. Nomura. Translational regulation by ribosomal protein S8 in E. coli: structural homology between rRNA binding site and feedback target in mRNA. *Nucleic Acids Res. 9*:1757-1764 (1981).
258. Munro, H. N., A. Aziz, E. A. Leibold, M. Murray, J. Rogers, J. K. Vass, and K. White. Ferritin gene structure, expression and regulation. *Ann. N.Y. Acad. Sci. 526*:113-123 (1988).
259. Hunt, T. False starts in translational control of gene expression. *Nature 316*:580-581 (1985).
260. Werner, M., A. Feller, F. Messenguy, and A. Pierard. The leader peptide of yeast gene CPA1 is essential for the translational repression of its expression. *Cell 49*:805-813 (1987).
261. Kahana, C., and D. Nathans. Nucleotide sequence of murine ornithine decarboxylase mRNA. *Proc. Natl. Acad. Sci. USA 82*:1673-1677 (1985).
262. Rosenthal, E. T., T. Hunt, and J. V. Ruderman. Selective translation of mRNA controls the pattern of protein synthesis during early development of the surf clam, Spistula solidissima. *Cell 20*:487-494 (1980).
263. Davidson, E. H. *Gene Activity in Early Development*. Academic Press, New York (1986).
264. Caldwell, D. C., and C. P. Emerson. The role of cap methylation in the translational activation of stored maternal histone mRNA in sea urchin embryos. *Cell 42*:691-700 (1985).
265. Ilan, J., and J. Ilan. Translation of maternal messenger ribonucleoprotein particles from sea urchin in a cell-free system from unfertilised eggs and product analysis. *Dev. Biol. 66*:375-385 (1978).
266. Jenkins, N. A., J. F. Kaumeyer, E. M. Young, and R. A. Raff. A test for masked message: the template activity of messenger ribonucleoprotein particles isolated from sea urchin eggs. *Dev. Biol. 63*:279-280 (1978).
267. Moon, R. T., M. V. Danilchik, and M. B. Hillie. An assessment of the masked message hypothesis: sea urchin messenger ribonucleoprotein complexes are efficient templates for in vitro protein synthesis. *Dev. Biol. 93*:389-403 (1982).
268. Richter, J. D., and L. D. Smith. Differential capacity for translation and lack of competition between mRNAs that segregate to free and membrane-bound polysomes. *Cell 27*:183-191 (1981).
269. Richter, J. D., and L. D. Smith. Developmentally regulated RNA binding proteins during oogenesis in Xenopus laevis. *J. Biol. Chem. 258*:4864-4869 (1983).
270. Richter, J. D., L. D. Smith, D. M. Anderson, and E. H. Davidson. Interspersed poly(A) RNAs of amphibian oocytes are not translatable. *J. Mol. Biol. 173*:227-241 (1984).

271. Miller, A. M., V. L. MacKay, and K. A. Nasmyth. Identification and comparison of two sequence elements that confer cell-type specific transcription in yeast. *Nature 314*:598-603 (1985).
272. Johnson, A. D., and I. Herskowitz. A repressor (MAT alpha2 product) and its operator control expression of a set of cell type specific genes in yeast. *Cell 42*:237-247 (1985).
273. Dranginis, A. M. Regulation of cell type in yeast by the mating-type locus. *Trends Biochem. Sci. 11*:328-331 (1986).
274. Hope, I. A., and K. Struhl. GCN4 protein, synthesized in vitro, binds HIS3 regulatory sequences: implications for general control of amino acid biosynthetic genes in yeast. *Cell 43*:177-188 (1985).
275. Fowlkes, D. M., N. T. Mullis, C. M. Comeau, and G. R. Crabtree. Potential basis for regulation of the coordinately expressed fibrinogen genes: homology in the 5′ flanking regions. *Proc. Natl. Acad. Sci. USA 81*:2313-2316 (1984).
276. Goodburn, S., K. Zinn, and T. Maniatis. Human beta-interferon gene expression is regulated by an inducible enhancer element. *Cell 41*:509-520 (1985).
277. Stanley, E., D. Metcalf, P. Sobieszcuk, N. M. Gough, and A. R. Dunn. The structure and expression of the murine gene encoding granulocyte-macrophage colony stimulating factor: evidence for utilisation of alternative promoters. *EMBO J. 4*:2569-2573 (1985).
278. Treisman, R. Identification of a protein-binding site that mediates transcriptional response of the c-fos gene to serum factors. *Cell 46*:567-574 (1986).
279. Caput, D., B. Beutler, K. Hartog, R. Thayer, S. Brown-Shimer, and A. Cerami. Identification of a common nucleotide sequence in the 3′-untranslated region of mRNA molecules specifying inflammatory mediators. *Proc. Natl. Acad. Sci. USA 83*:1670-1674 (1986).
280. Shaw, G., and R. Kamen. A conserved AU sequence from the 3′ untranslated region of GM–CSF mRNA mediates selective mRNA degradation. *Cell 46*:659-667 (1986).
281. Kruys, V., M. Wathelet, P. Poupart, R. Contreras, W. Fiers, J. Content, and G. Huez. The 3′ untranslated region of the human interferon-beta mRNA has an inhibitory effect on translation. *Proc. Natl. Acad. Sci. USA 84*:6030-6034 (1987).
282. Humphries, R. K., T. Levy, P. Turner, A. D. Moulton, and A. W. Nienhuis. Differences in human alpha-, beta- and delta-globin gene expression in monkey kidney cells. *Cell 30*:173-183 (1982).
283. Mellon, P., V. Parker, Y. Gluzman, and T. Maniatis. Identification of DNA sequences required for transcription of the alpha1-globin gene in a new SV40 host-vector system. *Cell 27*:279-288 (1981).
284. Treisman, A., M. R. Green, and T. Maniatis. Cis and trans activation of globin gene transcription in transient assays. *Proc. Natl. Acad. Sci. USA 80*:7428-7432 (1986).

285. Green, M. R., R. Treisman, and T. Maniatis. Transcriptional activation of cloned human beta-globin genes by viral immediate-early gene products. *Cell 35*:137-148 (1983).
286. Allan, M., J. de Zhu, P. Montague, and J. Paul. Differential response of multiple epsilon-globin cap sites to cis- and trans-acting controls. *Cell 38*: 399-407 (1984).
287. Harrison, P. R. Molecular analysis of erythropoiesis. *Exp. Cell Res. 155*: 321-344 (1984).
288. Chao, M. V., P. Mellon, P. Charnay, T. Maniatis, and R. Axel. The regulated expression of β-globin genes introduced into mouse erythroleukemia cells. *Cell 32*:483-493 (1983).
289. Young, K., M. Donovan-Peluso, K. Bloom, M. Allan, J. Paul, and A. Bank. Stable transfer and expression of exogenous human globin genes in human erythroleukemia (K562) cells. *Proc. Natl. Acad. Sci. USA 81*:5315-5319 (1984).
290. Allan, M., P. Montague, G. J. Grindlay, G. Sibbet, M. Donovan-Peluso, A. Bank, and J. Paul. Tissue-specific transcription of the human epsilon-globin gene following transfection into the embryonic erythroid cell line K562. *Nucleic Acids Res. 13*:6123-6134 (1985).
291. Rutherford, T., and A. W. Nienhuis. Human globin gene promoter sequences are sufficient for specific expression of a hybrid gene transfected into tissue culture cells. *Mol. Cell. Biol. 7*:398-402 (1987).
292. Townes, T. M., J. B. Lingrel, H. Y. Chen, R. L. Brinster, and R. D. Palmiter. Erythroid-specific expression of human beta-globin genes in transgenic mice. *EMBO J. 4*:1715-1723 (1985).
293 Magram, J., K. Chada, and F. Constantini. Developmental regulation of a cloned adult beta-globin gene in transgenic mice. *Nature 315*:338-340 (1985).
294. Chada, K., J. Magram, K. Raphael, G. Radice, E. Lacy, and F. Constantini. Specific expression of a foreign beta-globin gene in erythroid cells of transgenic mice. *Nature 314*:377-380 (1985).
295. Chada, K., J. Magram, and F. Constantini. An embryonic pattern of expression of a human fetal globin gene in transgenic mice. *Nature 319*: 685-689 (1986).
296. Wright, S., E. deBoer, F. G. Grosveld, and R. A. Flavell. Reguated expression of the human beta-globin gene family in murine erythroleukaemia cells. *Nature 305*:333-336 (1983).
297. Fordis, C. M., N. P. Anagnou, A. Dean, and A. W. Nienhuis. A beta-globin-gene, inactive in the K562 leukemic cell, functions normally in a heterologous expression system. *Proc. Natl. Acad. Sci. USA 81*:4485-4489 (1984).
298. Kioussis, D., F. Wilson K. Khazie, and F. Grosveld. Differential expression of human globin genes introduced in K562 cells. *EMBO J. 4*:927-931 (1985).
299. Karlsson, S., K. Van Doren, S. G. Schweiger, A. W. Nienhuis, and Y. Gluz-

man. Stable gene transfer and tissue specific expression of a human globin gene using adenoviral vectors. *EMBO J. 5*:2377-2385 (1986).
300. Fordis, C. M., N. Nelson, M. McCormick, R. Padmanabhan, B. Howard, and A. N. Schechter. The 5′-flanking sequences of human globin genes contribute to tissue specific expression. *Biochem. Biophys. Res. Commun. 134*:128-133 (1986).
301. Young, K., M. Donovan-Peluso, R. Cubbon, and A. Bank. Trans acting regulation of beta globin gene expression in erythroleukemia (K562) cells. *Nucleic Acids Res. 13*:5203-5213 (1985).
302. Koshe, K. A., C. Dobkin, and A. Bank. DNA sequences regulating human beta globin gene expression. *Nucleic Acids Res. 12*:7781-7793 (1985).
303. Emerson, B. M., and G. Felsenfeld. Specific factor conferring nuclease hypersensitivity at the 5′ end of the chicken adult beta-globin gene. *Proc. Natl. Acad. Sci. USA 81*:95-99 (1984).
304. Emerson, B. M., C D. Lewis, and G. F. Felsenfeld. Interaction of specific nuclear factors with the nuclease-hypersensitive region of the chicken adult beta-globin gene: nature of the binding domain. *Cell 41*:21-30 (1985).
305 Plumb, M. A., R. H. Nicolas, C. A. Wright, and G. H. Goodwin. Multiple sequence specific DNA binding activities are eluted from chicken nuclei at low ionic strengths. *Nucleic Acids Res. 13*:4047-4065 (1985).
306. Plumb, M. A., V V. Lobanenkov, K. H. Nicolas, C. A. Wright, S. Zavou, and E. H. Goodwin. Characterisation of chicken erythroid nuclear proteins which bind to the nuclease hypersensitive regions which bind to the nuclease hypersensitive regions upstream of the β^{A}- and β^{H}-globin genes. *Nucleic Acids Res. 14*:7675-7693 (1986).
307. Jackson, P. D., and G. Felsenfeld. A method for mapping intranuclear protein-DNA interactions and its application to a nuclease hypersensitive site. *Proc. Natl. Acad. Sci. USA 82*:2296-2300 (1985).
308. Frampton, J., I. Chambers, D. Conkie, and P. R. Harrison. The regulation of erythroid ce.l-specific gene expression, in *Molecular Approaches to Developmental Biology*, A. R. Liss, New York, pp. 533-545 (1987).
309 Frampton, J., D. Conkie, I. Chambers, W. McBain, M. Dexter, and P. Harrison. Changes in minor transcripts from the alpha1 and beta major and glutathione peroxidase genes during erythropoiesis. *Nucleic Acids Res. 15*:3671-3688 (1987).
310 Antoniou, M., E. deBoer, G. Habets, and F. Grosveld. The human beta-globin gene contains multiple regulatory regions: identification of one promoter and two downstream enhancers. *EMBO J. 7*:377-384 (1988).
311. Cone R. D., A. Weber-Benarous, D. Baorto, and R. C. Mulligan. Regulated expression of a complete human beta-globin gene encoded by a transmissible retrovirus vector. *Mol. Cell. Biol. 7*:887-897 (1987).
312. Wright, S., A. Rosenthan. R. Flavell, and F. Grosveld. DNA sequences required for regulated expression of beta-globin genes in murine erythroleukemia cells. *Cell 38*:265-273 (1984).

313. Kemper, B., P. D. Jackson, and G. Felsenfeld. Protein-binding sites within the 5′ DNase I-hypersensitive region of the chicken alpha-D globin gene. *Mol. Cell. Biol. 7*:2059-2069 (1987).
314. Thiele B. J., J. Fleming, K. Kasturi, J. O'Prey, E. Black, J. Chester, S. M. Rapoport, and P. R. Harrison. Cloning of a rabbit eythroid-cell specific lypoxygenase mRNA. *Gene 57*:111-119 (1987).
315. Yamamoto, M., N. S. Yew, M. Federspiel, J. B. Dodgson, N. Hayashi, and J. D. Engel. Isolation of recombinant cDNAs encoding chicken erythroid delta aminolevulinate synthetase. *Proc. Natl. Acad. Sci. USA 82*:3702-3706 (1985).
316. Schoenhaut, D. S., and P. J. Curtis. Nucleotide sequence of mouse 5-aminolevulinic acid synthetase cDNA and expression of its gene in hepatic and erythroid tissues. *Gene 48*:55-63 (1986).
317. Grandchamp, B., H. DeVerneuil, C. Beaumont, S. Chretien, O. Walter, an and Y. Nordmann. Tissue specific expression of porphobilinogen deamininase-two isozymes from a single gene. *Eur. J. Biochem. 162*:105-110 (1987).
318. Romana, M., A. Dubart, D. Beaupain, C. Chabert, M. Goosens, and P.-H. Romeo. Structure of the gene for human uroporphyrinogen decarboxylase. *Nucleic Acids Res. 15*:7343-7356 (1987).
319. Crick, F. H. C., and P. A. Lawrence. Compartments and polyclones in insect development. *Science 189*:340-347 (1975).
320. Morata, G., and P. A. Lawrence. Homoeotic genes, compartments and cell determination in Drosophila. *Nature 265*:211-216 (1977).
321. Lewis, E. B. A gene complex controlling segmentation in Drosophila. *Nature 276*:565-571 (1978).
322. Garcia-Bellido, A., P. A. Lawrence, and G. Morata. Compartments in animal development. *Sci. Am. 241*:90-98 (1979).
323. Brower, D. L. The sequential compartmentalization of Drosophila segments revisited. *Cell 41*:361-364 (1985).
324. Hogan, B., P. Holland, and P. Scholfield. How is the mouse segmented? *Trends Genet. 1*:67-74 (1985).
325. Pfeifer, K., B. Arcangoli, and L. Guarente. Yeast HAP 1 activator competes with the factor RC2 for binding to the upstream activation site UAS1 of the CYC1 gene. *Cell 49*:9-18 (1987).
326. Mantovani, R., N. Malgarett, B. Giglioni, P. Comi, N. Cappellini, S. Nicolis, and S. Ottolenghi. A protein factor binding to an octamer motif in the gamma-globin promoter disappears upon induction of differentiation and hemoglobin synthesis in K562 cells. *Nucleic Acids Res. 15*:9349-9364 (1987).
327 Behringer, R. R., R. E. Hammer, R. L. Brinster, R. D. Palmiter, and T. M. Townes. Two 3′ sequences direct adult erythroid-specific expression of human beta-globin genes in transgenic mice. *Proc. Natl. Acad. Sci. USA 84*:7056-7060 (1987).
328. Gehring, W. H. Homeo boxes in the study of development. *Science 236*: 1245-1252 (1987).

329. Akam, M. E., and A. Martinez-Arias. The distribution of Ultrabithorax transcripts in Drosophila embryos. *EMBO J. 4*:1689-1700 (1985).
330. Garber, R. L., A. Kuroiwa, and W. J. Gehring. Genomic and cDNA clones of the homeotic locus Antennapedia in Drosophila. *EMBO J. 2*: 2027-2036 (1983).
331. Ollo, R., and T. Maniatis. Drosophila Kruppel gene product produced in a buculovirus expression system is a nuclear phosphoprotein that binds to DNA. *Proc. Natl. Acad. Sci. USA 84*:5700-5704 (1987).
332. Plumb, M., Frampton, J., Wainwright, H., Walker, M., MacLeod, K., Goodwin, G., and Harrison, P. F. GATAG; a *cis*-control region binding an erythroid-specific nuclear factor with a role in globin and non-globin gene expression. *Nucl. Acids Res. 17*:73-92 (1989).
333. Harrison, P. R., Plumb, M., Frampton, J., Llewellyn, D., Chester, J., Chambers, I., MacLeod, K., Fleming, J., O'Prey, J., Walker, M., Wainwright, H., and Anton, I. Regulation of erythroid cell-specific gene expression during erythropoiesis. *Br. J. Cancer 58*(suppl IX):46-57 (1988).
334. Fleming, J., Thiele, B. J., Chester, J., O'Prey, J., Janetzki, S., Aitken, A., Anton, I. A., Rapoport, S. M., and Harrison, P. R. The complete sequence of the rabbit erythroid cell-specific 15-lipoxygenase mRNA: Comparison of the predicted amino acid sequence of the erythrocyte lipoxygenase with other lipoxygenase. *Gene 79*:181-188 (1989).
335. O'Prey, J., Chester, J., Thiele, B. J., Rapoport, S. M., and Harrison, P. R. The complete sequence and exon/intron structure of the rabbit reticulocyte 15-lipoxygenase gene. *Gene*, in press (1989).

16
Colony-Stimulating Factors: Future Prospects

JOHN M. GARLAND
University of Manchester Medical School, Manchester, England

T. MICHAEL DEXTER and NYDIA G. TESTA
Paterson Institute for Cancer Research, Christie Hospital and Holt Radium Institute, Manchester, England

Without question, colony stimulating factors and hematopoietic growth factors (HPGFs) are here to stay. It has taken nearly 20 years from the original observations (more in the case of erythropoietin) that lymphocytes can produce hematopoietic-cell stimulatory factors to reach the current stage where at least four of them are entering into clinical trials. The two most important questions being asked are: "How do they work?" and "Are they of any use?" The first question is directed to basic science in a rapidly expanding field in which techniques are only just becoming available to provide answers. In the last few years, an enormous stride has been taken in understanding how cells are controlled, which has impacts extending from basic differentiation processes to malignant change. The peptide growth factors, their receptors, and the manner in which oncogenes and viruses contribute to cellular control are all under intense scrutiny by molecular and cell biologists. What may be considered as lagging behind is the application of molecular biology to cell biology, biochemistry and the traditional approaches in systematic biological sciences such as immunology and pathology. Clearly, HPGFs inevitably attract an ever-increasing circle and range of disciplines, because the cell does not understand these convenient divisions and uses everything it has to generate a suitable response. In terms of basic science, HPGFs assume an important role throughout the spectrum of interests, and provide convenient, controllable models with which to analyze how cells work. At present, much attention is focused on signal transduction from the membrane to the nucleus, and how this results in inducing gene expression. This is clearly a major investigative activity. However, the relationship between this pathway and those of equal importance (upregulation of cellular metabolism, signals generating differentiation or proliferation decisions) is waiting for further exploration, and it is perhaps in this area that much progress may be expected. Unfortunately, the relevance of classic biochemistry has been somewhat subdued under the blossoming of molecular biology, and only now it may be realized that those who truly understand metabolic pathways (the "you are here" sign on those incredibly complicated wall charts of intermediary metabolism) are in short supply. When the growth-factor-stimulated new gene transcription has been initiated, what then? What happens before this to induce the cell to become "activated" or be capable of a completely different mode of existence? The gap in understanding of mechanisms generated between the cell membrane and nucleus is very large. Despite the enormous increase in knowledge about the structure of HPGFs, it is salutary to realize that the cell membrane is still a "black box" after nearly 70 years, and that how well-known growth factors, for example, insulin, work is still a mystery after 40 years.

On the usefulness of HPGFs, it may be said with considerable confidence that they will find an important place in many areas of clinical medicine. They have considerable advantages over many pharmaceuticals. They are "natural," short-lived, of low toxicity, highly effective, without (to date) any indication of lasting side effects, and (an important consideration) are likely to be economical. Interestingly, their clinical use has preceded their "scientific" evaluation, in that very little is known about their normal in-vivo function. However, medical application is often in advance of the "scientific" process, and it may be expected that the HPGFs will be applied to many situations clinically on a trial-and-error basis. The future for HPGFs appears to be exceedingly good, and is perhaps one of the major breakthroughs in clinical medicine equivalent to the antibiotic era. This is because the HPGFs have such a wide range of action, not necessarily restricted to the hematopoietic system, that many conditions may be expected to benefit either through a primary effect or secondarily (for example, in reducing infections). Thus, goals approached may be reductions in bed occupancy, reduced antibiotic usage, decrease in transfusion requirements, increased efficacy of vaccines, and increased efficiency of anti-tumor agents. At present, the use of HPGFs is limited because of their properties, which demand a hospital environment for infusion and monitoring. However, a new science is emerging, that of "designer" molecules, in which active epitopes can be combined in various controlled ways to produce multifunctional or function-specific artificial molecules. Although this is some way away, the combination of everincreasing computing power and ability to precisely manipulate genes is a combination likely to advance this concept very rapidly.

A word of caution may, however, be appropriate. The summary above indicates not only the positive progress that has been made, but some of the confusion in data obtained from experimental analysis. The total number of lymphokines elaborated by the immune system is doubtless set to rise substantially and all will be tested in various ways for activity on various cells, including those of the hematopoietic system. It would not be surprising, therefore, if many more effects of lymphokines on hematopoiesis are uncovered. The major problem is going to be how to evaluate their significance because much of the data are highly susceptible to the system in which they are evaluated. The HPGFs in fact, present the very real problems intrinsic in trying to dissect multifactor responses and cascade reactions. The secret to beginning an understanding is to define the cell populations and the receptors they express, and it is to this that considerable energy will doubtless be directed. Again, it is interesting to note that although interleukin-3 (IL-3) was the first hematopoietin to be cloned, its receptor is still elusive, furthermore, it took several years before the human equivalent was found, and the lack of homology between the murine and human IL-3s suggests that the receptors are going to be equally different. Not only that, but the highly useful murine IL-3-dependent lines and their ease of generation do not exist for human marrow and human IL-3; the best are lines whose characteristics are considerably different from murine IL-3-dependent cells.

Considerable work therefore still needs to be done. The isolation of cloned factors is only the first and essential stage in a much greater understanding of the cellular and molecular biology of these extraordinary proteins.

Index